PUBLICATIONS

OF THE

NAVY RECORDS SOCIETY

VOL. 163

THE MEDITERRANEAN FLEET,
1930–1939

The NAVY RECORDS SOCIETY was established in 1893 for the purpose of printing unpublished manuscripts and rare works of naval interest. The Society is open to all who are interested in naval history, and any person wishing to become a member should either complete the online application form on the Society's website, www.navyrecords.org.uk, or apply to the Hon. Secretary, The Mill, Stanford Dingley, Reading, RG7 6LS, United Kingdom, email address robinbrodhurst@gmail.com. The annual subscription is £40, which entitles the member to receive one free copy of each work issued by the Society in that year, and to buy earlier issues at much reduced prices.

SUBSCRIPTIONS should be sent to the Membership Secretary, 19 Montrose Close, Whitehill, Bordon, Hants GU35 9RG.

THE COUNCIL OF THE NAVY RECORDS SOCIETY wish it to be clearly understood that they are not answerable for any opinions and observations which may appear in the Society's publications. For these the editors of the several works are entirely responsible.

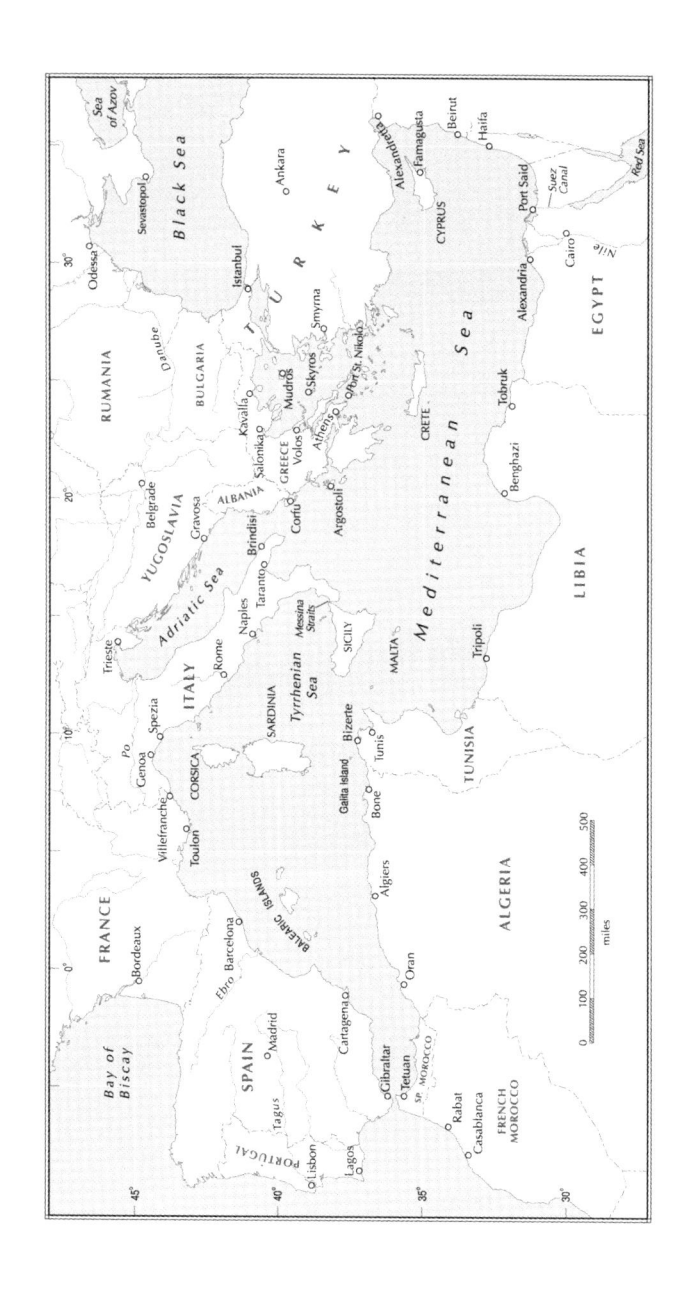

Map 1 The Mediterranean Sea

THE MEDITERRANEAN FLEET, 1930–1939

Edited by

PAUL G. HALPERN

Professor Emeritus, Florida State University

PUBLISHED BY ROUTLEDGE
FOR THE NAVY RECORDS SOCIETY
2016

Routledge
Taylor & Francis Group

LONDON AND NEW YORK

First published 2016
by Routledge
2 Park Square, Milton Park, Abingdon, Oxon OX14 4RN

and by Routledge
711 Third Avenue, New York, NY 10017

Routledge is an imprint of the Taylor & Francis Group, an informa business

© 2016 The Navy Records Society

Crown copyright material is reproduced by permission of The Stationery Office.

The right of Paul G. Halpern to be identified as the author of the editorial material, and of the authors for their individual chapters, has been asserted in accordance with sections 77 and 78 of the Copyright, Designs and Patents Act 1988.

All rights reserved. No part of this book may be reprinted or reproduced or utilised in any form or by any electronic, mechanical, or other means, now known or hereafter invented, including photocopying and recording, or in any information storage or retrieval system, without permission in writing from the publishers.

Trademark notice: Product or corporate names may be trademarks or registered trademarks, and are used only for identification and explanation without intent to infringe.

British Library Cataloguing in Publication Data
A catalogue record for this book is available from the British Library

Library of Congress Cataloging in Publication Data
Names: Halpern, Paul G., 1937– author.
Title: The Mediterranean Fleet, 1930–1939 / by Paul G. Halpern.
Description: London: Routledge, 2016. | Series: Navy Records Society publications ; 163 | Includes bibliographical references and index.
Identifiers: LCCN 2015022940
Subjects: LCSH: Great Britain. Royal Navy. Fleet, Mediterranean. | Great Britain. Royal Navy.
Fleet, Mediterranean – History –20th century. | Mediterranean Region – History, Naval – 20th century. | World politics – 1900–1945.
Classification: LCC VA457.M44 H3523 2016 | DDC 359.3/1–dc23 LC record available at
http://lccn.loc.gov/2015022940

ISBN: 978-1-4724-7597-8 (hbk)
ISBN: 978-1-315-55561-4 (ebk)

Typeset in Times New Roman
by Manton Typesetters, Louth, Lincolnshire, UK.

Printed in the United Kingdom
by Henry Ling Limited, at the Dorset Press, Dorchester DT1 1HD

THE COUNCIL OF THE NAVY RECORDS SOCIETY 2015–16

PATRON
H.R.H. THE PRINCE PHILIP, DUKE OF EDINBURGH,
K.G., O.M., F.R.S.

PRESIDENT
Vice Admiral Sir ADRIAN JOHNS, K.C.B., C.B.E., K.St.J., A.D.C.

VICE-PRESIDENTS
M. DUFFY, M.A., Ph.D., F.R.Hist.S.
Lt. Cdr. F.L. PHILLIPS, R.D., T.D., R.N.R.
R. MORRISS, M.A., Ph.D., F.R.Hist.S.
Professor N.A.M. RODGER, M.A., D.Phil., F.B.A., F.S.A., F.R.Hist.S.

COUNCILLORS

J.D. DAVIES, M.A., D.Phil., F.R.Hist.S.
Rear Admiral J. GOLDRICK, A.O., C.S.C., D.Litt., R.A.N.
Professor M.S. SELIGMANN, M.A., D.Phil., F.R.Hist.S.
T. BENBOW, M.A., D.Phil., P.G.C.A.P., F.H.E.A.
M.J. WHITBY, B.A., M.A., B.Ed.
J. ROBB-WEBB, M.A., Ph.D.
B. VALE, C.B.E., B.A., M.Phil.
N. BLACK, M.A., Ph.D., F.R.Hist.S.
E. CHARTERS, B.Hum., M.A., D.Phil.
G.J. COLE, M.A., Ph.D.
S. PRINCE, M.A.
O. WALTON, M.A., Ph.D.
Professor D. LAW, M.A., F.C.L.I.P., F.I.Inf.Sc., F.K.C., F.L.A.

J. DAVEY, M.A., Ph.D.
Professor B. GOUGH, M.A., Ph.D., F.R.Hist.S.
D. MORGAN-OWEN, M.A., Ph.D.
S. WILLIAMS, M.A.
Col. E. MURCHISON, D.S.O., M.B.E., R.M.
Professor E.J. GROVE, M.A., Ph.D., F.R.Hist.S.
Professor R. HARDING, M.A., Ph.D., F.R.Hist.S.
Professor R. KNIGHT, M.A., Ph.D., F.R.Hist.S.
E. MURPHY, M.Litt., Ph.D., F.R.Hist.S.
V. PRESTON, M.A., Ph.D.
Dr M. WILCOX, M.A., Ph.D.

HON. SECRETARY, R.H.A. BRODHURST, B.A.

HON. TREASURER, P. NORTHCOTT, M.A.

GENERAL EDITOR, B. JONES, M.A., Ph.D.

MEMBERSHIP SECRETARY, Mrs J. GOULD

WEB MANAGER, G.J. COLE, M.A., Ph.D.

SOCIAL MEDIA EDITOR, S. WILLIS, M.A., Ph.D.

CONTENTS

	PAGE
List of Maps	xi
Preface	xiii
Glossary of Abbreviations	xvii
Part I: The Mediterranean Fleet from 1930 to the Ethiopian Crisis	1
Part II: The Abyssinian Crisis, 1935–1936	61
Part III: The Spanish Civil War and the Nyon Agreements, 1936–1937	189
Part IV: The Approach of War, 1938–1939	379
Sources and Documents	559
Index	569

MAPS

1. The Mediterranean Sea . . . . frontispiece
2. The Western Mediterranean xxi
3. The Central Mediterranean xxii
4. The Eastern Mediterranean xxiii
5. Southern Red Sea and East Africa xxiv
6. Alexandria xxv
7. Zones Established by the Nyon and Paris Agreements xxvi

PREFACE

At the beginning of the 1930s the Mediterranean Fleet, Great Britain's premier naval force, was still operating in the atmosphere of what would soon seem to have been the halcyon days of the 1920s. The fleet in many of its training and exercises seemed to be preparing for a repetition of Jutland with squadrons of heavy warships and massed destroyer flotillas engaged against their enemy counterparts. At the same time, the Navy was endeavouring to integrate new weapons, notably aircraft, into its battle practices. It was, however, a period of world depression and the financial stringency was felt in the fleet, particularly in deferred spending on the defences of the bases of Malta and Gibraltar. It would be difficult to remedy these deficiencies. The world situation was also changing. In 1931 the Japanese, a former ally, embarked on aggressive expansion and in subsequent years became an apparent threat to the British position in the Far East. In the event of war with Japan, the Mediterranean Fleet was designated to move to the Far East. In 1932 the 'Ten Year Rule' – that Britain would not face a major war within ten years – was abandoned. In 1933 Hitler and the Nazis came to power in Germany and began policies that would lead to war in less than a decade. The technological progress of aviation also cast a growing uncertainty over traditional naval operations: how well were warships equipped to meet the new threat?

The first major crisis came in 1935 with the Italian invasion of Ethiopia, a member of the League of Nations. There was a very real possibility the Mediterranean Fleet would be engaged in hostilities against the Italians in support of League policies. The lasting effect of the crisis was that Italy went from being a traditional friend to a potential enemy. This had enormous consequences for the Mediterranean Fleet because of Italy's central position in the Mediterranean. The principal base of the Mediterranean Fleet at Malta was now in easy striking distance of the Italian air force. While the exact effectiveness of air power was still uncertain, and possibly underestimated, the potential danger to a fleet caught at anchor in harbour was apparent and the Mediterranean Fleet for the duration of the crisis shifted its major base far eastward to Alexandria. In the succeeding crises of the 1930s, notably Czechoslovakia in 1938 and the final months of peace, the Mediterranean Fleet would again leave

Malta for the more secure location of Alexandria. The Ethiopian crisis was no sooner over when the Spanish Civil War began and for the next few years portions of the Mediterranean Fleet were engaged on duties off the Spanish coast with the threat of mines, air and submarine attack. The end of the Spanish Civil War by the spring of 1939 brought little real relief when the Italian invasion of Albania in April 1939 began another crisis and a run-up to the events which ended in the outbreak of the Second World War. The challenges grew with not one but three potential enemies, plus a potentially hostile Spain where the victorious Nationalists were indebted to Germany and Italy for assistance during the war. In return, the hostility with Turkey so apparent in the first half of the 1920s was gone, although hopes that Turkey might prove an ally in a potential conflict were overly optimistic. The relationship with the French, relatively cool during the Ethiopian crisis, grew warmer with both navies engaged in conversations for at least distant cooperation in the impending war.

The naval archives for this momentous period are voluminous, continuing the mass of paper records that began with the outbreak of the First World War. It is a real challenge to choose documents to include in this volume from the many that are available and painful to have to make cuts when the manuscript threatened to become too long. Nevertheless, it remains frustrating to discover that, despite the apparent abundance of records, certain documents referred to in other records apparently no longer exist having been 'weeded'. I have tried to concentrate on naval activity in the documents chosen and resisted the temptation to include, for example, fascinating accounts of local conditions.

The Mediterranean Commanders-in-Chief now seem under tighter control from the Admiralty because of the widespread use of telegrams, mostly transmitted by wireless, although this method of transmission is no longer usually indicated. In the 1930s it became increasingly common to refer to the telegrams by time and date of transmission rather than their number which would also be different for each of the copies that were repeated to different authorities. These references were usually expressed in the form 'Your 2312/8', referring to the time and date of the message and usually located at the bottom of the telegram. The original forms and capitalisations have been retained when possible although the salutations and closings have been omitted. A glossary of the numerous abbreviations and acronyms has been included but for reasons of space well-known honours and decorations have not, as they are easily found in most dictionaries. There are also far too many ships mentioned to provide technical details of each; however, the reader will find these in standard references such as J.J. Colledge, *Ships of the Royal Navy*, *Conway's All the World's Fighting Ships, 1922–1946*, and the large and steadily growing number of monographs on ships and classes. When looking up these details it will also be sobering for the reader to learn how many of the ships mentioned in this volume would be lost during the war.

Sources identified as ADM [Admiralty], AIR and FO [Foreign Office] are located in The National Archives (formerly Public Record Office) at Kew, while those designated CHT refer to the Chatfield MSS. at the National Maritime Museum, Greenwich. Minor omissions are indicated by an ellipsis ... while more substantial cuts are indicated by a row of asterisks * * *. Documents are typescript unless otherwise indicated.

For their assistance over the years in the preparation of this volume I would like to thank: the late Roderick Suddaby, Keeper of Documents, the Imperial War Museum London; Robin Brodhurst; the staff at the National Archives, Kew; Captain Christopher L. W. Page, RN and Dr Malcolm Llewellyn-Jones of the Naval Historical Branch, Portsmouth; the archivists at the National Maritime Museum, Greenwich; and the archivists and staff at the Churchill College Archives Centre, Cambridge. Peter Krafft of the Florida Resources and Environmental Analysis Center was the cartographer and indefatigable in locating obscure maritime reference points whose names might have changed over the years. Dr Ben Jones, General Editor of the Navy Records Society, guided the manuscript through the press.

Documents that are Crown Copyright are reproduced by permission of The Stationery Office and I extend my apologies to any holders of copyright I have been unable to trace.

GLOSSARY OF ABBREVIATIONS

A/A	Anti-Aircraft
A/B	Anti-Boat (Boom)
AC	Admiral Commanding
ACNS	Assistant Chief of Naval Staff
ADGB	Air Defence of Great Britain
Adm	Admiral
Admy	Admiralty
AFO	Admiralty Fleet Order
ALO	Admiralty Liaison Officer (may also be Army or Air depending on context)
Amb	Ambassador
AOC	Air Officer Commanding
AoF	Admiral of the Fleet
ARP	Air Raid Protection
A/S	Anti-Submarine
AS	German Armoured Ship (*Deutschland* class, commonly known as 'pocket battleships')
ASDIC	Anti-Submarine Detection (later SONAR)
ASIS	Armament Store Issuing Ship
Asst	Assistant
A/T	Anti-Torpedo
BL	Breech Loading
BP	Battle Practice
BS	Battle Squadron
Capt	Captain
CAPTAIN (D)	Captain commanding Destroyer Flotilla (the flotilla number may also be included)
CAS	Chief of Air Staff
CB	Confidential Book
Cdr	Commander
Cdre	Commodore
Chllr	Chancellor
Chmn	Chairman
CID	Committee of Imperial Defence

CIGS	Chief of Imperial General Staff
C-in-C	Commander-in-Chief
CMB	Coastal Motor Torpedo Boat
CNS	Chief of Naval Staff
CO	Commanding Officer
COS	Chief of Staff or Chiefs of Staff Committee
CS	Cruiser Squadron
CSA	Chemical Smoke-making Apparatus
Cttee	Committee
DASO	Deputy Armaments Supply Officer
DC	Depth Charge
DCNS	Deputy Chief of Naval Staff
D/F	Direction-Finding
DF	Destroyer Flotilla
Dir	Director
DNC	Director of Naval Construction
DNO	Director of Naval Ordnance Department (Admiralty)
DNI	Director Naval Intelligence Division (Admiralty)
DOD	Director of Operations Division (Admiralty)
DoD	Director of Dockyards
DP	Director of Plans Division (Admiralty)
DPS	Director of Personal Services (Admiralty)
FAA	Fleet Air Arm
FB	Flying Boat
FO	Foreign Office (also Flag Officer)
FP & MS	Fishery Protection & Minesweeping (Flotilla)
G & T	Gunnery and Torpedo
Gen	General
GOC	General Officer Commanding
Govr	Governor
Govt	Government
G/R	General Reconnaissance
GREEN	Starboard (usually coupled with a bearing expressed in degrees)
HA	High Angle (anti-aircraft)
HACS	High Angle Control Station
HBM	His Britannic Majesty's
HE	High Explosive
HF	Home Fleet
HM	His Majesty's
HMG	His Majesty's Government
HMS	His Majesty's Ship

HSBPT	High Speed Battle Practice Target
IDC	Imperial Defence College
INA	Standing Instructions for H.M. Ships in connection with the Nyon Agreement
IPA	Standing Instructions for H.M. Ships in connection with the Paris Agreement
ISCW	Standing Instructions for H.M. Ships in connection with the Spanish Civil War
JDC	Joint Home and Overseas Defence Committee
KR	King's Regulations
LA	Low Angle
L/T	Land Telegraph
M	Military Branch (Admiralty)
Med	Mediterranean
Min	Minister
MNBDO	Mobile Naval Base Defence Organisation
M/S	Minesweeper
MSF	Mine Sweeping Flotilla
MTB	Motor Torpedo Boat
MTM	Mediterranean Tactical Memorandum
NAAGC	Naval Anti-Aircraft Gunnery Commission
NI	Non Intervention (Spanish Civil War)
NID	Naval Intelligence Division (Admiralty)
NLO	Naval Liaison Officer
OC	Officer Commanding
P & O	Peninsula & Oriental Steam Navigation Company
pom-pom	Short Range Rapid Fire Anti-Aircraft Gun
Pres	President
PVs	Paravanes
QF	Quick Firing
Q-Ship	Anti-Submarine warship disguised as innocent merchantman
RA	Rear Admiral
RAA	Rear Admiral Aircraft Carriers
RAC	Rear Admiral Commanding
RA (D)	Rear Admiral Commanding Destroyer Flotillas
RDF	Radio Direction Finding (later RADAR)
RED	Port (usually coupled with a bearing expressed in degrees)
RFA	Royal Fleet Auxiliary
RNAD	Royal Navy Armament Department
SAP	Semi-Armour Piercing

Sec	Secretary
S/M	Submarine
SO (I)	Senior Officer Intelligence
SL	Sea Lord (1 S.L. = 1st Sea Lord, 2 S.L. = 2nd Sea Lord, etc.)
SNO	Senior Naval Officer
SPs	Signal Publications
Sqdn	Squadron
TB	Torpedo-boat
TLs	Their Lordships (Board of Admiralty)
TSDS	Twin Speed Destroyer Sweep
TSMS	Twin Screw Minesweeper
USS	United States Ship
VA	Vice Admiral
VAA	Vice Admiral Aircraft Carriers
VAC	Vice Admiral Commanding
VACBCS	Vice Admiral Commanding Battle Cruiser Squadron
VACRF	Vice Admiral Commanding Reserve Fleet
VAQ	Vice Admiral Commanding Battle Cruiser Squadron
V/S	Visual Signal
WO	War Office
W/T	Wireless Telegraphy

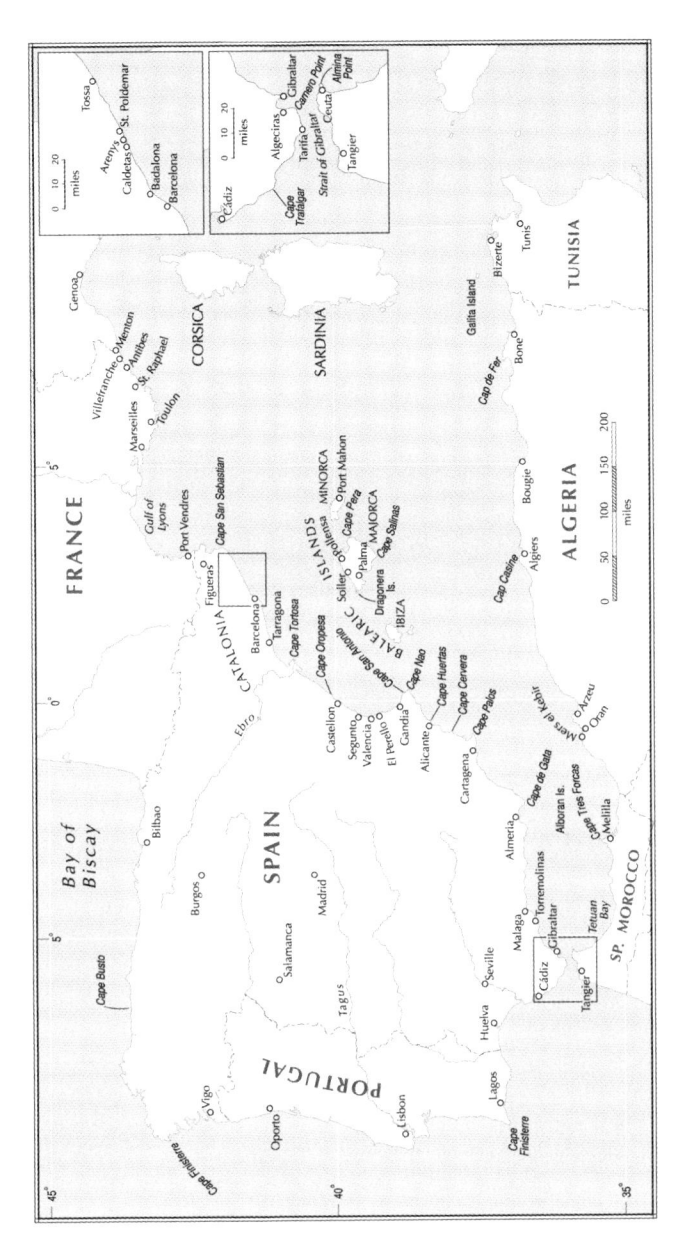

Map 2 The Western Mediterranean

Map 3 The Central Mediterranean

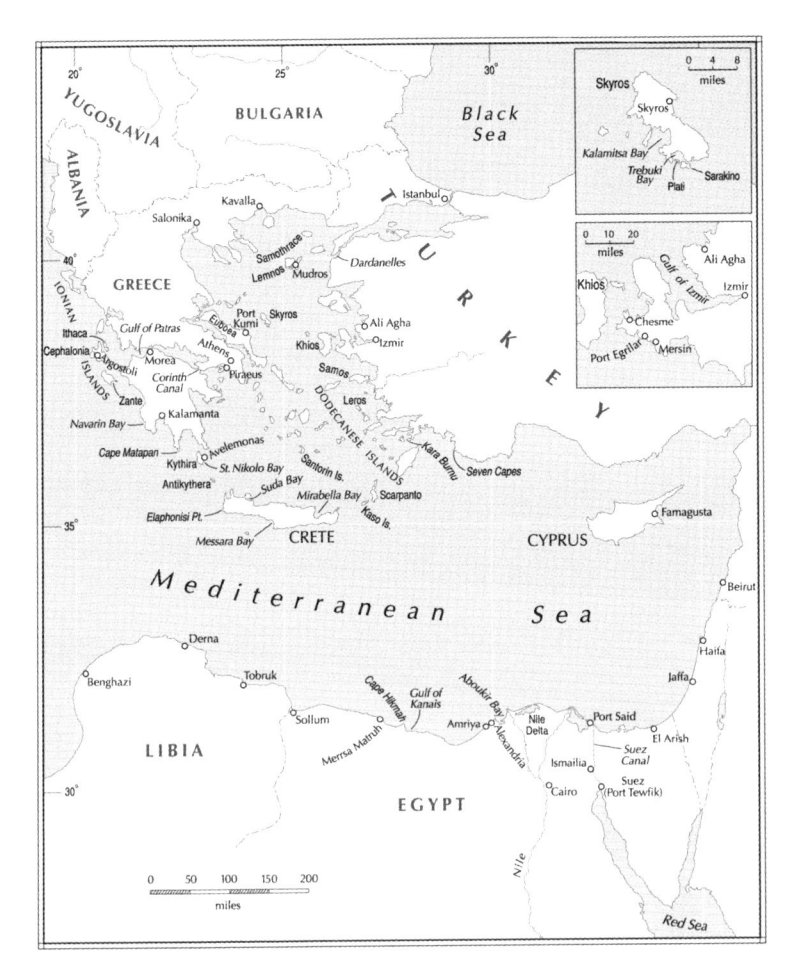

Map 4   The Eastern Mediterranean

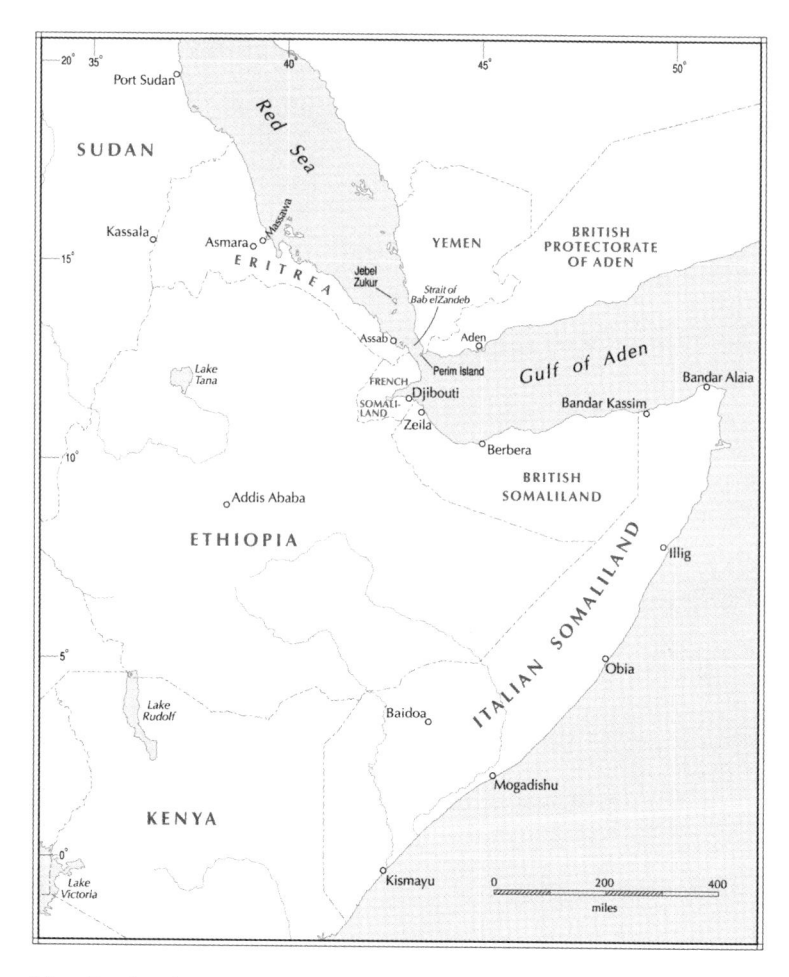

Map 5   Southern Red Sea and East Africa

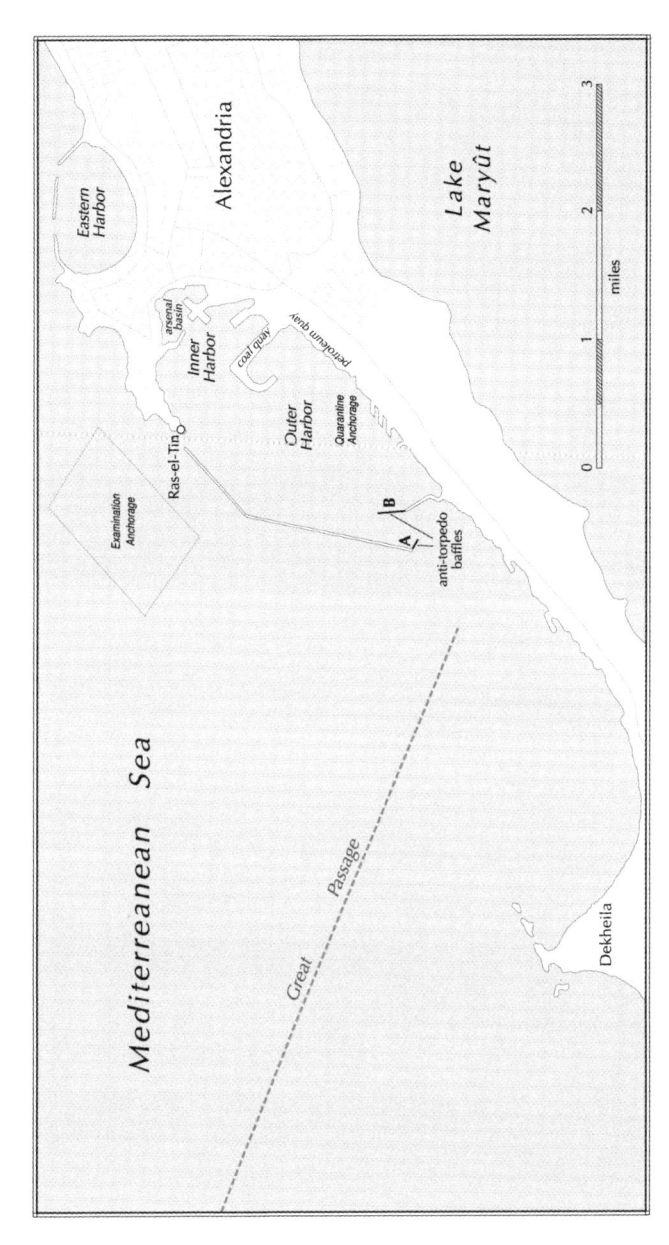

Map 6 Alexandria

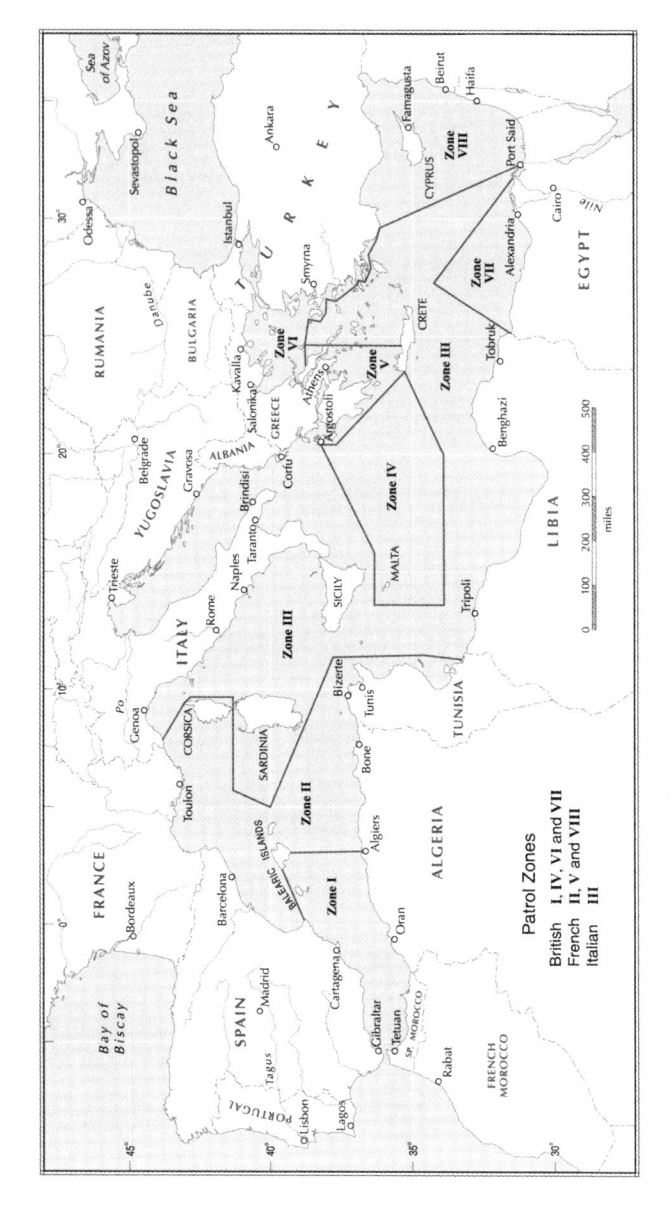

Map 7 Zones Established by the Nyon and Paris Agreements

PART I

THE MEDITERRANEAN FLEET FROM 1930 TO THE
ETHIOPIAN CRISIS

The Mediterranean Fleet began the 1930s while still under the command of Admiral Sir Frederick L. Field. He was succeeded in May 1930 by Admiral Sir Ernle Chatfield, who had been Admiral Beatty's flag captain during the First World War and still had an important career ahead of him as First Sea Lord and Minister for Co-ordination of Defence.[1] With its six battleships, one and often two aircraft carriers, two cruiser squadrons, four destroyer flotillas and a submarine flotilla, the Mediterranean Fleet was Great Britain's premier naval force.[2] It was still more powerful than the naval forces of the two major Mediterranean naval powers, France and Italy. Moreover, it was also destined to play a key role in case of war in the Far East against Japan, proceeding through the Suez Canal and Indian Ocean to Singapore [24]. With the exception of W.W. Fisher, who died prematurely, every Mediterranean Commander-in-Chief during the 1930s subsequently became First Sea Lord.[3] This was in contrast to the preceding decade when for a variety of reasons only Frederick Field (Mediterranean C-in-C June 1928–May 1930) eventually became head of the Navy.

At the beginning of the 1930s there was little to distinguish the activities of the Mediterranean Fleet from those of the late 1920s. The annual routines of training, drills and exercises, cruises and combined exercises with the Atlantic Fleet continued. Chatfield provides a rough outline of the annual routine. There would be a winter cruise in January and a spring cruise in March, generally culminating with combined exercises with the Atlantic Fleet. Rigorous fleet training took place in May and June, a period when the weather at Malta was at its best. The first summer cruise took place in July and August. A much anticipated fleet regatta usually took place during this cruise, often in Greek waters.

[1] For Chatfield's account of his time as Mediterranean C-in-C, see Admiral of the Fleet Lord Chatfield, *The Navy and Defence* (London, 1942), chap. xxiii.
[2] In contrast, the Atlantic Fleet (after 1931, Home Fleet) was: five battleships, three battle cruisers, a cruiser squadron, two aircraft carriers, two destroyer flotillas and a submarine flotilla. Admiralty, *The Navy List, July 1931* (London, 1931), pp. 201–2, 207–8.
[3] Adm Sir Roger Backhouse, the only 1st Sea Lord (November 1938–June, 1939) during the 1930s not to have been C-in-C Med, had been VA commanding First Battle Squadron and 2nd-in-command, Med Fleet, 1932–34.

This would be followed in late August by about a fortnight back at Malta when ships would be refuelled and re-stored. This would be followed by a second summer (later termed autumn) cruise until the end of October. Then came what Chatfield termed a 'self-refit' period back at Malta with the majority of ships laid up for the overhaul of hulls and machinery. The cruises were opportunities for drills and exercises and the fleet had a reputation for playing hard and working hard.[1] The routine could naturally be altered by periodic crises or external events.

Formal visits and relations with foreign fleets were an important part of the diplomatic role of the fleet. A report by the British naval attaché in France of a visit to Toulon in 1931 is included as an example of what was involved [4]. An indication that times were changing appears in the attaché's observation that the conduct of ships' companies ashore apparently reflected a higher standard of education than had been true in the past. Relations with Italy were still cordial and in April 1931 Chatfield could describe his visit to the Italian naval base at La Spezia as 'a happy and informal affair' [3]. He also noted the improvement in appearance of the Italian Navy over the past six years. Not all in the fleet were as sanguine about the Italians and after Mussolini had apparently gotten away with aggressive action against the Greeks at Corfu in 1923 a former officer described the years that followed as 'a period of slow deterioration of the influence exerted upon the Mediterranean Powers by the British Navy'.[2]

The Mediterranean Fleet could put on an impressive display. This was demonstrated in August 1932 when the fleet was concentrated at Corfu and visited by the heir to the throne, the Prince of Wales, and his younger brother Prince George [8]. The Royal Princes carried out what must have been a strenuous programme of inspections in which a high point, literally and figuratively, was their flight over the fleet in aircraft launched from a carrier. This might have had tragic consequences and there were some anxious moments when the sudden onset of a fog bank prevented Prince George's aircraft from landing on the carrier until the fog cleared.[3]

The fleet and combined exercises in 1930 in some ways still seemed to be looking back towards Jutland [1, 2]. There were problems, such as

---

[1]Chatfield, *The Navy and Defence*, pp. 233–9. Chatfield's account is generally echoed by the numerous memoirs, letters, diaries and midshipmen's journals by former members of the Mediterranean Fleet now in the Department of Documents, Imperial War Museum, London.

[2]Kenneth Edwards, *The Grey Diplomatists* (London, 1938), p. 92. Edwards was writing under the influence of the Ethiopian crisis and the Spanish Civil War then in progress and this may have coloured his remarks on the period preceding it. On the Corfu incident see James Barros, *The Corfu Incident of 1923* (Princeton, 1965).

[3]Chatfield's anxieties over the sudden fog are recounted in Chatfield, *The Navy and Defence*, pp. 247–8.

preventing the junction between enemy advanced forces and the enemy main fleet and operations when fleets met late in the day and encounters continued into the night. There was an admission that much still needed to be learnt in night fighting [2]. Chatfield later recalled that 'a large part of my time in the Mediterranean was consequently occupied in developing our night-fighting efficiency'.[1] Nevertheless, naval warfare was changing. Aircraft played an increasingly important role with air attacks executed from a distance of 90 miles and reconnaissance carried out at 'a record' 164 miles [1]. The fleet investigated other problems associated with aircraft, such as defending a fleet anchorage against air attack.[2]

The Mediterranean Fleet may have been imposing at first glance, but it was an ageing fleet. Ships needed refits and modernisation, particularly in regard to countering the developing threat from the air. However, this was also a period of financial stringency with the world entering the Great Depression of the 1930s. A round of pay cuts promulgated in an ill-considered manner resulted in disturbances in the Atlantic Fleet and a temporary refusal of duty that has since been known as the Invergordon Mutiny. The Mediterranean Fleet because of its distance largely escaped these disturbances, although certainly not the hardships caused by the cuts in pay.[3] Moreover, the reduction in monies available for the maintenance of the fleet meant that the number of ships in full commission would have to be reduced. In this context the Admiralty faced the problem of balancing the needs of the Atlantic Fleet with those of the Mediterranean Fleet. In destroyers, for example, each fleet was considered to require three flotillas. In October 1931 the Admiralty proposed a redistribution of strength that would reduce the Mediterranean Fleet by one battleship, an aircraft carrier and a destroyer flotilla [5]. Chatfield not surprisingly argued against this, and not merely because it meant a reduction in his command. With an eye towards recent events at Invergordon, he remarked that foreign service at the moment was 'more wholesome' than home service for personnel and, noting that the Atlantic Fleet was also to lose a battleship, added that he would not favour retention of a second carrier if it meant the loss of two battleships instead of the one proposed [6]. He

---

[1]Ibid., p. 240.
[2]On the growing appreciation of naval air power see Geoffrey Till, *Air Power and the Royal Navy, 1914–1945* (London, 1979), pp. 161–4.
[3]On the Invergordon Mutiny see: Stephen Roskill, *Naval Policy Between the Wars.* Vol. II: *The Period of Reluctant Rearmament, 1930–1939* (London, 1976), chap. 4; Anthony Carew, *The Lower Deck of the Royal Navy, 1900–39: The Invergordon Mutiny in Perspective* (Manchester, 1981), chap. 8. An excellent short account is Christopher M. Bell, 'The Invergordon Mutiny, 1931' in Christopher M. Bell and Bruce A. Elleman (eds), *Naval Mutinies of the Twentieth Century: An International Perspective* (London, 2003), pp. 170–92.

feared this would give strength to the 'false theories' expounded in the press that battleships were of 'questionable utility'. Chatfield recognised that in the confined waters around Europe against minor powers aircraft carriers 'would appear to be even more valuable than battleships' but nevertheless, in this difficult choice he apparently opted for the battleship, to all appearances still the ultimate weapon in his eyes. The Admiralty was quick to point out that the proposed reduction did not imply any diminution of the importance of capital ships and that the proposals really reflected the existing situation when ships would be away on long refits [7]. It was also a question of balancing Home Service with Foreign Service and the fact that specialists such as telegraphists were reluctant to re-engage if it meant long service abroad away from their families. This was another indication of the changing nature of the Navy. Chatfield would later write that he considered the Home Fleet more difficult to command than the Mediterranean Fleet. In the Mediterranean there were long commissions, infrequent leaves and a smaller turnover in personnel. At home there were frequent leaves, changes in personnel occurred more often, and the proximity of the sailor's homes meant that domestic life was apt to 'over-dominate' the life of ships.[1]

Admiral Sir William W. Fisher succeeded Chatfield as Mediterranean Commander-in-Chief in October 1932. Known in the service as 'W.W.', he was one of those flamboyant and commanding personalities that inspire legends.[2] But for his premature death at the age of 62 in 1937, he might well have succeeded Chatfield as First Sea Lord. The eighth child of eleven in a distinguished and intellectual family – his eldest brother was the famous historian H.A.L. Fisher – he was recorded as able to conduct a correspondence in Latin with a Yugoslav professor he had met at Split.[3]

Fisher was soon conscious of the vulnerability of the Mediterranean Fleet to air attack and even more aware of the lamentable state of the defences against air attack of the fleet's bases at Gibraltar and Malta [9]. The problem at Gibraltar had in theory been under study and in November 1931 the Committee of Imperial Defence had made recommendations concerning Gibraltar. There was, however, a big difference between recommendations and action, and while the Admiralty assured Fisher in June 1933 that they understood his concerns, the order of priority to the allocation of scarce funds was to be decided by the Committee of Imperial Defence [11]. Concern about a potential air attack on Gibraltar from North Africa, meaning of course by the French or Spanish, was highly

[1]Chatfield, *The Navy and Defence*, pp. 245–6.
[2]There is a short admiring biography by an officer who served under him. See Admiral Sir William James, *Admiral Sir William Fisher* (London, 1943).
[3]Ibid., p. 3.

theoretical and part of the common defence planning against all contingencies and vulnerabilities in a future that could not be foreseen with certainty. France had been seen as a potential enemy in a so-called 'Locarno War', that is a war in which the French had been the aggressor against Germany and Britain had been obligated under the provisions of the Locarno Treaty to go to the assistance of the victim of that aggression. By the spring of 1933 Hitler and the Nazi regime were in power in Germany and planning of this sort was highly unrealistic, although within a few years a German-backed Nationalist Spain would become a potential threat.[1] Relations with Italy were still good in 1933, but geographical realities could not be ignored and the base at Malta was in easy range of a number of Italian air bases on the island of Sicily or southern Italy. Mussolini's Fascist regime had made much propaganda over its interest in air power and there were some spectacular mass formation flights by the Italian air force. Malta's defences against air attack in June 1933 consisted of a mere four anti-aircraft guns and an RAF squadron of six Fairey IIIF float planes [12]. Fisher was not satisfied with existing plans for improving the islands A/A defences and urged the Admiralty to emphasise to the Air Council the importance of the island in Mediterranean naval strategy and ask them to at least consider the provision of ground facilities necessary for an 'adequate' force of fighter aircraft.[2] He was afraid Chatfield would be inclined to put Malta and Gibraltar 'in cold storage for a bit' after the First Sea Lord's recent successful efforts over the Singapore base and stated succinctly that as Mediterranean C-in-C, 'I can't' [13].

The potential danger to Malta did not come solely from the air. The Governor of the island arranged for a combined exercise in which a landing force was able to inflict severe damage on the dockyard and other key points on the island. Although Fisher considered the air threat the greatest, he pointed out that the exercise had demonstrated that the danger of a military attack also existed and the necessary countermeasures that were then lacking had to be taken [14].

Fisher could not have been pleased by the eventual Admiralty reply to his concerns in October that the question of the scale of attack to be guarded against and the scale of defences to be provided were 'still under examination'. This was coupled with a warning that, in view of even more

---

[1] However, as proof that one must prepare against all contingencies, in July 1940 Gibraltar was the objective of air raids by Vichy French aircraft in retaliation for the British attack on the French Fleet at Mers-el-Kébir.

[2] In the period between the Ethiopian crisis and the outbreak of the world war the Navy and the RAF had differing views on the possibilities of defending Malta from air attack. See Michael J. Budden, 'Defending the Indefensible? The Air Defence of Malta, 1936–1940', *War in History*, vol. 6, no. 4 (November, 1999), pp. 447–67.

serious defence commitments elsewhere, 'financial considerations' might 'preclude the installation of defences at Malta on the full scale at present' [15]. This discouraging reply came virtually at the same moment as the results of a strategical investigation carried out by the fleet. The problem involved a hypothetical situation in which the British would be obliged through the Covenant of the League of Nations to intervene in the event of Italian aggression against Jugoslavia (later Yugoslavia). The conclusions were that Malta would be untenable as a main fleet base in the event of a war with Italy because of the air menace and that the battleships of the fleet would be unfitted for operations where they would be exposed to frequent or heavy attack from the air [16]. Fisher appreciated that Chatfield as First Sea Lord had to weigh the urgency of the Far East as compared to the Mediterranean but argued that 'we are taking an impossible risk about Malta' [17]. At this time there were far too many unfulfilled British defence requirements and far too little available to meet the demands. Fisher reflected this in his rhetorical question: must they leave Malta defenceless so as not to retard the mechanisation of the army?

Approximately fifteen months after assuming his command Fisher reported in January 1934 that he was convinced his flag officers and commanding officers had done and were doing all that was possible to make their squadrons and ships ready for war. There was, however, a serious weakness that owing to a lack of material it was beyond their power to remedy. Chatfield and the Admiralty could not have been surprised after previous reports when Fisher identified the weakness as the fleet's inability to repulse an attack in force from the air [18]. The fleet needed at least double the number of anti-aircraft guns, suitable high angle and anti-aircraft control positions, more anti-aircraft ammunition both aboard ships and in reserve at Malta and a greater allowance of practice ammunition and more suitable targets for practice. The preceding spring the fleet had conducted an interesting trial with an experimental radio-controlled drone, the 'Fairey Queen', as a target for a cruiser and two battleships, but the drone had been shot down by the cruiser relatively quickly and before the battleships had their chance to shoot [10]. Now Fisher argued that until these anti-aircraft requirements were met he could not say the fleet was as ready for war as it should be for, in his view, air attack was the only form of attack that could really harm them. Fisher also pointed out that the Royal Air Force still had no aircraft based at Malta capable of participating in either the defence of the island or an attack on enemy air bases. The questions Fisher had raised about anti-aircraft armament of ships and the defences of Malta were 'under consideration' by the Admiralty at the beginning of 1935, less than a year before the crisis over Ethiopia [23].

The potential role of the Mediterranean Fleet in a war in the Far East coupled with the modernisation programme of the Royal Navy's ageing battleships promised to bring about another redistribution of naval forces. The modernisation programme included extensive modifications – not all completed by the outbreak of war – to the *Queen Elizabeth* class [19].[1] Chatfield proposed transferring the Royal Navy's battle cruisers to the Mediterranean Fleet. These fast ships were intended to act as an advance guard and from a central position in the Mediterranean could operate either in the Far East or in Home Waters [19]. Initially, the manning situation would require the Mediterranean Fleet to lose a battleship in exchange for two – and eventually in 1938, all three – of the Navy's battle cruisers. Chatfield also indicated the Mediterranean Fleet would eventually have two aircraft carriers. The Mediterranean flagship *Queen Elizabeth* was due for a long refit and modernisation and, since none of the *Royal Sovereign* class seemed suitable for conversion to a fleet flagship, the Board of Admiralty finally decided in October 1934 that for reasons of homogeneity all of the *Queen Elizabeth* class would be transferred to the Mediterranean and the slower *Royal Sovereign*s gradually transferred to the Home Fleet between the summer of 1935 and the autumn of 1936 [22]. The large repair and modernisation programme meant, however, that for the next five years there would actually be only four battleships on the station [22].

While relations with Turkey had improved significantly since the crises of the 1920s, incidents could still occur. One of the nastiest happened in July 1934 when Turkish border guards fired on a small skiff carrying British naval officers that had sailed too close to the Turkish coast near the Greek island of Samos. A Surgeon-Lieutenant was killed, his body never found. Fisher interrupted his stay on Cyprus to rush to the scene with other ships of the fleet. In the end the incident was smoothed over, the Turkish government eventually agreeing to pay compensation to the family and render honours – possibly with bad grace – in a memorial ceremony [21]. This was one of these incidents when conflicting emotions were likely. A show of force would be the natural reaction but there were also higher considerations likely to be advanced by the Foreign Office. These included Turkey's position in the diplomatic scene and the possibility of Turkish defence orders from British industry. There were apparently elements within the Turkish government anxious for better relations and the situation ought not to be compromised by the actions of unnamed and possibly trigger-happy frontier guards. Chatfield accordingly

---

[1] For details of these modifications, see Alan Raven and John Roberts, *British Battleships of World War II* (London, 1976), chap. 11.

urged Fisher to put aside the bad taste left by recent incidents and welcome an anticipated Turkish proposal for a visit by the Turkish Fleet to Malta [25]. The very successful visit by the Mediterranean Fleet to Istanbul in 1929 had for varying reasons never been reciprocated by the Turks.[1] This would be the opportunity to do so, although the onset of the Abyssinian crisis delayed the actual visit until the following year.

The majority of the Mediterranean Fleet returned home in the spring of 1935 for the naval review held in honour of King George V's Silver Jubilee. Only a few cruisers and a flotilla of destroyers were left in the Mediterranean. Chatfield was anxious to put on a good show for the first real naval review since 1914 and the Home Fleet alone would not suffice without temporarily mobilising ships from the reserve [20].

In the event of a war in the Far East the security of the Suez Canal was of vital importance and plans were made for the seizure and control of the canal in case of an emergency. Chatfield informed Fisher it was important that the Mediterranean Fleet had forces which could be sent immediately to the canal once a state of emergency was declared [19]. A major concern was the possibility the Japanese would attempt to block the canal with blockships thereby dividing the British forces or, if the fleet had already passed through, severing its lines of communication [24]. Fisher proposed that two cruisers of the Third Cruiser Squadron and the Third Destroyer Flotilla act as an advanced Canal Control Force, sailing from Malta as soon as possible. The remaining two cruisers of the squadron would proceed to Port Said as soon as they were ready and strengthen the defences of the canal [26].

The Mediterranean Fleet during the first half of the 1930s continued its usual activities: namely training, exercises, and showing the flag. When necessary it could also act as an Imperial police force, providing ships and men when required to reinforce the civil and military authorities in Egypt or Cyprus.[2] At sea the fleet and its commanders were confident they could handle any rival naval force likely to face them. But the fleet was ageing. The Mediterranean was becoming something of a 'closed lake' as the performance of aircraft improved and this represented a threat the ships were not completely prepared to face. This was particularly evident to Fisher who regarded air attack as the only thing that could really hurt them. Unfortunately, as Chatfield made clear, it was a hard struggle in the era of financial stringency to prise sufficient funds from the

---

[1] On the visit in 1929 by the Mediterranean Fleet see Paul G. Halpern (ed.), *The Mediterranean Fleet, 1919–1929* (Farnham, 2011), Docs Nos 327 and 328.

[2] For example, Chatfield in his flagship accompanied by the battleship *Ramilles* proceeded to Alexandria after rioting occurred there in September 1930 and ships of the fleet were also sent to Cyprus following severe disturbances in the autumn of 1931.

Treasury for modernisation, let alone for a naval construction programme, at a time when the other services had equally pressing needs. In the spring of 1935 the threats had been to a certain extent theoretical, planning for all contingencies, some highly unlikely. With the Italian invasion of Ethiopia, a fellow member of the League of Nations, the threats of air attack became all too real if the British were drawn into conflict as a result of their obligations under the Covenant of the League. The year 1935 became one of crisis for the Mediterranean Fleet and the relatively peaceful decade which had preceded it would not return until years after the Second World War when Great Britain would be in very different circumstances. It is not surprising that many looked back on the late 1920s and early 1930s with a great deal of nostalgia.

1. Summary of 1930 Mediterranean Fleet[1] Exercises (1)

[ADM 186/147]
Tactical Division, Naval Staff
November 1930

C.B. 1769/30 (1)

* * *

No. 1. Exercise O.P. [9–11 March 1930, Atlantic and Mediterranean Fleets] *A strategical and tactical exercise in which an attempt is made to prevent a junction between enemy advanced forces and their reinforcing main fleet.*

The principle of drawing lots for certain alternative conditions was again employed satisfactorily. Aircraft and submarines played particularly important parts in the exercise. The depth of successful air reconnaissance reached a record of 164 miles, and air attacks took place at a distance of more than 90 miles from the point of departure.

No. 2. Exercise Z.C. [17–18 March 1930, Atlantic and Mediterranean Fleets] *A tactical exercise with two fleets meeting late in the day without decisive results. Operations continue during the night and action is renewed next morning.*

This exercise demonstrated the two-edged nature of a smoke screen. Flotillas were able to advance under the cover of the enemy's screen, which also hampered their own secondary armament fire, and made their attacks on the battle fleet, which was then still in the process of re-forming for the night. During the night, both Commanders-in-Chief were placed out of action and subsequent operations were conducted by the Seconds-in-Command.

* * *

No. 4. Exercise D.A. [2 April 1930] *A tactical exercise in which the chasing fleet endeavours to delay the enemy's retirement by means of destroyer and air attacks.*

Destroyers attacked in the two-divisional formation – torpedoes being fired. The torpedo bomber, fighter and bombing attacks by aircraft were well synchronised with the arrival of destroyers' torpedoes.

* * *

---

[1] Exercises conducted by the Atlantic Fleet are not reproduced.

2. Summary of 1930 Mediterranean Fleet Exercises (2)

[ADM 186/148]　　　　　　　　　　　Tactical Division, Naval Staff
　　　　　　　　　　　　　　　　　　　　November 1930

C.B. 1769/30 (2).

* * *

No. 7. Exercise O.X. [29 October 1930] *A tactical exercise with two fleets meeting late in the day without decisive results. Operations continued during the night., and included a night action between heavy ships.*

This exercise showed how much there is to be learnt in night firing. An interesting feature was the illumination of the enemy battle fleet by the searchlights of a detached force.

This exercise was framed similarly to Exercise Z.C. (C.B. 1769/30 (1)) but in that exercise no night action materialised.

No. 8. Exercise O.Y. [28 October 1930] *To investigate the methods of defending a fleet in harbour against air attack. The exercise was arranged to continue the investigation begun in Exercise O.L. (C.B. 1769/29 (2)).*

A satisfactory anchorage plan was arranged and separate arcs of fire for H.A. and L.A. were allocated

Seventeen aircraft took part.

Several targets were selected by the attackers, with satisfactory results. In 'O.L.' the attack was concentrated on one ship.

Several points requiring further investigation were brought out.

* * *

3. *Admiral Sir Ernle Chatfield[1] to Admiralty*

[ADM 1/8752/198] *Queen Elizabeth* at Villefranche
24 April 1931

CONFIDENTIAL

No. 676/561/114.

Be pleased to lay before Their Lordships the following report on my visit to Spezia in *Queen Elizabeth*, 8th–10th April, 1931.

2. My visit was intended primarily as a personal one on Vice-Admiral F. Gambardella, Commanding the First Squadron, who had been good enough to come to Rhodes to meet me in October last.

* * *

4. On the evening of the same day [8 April] an official dinner was given by Vice-Admiral Gambardella on board his flagship, *Trieste*, to myself, my Staff and ten officers of *Queen Elizabeth*. It was also attended by the principal civil authorities. Although this was an official function and speeches were made, the very friendly attitude of our hosts made it, in actuality, a happy and informal affair and quite free from the initial reserve that officers of two different nations usually experience when meeting for the first time. I was very impressed by the evident desire on the part of the Italian Admirals and their officers of taking that earliest opportunity of displaying such complete friendliness and appreciation of our visit.

Vice-Admiral Gambardella in a speech lasting some five minutes, warmly welcomed myself and my flagship. He referred particularly to the comradeship between the two navies during the War and stressed the pleasure felt by himself and the Italian Navy at the visit of one of H.M. Ships. He spoke in Italian which was translated paragraph by paragraph into English by Lieutenant Mimbelli.[2] I replied in English and my speech was translated into Italian by Captain Bevan.[3]

* * *

[1] Adm [later AoF] Sir Alfred Ernle Montacute Chatfield (1873–1967). Created Baron, 1937. Flag Capt to Beatty in battlecruiser *Lion*, 1914–16; Flag Capt and Fleet Gunnery Officer to Beatty when latter was C-in-C Grand Fleet, 1917–19; 4th Sea Lord, 1919–20; ACNS, 1920–22; commanded 3rd LCS, 1923–25; 3rd Sea Lord and Controller of the Navy, 1925–28; C-in-C Atlantic Fleet, 1929–30; C-in-C Med, 1930–32; 1st Sea Lord and CNS, 1933–38; Min for Co-ordination of Defence, 1939–40.
[2] Attached to Chatfield and his staff as personal liaison officer.
[3] Capt [later RA Sir] Richard Hugh Loraine Bevan (1885–1976). Flag Capt and COS to C-in-C Africa Station, 1924–26; Naval Attaché in Italy and Greece, 1928–31; commanded Signal School, Portsmouth, 1934–35; retired, 1935; recalled and Naval Attaché, Rome, Feb–June 1940; Senior British Naval Officer, North Russia, 1941–42; Flag Officer-in-charge, Belfast, 1942–45; KBE, 1946.

7. I am glad to report that this visit was an undoubted success and has certainly strengthened the good feeling between the two navies. The good will of the two Commanders-in-Chief[1] and of all officers in their commands was unmistakably sincere and the happiest relations existed throughout our stay between the personnel of the two nations. It was a great personal pleasure to me to meet Vice-Admiral Gambardella again and I trust that the Mediterranean Fleet will have the honour of receiving an early visit from him and his squadron at Malta.

One cannot fail to be impressed by the fine bearing and smartness of the officers and men of the present-day Italian Navy. The improvement in appearance of both the personnel and their ships in the last six years has been striking and the state of the units at Spezia and their officers and ships' companies was very creditable.

The references in the principal Italian newspapers to the presence of the *Queen Elizabeth* at Spezia were lengthy and were expressed in very cordial terms, particular reference being made to the 'comradeship and friendship between the two navies'.

* * *

4. *Captain Guy W. Hallifax[2] to Lord Tyrrell[3]*

[ADM 1/8752/198]   British Embassy, Paris
8 May 1931

[Carbon]

France No. 8/31.

I have the honour to report that the reception accorded to Admiral Sir Ernle Chatfield, K.C.B., Commander-in-Chief of the Mediterranean Fleet, at Toulon recently [25–27 April] was very cordial.

The *Queen Elizabeth*, the Commander-in-Chief's flagship, only stayed 3 days in the port, but they were crowded days and the festivities arranged included an official dinner and a ball on separate nights at the Prefecture Maritime, an official lunch on board the *Provence*, flagship of the French

---

[1]VA Gambardella (Commanding 1st Sqdn) and VA Monaco Duca di Longano (Commanding Upper Tyrrhenian).

[2]Capt [later RA] Guy Waterhouse Hallifax (1884–1941). Naval Attaché in France, 1928–31; retired 1935; commanded South African 'Seaward Defence Force', 1940–41.

[3]William George Tyrrell (1866–1947). Created Baron, 1929. Sec of Cttee of Imperial Defence, 1902–04; Private Sec to Sir Edward Grey, Sec of State for Foreign Affairs, 1907–15; Asst Under-Sec of State for Foreign Affairs, 1919–25; Perm Sec of State for Foreign Affairs, 1925–28; Amb at Paris, 1928–34.

Commander-in-Chief of the 1st Squadron,[1] and a gala performance at the opera which was preceded by a dinner given by Admiral Chatfield on board the *Queen Elizabeth*, to which all the local naval high authorities were invited. There were also 'matinées dansantes' on board both the *Queen Elizabeth* and the *Provence*.

Sports and entertainments were arranged for the ship's company, and it was gratifying to see in the streets a considerable amount of sober fraternising between the sailors of the two Countries. It was not so many years ago that our men seldom fraternised with foreign sailors, except under the influence of drink. I think it is one of the beneficial results of the higher educational standard attained in the Navy that the men not only realise the importance of these visits, but try to ensure their success by fraternising with the local sailors.

As usual the two Navies got on very well together, but I fancy that the local population were not so enthusiastic about the visit – or at least not so demonstrably so – as was the civilian element at Brest on the occasion of the *Nelson*'s visit last year.

5. *Admiralty to Admiral Sir Ernle Chatfield*

[ADM 116/2860]   23 November 1931

SECRET.

[Carbon]   M.02804/31.

I am to acquaint you that in connection with the probable redistribution of the Fleet mentioned in Confidential Admiralty Fleet Order 2572/31 of the 30th October, 1931,[2] the following alterations in the strength of the Mediterranean Fleet are tentatively under consideration, and it is requested that you will forward your remarks on the subject.

Capital Ships.

2. It is proposed that the number of Battleships should be reduced by 1. This would mean that H.M.S. *Ramilles*, on her return from the Mediterranean during 1932, to pay off and to be taken in hand for large repair, would cease to belong to the Mediterranean Fleet. The ship would,

---

[1]Vice Amiral Georges Robert (1875–1965). Second and subsequently first sous-chef of naval general staff, 1924–28; commanded torpedo boat flotilla of 1re escadre, 1928–29; Dir of Military Personnel, 1929–30; C-in-C, 1re escadre, 1931–32; Inspector Gen of Maritime Forces in the Med, 1932–35; head of French delegation at London Naval Conference, 1935; retired 1937; recalled to service on outbreak of war and named C-in-C West Atlantic Zone and High Commissioner for the Antilles, September 1939–July, 1943; dismissed in September 1944 and subsequently tried and condemned by the High Court for his support of the Vichy regime, but the punishment was set aside.

[2]Not reproduced.

however, be available subsequently as a relief for another ship of the class (probably H.M.S. *Royal Oak*) when the time arrives for the latter to undergo large repair.

Aircraft carriers.

3. It is proposed that the number of aircraft carriers should be reduced by 1. This would mean that H.M.S. *Eagle*, now paid off for repairs, would not rejoin the Mediterranean Fleet, which would therefore have only one Aircraft Carrier.

4. As regards Destroyer Flotillas, it has been decided that the 2nd Destroyer Flotilla, due to return home in 1932, will not return to the Mediterranean, but will be attached to the Atlantic Fleet. Each of these Fleets will then have 3 Destroyer Flotillas. Separate instructions have been given in Admiralty Letter M.02672/31 of the 23rd instant[1] to this effect.

5. An early reply to this letter is requested.

6. A copy of a letter addressed to the Commander-in-Chief, Atlantic Fleet, on the subject of alterations proposed in the strength of the Atlantic Fleet is enclosed.

6. *Chatfield to Admiralty*

[ADM 116/2860]   at Malta, 21 December 1931

SECRET.
COMMANDER-IN-CHIEF, MEDITERRANEAN STATION
No. 2060/596/88.
STRENGTHS OF MEDITERRANEAN AND ATLANTIC FLEETS.

Be pleased to lay before Their Lordships the following remarks on Admiralty letter, M.02804/31, of 23rd November, 1931, on the above subject.

2. I fully appreciate that the reduction of the money granted for the maintenance of our Naval strength renders inevitable measures of reduction in the number of ships in full commission; at the same time, it appears desirable, in considering the best steps of this nature to be taken, that it should be guided partly by any action of a similar nature to be taken by the navies of our possible enemies.

I have no information as to this and, therefore, such remarks as I am able to make lack this basis of consideration.

3. Although Their Lordships have only requested my remarks on the proposals affecting the Mediterranean Fleet, they have been pleased to send me a copy of the corresponding proposals for the Atlantic Fleet. I

---

[1] Not reproduced.

find it impossible to deal suitably with the proposals for one fleet only, in view of our possible dependence in war on a two-fleet combination (wholly or partial) and I have therefore ventured to deal with the problem in a more general manner than was perhaps Their Lordships' intention.

4. The reduction of three capital ships is one that I know Their Lordships will have proposed only with great regret and anxiety.[1] Not only is it a 20 per cent reduction of immediate fighting strength but it is a 20 per cent reduction in capital ship training for Captains, officers and men. The capital ships are the most important schools afloat for all ranks and ratings; in them every branch of naval warfare and weapon technique can be seen and taught.

On the correct handling of the British main fleet much may in the future depend. Adequate practice is sufficiently difficult to obtain already; with the reduced number of ships it will be still rarer.

Further, in view of the false theories now expounded in the press, any impression that might be gained by the paying off of three capital ships, such as that battleships are of questionable utility, would be specially unfortunate.

5. Experience has already been obtained both in the Mediterranean Fleet, in 1920–23, and in the Atlantic Fleet in later years with the Third Battle Squadron, of having a fleet with some ships fully and others partly manned. It did not prove satisfactory, as the fleet life and training were much handicapped by the inequality of what could be expected from complements of such different strength. I suggest, therefore, that a general and equal reduction of complements in one fleet (as proposed in paragraph 3(b) of Admiralty letter, M.02804/31, of 23rd November, 1931, addressed to the Commander-in-Chief, Atlantic Fleet) is far preferable to any unequal reduction.

6. Further, I am of opinion that a general reduction of complements in one fleet is a better step than paying off ships. If war were to break out unexpectedly, the reduced ships would only need filling up, they would be in running order and have onboard a strong nucleus of trained personnel.

I would, therefore, favour an even larger reduction in complements in the Atlantic Fleet, i.e. by, say 30 per cent and the maintenance of a correspondingly larger number of ships in commission.

7. If this can be accepted in principle, I am in favour of the maintenance of the Mediterranean Fleet at its present strength and any unavoidable reduction in big ships in commission being made in the Atlantic Fleet.

---

[1] The Admiralty, in addition to reducing one battleship in the Mediterranean Fleet, had also proposed reduction of the Atlantic Fleet by one battleship and one battle cruiser.

There are indications that foreign service is, at the moment anyhow, more wholesome than home service. For that reason, any reduction of strength in the Mediterranean seems inadvisable; the tendency should be the other way.

I believe there is generally a desire among the younger men for foreign service, the older and married preferring to stop in Home waters. I am of the opinion that the pressure of the married men to increase Home service should be resisted. A stay-at-home navy is undesirable; joining the Navy should be synonymous with going abroad and those who do not wish to do so are not wholly fitted for a sailor's career. A scheme of complement that could differentiate between home and foreign service could, I believe, be accepted on stations abroad but would doubtless involve a reduction in the number of higher ratings (to avoid superfluity at home) but some such reduction would be welcome, to stimulate the flow of promotion.

8. I hope it will not be considered that in expressing this opinion I am in any way prejudiced by being myself in the Mediterranean Fleet. I have considered the matter solely from the broadest aspect of our naval policy.

9. If these suggested slight modifications of Their Lordships' proposals are not acceptable, I concur in the reductions in the Mediterranean Fleet proposed in the Admiralty letter under reply. The reduction of an aircraft carrier is, of course, regretted. This question is again largely governed by the nature of any war menace for which the Mediterranean Fleet may be called on to train. If such may be expected to occur in the confined waters around Europe against minor Powers, aircraft carriers would appear to be even more valuable than battleships, but I should not advocate the retention of a second carrier if it necessitated the reduction of two battleships instead of the one now proposed.

10. The decision of Their Lordships regarding the flotillas is noted and, I admit, follows on the combined recommendations of the Commanders-in-Chief, Mediterranean and Atlantic Fleets, that the Atlantic Fleet needs three flotillas. Nevertheless, I submit that the reduction of the Mediterranean Fleet flotillas to three will decidedly affect the development of destroyer tactics, particularly at night. Both by day and night three flotillas are the minimum required to form a proper striking force and they must be faced with opposition if incorrect lessons are to be avoided. The progress that I feel may be claimed in night destroyer work in the Mediterranean Fleet in the past two years has only been possible by the possession of four flotillas.

11. If, therefore, only six flotillas in all can be afforded, the present distribution appears to me soundest. If, however, Their Lordships are unable to approve this, I consider it most desirable that the flotillas of the

two fleets shall meet annually for longer periods, to continue the work already begun. Skeleton forces can, of course, be used, but to trust too much to the lessons drawn from skeleton forces is undesirable, apart from the breaking up of the flotillas that is entailed in this method.

7. *Admiralty to Chatfield*

[ADM 116/2860]   1 March 1932

CONFIDENTIAL.
[Carbon]   M.0308/32.

Their Lordships desire me to refer to your submission No. 2060/596/88 of the 21st December, 1931, concerning the strength of the Mediterranean and Atlantic Fleets.

2. The decisions of Their Lordships as to the future constitution, etc., of these Fleets are being communicated separately to yourself and to the Commander-in-Chief, Atlantic Fleet, and are also being promulgated in A.F.O.'s ...

3. Before Their Lordships' decisions were reached, the various relevant factors, including the present and prospective naval strength of other Powers, the maintenance of our prestige, and our possible dependence in war upon a two-fleet combination, had received the fullest consideration.

4. Their Lordships' decision to reduce the number of Capital Ships in the Mediterranean from six to five, and to treat three of the total number of Capital Ships, viz. fifteen, as being in reserve, was based partly upon the fact that already at any given date two, and sometimes three of our Capital Ships, including one and sometimes two of those allocated to the Mediterranean Fleet, are undergoing Large Repair or refit (for which they may be paid off or reduced to special complement) and would on this account be unable to proceed on active service at short notice.

5. In these circumstances, the total number of Capital Ships available at any given time, either for service or for the training of personnel, does not exceed twelve or at most thirteen; nor does the number so available in the Mediterranean exceed five, and it may not, on occasions, exceed four. The decisions now taken are, therefore, little more than a recognition of a situation which already exists and is inevitable.

6. I am to state in this connection that Their Lordships intend that during the prolonged absence for Large Repair of any one of the five Capital Ships to be allocated in future to the Mediterranean, one of the three Capital Ships in reserve shall be commissioned for temporary service in the Mediterranean in her place, should this appear to them to be necessary.

7. Your suggestion that, as an alternative to the course which is being taken, there should be a general and equal reduction of complements (say by 30%) in the Atlantic Fleet, would not have similarly have accorded with the actual situation, would have impaired the effective strength of the Fleet affected, and would have involved serious difficulties on account of shortage of personnel both for training and for the ordinary running and maintenance of the ships. Their Lordships have, however, decided that the complements to be carried by the seven Capital Ships which are to be kept in commission at Home are to be on the same basis for all.

8. In reaching their decisions concerning Capital Ships, Their Lordships have been far from overlooking the importance of Capital Ships to any great Naval Power – they are mindful of the inestimable value of our Capital Ships to the allied cause in the European War, and their present decisions are not to be taken as implying any lack of conviction that in any future Naval War in which this country may be engaged an adequate number of modern Capital Ships will be a prime necessity. These decisions are related solely to the distribution and status of our Capital Ships at the present time.

9. The decision to reduce the number of destroyer flotillas in the Mediterranean from four to three was made after the fullest consideration and Their Lordships are satisfied that the change made is in the best interests of both Fleets. Your suggestion that the Flotillas of the two Fleets should meet annually for longer periods will receive due consideration.

10. As regards the incidence of Home and Foreign Service, the question is one of practical expediency. Before and during the war and for a few years after it, policy demanded a large concentration of ships in Home Waters, and the portion of ships abroad was in consequence comparatively small. Latterly, there has been a reversal of this distribution, and a resulting large alteration in the percentage of Home and Foreign Service. The transfer of four '*Queen Elizabeths*' from the Mediterranean to the Atlantic Fleet in 1929 modified the position but against this must be placed recent policy to station abroad a greater proportion of large new cruisers. The percentages of Home and Foreign Service in 1921 and 1931 respectively are as follows:

	Home	Foreign
1921	78.7%	21.3%
1931	56.7%	43.3%

Great difficulty in manning is now resulting from the unwillingness of ratings in certain branches, e.g., Telegraphists, to re-engage. This is said to be due partly to their long hours when serving in the large Fleets and partly to the reduced amount of their Home Service. Moreover, without a proper proportion of men at home it is not possible to arrange for the requisite numbers to undergo the qualifying courses for promotion. The whole matter of Home and Foreign Service, however, together with kindred questions, is at present under consideration.

8. *Chatfield to Admiralty*

[ADM 1/8761/245] H.M.S. *Queen Elizabeth*, at Malta
25 August 1932

No. 1371/109/14.
VISIT OF THEIR ROYAL HIGHNESSES
THE PRINCE OF WALES AND THE PRINCE GEORGE
TO THE MEDITERRANEAN FLEET.

Be pleased to inform Their Lordships that the Mediterranean Fleet under my command assembled at Corfu[1] on Saturday, 13th August, in preparation for inspection by Their Royal Highnesses the Prince of Wales[2] and Prince George.[3]

* * *

2. Their Royal Highnesses arrived at Corfu at 2115 (Corfu time – Zone minus 3) on Saturday, 13th August by Imperial Airways Flying Boat *Satyr*, which was accompanied by the R.A.F. Southampton flying boats *S.1058* and *S.1302*.[4]

As Their Royal Highnesses arrived after dark and therefore too late to permit them to be received with the customary ceremony that day, they came onboard H.M.S. *Queen Elizabeth* in plain clothes, the Standard of H.R.H. The Prince of Wales being broken in that ship.

In the evening I gave an official dinner at which the Flag Officers and a number of Captains and other Senior Officers of the Fleet had the honour of meeting Their Royal Highnesses.

The official dinner was followed by a searchlight display by the Fleet and a tattoo performed by the Royal Marine Band of the Fleet Flagship.

[1] The ships present during the period of inspection included: 4 battleships; 1 aircraft carrier; 3 8-inch gun cruisers; 4 6-inch gun cruisers; 22 destroyers and destroyer repair ship *Sandhurst*; 5 submarines and submarine depot ship *Cyclops*; fleet repair ship *Resource*; despatch vessel *Bryony*; hospital ship *Maine*; target service ship *Chrysanthemum*; and RFA auxiliary (store carrier) *Perthshire*.
[2] Edward, Prince of Wales (1894–1972). Eldest son of King George V. Succeeded father as King Edward VIII, 20 January 1936; abdicated 10 December 1936 and given title Duke of Windsor; liaison officer with French army, 1939–40; Govr of the Bahamas, 1940–45.
[3] George, fourth son of King George V (1902–42). Created Duke of Kent, 1934. Served in RAF rising to rank of Air Cdre, 1939–42; killed in air crash, August 1942.
[4] The Princes had flown from Brindisi to which they had travelled by train from Venice.

3. On Sunday, 14th August, at 0800, ships dressed overall and a Royal Salute was fired.

During the early part of the forenoon the officers of my Staff and of H.M.S. *Queen Elizabeth* were presented to Their Royal Highnesses, and the Prince of Wales then inspected H.M.S. *Queen Elizabeth* and attended Divine Service.

Prince George inspected *Chrysanthemum* and *Sandhurst* and attended Divine Service in the latter ship.[1]

* * *

6. On Monday forenoon, 15th August, H.R.H. The Prince of Wales inspected H.M. Ships *Revenge, Coventry, London, Colombo, Cyclops* and the submarines were berthed alongside her, *Douglas, Codrington* and *Keith*.

H.R.H. The Prince George inspected R.F.A. *Perthshire*, Hospital Ship *Maine, Bryony* and the eight destroyers of the First Flotilla.

* * *

By His Royal Highness's command ships' companies were given a 'make and mend clothes' in the afternoon and spliced the main brace in the evening.

7. In the evening an official dinner was given on board *Queen Elizabeth*, to which the Greek Minister of Marine, the A.D.C. to the President and the principal local authorities at Corfu were invited in addition to a number of officers of the Fleet.

After dinner Their Royal Highnesses attended an entertainment and an 'At Home' given on board *Queen Elizabeth*, when opportunity occurred to present a number of officers of all Branches.

From 2130 to 2330 H.M. Ships were illuminated.

8. On Tuesday forenoon, 16th August, H.R.H. The Prince of Wales inspected H.M. Ships *Royal Oak, Resolution, Resource, Sussex, Curlew, Ceres* and *Shropshire*. H.R.H. The Prince George inspected the destroyers of the Third and Fourth Flotillas.

9. At 1330 on Tuesday, 16th August, the Fleet weighed and proceeded to sea. When clear of the channel H.M.S. *Glorious* closed and Their Royal Highnesses were transferred to her to witness from the air a Torpedo Aircraft attack on the Fleet.

As reported to Their Lordships in my message 0025 of 17th August, a sudden bank of fog developed after the aircraft had completed their attack

---

[1] The remainder of the day involved activities arranged by the Greek authorities on shore.

and were returning to the Carrier. The Fairey IIIF carrying the Prince of Wales landed-on immediately before the fog enveloped H.M.S. *Glorious* but the remaining machines, including the Fairey IIIF carrying Prince George, had to delay landing-on until *Glorious* had passed through the fog bank.

* * *

10. It had been intended to carry out a submarine attack on the Fleet on conclusion of the air exercise but, owing to the delay reported in the previous paragraph, the re-embarkation of Their Royal Highnesses in H.M.S. *Queen Elizabeth* could not be effected in time to carry out the submarine exercise.

After dark a Night Destroyer Attack on the Battlefleet was carried out, but, owing to the risk of fog, Air Shadowing was not carried out as had been intended.

11. On Wednesday forenoon, 17th August, an exercise was carried out designed to continue the investigation of chasing and retiring tactics and the employment of smoke by a retiring fleet.

In the afternoon the Fleet was exercised at Junior Officer Manoeuvres and subsequently H.M.S. *Revenge* carried out a demonstration firing from her Mark 'M' Pom-pom.

After dark a night encounter exercise, designed to demonstrate the use of all types of illuminants including aircraft flares, took place.

12. On Thursday forenoon, 18th August, a Day Destroyer Attack on the Battlefleet, combined with target and throw-off firings by the Flotillas, was carried out off Malta.

13. Very keen interest in all the exercises at sea was taken by Their Royal Highnesses. The whole sea programme had been designed before the visit of Their Royal Highnesses was known and was not altered in any respect.

14. At 1100 the Fleet entered the Grand Harbour, H.M.S. *Queen Elizabeth*, wearing the Standard of His Royal Highness, following H.M. Ships *London* and *Revenge* into harbour.

15. A Royal Salute was fired by the Army as H.M.S. *Queen Elizabeth* entered, and ships were dressed overall on securing.

The Rear-Admiral, Malta, the Surgeon Rear-Admiral, R.N. Hospital and the Captains commanding Establishments and H.M. Ships which had not been present at Corfu, came on board H.M.S. *Queen Elizabeth* and were received by Their Royal Highnesses.

At 1220 Their Royal Highnesses embarked in the Royal Barge to proceed on shore. A royal salute was fired and Their Royal Highnesses were cheered as they passed each ship. On landing at the Customs House

they were received by His Excellency the Governor[1] and proceeded to San Antonio Palace where they stayed until their final departure on Saturday night, 20th August, in H.M.S. *Shropshire*, for Cannes.
16.

* * *

I should like to take this opportunity of submitting the deep appreciation of myself and the personnel in my command at the generous action of Their Royal Highnesses in coming so great a distance to visit this Fleet and gladly undertaking a heavy programme of inspections during the very hot weather experienced at Corfu.[2] The visit has been most impressive and has been of immense value and encouragement to every officer and man under my command.

[*Enclosure*]

COPY OF MESSAGES.[3]

* * *

From....His Majesty The King.
To........The Commander-in-Chief, Mediterranean Station.
Date.....24th August, 1932.
  On the conclusion of the visit of my two sons to the Mediterranean Fleet, I wish to express my warmest appreciation of your kindness to them. It was especially good of you to send them to Cannes in the *Shropshire*. I am glad that you were able to arrange for their visit to so many ships at Corfu and that on the way to Malta they witnessed the normal fleet training at sea. From all I hear the Fleet under your command is in a high state of efficiency and the tradition not only of the Mediterranean Fleet but of the whole Service is worthily maintained. George R.I.

REPLY.

  With my humble duty I thank Your Majesty for your most gracious message which is deeply appreciated. That Their Royal Highnesses should have visited the Fleet and also have watched its work at sea has given the greatest encouragement to us all. It will be our continuous endeavour to be worthy of the confidence Your Majesty places in us.

[1] Maj Gen Sir David Graham Campbell (1869–1936). Commanded 21st Division, 1916–18; Military Sec, 1926; GOC, Aldershot, 1927–31; Gov of Malta, Oct 1931–Mar 1936.
[2] A midshipman in the *London* thought the Prince of Wales looked very tired and considered the programme a lot for him to do in one forenoon. Midshipman H.H.E. Kemball to his parents, 22 Aug 1932, Imperial War Museum, Kemball MSS.
[3] The telegrams exchanged between the Prince of Wales and Chatfield are not reproduced.

9. *Admiral Sir William W. Fisher[1] to the Admiralty*

[ADM 116/3473]   10 April 1933

SECRET
THE COMMANDER-IN-CHIEF, MEDITERRANEAN STATION.
Med. 081/040.
Subject: GIBRALTAR – DEFENCES.

Be pleased to lay before the Board a request, that I may be informed what steps have been taken by the War Office to provide and install the military material necessary to render Gibraltar secure as a Naval Base in war.

2. It appears that to date little, if any, of the recommendations by the Committee of Imperial Defence set out in C.I.D. Paper 365-C entitled 'The Defence of Ports at Home and Abroad – Gibraltar' and dated November, 1931, have been put into effect.

3. Other forms of defence will have to be provided before Gibraltar can be considered in a proper state to discharge its function in war, but I regard the lack of adequate defence against air attack launched from North Africa as the most important deficiency [of] the existing defences.

4. If there is any chance of hostilities in Europe, the proper defence of Gibraltar, as well as Malta, does not brook indefinite postponement, and I consider the early provision of the necessary anti-aircraft guns and anti-aircraft lights at Gibraltar [to] be a matter of great importance.

10. *Fisher to Chatfield*

[CHT/4/5]   Malta, 29 May [1933]

[Holograph]
We were to have gone out today for the Fairey Queen[2] trials but it is blowing hard from the westward – quite a sea so I have postponed it 24 hours. The F.Q. did a successful trial flight from Halfar – pilot controlling and I hope that both 'operations' – firing will be successful.

---

[1] Adm Sir William Wordsworth Fisher (1875–1937). Commanded battleship *St Vincent* at Jutland, 1916; Dir Anti Submarine Division, Admy, 1917–18; COS in Med Fleet, 1919–22; COS in Atlantic Fleet, 1922–24; RA in 1st Battle Squadron, Med Fleet, 1924–25; Dir of Naval Intelligence, 1926–27; 4th Sea Lord, 1927–28; Dep CNS, 1928–30; VA commanding 1st BS and 2nd-in-command, Med Fleet, 1930–32; C-in-C Med, Oct 1932–Mar 1936; C-in-C Portsmouth, 1936–37.

[2] An experimental radio-controlled, auto-stabilised gunnery target version of the Fairey IIIF. Only three were built: see H.A. Taylor, *Fairey Aircraft since 1915* (Annapolis, 1974), pp. 154–7.

Sussex   will tackle the High bombing mission.
Revenge   "   "   "   Medium height ".
R.[oyal] Sov[ereign]   "   "   "   Mark M. Pom-pom

I am satisfied that the ships have been well trained – but to produce hits with such a small volume of fire as is given by a pair of guns only in the bombing runs may prove difficult. If we can't I think that we shall probably be able to say that it is not the system which is at fault but the inadequate number of guns.

There is lots of interest going on. *Adventure*'s mining trials – TSDS[1] sweeps – which went very well, Combined Operations with the Army 2 nights ago, CMB firings, Mine bumping trials & 15″ Day concentration next week.

The Combined Operations were great fun. Landing RMSF[2] + Worcesters to attack vulnerable points in the Dockyard. Invading Forces were put ashore anywhere between [?] Marsa Scala and Vico Nunico – about 8 different parties in all of which two got right through with their demolition parties into the yard.

Our first summer cruise brings all ships into the Adriatic and we are going to have a H.A. competition in the middle of it and the Regatta at Kotor at the end. Everyone says yours last year was such a great success there.

The 2nd summer cruise will be to the Levant with another HA competition in the middle of it in order to keep main & secondary armaments 'alive' during the 2nd part of the cruise. I am making the experiment of doing the 15″ & 6″ competition shoots at the end of the 2nd summer cruise before ships go into harbour. That provides an incentive & a shoot to break up the long no-gunnery period July–February though the results will not be comparable to those that we should get at the end of the training period.

Tue., May 30., 7 p.m.

Just back from Fairey Queen trials. Started off at 8.30 but low cloud & then no wind prevented a start. At last at 2.30 p.m. the conditions became well nigh ideal & to our relief. *London* shot the F.Q. off & she sailed away without a falter. An hour after she had climbed as high as they could get her but it was short of 10,000 ft. (nearer 9000 I believe) and Run 1 was started, F.Q. coming in at roughly right angles from the port beam of *Sussex*.

She managed to pick her way through all the bursts the first run which was average shooting, not more than that, 1 or 2 very close but not

---
[1] Twin Speed Destroyer Sweep.
[2] Royal Marine Striking Force.

consistent fine [?] rate of fire. The 2nd run started about 10 mins. later and I think *Sussex* had got over stage fright for the bursts were close from the start and before it was ⅓ over the F.Q. was hit, threw her nose up into the air, altered course right round & came spinning down into the water. I'm afraid she sank before anyone could get to her. The *Sussex*'s gave a great cheer but the *R. Sov.*s and *Revenge*s were very glum about being baulked of their go at her. One can't help feeling glad that she only lasted about 3 minutes altogether (this is a guess only, I not sure really[?]) and at the same time sorry that so splendid a target couldn't have had more ships firing at it. The shoot was witnessed by the Governor, all Military & Air Force chiefs, etc.

The First Lord will have told you so much of what I could otherwise write that I'll finish this up now & just catch the mail.

[P.S.] It was good to read of your going up to Invergordon. I'm sure the H.F. [Home Fleet] appreciated it tremendously. Come to us soon when fed up with Geneva or Parliament.

11. *Admiralty to Fisher*

[ADM 116/3473] 2 June 1933

SECRET.

[Carbon] M.01354/33.

With reference to your submission No.081/040, of 10th April last,[1] I am to inform you that some progress has been made with the modernisation of the defences of Gibraltar, three of the eight 9.2-inch guns being in process of conversion to 35° mountings and half of the A/A guns and lights (eight guns and six lights) having been provided.

2. Their Lordships fully appreciate the inadequacy of the defences of many ports at home and abroad and the necessity for augmenting the defences at Gibraltar in conjunction with those elsewhere, and quite understand your anxiety in the matter. The order of priority in the allocation of available funds to rectify the position is now under the consideration of the Committee of Imperial Defence.

3. In this connection, Their Lordships fully realise the importance of our bases in the Mediterranean in the event of hostilities in Europe.

---

[1] Doc. No. 9.

12. *Fisher to the Admiralty*

[ADM 116/3473] at Malta, 20 June 1933

MOST SECRET.
THE COMMANDER-IN-CHIEF, MEDITERRANEAN STATION.
Med.0163/040.

AIR DEFENCE – MALTA

Be pleased to inform Their Lordships that I have recently had under consideration the Air Defences for the Naval Establishments in Malta and to lay before them the following remarks and recommendations on the subject.

2. Estimated Air Threat – Malta is within air striking distance of Sicily, South Italy and North Africa. In estimating the probable scale of attack, the case of aircraft operating from Sicily has been taken since, due to the proximity of its Air Bases to Malta, it would be possible to develop a greater weight of attack from there than from either of the other two areas.

3. The scale of attack calculated for the Malta Command Defence Scheme, 1933, from information available in 1932, outlines this threat as follows:–

'Syracuse, Augusta and Marsala are already established as Seaplane Bases and there are two Civil Bases at Palermo and Milazzo, which could be taken over.

It is estimated that the Italians would have no difficulty in rapidly concentrating among the above bases, 50 land-planes and 50 seaplanes, even supposing that the strategical situation called for considerable air commitments elsewhere. With these aircraft, excluding any subsequent expansion accommodated at improvised landing grounds, the total weight of attack that could be launched would be some 30–35 tons per day.'

4. From this statement and from the fact that there is an extensive plain in the neighbourhood of Catania on which many temporary aerodromes and refuelling bases could be secretly constructed within a month, it is clear that the total of 30–35 tons per day is the minimum that may be expected at the present time; and that, unless Italy has heavy air commitments in other spheres, this total can be largely exceeded.

5. Counter measures – The counter measures at present available consist only of four 3" H.A. Guns and one R.A.F. Squadron of six III.F. Floatplanes, but I am given to understand that the Military Authorities have had the A/A Defences of Malta under review for some considerable time and that an A/A re-armament scheme is in existence though its details have not yet received final War Office approval.

This scheme allocates sixteen 3" H.A. Guns to the defence of the Grand Harbour and Dockyard Area and is intended to meet the threat outlined in paragraph 3 above.

6. There are three important points in connection with this proposed re-armament scheme to which I desire to draw Their Lordships' attention:–

(a) The scale of attack which it is designed to meet is the minimum which may reasonably be expected at the present time (vide paragraph 4).

(b) It assumes that the total of 30–35 tons per day will be arrived at by comparatively small formations of aircraft attacking at frequent intervals, so that the A/A Guns will not be called upon to engage more than, say four separate formations at any one time. No provision is made to counter attacks delivered by really large formations of say 50 aircraft such as, it is known from Press Reports, have been practiced in the Italian Air Force.

(c) The most probable direction from which an air attack on the Grand Harbour will come is the North Eastward and the 3" Gun has insufficient reach to bring hostile aircraft under effective fire before they reach their bombing positions.

7. For these reasons it is my considered opinion that the proposed A/A re-armament scheme cannot be accepted as being adequate to meet the existing air threat and that a considerably greater number of guns is required and that they should be of larger calibre. If, as I am given to understand, there is no Army A/A Gun of a larger calibre than 3" at present in production, I would suggest that Naval 4" H.A. Guns should be adapted for the purpose and mounted at Malta without delay.

8. <u>Aircraft for Defence</u> – It appears that when the substitution of three Iris Flying Boats for the existing floatplanes of No. 202 Squadron is completed, there will be no aircraft permanently based on Malta capable of acting in a defensive role.

It is suggested therefore, that this state of affairs should be brought to the notice of the Air Council and that they should be invited to consider the provision of at least the ground facilities necessary for an adequate force of fighter aircraft. Also I consider that both the War Office and the Air Ministry should be urged to investigate the possibility of employing any ancillary devices such as sound mirrors, protective curtains, smoke screens, etc., which might hamper the tactics of the enemy, or adversely affect his morale.

9. Another consideration to which I desire to draw attention is that, in the event of the present Disarmament Conference failing to come to some agreement as to the future limitation of military aircraft, it must be supposed that the Italian Air Force, which has been steadily growing in

the past, will continue to do so. Also, improvement in aircraft construction can be looked forward to so that each individual bombing aircraft must become an increasing potential menace.

Hence, should the Nations still retain their freedom of action in these matters at the conclusion of the conference, I am of the opinion that the estimated scale of attack from which the necessary counter measures are deduced, should be greater than that which actually exists at the moment when the estimate is made. The provision of defence today must take account of the increasing power of offence tomorrow.

10. Strategical Importance of Malta – It is certain that the British Fleet will require to use Malta in war time, frequently, if not continuously, and whilst the ships are there they must be reasonably protected, and when they are not there, the services on shore on which their operation and maintenance depend must, as far as possible, be made impervious to attack. Inability to use such a large Central Fleet Base would most seriously affect strategy and have political effects of far reaching importance.

11. These facts are admittedly obvious but are set forth since they must be equally obvious to Foreign strategists. It must be assumed that the state of the Anti Aircraft Defences of Malta is known abroad, and that the certainty of a large measure of success in the air when none can be looked for on the sea, will not only dictate the conduct of war operations but influence policy during peace.

12. Conversely, were the Anti Aircraft Defences of Malta brought up to a proper standard, this knowledge might tend to curb a possible hostile attitude.

13. I am of the opinion, therefore, that to remedy the comparatively defenceless state of this most important Naval Base from the Air is essential and that the mounting of a sufficient number of H.A. Guns, 4″ or above is a matter of urgent priority. The cost of installing the necessary number of guns and lights is considered to be trivial when set against the value of all they would protect and I would submit that the necessary measures to provide them and the ammunition, should not be delayed by considerations as to how they should be manned or maintained.

14. Though a matter outside my province I cannot help thinking that the making of Malta as impregnable as Malta should be, would have a wholesome effect in that Island as well as elsewhere.

13. *Fisher to Chatfield*

[CHT/4/5] *Queen Elizabeth*, at sea, Abbazia to Brioni
21 July 1933

[Holograph]

* * *

I had a good long talk with im Thurn[1] when he arrived[2] but I fear I have not quite met your wishes in raising the question of Malta air defences so soon after your successful efforts over Singapore. I felt, however, that (a) you would put Malta (& Gib.) In cold storage for a bit if you thought it best whereas I can't., and (b) with a National Govt. awakened by you to a proper sense of our vulnerability as they have been, <u>now</u> might be the time to 'go on hitting'.

I do feel it is a futility to put 20– 3″ guns into Malta. When they are there it seems to me it will be more difficult to get the proper thing done than it will be when <u>none</u> are there (or only 4 as at present). And so Malta will remain without proper AA protection for years to come unless a great effort is made now.

I suppose few tears have been shed over the demise of the Disarmament Conference. My view has always been however that as the U-Boat was our greatest source of anxiety in the last war so the aeroplane (attacking merchantmen) will be our greatest in the next. It may be that whatever we do in peace beforehand you will not <u>safeguard</u> against such attack in war but if you delay the initiation of such attack or reduce its scale it is of paramount importance for us and outweighs the 'police conveniences' in disturbed areas, etc.[3] For some time I had hoped that British reluctance to subscribe to total abolition of military aircraft was in the nature of a smokescreen to hide what I should call the incalculable benefit that we should receive – e.g. other nations saw how great it was to our advantage to abolish the submarine and therefore cold shouldered the project – but I'm afraid that that is not the case. Every extension of the military side of the Air can be said to discount the value & power of a Navy. Par contra, the elimination of military aircraft enhances all military waterborne craft.

---

[1]VA John Knowles im Thurn (1881–1956). Dir Signal Dept, Admy, 1920–21; commanded cruiser *Ceres*, 1921–23; commanded battlecruiser *Hood*, 1923–25; commanded Signal School, Portsmouth, 1925–28; COS in Med Fleet, 1928–30; ACNS, 1931–33; commanded 1st CS, Med Fleet, 1933–35; retired list, 1935; Cdre RNR (Convoys), 1940; served at Admy, 1941–44.

[2]Probably to convey the news that this would be his final appointment before being replaced on the retired list.

[3]A reference to the use of the RAF in so-called 'air control' of areas such as Iraq or the Northwest Frontier of India.

Balbo[1] & his menagerie will give a great impetus, I fear, in an undesirable direction. In 10 years time will the Navy be to England what it was! A different attitude at Geneva might, I feel, have provided an answer in the affirmative!

I have very carefully absorbed your remarks about the Fleet's attitude if there is agitation in other quarters to remove the 10% cut.[2] I entirely agree that the Lower Deck must be silent. As you say, they have to regain their good name & can't possibly lose by standing aloof. I am proposing to say this to Flag Officers but think it best to withhold more active measures or propaganda till the time appears ripe. At the moment, <u>out here</u>, interest in these matters seems to have subsided and it would be a pity to revive it. If your information leads [?] you to think otherwise I know you will have me informed.

* * *

14. *Fisher to the Admiralty*

[ADM 116/3473] at Malta, 29 August 1933

SECRET
THE COMMANDER-IN-CHIEF, MEDITERRANEAN STATION.
Med.0231/181/17.
COMBINED OPERATIONS, MALTA, 1933.

Herewith is forwarded for the information of Their Lordships a copy of the report by His Excellency the Governor on the Combined Operations, 1933, together with a report[3] by Vice Admiral Chetwode,[4] then Commanding 1st Cruiser Squadron.

---

[1] Italo Balbo (1896–1940). Served with Alpini [Mountain troops] during First World War; joined Fascist movement, 1921; Under-sec of Aviation, 1926–29; Min of Aviation, 1929–33 and famous for leading a series of mass flights including one to Brazil and another across the North Atlantic to Chicago; Govr of Libya, 1934–40; killed when his aircraft was mistaken for an enemy and shot down over Tobruk by Italian A/A batteries, June 1940.

[2] A reference to the pay cuts imposed in September 1931 as a result of the financial emergency. The initial scheme had been far more drastic and resulted in disturbances in the Atlantic Fleet, commonly known as the Invergordon Mutiny. The Government cancelled the original order and placed a limit of 10 per cent on the cuts imposed with the exception of higher-ranking officers. The Mediterranean Fleet had been dispersed on manoeuvres at the time of these events and was not seriously affected. See the full discussion in Stephen Roskill, *Naval Policy between the Wars* Vol. II: *The Period of Reluctant Rearmament, 1930–1939* (London, 1976), chap. 4.

[3] The reports are too lengthy to reproduce.

[4] VA [later Adm Sir] George Chetwode (1877–1957). Capt (D), 5th DF in Med, 1917–19; Dep DNI, 1923–25; commanded battleships *Queen Elizabeth* and *Warspite*, 1925–27; Naval Sec to 1st Lord, 1929–32; commanded 1st CS, Med Fleet, 1932–33; Adm commanding Reserves, 1933–36; retired list, 1936.

2. This operation was designed by His Excellency the Governor to test the Malta Defences as they now exist against a raiding force of picked men, landed, during the precautionary period as the first act of war, with the object of obtaining control of the dockyard for a period sufficient to render it incapable of fleet repair work.

3. This form of attack has not been tested hitherto, and has recently been added to the forms of attack set out in the Malta Defence Scheme, being considered as constituting the most likely threat of land attack to which Malta is exposed.

4. The naval side of the Combined Operation presented no unusual aspect.

The outcome of the operation ashore, namely, that a comparatively small force of 1138 officers and men who succeeded in leaving the raiding squadron off shore, were adjudged to have been able, in spite of heavy casualties on some beaches and elsewhere, to damage beyond local repair important installations in the Yard, evidently requires serious consideration.

5. Although the exercise showed that the Military Defence with existing strength of personnel is capable of some improvement, it appears that a certain defence against this threat is not possible without some increase in infantry of the garrison.

6. In the revised Malta Command Defence Scheme – as tested in this exercise – all guards at vulnerable points have been cut down, to provide a maximum of troops to defend the beaches. In spite of the increased strength thus obtained, at the cost of security elsewhere, landings were not prevented. Not only was the Dockyard crippled by the raid, but other vulnerable points (e.g. Rinella W/T Station, etc.) were damaged, or lay open to damage in the course of the exercise.

7. Although I consider that Malta as at present defended offers a better target for an air offensive than for Military attack the threat of the latter exists and the necessary counter measures now lacking are requisite to the security of Malta in its role as a Naval Base.

15. *Admiralty to Fisher*

[ADM 116/3473] 3 October 1933

<u>MOST SECRET.</u>
[Carbon] M.01930/33.

With reference to your submission No. 0163/040 of 20th June last,[1] I am to acquaint you that the whole question of the defences of Malta,

---
[1] Doc. No. 12.

including A/A defences, is still under examination by the Committee of Imperial Defence and a decision has not yet been reached as to the scale of attack on which the defences are to be based, nor as to the scale of the defences.

2. With reference to your remarks on the calibre of A/A guns, the Army have a 4.7" gun now under trial. Although Their Lordships are fully in agreement with your views as to the inadequacy of the 3" A/A gun, after full consideration of the circumstances they do not see their way to suggest to the Army Council that Naval 4" A/A guns should be mounted at Malta.

3. The provision of additional aircraft and the necessary ground facilities, as well as the use of ancillary devices, are being considered in the course of the examination referred to in paragraph 1.

4. The general situation in regard to the defence of our bases in the Mediterranean and Their Lordships views thereon were communicated to you in Admiralty Letter M.01354/33 of 2nd June, 1933.[1] Their Lordships and indeed the other Service Departments fully realise the total inadequacy of the existing A/A defences at Malta but, in view of even more serious defence commitments elsewhere, financial considerations may preclude the installation of defences at Malta on the full scale at present.

16. *Fisher to the Admiralty*[2]

[ADM 116/3473]   at Milo, 6 October 1933

MOST SECRET
THE COMMANDER-IN-CHIEF, MEDITERRANEAN STATION
Med. 0274/721/17.
STRATEGICAL INVESTIGATIONS CARRIED OUT BY MEDITERRANEAN FLEET.

Be pleased to inform Their Lordships that among other Strategical Investigations carried out by H.M. Ships of the Mediterranean Fleet during the winter 1932–1933 was set the following problem:–

PROBLEM A.

Feeling has been running high between the Jugo-Slavs and the Italians, which has culminated in an anti-Italian outburst at Zagreb on September 9th. The Italians have demanded a large indemnity and an official apology, and have threatened to bombard Split if they are not given satisfaction within 48 hours, which will be up at 1200, September 13th.

---
[1]Doc. No. 11.
[2]Copy to VA Commanding, 1st Battle Squadron.

The Jugo-Slavs have put their case before the League of Nations and in any case do not propose to accede to this demand.

The Italians claim that it is entirely their own private affair, and they will not necessarily abide by any decisions of the League.

France is in the middle of a general election and is having trouble in Morocco. She is not prepared to give active support to the League.

Great Britain proposes to act in accordance with Article 16 of the Covenant of the League[1] should the Italians carry out their threat.

DISPOSITION OF THE ITALIAN FLEET.

Battleships were at Naples and Spezia on 11th September, but proceeded to sea at 1200 on that day.

Trento and *Trieste* are at Otranto.[2] The exact whereabouts of the remainder of the Italian Fleet at the moment is not known, but it has been ascertained that the majority of their light craft are in the Adriatic.

The Air Forces are disposed in accordance with the latest intelligence report.

THE BRITISH FLEET.

The Mediterranean Fleet (constituted as on 1st August, 1932) is disposed as follows:–

At Argostoli – Third Cruiser Squadron.
　　　　　　　Fourth Destroyer Flotilla.
At Malta – 　Remainder of the Fleet.

Commander-in-Chief, Mediterranean has been kept informed about the situation by the Admiralty, and receives the following signal at 1500 on September 11th 1932:–

'War between Italy and Jugo-Slavia probable. Make the necessary dispositions to render every possible assistance to the Jugo-Slavs, if an attack is made on them. Full use of Jugo-Slav bases may be made.

Battle Cruiser Squadron, Second Cruiser Squadron, *Courageous*, Fifth Destroyer Flotilla and Second Submarine Flotilla will leave England to reinforce on 13th September. Squadrons and Flotillas proceeding independently at high speed.'

REQUIRED.

An appreciation of the situation from the point of view of the Commander-in-Chief, Mediterranean.

2. Well reasoned and carefully thought out solutions of this problem were received from H.M. Ships *Revenge* and *London*.

---

[1] In case a member of the League resorted to war it shall *ipso facto* be considered to have committed an act of war against all other members of the League and would be subject to trade and financial sanctions as well as possible common military action by the several governments concerned.

[2] *Trento* and *Trieste* were 10,500-ton cruisers armed with 8″ guns.

3. In forwarding these solutions the Vice Admiral Commanding, 1st Battle Squadron, remarked as follows:–

X   X   X   X   X   X

3. I desire to draw your attention to the following special points:–

(a) Both these ships and others also which have undertaken strategical investigations in 1932/33, express the opinion that Malta would be untenable as our main Fleet base in the event of war with Italy, on account of the air menace.

(b) Both ships, again, imply that, on account of insufficient anti-aircraft armament, our battleships are unfitted for operations where they are liable to frequent or heavy attack by aircraft.

4. I submit that, at the present time, there is reason in both these opinions, which I believe to be held by a number of other officers besides.

I do not consider that this can be regarded as a satisfactory state of affairs, as it encourages an attitude of mind the opposite to what is desired, namely a firm belief in our own strength.

5. I am, of course, aware to some extent of what is being done in these directions, but it appears to me that for moral reasons alone it would be advantageous to quicken up the provision of material which is well known to be necessary and about which no real doubt exists.

6. As regards the battleships, I consider that they will not be reasonably well equipped to resist aircraft attack until they are provided with a set of H.A.C.S.[1] for use on each side and two mark 'M' pom-poms. The addition of a third or fourth 4-inch gun each side is extremely desirable.

X   X   X   X   X   X

4. I fully concur with these remarks and, in particular, with the opinions expressed in paras. 5 and 6.

5. I have already put before Their Lordships, in my letter Med. 0163/040 of 20th June, 1933,[2] my views as regards the Anti-Aircraft defences of Malta. The effect of the inadequacy of the present armament on the morale of the Fleet lends added force to the arguments there advanced.

6. Marked improvement has been shown in recent years in the H.A. gunfire of the Fleet. The impression left, nevertheless, by every exercise in which the Fleet is attacked by aircraft is that strengthening of the H.A. armament of the battlefleet is required before the Fleet can face such an attack free from anxiety.

---

[1] High Angle Control System.
[2] Doc. No. 12.

17. Fisher to Chatfield

[CHT/4/5]　　　　　　　　　　　　　　　　　　Malta, 17 November 1933

[Holograph]

* * *

The H.A. competition showed good progress. The broad result was that the time from 'Alarm' to the first close burst this year exactly equals the time from 'Alarm' to 'Open Fire' last year and the rate of fire has doubled, approximately, e.g. 10–20 rds. per gun per min.; as against 10 in 1932. With harder conditions average points of all ships has advanced from 490 to 796 with the same scale of marking. I hope you will see the British Movietone film – scenes at Navarin, Corfu & Malta. Lots of mistakes to pickup I'm sure as the cameramen take their selected moments and not ours. I took 6 Greek naval officers to sea from Athens to Navarin & that was appreciated as a return for all they do for us.

The great political crisis in Malta has so far caused no popular excitement.[1] I notice that people of all sorts are more polite in the streets than they were. In fact it is almost embarassing.

I was afraid you could not help me over Malta A/A defences. You can weigh the urgency of 'Far East' & 'European' and I can't but I do really feel that we are taking an impossible risk about Malta.[2] The cost of a proper defence is not great. Spread over 2 or 3 years it is negligible. Because Singapore has (admittedly) to be dealt with must Malta stand down? Can't both be tackled? By the War Office? Would 20 4″-guns + emplacements cost more than £100,000? Would searchlights, wiring, control pos[ns], directors, predictors, cost more than £30,000? And with an equipment of ammunition would the total cost exceed £150,000? Or £50,000 p.a. for 3 years?[3]

---

[1]Elections in June 1932 resulted in the victory of the pro-Italian Maltese Nationalist Party and after a period of confrontation with the authorities, particularly over the teaching of Italian in schools, the ministry was dismissed and the constitution suspended leaving the island under the direct control of the Governor. For brief summaries of the Maltese political situation, see Douglas Austin, *Malta and British Strategic Policy, 1925–1943* (London, 2004), pp. 18–19, 46–8.

[2]Although Fisher probably did not know yet, the Committee of Imperial Defence on 9 November decided to set the priorities for defence spending in the order: (1) defence of British possessions and interests in the Far East; (2) European commitments; and (3) the defence of India. No expenditure was to be incurred on measures of defence required to provide exclusively against attack by the United States, France or Italy. Minute by Director of Plans, 13 Nov 1933, ADM 116/3473, Docket No. M.03188/33. The Defence Requirements Committee might modify the CID view and on 19 December Chatfield noted that he would inform Fisher personally about the Defence Requirements Committee and its deliberations. Minutes by Deputy Chief of Naval Staff, 13 Dec and Chatfield, 19 Dec 1933, ibid.

[3]Chatfield marked this paragraph for the attention of the DCNS.

Europe is pretty tender now. Malta is the key position for us. Must it be left defenceless because mechanisation of military units must not be retarded?

Im Thurn asked if I would suggest *London* going to Gravesend when she is o.k. for recommissioning, so that Lord Mayor, etc. might see over her but I said I did not feel inclined to take the initiative in the matter as we shall be so short of cruisers and the sooner she is back on the station the better.

Boyle[1] & I are in communication over Combined Exercises. I hope they will come off but I expect I shall have to draw on some of his fuel surplus. Is there a chance of your being able to be with us? Outline scheme is Home Fleet SW of Gib. making for a N or NE port with Troop Convoy. Med. Fleet to intercept. Operations last 4 days. Combined Fleets to Gib. on conclusion. Periods roughly March 9–14 sea, [March] 14–22 Gib.

Very glad to get *Resolution* back & she has made good use of her working up time. She looks smart and her men 'on their toes' to do well.

There was a rumour that the Maltese & especially the University undergraduates (following Oxford or Cambridge?) were going to boycott Poppy Day – but the receipts were bigger than ever and I'm told that not a single Dockyard man entered the gates without paying & generally overpaying for his poppy.

* * *

18. *Fisher to the Admiralty*

[ADM 116/3473] 23 January 1934

MOST SECRET
THE COMMANDER-IN-CHIEF, MEDITERRANEAN STATION.
Med.024/0015.
MEDITERRANEAN FLEET – STATE OF PREPAREDNESS FOR WAR

Be pleased to inform Their Lordships that I am satisfied as a result of inspections, firings and other practices that Flag and Commanding Officers have done and are doing all that is possible to make their Squadrons and Ships ready for war.

---

[1] Adm [later AoF] Sir William Henry Dudley Boyle (1873–1967). Succeeded cousin as 12th Earl of Cork and Orrey, 1934. Naval attaché in Rome, 1913–15; SO, Red Sea Patrol during First World War; commanded 1st CS, Med Fleet, 1926–28; Pres, RN College, Greenwich, 1929–32; C-in-C Home Fleet, 1933–35; C-in-C Portsmouth, 1937–39; commanded combined expedition to Narvik, Norway, 1940.

2. There is, however, one serious weakness which, owing to lack of material, is beyond their power to remedy and that is the inability of the Fleet to guarantee the successful repulse of an air attack in force.

To deal with such an attack the Fleet requirements are considered to be:–

(a). At least double the number of H.A. Guns, both for long range and close A/A defence, in capital ships and cruisers,

(b). A second H.A. Control Position in all capital ships and large cruisers,

(c). The provision of suitable A/A control for the Low Angle armament of destroyers to enable them to play their part in the defence of the Fleet with their 4.7″ guns,

(d). More A/A ammunition on board all ships and in reserve at Malta. The outfits of A/A ammunition on board are expended by ten minutes continuous fire. Experience in recent air attack exercises in the Mediterranean Fleet prove that determined attacks by an air striking force of moderate strength, operating during daylight hours of one day only, call for, in the aggregate, at least ten minutes gunfire from both batteries of the H.A. armaments. The reserve stored at Malta is only 75% of the outfits carried on board,

(e). More suitable targets and a larger annual allowance of practice ammunition.

Naval operations in the Mediterranean are seldom likely to be beyond the range of shore based aircraft, and until the above requirements have been met, I am, therefore, unable to say that the Mediterranean Fleet is as ready for war as it should be.

3. I am aware that these considerations are constantly in the minds of Their Lordships and that in particular steps are being taken to deal with (b), (d) and (e). I would submit, however, that they are all such as to deserve – in the Mediterranean Fleet anyhow – priority over any other considerations.

4. When intelligence is insufficient to assess with confidence the ability of an enemy it is generally assumed to be roughly equivalent to our own. This may or may not provide a large margin of safety vis à vis European Naval Powers.

I am well aware of the powers of the air striking force of *Glorious* to inflict serious damage on enemy capital ships. I am not able to say, with positions reversed, that the First Battle Squadron would not suffer equally from an attack by shore based aircraft which, further, might be on a much larger scale.

If the requirements (a) to (e) above, are fully met, then air attack on large warships may possibly be brought into the same category as the attack by ships on forts, an operation attended by grave risks and with only a problematic chance of success.

5. Complementary and essential to the operations of the Mediterranean Fleet at sea is the security of its principal base at Malta.

In the Dockyard, damages sustained by air attack, mine or torpedo will have to be made good and in the harbour ships will have to be refuelled, re-ammunitioned, revictualled, etc.

I am unable to report that, today, the functioning of these essential services can be relied upon with any certainty under conceivable war conditions.

6. Today the Island A/A defences can only engage one target at a time with a few guns of low power and poor reach, consequently the deterrent to an air attack at the moment is practically NIL. It is understood that some slight improvement will be achieved within the next two years by the drafting of more men and ammunition to Malta under the re-armament scheme, but even then the deterrent to an air attack in force will be extremely small. The personnel of the Dockyard is without protection from gas, so that even were the material damage to docks, warships, oil tanks, etc., insignificant, the Yard's services would be seriously affected by casualties or the fear of casualties.

7. It is universally recognised that aircraft play an essential part in the defence of land objectives against air attack, and are complementary to Military A/A defence weapons or measures. Yet with the existing distribution of R.A.F. squadrons, there are no aircraft based at Malta capable of co-operating in A/A defence or of operating against enemy Air Bases.

8. In short, both at the base and at sea, I am of opinion that immediate steps are necessary to counter the only form of attack which can really harm us and which is, therefore, the only form of attack likely to be adopted by an enemy.

9. I realise that politically it may be argued that the contingency envisaged is not sufficiently probable to warrant expenditure when there are other more likely eventualities involving still more vulnerable points. I am, of course, not competent to express any opinion as to that, but the state of Europe appears to be such that any steps that will ensure the Fleet being able to operate in any waters to fullest effect should now be taken.

10. I am of opinion that development in the air and perception of the use of aircraft against the British Fleet and Merchant Navy may increase in the near future and the provision of counter measures will take longer than the formulation, organisation and initiation of a proper air offensive. For this reason I have come to the opinions expressed in paras 3 and 8.

11. Attached hereto are:–
(a). Drawings showing possible positions for extra 4″ H.A. guns in *Queen Elizabeth* and *Royal Sovereign* classes. (Enclosure 1).

(b). A computation of the air menace when within reach of French or Italian shore concentrations. (Enclosure 2).[1]

12. It is considered that in view of the age of our Battleships and the requirements of new construction it will be difficult and costly to supply armoured protection that will confer a satisfactory degree of immunity and anyhow the process must be a slow one.

Pending the development and installation of twin mountings of satisfactory output, which it is understood was recommended by the report of the N.A.A.G.C., recently before Their Lordships, I consider that much can be done in the immediate future by making use of 4" H.A. mountings of existing design mounted in the best of the various positions shown in Enclosure 1 hereto. The A/A broadside could thus be increased to 4 or 5 guns without encountering any structural difficulties and with only slight inconvenience in other directions. The selected gun positions can be prepared in peace and the necessary wiring run, those of the extra gun mountings inconvenient to the ordinary work of the ship need not be kept mounted in peace, but be held immediately available at the base.

When prepared for war, boats interfering with the service of these extra guns could be left behind on the Fleet leaving Malta as was the case in the Grand Fleet. Periodically the guns could be mounted in a selected ship to exercise the scheme and provide experience with the heavier armament.

19. *Chatfield to Admiral W.W. Fisher*

[CHT/4/5] 11 May 1934

[Carbon]

I am sending you with this letter, for your own information, a copy of a tentative programme for the modernisation and large repair of the ships of the Battle Fleet.

The present decision is that all the ships of the *Queen Elizabeth* class and also the *Royal Oak* (the first of the 'R' Class now coming in hand) shall be fitted with additional horizontal armour in addition to other modern improvements. It has now been decided to re-engine and boiler *Warspite*, *Queen Elizabeth*, *Valiant* and *Renown*.

The latter proposal has been come to on account of the very great ages some of our ships will have run to before they are replaced by new construction. For instance, with a replacement programme of one ship a year commencing in 1937, the ages will come up to as much as 34 and 36

---

[1] The two enclosures are not reproduced.

years, and even with a replacement of three ships every two years an age of 32 years will be reached. It is impossible to guarantee that their present machinery will remain entirely reliable for another 15 years or so, and the intention is to retain the ships which are being re-engined to the last.

2. To turn to another subject, you will remember that Admiralty strategic policy has always visualised the possibility of stationing the Battlecruisers in the Far East when the facilities at Singapore would admit.

I brought this question to the Committee of Imperial Defence a year ago and it was decided that, principally for political reasons, it would be undesirable to make this redistribution. I think it is also undesirable, now that our Battle fleet is reduced to the minimum number of units, to divide our Capital Ships so much.

On the other hand it seems to me that the presence of the Battlecruisers in the Home Fleet is open to criticism. The Battlecruiser force is essentially an advanced force and for this reason I think that if it was stationed with the Mediterranean Fleet the Battlecruisers would be in a central position from which they could operate in either direction.

Owing to the decision to take a fourth Capital Ship in hand, namely, *Malaya* this autumn, the manning situation would admit of the 2 Battlecruisers being stationed in the Mediterranean at any time after this, providing one Battleship is withdrawn from the Mediterranean to join the Home Fleet.

Other complications arise, however, such as the commissioning dates of the ships concerned and it appears to me that the best time to make this redistribution, if it is decided upon, would be almost immediately after you give up your command towards the end of 1935. *Hood* and *Repulse* would join the Mediterranean Fleet as soon as *Repulse* is ready on the completion of her large repair.

There is the question of berthing in the Grand Harbour to be considered. In this connection we should anticipate, in years to come, two aircraft carriers being included in the Mediterranean Fleet in addition to the Battlecruisers. With the present arrangement of moorings it seems possible to berth all these ships and four battleships, making use of Parlatorio Wharf.

I should like to have your comments on any or all of the above proposals and also in regard to the position of the Vice Admiral Second-in-Command, Mediterranean Fleet, in the new arrangements and in regard to which Battleship of the Mediterranean Fleet should be withdrawn.

In regard to the Vice Admiral, I think it essential that he should remain in the Mediterranean and that in order to carry out his functions of Second-in-Command he should remain in the Battle Fleet notwithstanding that it has been reduced by one ship. The alternative which you might consider is for him to be in the Battlecruiser Force and a Rear Admiral in the Battle Fleet.

As regards the Battleship to be withdrawn frrom the Mediterranean Fleet, for the sake of homogeneity it would be preferable to withdraw *Queen Elizabeth*, in which case it would possibly be convenient to take *Ramilles* in hand for refitting as Fleet Flagship when she is due for docking at Malta in 1935. Estimated time required at least 2 months.

I have it in mind that *Queen Elizabeth* would bring your Flag home from the Mediterranean to pay off at the end of her present commission, extending it to October 1935. This would make her commission 2 years and 8 months. She would then have a short commission in the Home Fleet before being taken in hand for modernisation. The alternative which you may consider is to withdraw one of the 'R' Class to the Home Fleet, sending *Queen Elizabeth* out to the Mediterranean again, after recommissioning to resume her duties as Fleet Flagship, being subsequently replaced by *Warspite*.

The one proposal gives homogeneity at the cost of £13,000 to fit out *Ramilles*. The second saves the money at the expense of an odd ship in both Fleets.

But the latter alternative would involve the Mediterranean Fleet being without a Fleet Flagship at the following periods:–

(i). During change of Flag in October 1935.

(*Resolution* would be available as a Temporary Flagship).

(ii). Possibly during relief of *Queen Elizabeth* by *Warspite* March 1937. It is not yet clear whether the manning situation will permit of *Warspite* proceeding to the Mediterranean before *Queen Elizabeth* returns to the United Kingdom. (Vice Admiral will then be in *Resolution* since *Revenge* will be in hand for Large Repairs).

3. I have now got the question of the defence of the Suez Canal before the Chiefs of Staff Sub-Committee with a view to:–

(a). Representing to the C.I.D. the extreme importance of seizing and controlling the canal with British Forces when an emergency arises, as I am convinced that this is the only proper way of avoiding what would be a serious blow to our strategy in the event of the canal being blocked.

(b). Sending out instructions to you and the G.O.C., for the officers carrying out the reconnaissance which is shortly to be made, in order to produce a defence plan for the Canal.

In regard to (a) it is important that you should have forces which can at any moment be immediately sent to the Canal, without previously calling at Malta, in the event of 'a state of emergency' being declared in the Canal area. The Red Sea Sloops should be available but more would be required and it seems that a Destroyer Flotilla would possibly be the most valuable unit for such a purpose. You might consider whether you could bring on the personnel and war stores for the flotilla from Malta and transship them at Port Said or Suez.

20. *Chatfield to Fisher*

[CHT/4/5] 14 June 1934

[Carbon]

* * *

Now there is another point that I want to write to you about. Next year is the 25th anniversary of the King's Accession and it is probable that considerable ceremonies and national rejoicings will be staged for that important occasion, lasting for a considerable period in London, including various important functions, reviews of the Army and Air Force and a Naval Review is also being considered at Portsmouth.

I feel it would be a poor show for a Review of the Home Fleet to be all that we could put up and it would be compared adversely with the recent Review of the American Combined Fleets by the President of the United States. I can, of course, take steps to mobilise a certain number of the Reserve Fleet cruisers and destroyers, but to my mind if we are going to have a Naval Review it ought to be a good one, being the first real Review since 1914.

This leads me to consider bringing home the Mediterranean Fleet under your command for the occasion. I have things from the strategical situation and provided the situation is peaceful I can see no real objection to it. It would mean that in the event of trouble the two Fleets would be together instead of separated, but of course I should have to leave in the Mediterranean a few cruisers and perhaps a flotilla of destroyers as an out-post and for general requirements.

The other point of view is the point of view of welfare. It would not be possible to bring your Fleet home unless it was at home for several weeks so that some days leave could be given to each Watch of officers and men either before or after the Review.

Further, it would affect the Spring Cruise because it would be best in that case to endeavour to have the Combined Fleet Exercises then rather than as at present in March. It would perhaps be a little hard on the ships who were left in the Mediterranean not coming home but I do not think that can be avoided though it might be possible to give preference to the ships which had been least long away from England as those to be retained on the Station.

I shall be very glad if you will think it over and let me have your very confidential views, as early as possible, as decisions are now being come to and a Committee has been set up to consider the whole matter. I can see it would be disturbing to you and your work, refits, etc., but I do not think that these are outstanding objections compared to the importance to the Navy of making a bad show on such a big occasion.

* * *

21. *Chatfield to Fisher*

[CHT/4/5] 2 August 1934

[Carbon]

Very many thanks for your letter of the 25th June about the future composition of the Mediterranean Fleet and the Silver Jubilee Celebrations. Your remarks are most helpful and are now being considered by the Admiralty Departments, and I hope we shall be able to meet all your views. However it may be some little time before you get any definite decisions, so I am just writing this to let you know that everything is under consideration.

In any case you can put out of your mind that there would be any chance of my accepting your generous offer to transfer your Command to your successor after the Review. It would be entirely against my wishes. I want you to do your full 3 years. Who your successor will be is not yet decided, but it probably lies between Pound[1] and Backhouse.[2] My inclination is that Pound should relieve you at the end of your 3 years, Backhouse having previously relieved Boyle in the Home Fleet next Summer. Boyle has only been given the appointment for 2 years in view of the somewhat exceptional circumstances under which he was appointed. I feel that Backhouse, having just done 2 years in the Mediterranean, it would be better for him to go to the Home Fleet where his solid qualities of administration in that difficult Command would be best used, whereas Pound, being younger and probably rather more active and having already been in the Home Fleet, would be better placed in the large and longer Command. I wonder whether you will agree with this!

[1] VA [later AoF Sir] Alfred Dudley Pickman Rogers Pound (1877–1943). Commanded battleship *Colossus* at Jutland, 1916; DP, Admy, 1922–25; COS in Med Fleet, 1925–27; Asst CNS, 1927–29; RA commanding Battle Cruiser Sqdn, 1929–31; Admy rep on League of Nations Advisory Commission, 1932; 2nd Sea Lord, 1932–35; temporary COS, Med Fleet, 1935–36; C-in-C Med Fleet, 1936–39; 1st Sea Lord and CNS, 1939–43.
[2] VA [later Adm Sir] Roger Backhouse (1878–1939). DNO, 1920–22; RA commanding 3rd BS, Atlantic Fleet, 1926–27; 3rd Sea Lord and Controller of the Navy, 1928–32; VAC 1st BS, Med Fleet, 1932–34; C-in-C Home Fleet, 1935–38; 1st Sea Lord and CNS, 1938–39.

Very many thanks for your letter written from Samos about the horrible Turkish affair.[1] I want you to realise how much I sympathise with you and how much I admire your quick and decisive action, and how maddened you all were at the outrage.[2] I had to steer between my natural wish to support you in every way, and to approve all your proposals, and the larger question, as I saw it at a distance, of getting national satisfaction from the Turks. To my mind the incident gave us an opportunity to turn the tables on them, more especially as at that moment they were actually doing all they could to ingratiate themselves with us for various political reasons, that was why I thought it was best to restrain you from your natural wish to act locally.

On the whole, I feel that we have accomplished our object. We have demonstrated in force at the place saying that we were ready to avenge the insult if we could not get redress and the presence of your ships there undoubtedly strengthened the hand of our representative at Angora.

I have not heard any more about the incident you reported, namely, that the Turkish Captain never arrived for the memorial service, but I do hope that you are satisfied with the final outcome of the situation, including the £2000 compensation which we insisted on, the Turks having originally offered £1000. The 4th Sea Lord[3] came in there and was very helpful, and we said that if they gave us only £1000 it would not be called compensation but an ex gratia payment as a practical proof of their policy, but if they wished to call it compensation £2000 was the minimum.

---

[1] On 14 July Turkish patrols fired on a small skiff carrying three British naval officers from the cruiser *Devonshire*. The ship was on a visit to the Greek Island of Samos, close to the Turkish coast. A Surgeon-Lieutenant was killed and another officer wounded. The Turks claimed the men had been regarded as smugglers and in an apparent misunderstanding over the meaning of signals had refused to stop. The British Government eventually decided to accept the Turkish 'expression of regret' over the 'genuine misunderstanding'. Although the body was never recovered, on 21 July a memorial service was held at sea with *Queen Elizabeth*, *London* and *Devonshire* as well as a Turkish destroyer taking part. The Turkish Government offered a compassionate grant of £2,000 to the next of kin. A summary account is in *Turkey: Annual Report, 1934*, p. 6. The National Archives, FO 371/19037.

[2] Fisher had been staying at a hotel in Cyprus when the incident took place, but wrote that he had 'gathered every ship I could lay my hands on' and within 36 hours had 2 battleships, 3 cruisers and 7 destroyers at Samos. The Turks, who had initially refused to allow the British to search for the body, agreed and later at the memorial service Fisher made the Turkish destroyer anchor close to the *Queen Elizabeth*. As the British fired a ceremonial volley, the Turkish destroyer dropped a large wreath. Admiral Sir William James, *Admiral Sir William Fisher* (London, 1943), pp. 128–9.

[3] RA [later VA Sir] Geoffrey Blake (1882–1968). Gunnery Cdr in *Iron Duke* at Jutland, 1916; Naval Attaché to the US, 1919–21; commanded *Queen Elizabeth*, 1921–23; Dep Dir RN Staff College, 1925–26; Dir RN Staff College, 1926–27; COS in Atlantic Fleet, 1927–29; Cdre in Command of NZ Station and First Naval Member, NZ Naval Board, 1929–32; 4th Sea Lord, 1932–35; VACBCS and 2nd-in-Command, Med Fleet, 1936–38; placed on retired list for reasons of health curtailing a career that may well have led to his being chosen 1st Sea Lord, 1938; Asst CNS (Foreign), Apr–Dec 1940; Flag Officer Liaison US Navy in Europe, 1942–45; Gentleman Usher of the Black Rod, 1945–49.

* * *

By the by, there is another point about the Samos incident and that is that I cannot help feeling the officers were ill-advised to go to the Turkish coast as when I was out there and used to fish in those waters the Greeks and British residents always told me the Turks fire at you if you went near there. I never went within 3 miles of the Turkish coast in my perambulations in a motor boat.

The subsequent incident of their firing on an Italian fishing boat shows that they are not to be trusted anywhere and you may consider it desirable, therefore, to lay down that, for the present at any rate, the Turkish coast should be given a wide berth. I rather fancy the Foreign Office may ask us to ensure that there is no further opportunity given to the Turks to repeat their murderous propensities, so if you do get a letter from the Admiralty about it you will understand the reason.

I have been having a very difficult time over monies for the Navy. As you know, what was called the Defence Requirements Committee recommended large increases for deficiencies of the 3 Services, mainly at my instigation and laid down a large programme of expenditure over a series of years for all 3. They have, however, been so panicked by the Press and Parliament about the Air that they have decided on a large Air expenditure and a considerable Army expenditure also, and consequently they have not got enough for me also, or rather the Chancellor[1] says he has not. The Treasury are frightened to death of having to fight on two fronts, i.e. to maintain an Air Force and Army ready to be of some value in Europe and in the defence of these Islands, including the Low Countries, and also keeping up the Navy for Imperial Defence. Probably the only way to do both is by a loan, but this they are at present opposed to. I am going, consequently, to give up the idea of a Naval Programme based on a number of years, and to concentrate on getting an increase of some £4 million in next year's Estimates, which I think I shall succeed in doing. We are hurrying on our sketch Estimates so as to be able to put a figure to the Chancellor in October, and if he refuses we shall then have time to go to the Cabinet and fight it out.

The further complication is, of course, the Naval Conference, and our initial preliminary negotiation lead[s] us to think it would not be very easy to get agreement. Consequently, our whole new Construction Programme is very much in the melting pot, and that is another reason against a

[1]Neville Chamberlain (1869–1940). Lord Mayor of Birmingham, 1915–16; Dir Gen of National Service, 1916–17; MP (U) Ladywood divn of Birmingham, 1918–29; Edgbaston divn, 1929–40; Postmaster Gen, 1922–23; Paymaster Gen, 1923; Min of Health, 1923, 1924–29, Aug–Nov 1931; Chllr of the Exchequer, 1923–24, 1931–37; PM and 1st Lord of the Treasury, May 1937–May 1940.

programme being agreed to over a number of years. We shall have to wait and see what happens next year. I have, however, got them to agree to an increase in the Fleet Air Arm.

Although I have got the Prime Minister[1] to agree to the replacement of the Battle Fleet, the Treasury are still opposed to it, especially as at the same time I am making heavy demands for the modernisation of the Battle Fleet, including the complete modernisation of the *Warspite*, *Queen Elizabeth*, *Valiant* and *Renown*, the 4 ships that will have to last the longest.

In the *Warspite* I am taking out four 6-inch guns, removing the 6-inch battery armour, and putting it on the engines and boiler rooms, re-engining and re-boilering the ship, doubling the number of 4-inch guns, fitting 4 Mark M Pom Poms, and a catapult on the boat deck with 2 Observation and 2 Fighting Aircraft.

In the other ships I intend to remove the 6-inch batteries altogether and substitute the 4.7-inch twin mountings, 8 guns a side, H.A. and L.A. This mounting will not be ready for the *Warspite*, which is unfortunate, and after discussing it with Backhouse I have to effect a compromise.

All this means a lot of money, something like £6 million for the 4 ships for alterations alone. It is a big problem. We are so much behind hand and we have such an accumulation of deficiencies as regards personnel, fuel, defence of bases, and ammunition that I am very anxious.

I daresay you will have seen in to-day's papers the action of Keyes[2] in the House of Commons in re-raising the whole question of the Mutiny and the supersession of Tomkinson.[3] I did all I could to dissuade him in

---

[1] James Ramsay MacDonald (1866–1937). Labour MP for Leicester, 1906–18, Aberavon, 1922–29 and Seaham, 1929–31; Leader of the Labour Party, 1911–14; PM and Sec of State for Foreign Affairs, Jan–Nov 1924; PM, 1929–35; Lord Pres of the Council, 1935–37.

[2] AoF Sir Roger John Brownlow Keyes (1872–1945). Created 1st Baron, 1943. Inspecting Capt and subsequently Cdre commanding Submarines, 1912–15; COS in Eastern Med Sqdn, 1915–16; DP, Admy, 1917; VA Dover (led raids on Zeebrugge and Ostend), 1918; commanded Battlecruiser Sqdn, 1919–21; DCNS, 1921–25; C-in-C Med, 1925–28; C-in-C Portsmouth, 1929–31; MP (C) Portsmouth North, 1934–43; Dir of Combined Operations, 1940–41.

[3] VA Wilfred Tomkinson (1877–1971). Closely associated with Keyes during the latter's early career including service as First Lieutenant in Keyes's destroyer *Fame*, 1900; commanded '*B*' class submarines at Venice, 1915–16; COS in Med Fleet, 1927; Asst CNS, 1929–31; commanded Battlecruiser Sqdn, 1931–32; retired list, 1935; Flag Officer-in-charge, Bristol Channel, 1940–42. Tomkinson had the bad luck to be senior naval officer in the Atlantic Fleet when the C-in-C had recently been taken ill and therefore became the scapegoat for the refusal of duty in certain ships in September 1931 that has become known as the Invergordon Mutiny. The incidents followed the Admiralty announcement of pay cuts because of the economic crisis. Although his conduct was initially approved by the Admiralty, he was superseded a few months later and several months before his appointment was due to end. It was the effective end of his naval career although there had been no court martial or formal enquiry. Keyes championed Tomkinson's cause but in the long run was unsuccessful. There are numerous documents on the subject in Paul G. Halpern (ed.), *The Keyes Papers*. Vol. II: *1919–1938* (London, 1980), Part III.

the interests of the Service but he would not be dissuaded. I know that the Service as a whole is with me about it, and I am thoroughly disgusted with the whole matter, as I am sure you will be also.

This is a very long letter and needs no answer. I am very glad to hear that you are very fit and your Command in splendid fettle, as is natural.

I think the Home Fleet is in good fettle too, and I send you my very best wishes and reassure you of my continued support in everything you do.

22. *Minutes of the Board of Admiralty*

[ADM 167/90] 4 October 1934

SECRET.

* * *

Redistribution of the Fleet.

3230. The Board considered a Memorandum dealing with the proposed redistribution of the Fleet whereby the five Battleships of the '*Queen Elizabeth*' Class and the Battle Cruisers would be stationed in the Mediterranean and the five Battleships of the '*Royal Sovereign*' Class, now in the Mediterranean, would join the *Nelson* and *Rodney* in the Home Fleet. It was proposed to effect this redistribution gradually between August, 1935, and October, 1936, the ultimate result being as follows:–

Mediterranean.	*Home.*
Queen Elizabeth (Flag).	*Nelson* (Flag).
Barham (Flag).	*Rodney.*
Warspite.	*Revenge.*
Valiant.	*Royal Sovereign.*
Malaya.	*Ramilles.*
Hood.	*Resolution* (Flag).
Repulse.	*Royal Oak.*
Renown.	

It was considered that strategically it would be an advantage for the Battle Cruisers, which were essentially an advance force, to be stationed in a central position whence they could operate either to the East or West. The manning situation, however, necessitated one Battleship being withdrawn from the Mediterranean Fleet in order to compensate for the transference of the Battle Cruisers, and as none of the '*Royal Sovereigns*' were suitable for use as Mediterranean Fleet Flagship, it appeared desirable for the sake of homogeneity to exchange the battleships as proposed.

Apart from strategical reasons, which political considerations might render it undesirable to emphasise in public, it was pointed out that the proposed change was desirable in view of the effect which the contemplated modernisation programme would have on the numbers and homogeneity of the Mediterranean Battle Squadron.

It has been reported that no important administrative difficulty attached to the proposal. The financial effect was estimated to be as follows:–

(a) Non-recurrent Expenditure involved in fuel for the
'reshuffle' £6,000

(b) Annual Expenditure.
   Personnel –
   Pay.. £28,000 ⎫
   Non-effective liability .... 7,500 ⎬ 35,500
   Docking   2,500
   Fuel –
   (i) Extra allowance for a Mediterranean ship.... £1,380 ⎫
   (ii) Recommissioning voyage every 2½ years   ⎬ 2,580
   (At £3,000) .... 1,200 ⎭
   £40,580

The Board approved these proposals and decided to defer the question of any public announcement on the subject pending consideration of the draft of the First Lord's Statement on Navy Estimates, 1935.

23. *Admiralty to Fisher*

[ADM 116/3473] 5 January 1935

SECRET.

[Carbon] M.0400/34.

With reference to your submission Med.024/0015 of 23rd January 1934,[1] I am to inform you that the improvement of the A.A. armament of 15" capital ships and 8" cruisers is under consideration and you will be informed in due course of the conclusions reached.

2. The defences of Malta are now being reviewed by the Committee of Imperial Defence, a copy of J.D.C. Paper No. 184 which contains recommendations by the Joint Oversea and Home Defence Sub-Committee

---

[1] Doc. No. 18.

is enclosed for your information.[1] This document should be taken on charge in the C.B. account and its receipt should be acknowledged.

3. Their Lordships propose in due course to give consideration to the provision of extra storage space for explosives at Malta, but whatever decision is reached the additional accommodation will not be available for at least two years and possibly much more, therefore if you wish more reserves of A.A. ammunition to be held at Malta meanwhile, other reserve ammunition must be moved elsewhere. Their Lordships request that you will forward any proposals you wish to make on the subject. For your information, I am to state that a total of 50,000 cubic feet of storage space for group 3 (B.L. cartridges) and/or for Group 6 (Q.F. ammunition) is available at Gibraltar.

24. *Chatfield to Fisher*

[ADM 116/3489] 24 January 1935

MOST SECRET.

[Carbon]

The Committee of Imperial Defence has now approved the recommendation of the Chiefs of Staff that H.M. Government should be prepared to assume control of the Suez Canal at any time at which the situation appeared to warrant such action.

This gives us a firmer basis for our defence of the Canal in emergency, and I feel that the time has now come to consider these in relation to the broader strategical aspects of the question.

2. From this point of view there are three possibilities to be considered:–
That the Canal may be blocked, or attempts be made to block it,
(a) Before the arrival of the Mediterranean Fleet at Port Said,
(b) At some time during the passage of the two Fleets through the Canal,

---

[1] The committee agreed that Malta was 'perilously close' to falling to so low a standard of efficiency that it would be impossible to bring it up to a suitable standard within a reasonable time. They recommended a series of initial measures. For local naval defences they were: the provision of anti-torpedo booms and anti-coastal motor boat booms for the Grand Harbour; and for Marsamuscetto Harbour, anti-torpedo nets and a loop mine field for the entrance. The coast and anti aircraft defences should be improved by reconditioning and modernising defence electric lights, range finding and communications systems; the reconstructing and re-arming with modern guns and mountings of one 6" gun battery; the conversion to high angle mountings of one 9.2" battery and the provision 'of the proper scale' of anti-aircraft guns, searchlights and acoustic mirrors. The air forces stationed at Malta initially would be primarily for training purposes but eventually expanded to a composite squadron consisting of one flight of fighter bomber aircraft, two flights of torpedo bombers and one flight of spotter aircraft. The construction of a second airfield was also most desirable. J.D.C. No.184, *The Defence of Ports at Home and Abroad: Malta*, 1 Oct 1934, ADM 116/3473.

(c) After the fleets have passed through.

3. The timetable issued by the Admiralty for the passage of the fleets (Appendix 6 of the War Memorandum – Eastern) is only a typical one, and your emergency scheme for the dispatch of the Mediterranean Fleet from Malta is elastic in that it is designed to meet various alternative initial dispositions of the Fleet. Similar considerations apply to the Home Fleet, and it is in fact impracticable to base our plans upon a rigid timetable for the passage of the fleets through the Canal.

4. Recent investigations of defence measures have made it clear that the Suez Canal commitment must be regarded primarily as a naval one, and I am strongly of the opinion that the measures to be taken for the security of the canal must be viewed as one continuous naval operation.

5. With regard to the possibility mentioned in paragraph 2(a), consideration must be given to the correct moment to take action. It may be assumed that if strained relations developed, the Mediterranean and Home Fleets would be ordered to their bases to prepare for passage to the Far East and that the Mediterranean Fleet would not be required to leave Malta within about seven days from the time that, on the emergency first arising, the secret tanker movements were started. If the Japanese had any plan to block the Canal they would probably refrain from doing so until the critical moment, and this would be the departure of the Mediterranean Battle Fleet from Malta. It may be anticipated, therefore, that the Admiralty would advise H.M. Government to issue instructions to take control of the Canal at the latest not less than 24 hours before the sailing of the Mediterranean Battle Fleet.

The Canal could not, however, be seized unless the ships requisite for the purpose were already there and the early and unobtrusive departure of this force from Malta or elsewhere must therefore be an essential part of our plans.

6. A considerable portion of the Mediterranean Fleet is already earmarked for departure in advance of the Main Body, and this new commitment is likely to represent a further substantial addition. It may be necessary to consider whether reductions in the force now detailed for special duties ahead of the Fleet, and in this connection your investigation should visualise the Battle Cruisers being stationed in the Mediterranean. The despatch of the Suez Canal Force, referred to in paragraph 9 below, must necessarily be given a high priority, and I envisage that, in certain circumstances, docking and other non-essential preparations would have to be dispensed with.

7. With regard to the second possibility, if our Battle Fleet was divided by the blocking of the Canal during the transit of the fleets, the strategical consequences would be so serious that it is for consideration whether our plans should not be altered so as to provide for the virtual concentration

of the Battle Squadrons before passing the Canal. Your opinion as to the certainty of being able to assure the security of the Canal will influence the decision on this important point.

8. The foregoing considerations will necessitate a revision of the typical timetable for the passage of the Fleet, and also for your arrangements for docking and storing the Mediterranean Fleet in emergency.

Both these matters will, in any event, need consideration in connection with the re-distribution of the Fleet.

9. You will have received Admiralty Letter M.03555/34, containing a suggested detailed scheme for defending the Canal against blockship attack by means of destroyers stationed at Port Said and Suez. This scheme is the outcome of the conviction that guns and searchlights, even in the unlikely contingency of these being installed in peace and manned, cannot be relied upon as the primary means of defence. We must, therefore, be prepared to rely solely on naval measures. In addition to the Flotilla, cruisers will be required at Port Said and Suez to organise the counter-blocking measures. I am of opinion that an Officer of Captain's rank should be in charge at each end of the Canal, and it is for consideration that a Flag Officer should be stationed at Ismailia to take charge of all operations in the Canal Zone in close collaboration with the Canal Company's Headquarters.

10. It will be imperative that the Senior Officers principally concerned should have local knowledge, and be thoroughly acquainted with the principles upon which the defence scheme is based. They should also be prepared to assume complete responsibility in whatever action, however drastic, they consider necessary for the safety of the Canal.

11. If the scheme outlined in the Admiralty letter referred to above is accepted as sound and practicable, the corollary would be for the Senior Officers from H.M. Ships, which may be called upon to defend the Canal in an emergency, to visit the Canal in peace time. In this connection it might be possible to make use of the services of selected British pilots in an advisory capacity, both at Suez and Port Said, but you will probably consider it inadvisable to take such pilots into our confidence in peace time. The Senior Officers could, during such visits, discuss the defence scheme and become acquainted with local conditions both at Port Said and Suez, in co-operation with the N.L.O. Port Said and Suez, who is best qualified to advise them on all aspects of the problem. Without such prior knowledge I think it would be very difficult for the Commanding Officers of ships detailed for the defence of the Canal to act with energy and judgement in a delicate and dangerous situation, upon the correct and efficient handling of which so much would depend.

12. With regard to the possibility that the Japanese would attempt to block the Canal after the passage of the fleets this, though perhaps less

likely, cannot be ignored, since the maintenance of our communications, both naval and commercial, through the Canal would continue to be of primary importance. In order that there should be no break in the continuity of our measures in the Canal Zone, it will be necessary to arrange for the relief, at the Canal, of the units detached from the Mediterranean Fleet by a force despatched from the United Kingdom for this purpose, and this is being investigated.

13. In view of the importance of these questions, including the defence scheme, their early solution is very desirable.

25. *Admiralty to Fisher*

[ADM 116/3536] 15 March 1935

SECRET.

[Carbon] M.01344/35.

I am to acquaint you that My Lords have been informed semi-officially by the Foreign Office that it is likely that later in this year the Turkish Government will propose an official visit to Malta in the autumn in return for the official visit paid to them by the then Commander-in-Chief, Mediterranean in 1929.

2. The Secretary of State for Foreign Affairs[1] hopes that it will not be considered necessary, as a consequence of various recent incidents,[2] to reject such a proposal if it is made, and he has asked for an assurance that a visit by some Turkish warships will be welcome.

3. In considering this question My Lords have not overlooked the view expressed by you in your telegram 1148 of 19th November, 1934, nor the recent reception of H.M.S. *London* during her visit to Istanbul.[3]

---

[1] John Allsebrook Simon (1873–1954). Created Viscount, 1940. MP (Lib.) Walthamstow div, Essex, 1906–18; Spen Valley div, Yorkshire, 1922–31; (Liberal National), 1931–40; Solicitor-Gen, 1910–13; Attorney-Gen, 1913–15; Sec of State for Home Affairs, 1915–16; Sec of State for Foreign Affairs, 1931–35; Sec of State for Home Affairs and Dep Leader of the House of Commons, 1935–37; Chllr of the Exchequer, 1937–40; Lord Chllr, 1940–45.

[2] Notably the Samos incident; see Doc. No. 21.

[3] In November 1934 Fisher had been annoyed by the Turkish refusal to permit HMS *Frobisher* to visit Khelia Liman and Imbros and in view of this, as well as the Turkish reluctance to return the call made by the Mediterranean Fleet in 1929, Fisher cabled: 'I personally deprecate any visits, however brief, to Constantinople.' Fisher to Admiralty, Telegram No. 918, 19 Nov 1934, ADM 116/3536. In January Rear Admiral im Thurn reported that, except for the initial exchange of calls with the Vali and the General Commanding, Istanbul, the Turkish authorities had entirely ignored the presence of the *London* and that was 'so contrary to the welcome invariably extended by other countries we visit that I felt that we were an embarrassment to the Turks and our presence unwelcome to them'. First Cruiser Squadron, Report of Proceedings, 22 Jan 1935, ibid.

On the other hand, there have recently been indications of a genuine desire on the part of Turkey to improve relations with this country, notably their desire to place substantial orders for warships in Great Britain. Their Lordships note also the statement of the Rear Admiral Commanding, 1st Cruiser Squadron, that the Turkish officers themselves are far from being unfriendly.

They feel that from the national point of view there would be every advantage in accepting the proposal of the Turkish Government, while from the naval point of view, if the question is to be judged by the ungracious reception of H.M.S. *London* by the Turkish authorities the chance of any friendly intercourse between the two Navies may well be indefinitely postponed, which would be highly undesirable. The best cure for the present state of affairs is, in Their Lordships' opinion, to encourage the Turks to visit the Royal Navy where they may learn how one Navy ought to receive another.

4. My Lords propose, therefore, to inform the Secretary of State for Foreign Affairs that a visit will be welcome if put forward, but before doing so they would be glad to receive your early views.[1]

26. *Fisher to Chatfield*

[ADM 116/3489] 30 March 1935

MOST SECRET.

Commander-in-Chief, Mediterranean Station.

My views on the points raised in your letter of 24th January, 1935,[2] are as follows:–

Paragraph 3. I concur that it is impractical to base the plans for the passage of the fleets through the Suez Canal upon a rigid time table. I feel confident, however, that if the Canal has not already been blocked before the arrival of the advanced Canal Control Force, and the Government instructions to take all measures necessary are not withheld, the fleets will make the passage in safety.

Within the general framework of the War Memo. – Eastern, the Commander-in-Chief, Mediterranean, will of course feel himself free to arrange the plan in detail.

2. Paragraphs 5, 6 and 9. I propose that the advanced Canal Control Force should consist of the flagship and one other cruiser of the Third Cruiser Squadron and the Third Destroyer Flotilla.

---

[1] While Fisher agreed to the invitation, the events surrounding the Ethiopian War delayed the visit of the Turkish Fleet to Malta until November 1936.
[2] Doc. No. 24.

This force should leave Malta as early as possible, and not later than Z-5 day, so as to be at its station in the Canal Zone by Z-3 day. The remaining two cruisers of the Third Cruiser Squadron would proceed to Port Said as soon as they are ready so as to be under the orders of their own Flag Officer and to strengthen the defence of the Canal at the critical time during and just after the departure of the Mediterranean Fleet from Malta.

These two cruisers would proceed in company with the Fleet from Suez as it is assumed from Admiralty Letter M.00556/34 of 12th October, 1934,[1] that they will not now be required for the convoy of tankers.

3. As regards the protection en route and safe arrival of tankers at Secret Anchorage 'T',[2] I think it improbable that the Japanese would send cruisers so far, for the specific purpose of intercepting tankers. A surface ship operating at such a distance from a home base is beset with difficulties. The Japanese might reasonably expect the Main Fleet itself, or parts of it, to be convoying its tankers, which would therefore be inaccessible.

4. For these reasons I do not recommend the detachment of any important units at any great distance from the main fleet. The most likely attack on tankers when in open waters would be from disguised armed merchant vessels prepared and sailed before a declaration of war, and with this a destroyer could deal, and would be able to intercept reports and route the tanker accordingly.

5. In view of the alteration of plans, by which all the initial supply of fuel for Secret Anchorage 'T' will be coming from Aden or Trincomali [*sic*], the task of safeguarding the tankers is simplified.

I would therefore prefer that the First Cruiser Squadron and *Glorious* should go to Aden and cover the movements of tankers to Secret Anchorage 'M';[3] thence proceeding ahead of the battle fleet to the northward of the tanker route from Aden to Secret Anchorage 'T' until about the longitude of 70° East; and should proceed thence to Trincomali via the One and a half degree channel, to cover the passage of tankers from that port to Secret Anchorage 'T'.

The individual tankers[4] to be escorted from Aden to 'T' by one or more destroyers which would subsequently proceed to Trincomali to fuel and rejoin the First Cruiser Squadron.

6. Notwithstanding the foregoing remarks, the arrival of sufficient fuel at Secret Anchorage 'T' is of course of transcendent importance and I do

---

[1] Not reproduced.
[2] Addu Atoll in the Maldives.
[3] Kamaran Bay in the southern portion of the Red Sea.
[4] Marginal note by Chatfield: '3 tankers.'

not know what International consideration (vide Note on Appendix 6 A to the War Memorandum – Eastern) could stand in the way of the despatch of whatever tankers were available in the Persian Gulf direct to Secret Anchorage 'T' as an additional and precautionary reinforcement of supply. In any case, whatever the difficulty in this connection, it is believed to be essential that not only must all available supplies be drawn from Persian sources but that these must be expanded to their full productive capacity in order to prosecute a war in the Far East.

7. If not on the outbreak of war, then not long after it, it seems certainly necessary to have vessels in the Strait of Ormuz ready to deal with enemy submarines whose best chance of intercepting the Fleet's fuel supplies would appear to lie in that neighbourhood. Such vessels must of course be equipped with Asdics.

8. I consider that the Battle cruisers should form an integral part of the Battle Fleet during the passage to the Far East.

9. Paragraph 7. I am of the opinion that the programme for the passage of the Mediterranean Fleet through the Suez Canal should remain unaltered. There are no really valid grounds for awaiting the arrival of the Home Fleet, and such a course would involve congestion at Secret Anchorage 'M'.

The possibility of the Canal being blocked at any time before or after the passage of the Mediterranean Fleet will always be present, in all probability can never be foreseen and must be guarded against as far as is humanly possible. To have half the Main Fleet successfully through the Canal is better than to have none through.

10. Paragraph 8. I propose that the Battle cruisers should dock at Malta after the Battleships and, by virtue of their superior speed should partially fill the gap between the departure from Port Said of the Mediterranean Battleships and the arrival there of the first ships of the Home Fleet.

The programme of storing and docking the Fleet as at present constituted, at Malta, has been drawn up; the amendments necessary to include the Battle cruisers are under consideration, but should not greatly affect the existing programme in view of the remarks at the beginning of this paragraph.

11. Paragraphs 9, 10 and 11. Until relieved by Officers from England, Rear Admiral Commanding, Third Cruiser Squadron, should assume command in the Canal Area, being stationed ashore at Port Said. His Flag Captain, in the flagship, should be in charge of the approaches to Port Said. The Captain of the other cruiser in charge at Port Tewfik. If necessary, Captain (D) of the Flotilla should be at Ismailia to start with, as liaison between the Flag Officer Commanding and the Canal Company. Four destroyers at each end of the Canal for guard duties.

12. It is proposed that Rear Admiral G.H.D'O. Lyon[1] should be made acquainted with the duties he might have to carry out, as also the Captains (D), First and Third Flotillas. During a Summer Cruise this year opportunity will be made for Rear Admiral Lyon and Captains (D) One and (D) Three to study Canal problems on the spot.

13. Paragraph 11. As to taking British Pilots into our confidence in peace time, I am of the opinion that this would be most inadvisable and unsafe. It is the very exceptional and rare man who can resist moments of expansion and thus become a danger to the State.

14. Paragraph 12. It is recognised that only a slightly lesser degree of control of the Canal will be required even after the passage of the Fleet in view of the necessity for maintaining our communications. At the same time the Mediterranean destroyer flotilla charged with this duty will be required in the Far East at the earliest moment, so that its relief by a force despatched from the United Kingdom will be a matter of great urgency.

---

[1]RA [later Adm Sir] George Hamilton D'Oyly Lyon (1883–1947). Dir of Physical Training and Sports, Admy, 1927; Head of British Naval Mission to Greece, 1929–31; Cdre (D) commanding Home Fleet Destroyer Flotillas, 1932–34; RAC 3rd CS, Med Fleet, 1935–37; C-in-C Africa Station, 1938–40; C-in-C The Nore, 1941–43; retired list, 1943.

PART II

THE ABYSSINIAN CRISIS, 1935–1936

On 5 December 1934 Italian and Ethiopian forces clashed at the desert town of WalWal. The location was so obscure that there was doubt as to whether it was located in Ethiopia or Italian Somaliland. The Italian government demanded both an apology and reparations and initially refused an Ethiopian proposal to refer the dispute to arbitration. Ethiopia was, however, a member of the League of Nations and requested an investigation of the responsibility. A long series of complicated diplomatic machinations began and by the winter of 1935 it was apparent the Italians were building up their military strength in East Africa. In May 1935 the Council of the League of Nations established an arbitration commission to examine the question, but in the face of Italian opposition there was no agreement on procedure. The League then called in an additional arbitrator in July with the more restricted charge of establishing responsibility for the clash, but not the ownership of the disputed territory. Italian military preparations continued and from the bellicose statements of the Italian leader Benito Mussolini there was little doubt the Italians were bent on war. A three-power conference in Paris between France, Great Britain and Italy on 16 August submitted compromise proposals, subject to Ethiopian approval, that were scornfully rejected by Mussolini on the 18th. The Commission of Inquiry into the WalWal incident delivered an anodyne verdict on 3 September. Neither side was to blame for the clash since both the Italians and Ethiopians considered WalWal to be in their own territory. The Italian preparations for war continued. Hostilities could therefore lead to a situation where other members of the League might be called upon to enforce sanctions against Italy in fulfilment of their obligations under Article XVI of the Covenant of the League.[1] On 11 September the British Foreign Secretary, Sir Samuel Hoare, delivered a strong statement at the League that Britain would honour her obligations under the Covenant, this despite the claim made by the French Foreign Minister Pierre Laval that the day before, in a

[1] Surveys of the crisis are: George W. Baer, *The Coming of the Italian–Ethiopian War* (Cambridge, Massachusetts, 1967); *id edn, Test Case: Italy, Ethiopia and the League of Nations* (Stanford, 1976); and the short summary in G.M. Gathorne-Hardy, *A Short History of International Affairs, 1920–1939*, 4th edn (London, 1950), pp. 402–18.

private meeting, Hoare had ruled out military sanctions, a naval blockade, closure of the Suez Canal or anything likely to lead to war.[1] In practice, a major role would be played by the two leading Mediterranean Powers, Great Britain and France. The French Government had recently been seeking a rapprochement with their Mediterranean rival Italy in the face of the growing power and aggressive nature of France's eastern neighbour Germany.[2] With a confused and weak political situation at home, the French were loath to act. This in turn led to the situation where Great Britain might have to act against Italy alone and the danger the unpredictable Mussolini might even launch a pre-emptive strike against the British.

The situation had serious implications for the Royal Navy.[3] The Government had begun the painful process of rearmament including modernisation of the fleet, but there were serious deficiencies, especially in regard to air power, that would take time to correct. The British reached a naval agreement with Germany in June 1935 permitting the Germans to build ships up to a certain percentage of the British Fleet.[4] The rationalisation behind it was that the Germans were going to rearm anyway and this was a means of avoiding the type of naval race that had taken place before the First World War. In the Far East, Japanese aggression in China was evident and Britain's Far Eastern possessions required reinforcement.[5] However, reinforcement of the Far East presupposed a secure Mediterranean and the possible rupture in the traditionally friendly relations with Italy threatened this. A conflict with Italy was not something British military leaders were enthusiastic about but the Government also had to take into account the strong pressure of idealistic public opinion that the League must be supported and Fascist aggression resisted. The First Sea Lord expressed these reservations forcefully in a private letter to the Commander-in-Chief of the China Station [58].

The Mediterranean Fleet would obviously be in the forefront of any conflict with Italy.[6] On 7 August the Admiralty warned Admiral Fisher

---

[1] Gathorne-Hardy, *International Affairs*, p. 411.
[2] Examined in detail in Jean-Marie Palayret, *L'Alliance Impossible: Diplomatie et outil militaire dans les relations franco-italiens* (Paris, 2004), pp. 271–80.
[3] For the thoughts of the First Sea Lord, see Admiral of the Fleet Lord Chatfield, *It Might Happen Again* (London, 1947), pp. 87–90.
[4] On the Anglo-German Naval Agreement of 1935, see Joseph A. Maiolo, *The Royal Navy and Nazi Germany, 1933–1939* (London, 1998), chap. 1.
[5] Great Britain's strategic dilemmas are examined in Christopher M. Bell, *The Royal Navy, Seapower and Strategy between the Wars* (Stanford, 2000).
[6] The basic accounts of the role of the Mediterranean Fleet in the Abyssinian Crisis are in Stephen Roskill, *Naval Policy between the Wars*. Vol. II: *The Period of Reluctant Rearmament, 1930–39* (London, 1976), chap. 9; and Arthur J. Marder, 'The Royal Navy and the Ethiopian

that strained relations with Italy were possible leading to a potential danger to the Mediterranean Fleet in Malta and that the fleet should be prepared to move to Egypt to avoid surprise and to secure the Suez Canal [27]. The Mediterranean station was not at full strength. Fisher, in the temporary absence of his flagship *Queen Elizabeth*, flew his flag in the battleship *Resolution*. Moreover, at the moment there were no carriers on station. The *Glorious*, then just ending a long refit at home, hurriedly embarked aircraft, wound up the final stages of refitting and made what has been described as a fast passage to Malta arriving on the 28th. Here she embarked more aircraft and a day later sailed for Alexandria. The ship now had five different types of aircraft onboard: Blackburn Baffins; Fairey Seals; Fairey IIIFs; Hawker Nimrods; and Hawker Ospreys.[1]

The first portion of the planned summer cruise was to be quietly cancelled. The Admiralty realised that at the moment the Mediterranean Fleet would not be strong enough to undertake an immediate general offensive but assured Fisher he would be reinforced with units from the Home Fleet as soon as possible. Fisher was also ordered to begin unobtrusive preparations in the form of anti-submarine and anti-torpedo nets and booms as well as preparations for the Mobile Naval Base which was to be sent to the Mediterranean and established at a bay on the Greek coast in the southern Ionian Sea. The location was designated Port 'X', the site Navarino Bay. Fisher was told that the policy of the government was to support the League and enforce the Covenant, notably Article 16, provided France and the other powers also gave active support. Nevertheless, the Chiefs of Staff were anxious to avoid precipitating any crisis in order to give the services time to prepare and coordinate plans with other powers [28]. The British preparations were to be made as unobtrusively as possible. The Mediterranean Fleet was to remain at Malta until 29 August when the summer cruise had been scheduled to begin. The first cruise was cancelled, and the fleet would concentrate at Alexandria [32]. The objective was to move the most important units of the fleet beyond the danger of large-scale air attack from Sicily and Italy and to occupy a strategic position near the Suez Canal and provide additional security to Egypt. A sufficient number of destroyers of low endurance and submarines would remain at Malta to repel a seaborne

---

Crisis of 1935–1936' in *From the Dardanelles to Oran: Studies of the Royal Navy in Peace and War, 1915–1940* (London, 1974), chap. 3. Among unpublished documents an important source is the lengthy history of the crisis prepared in the Mediterranean Fleet in 1937. This study, too lengthy to reproduce, contains references to some material which may have been 'weeded' and no longer exists. See 'History of the Italian–Abyssinian Emergency, August 1935 to July 1936', 20 Dec 1937, ADM 116/3476.

[1] John Winton, *Carrier Glorious: The Life and Death of an Aircraft Carrier* (London, 1986), pp. 50–51.

attack against the island [97]. Preparations for the defences at Malta and Gibraltar continued during the last half of August [31, 33, 34, 38–41]. Fisher was confident that after receiving reinforcement from the Home Fleet the Mediterranean Fleet would be able to deal with any situation at sea without the assistance of other powers, provided the French were friendly. He advocated meeting any hostile acts by the Italians with the strongest possible countermeasures within 24 hours if possible [29]. A major offensive, however, would have to await the junction of the two fleets. He assumed, correctly as the Admiralty informed him, that any air attack on Malta would mean a state of war existed and that he could undertake a counter offensive without further instructions [30, 42]. This did not include, however, retaliatory attacks on Italian bases by air if the Italians did not initiate air attacks themselves [48, 49]. The Admiralty gave additional instructions on the subject of objectives in January. The government as a general principle intended to avoid taking the initiative in any air or naval bombardment directed against the civilian population in order to avoid providing the Italians with justification for retaliatory action. Only counter-attacks of a corresponding nature were allowed in the case of Italian attacks against legitimate military targets. In addition, even if the Italians attacked non-military objectives, retaliatory attacks of the same nature would require specific authorisation. Fisher's proposed operations against air bases at Port Augusta and possibly Catania were considered legitimate [91].

Fisher and the Admiralty differed about the air defence of Malta. Fisher wanted to leave the majority of the carrier *Glorious*'s aircraft at Malta when the ship sailed for Alexandria. Only a half squadron would be left aboard the ship for reconnaissance. In the light of RAF weakness on the island, he considered the aircraft to be retained at Malta necessary for defence as well as an immediate counter-offensive. Should the fleet return to the central area of the Mediterranean, the aircraft would be re-embarked [35, 37]. The Admiralty disagreed. The Fleet Air Arm was weak in numbers and loss of the aircraft would immobilise the carrier [36]. Fisher disliked Malta 'having no reply to make to air attack' and regarded the Air Ministry as having failed them. He could see no use for *Glorious*'s aircraft far away in Egypt [44]. Fisher also believed that with additional high-angle guns, fighters and bombers Malta could be made capable of being used as a base [57]. Andrew Cunningham, then Rear Admiral (Destroyers) in the Mediterranean Fleet, recalled Fisher was 'greatly incensed' by an appreciation of the situation from the Chiefs of Staff Committee in London. Fisher rose to his feet and said in his most impressive manner: 'Cunningham, I have sent a signal to Their Lordships telling them I disagree with every word of this pusillanimous document. The Mediterranean Fleet is by no means so powerless as is here set out.'[1]

In a private letter to the First Sea Lord, Fisher also revealed his dissatisfaction with the quality and quantity of intelligence available to him, complaining that on this subject there was no one in the island of Malta who had imagination or initiative. It is interesting to note that he asked for Rear Admiral James Somerville to be sent out to the Mediterranean because Somerville 'had a mind which could function outside the ordinary grooves and make war à la 1935 instead of à la 1914'. Somerville was then Director of Personal Services at the Admiralty and would not arrive in the Mediterranean Fleet as Rear Admiral (Destroyers) until the following April.[2] Fisher also hinted at small operations he had in mind which apparently included 'to use and lose' the old battleship *Centurion*, now a radio-controlled target ship [44]. Treaty provisions, however, precluded use of the ship.[3]

At the end of August the Admiralty reinforced the Mediterranean Fleet at Alexandria with a number of ships from the Home Fleet, including the carrier *Courageous*, one and a half destroyer flotillas, a submarine flotilla and the netlayer *Guardian*. The remainder of the Home Fleet would be concentrated at Portland ready to sail for Gibraltar at short notice if the situation became dangerous [43, 50]. Alexandria, however, was not large enough to handle the concentration. Consequently, the Third Cruiser Squadron, a destroyer flotilla and the minelaying organisation concentrated at Haifa. The major threat would probably come from the air. The Italians had developed a reportedly powerful air force and the effect of air power on the attempt to establish the MNBDO at Port X was an unknown factor. Fisher did not take the extreme view that Italian air attack would make the task of establishing the base impossible and the anchorage untenable, but acknowledged that until the Italian air threat was reduced it might be too hazardous for refuelling and the latter would have to take place at a location further from the sources of air attack [45, 46]. The availability of anti-aircraft ammunition, likely to be consumed at a high rate, was another source of worry. After the Mediterranean Fleet received additional stocks of this ammunition, there would be little left in reserve at home [51, 52]. Furthermore, as Chatfield explained in a memorandum on the strategical position in the Mediterranean, the reserve stocks of this ammunition were being built up over a period of years and, since no war in the Mediterranean had been envisaged, they were not based on a fleet

---

[1] Admiral of the Fleet Viscount Cunningham of Hyndhope, *A Sailor's Odyssey* (London, 1951), pp. 173–4.
[2] Michael Simpson (ed.), *The Somerville Papers* (Aldershot, 1995), p. 6.
[3] See below, p. 80, note 1.

having to operate in narrow waters solely dependent on itself for AA protection. Chatfield went so far as to say that the whole reserve of AA ammunition in the fleet might not be enough for it to meet the sustained attack for even as short a period as a week [50]. Chatfield was confident of the long-term result of a war with Italy, but in the short term the Italians had certain advantages based on their central geographical position, British unpreparedness, and their relative strength in cruisers, destroyers and submarines which Britain, with worldwide responsibilities, found it hard to match in the Mediterranean. Moreover, Britain's only first-class base in the Mediterranean at Malta was only 60 miles from Sicily with its Italian air bases and there was also a lack of docking and repair facilities in the eastern Mediterranean where the British fleet was now concentrated. Victory, while not in doubt, would take time and could be more costly than anticipated [50].

Fisher planned either when ordered or after an act of war to move westward from Alexandria with his fleet. He estimated the latter would comprise (before reinforcement from the Home Fleet) three battleships, three 8-inch cruisers, four 6-inch cruisers, two aircraft carriers, twenty-five or twenty-six destroyers and two oilers. A battleship, destroyer leader and division of destroyers would be left at Port Said. The oilers would be detached to a point 17 miles off Cape Matapan. On arrival in the central area of the Mediterranean his light forces would carry out sweeps by night between Syracuse and Cape Spartivento, covering a bombardment of the railway near Taormina and possible attacks on ships in Port Augusta. Aircraft would carry out dawn attacks on Italian air bases at Port Augusta and Catania if such retaliatory action had been authorised. After 48 hours the light forces would fuel from the oilers off Scutari. A cruiser squadron and destroyer flotilla would return to Alexandria to convoy the depot ship *Woolwich* and netlayer *Guardian* and the MNBDO to Port X covered by the main forces. Fisher would try to keep Italian forces occupied by harassing the Taranto area after the MNBDO had arrived and British submarines would be deployed to intercept Italian forces at likely places [53, 54]. The Admiralty also took steps to counter Italian forces in the Red Sea, an essential step since it was anticipated through-Mediterranean traffic as well as the supply lines of the British armed forces would have to be routed around the Cape of Good Hope. The limits of the East Indies Station were temporarily extended up to the southern entrance of the Gulf of Suez and contraband control stations would be established at Aden and Port Said [55, 56].

On 18 September the Admiralty ordered Fisher to cancel the second summer cruise so as not to weaken Fisher's strategic position by dispersion [59]. It was by this time becoming increasingly likely that the threat of sanctions would be unlikely to deter the Italians from war against Abyssinia. However, it was also probable that the Italians would launch the war without actually attacking another member of the League. The support that the British could expect from other League members was therefore questionable and the Chiefs of Staff had to prepare for a war with the British facing Italy alone. Their general conclusion was that solely economic and naval pressure could not be relied upon to defeat the Italians, although the Italian army in East Africa could probably be cut off and forced to surrender. Control of the central Mediterranean would be even more important and to achieve this the use of Port X would be vital. The establishment of this base must therefore be Fisher's chief preoccupation at the beginning of hostilities. The Admiralty also envisaged a two-pronged attack, that is the forces from Gibraltar would sweep the Italian west coast in conjunction with the Mediterranean Fleet acting from the east. Nevertheless, the Admiralty still maintained an important restriction on British actions. There would still be no bombardment of shore objectives at this stage if the Italians had not already attacked Malta by air. If they had, then retaliatory action was authorised [60, 62]. Fisher would also be in general command of all forces operating in the Mediterranean [61]. Although the Cabinet in the period 20–26 September decided the Home Fleet would not proceed to Gibraltar, Fisher nevertheless received further reinforcements. The minelaying submarine *Porpoise*, cruisers *Berwick*, *Sussex* and *Galatea*, and the minelayer-cruiser *Adventure* were ordered to the Mediterranean, where *Galatea* subsequently became the flagship of the Rear Admiral (D).[1]

The Italians began their long-expected invasion of Ethiopia on 3 October. On 11 October the Assembly of the League voted to impose sanctions. These included an arms embargo and the prohibition of exporting certain raw materials to Italy. Significantly, oil was not included. While the diplomatic manoeuvring went on at Geneva and in the capitals of Europe, the Admiralty continued its preparations for the possibility of a single-handed war. The Home Fleet would have a role in the western Mediterranean. Admiral Sir Roger Backhouse, C-in-C Home Fleet, was worried about the planned mission for his battlecruisers and the aircraft carrier *Furious* in operations against the north-west coast of Italy, given the vulnerability of the old carrier to air attack or Italian surface warships. He did not think the carrier could be left 'unguarded for a moment' and also advocated a dawn attack for the projected strike against Genoa. He suggested possible objectives for his force, notably the Ansaldo works at Genoa, the naval station at Maddalena and the seaplane base near Cagliari

[1] 'The Italian–Abyssinian Emergency', p. 10. ADM 116/3476.

on the island of Sardinia [65, 67]. Fisher also favoured a dawn attack but also considered a personal conference with Backhouse necessary to work out the details of the Gibraltar Force operations [66]. The Admiralty considered the question concerning the timing of an air attack must depend on the actual circumstances at the moment and the strength of the anti-aircraft defences of the escorting ships [75]. The meeting between Fisher and Backhouse finally took place at the beginning of February when Backhouse proceeded to Alexandria in the destroyer leader *Faulknor* [92, 93].

In October Admiral Sir Dudley Pound arrived in the Mediterranean Fleet to serve as Fisher's Chief of Staff [68, 69]. Pound had been designated to succeed Fisher but the situation in which a full admiral served as Chief of Staff to his predecessor was unusual.[1] According to Pound, on his arrival '[Fisher] said, "We'll run this show together like two C.-in-C.s" but I don't think he imagined it was possible even more than I did'. Surprisingly, the combination seems to have worked well during the few months before Pound assumed command on 20 March 1936 [98].

There were a number of issues to consider in case of hostilities. Among them, the defence of Egypt's western frontier with Italian-ruled Libya was examined by the land, sea and air commanders in Egypt. The position at Sollum on the frontier was vulnerable and would probably have to be evacuated [64]. The voluntary evacuation of families from Malta was also encouraged by the Government [70]. The security of British undersea cables in the Mediterranean and Red Sea was another concern. These were considered relatively secure as long as secrecy as to their exact location was maintained, although absolute security could not be guaranteed [76]. The concentration of ships at Alexandria would have presented a lucrative and appealing target to Italian aircraft or warships. Fisher as a result devoted a good deal of attention to air defence and establishing coastal batteries. There was, however, competition between the requirements for manning Alexandria's defences and providing the necessary manpower for the establishment of the base at Port X. On 2 November the Admiralty agreed with Fisher's priorities, notably that delay in establishing the base at Port X would be inadmissible [71, 74].

Fisher attempted to have frequent exercises to keep the fleet at Alexandria from getting 'rusty'. He also signalled his intention if the fleet eventually arrived at Port X to transfer his flag to the light cruiser *Ajax* to control operations at sea, leaving the battleships inside. The latter would

---

[1] Robin Brodhurst, *Churchill's Anchor: The Biography of Admiral of the Fleet Sir Dudley Pound* (Barnsley, 2000), pp. 88–90.

be withdrawn to Alexandria once the defences at Port X were established. This would present the enemy with fewer targets, provided of course that the shore defences could guard the oilers and repair ships [77]. Fisher subsequently felt obliged to reassure Chatfield on his somewhat surprising intent to control operations from the smaller cruiser *Ajax* instead of one of his larger ships [95].

The diplomatic question of according Italy belligerent rights in the Ethiopian War was a complicated one. The British Government eventually decided that it was best not to raise the dangerous question and no forcible measures would be taken to withhold them except in pursuance of a policy unanimously adopted by members of the League [63, 72]. The potential support from France, the member of the League most concerned, remained problematical and the preliminary and limited talks with the French that began revealed that, in the early stages of any conflict, they would be in no position to render aid but would, on the contrary, need British assistance until French mobilisation was complete. This was especially true in regard to the transport of troops from North Africa to Metropolitan France, a traditional feature of French mobilisation plans [73].[1]

As the crisis dragged on into November with no end in sight, it became necessary to draw up a schedule for leave, refitting and recommissioning of Home Fleet ships serving in the Mediterranean [78]. Fisher believed that, as long as the emergency remained acute, it was necessary to keep the fleet at full strength and instantly ready. He was concerned that the Admiralty schedule for leave would cause the eastern Mediterranean to be weak in carriers and cruisers whose crews were highly trained [79]. The Admiralty reassured him that they did not contemplate weakening the forces in the eastern Mediterranean until the situation improved and had drawn up their schedule to avoid, once the situation returned to normal, having to give leave to all Home Fleet ships at the same time. They also authorised the respective commanders in the Mediterranean, the Red Sea and Gibraltar to communicate verbally to the ships' companies the reasons why they would be away from home at Christmas. The core of the argument was that the perception of the complete readiness of British forces in the Mediterranean and Red Sea was 'the greatest, if not the only, guarantee that Peace will be preserved' [81, 82]. Given the reluctance of the League powers to apply really serious sanctions, and the readiness of the British and French governments to meet a substantial portion of Italian demands as demonstrated in

---

[1] The French perspective is in M.A. Reussner (ed.), *Les conversations franco-britanniques d'État-Major (1935–1939)* (Paris, 1969), pp. 46–52.

December by the so-called 'Hoare–Laval Pact', this claim is an exaggeration.[1]

The question of French assistance also concerned Fisher as 1935 entered its final month. He asked for and received permission from the Admiralty to send a destroyer to Bizerta to gather information from the French naval authorities on a number of topics such as the availability of fuel oil for his fleet [80, 84]. The Admiralty raised the subject through the naval attaché in Paris before giving their approval. The French, worried over the large number of Italians resident in Tunisia, insisted on absolute secrecy. The mission was entrusted to then Commander Louis Mountbatten and his destroyer *Wishart*. Mountbatten found Vice Amiral Laborde, the Préfet Maritime, decidedly unwelcoming and gained the impression the French had few if any real plans for mobilisation and war. It was not an encouraging visit [86].

By the end of November Fisher had serious doubts about the advisability of establishing the advanced base at Port X at the beginning of hostilities. It would be a great advantage to have an advanced base in such a central position but it was not possible to estimate the strength of the Italian air threat and the risk was therefore not justified. Alexandria would remain the main base.[2] The Chiefs of Staff in London approved. The anti-aircraft guns and searchlights destined for Port X, would now be used at Alexandria but re-embarked if it was decided to establish the base in the future. The Chiefs of Staff also recommended consideration of a combined operation to drive the Italians back from the Egyptian frontier in order to deprive them of advanced air bases [83, 85]. They objected, however, to plans to establish an elaborate advanced base at Sollum.[3] The position of the latter right on the frontier was too vulnerable and a concentration here politically provocative. They recommended that the base be established further east at Sidi Barrani. The latter would serve as the starting point for the offensive to drive the Italians back from the frontier and to clear the Sollum area for a base in the future, followed by a further advance when preparations were complete [88, 89].

The abandonment of Port X caused Fisher to alter his plans somewhat. They remained fundamentally the same but the fleet could not occupy the central area of the Mediterranean continuously in sufficient strength because of the necessity of the destroyer flotillas to boiler clean after 21

---

[1] The public outcry when news of the peace plan was divulged led to Hoare's replacement as Foreign Secretary by Anthony Eden. Gathorne-Hardy, *Short History of International Affairs*, pp. 414–17.

[2] Steven Morewood, *The Defence of Egypt, 1935–1940: Conflict and Crisis in the Eastern Mediterranean* (London, 2005), pp. 56–7.

[3] On the British army in Egypt see Morewood, *The Defence of Egypt*, pp. 66–9.

days of continuous steaming. This meant a permanent reduction of one flotilla from the fleet. Fisher now intended to occupy the central area for only about a week and then return with the main force to Alexandria to fuel. He hoped, however, that it might be possible to use both Malta and Greek harbours for cruisers and destroyers to refuel [87, 93, 94].

As his period of command drew to a close, Fisher was confident that the fleet in Alexandria was secure against surprise attack and staged frequent exercises to make sure it remained so. He was also confident of his anti-aircraft defences and apparent success against the radio-controlled 'Queen Bee' targets [95–97]. Fisher, after his return to England, voiced similar sentiments to the Sub-Committee (of the CID) on the Vulnerability of Capital Ships to Air Attack, although he did acknowledge that their battleships had not yet been provided with sufficient anti-aircraft defences [108]. The impression remains of overconfidence: that is, if one only put enough AA weapons in a capital ship all would be well. He had formed his impressions from the fragile 'Queen Bees' and the lumbering three-engine Italian Savoia Marchetti S81s with their fixed undercarriages. A scant four years later the far more potent German Ju 87s and Ju 88s would prove to be a different story.

Shortly after assuming command Pound took the fleet to sea for two short exercises. They included a scenario in which a skeleton force represented Italian cruisers which had been intercepted but did not want to fight. The exercise convinced Pound that the old 'D' class cruisers of the 3rd Cruiser Squadron under his command were out of place in the eastern Mediterranean force. Their low endurance meant they would hardly ever be available in the central area and if present their low speed would prevent them coming into action. He accordingly requested that if the situation deteriorated the 3rd Cruiser Squadron should immediately change places with the newer ships of the 2nd Cruiser Squadron at Gibraltar [98].

By the spring of 1936 it was becoming more and more apparent that there was no will to impose serious sanctions on the Italians who were well on their way to achieving their objective, the conquest of Ethiopia. This also meant that the likelihood of their attacking British forces was slight. After all, they had achieved what they wanted. Vice Admiral James, the Deputy Chief of Naval Staff, pointed out that this affected the reasoning for keeping the fleet on a war footing. After eight months this was having serious effects on the well-being of the fleet in questions such as leave, docking and refitting. Chatfield agreed, and asked the First Lord to pay attention to the serious consequences on the Navy that present British policy in the Mediterranean created [99]. The appeal was partially successful, the Cabinet decided that pending further meetings and decisions at Geneva there would be no major redistribution of the fleet but that this would not preclude some relaxation of the present state of instant readiness nor unostentatious movements of ships provided they were not on a scale to reflect on foreign policy [101, 102]. This was the beginning of the end of the crisis and, while Pound welcomed the relaxation, he also warned about frequent alterations in the leave conditions as 'they tend to become an international barometer' [100].

The diplomatic discussions at Geneva dragged on through May and the Admiralty ordered Pound to make preparations, in an emergency, to reinforce British destroyers at Aden with ships drawn from the Mediterranean Fleet. They were to counter Italian naval forces in the Red Sea [103]. A week later, reflecting eased diplomatic tension, Pound was authorised to keep the fleet at only 72 hours' readiness [104]. Pound welcomed this opportunity to send the four 8-inch cruisers that had been longest out of dock to Malta but like many in the fleet at home and abroad he wished the League would end the Abyssinian affair. Nevertheless, he assured Chatfield 'we are quite happy to stay at Alex. for as long as we are required' [105]. In June Pound and the Admiralty made plans for actions to be taken at the end of the crisis [106, 107].

The Foreign Secretary in late June stated publicly that Government policy would be to maintain larger forces in the Mediterranean than had been the case at the beginning of the dispute. Sir Samuel Hoare, the new First Lord, had to remind him that, while the statement was literally true, this was only because the Mediterranean Fleet at the end of the preceding summer was not up to the strength announced in the 1935 Estimates. That standard could not be achieved 'at the present time or in the near future'. The Mediterranean Fleet would only see an appreciable increase in 1939 when the modernisation of old capital ships was nearly complete and new construction began to enter service [109].

Chatfield continued to press for permission from the Foreign Office to allow a redistribution of the fleet and a return to normality. He finally succeeded on 8 July when the Foreign Secretary agreed provided the withdrawals were not so large as to nullify his undertaking in the House of Commons that the British would maintain larger forces in the Mediterranean than previously [110]. With evident relief, the Admiralty orders were issued to revert to normal conditions [111, 112]. Fixed naval defences at Haifa were evacuated, booms and underwater defences at Haifa and Alexandria were recovered, the base organisation at Aden was reduced, and the MNBDO equipment returned to the United Kingdom. Most Home Fleet ships returned to England and cruisers, destroyers and submarines from the distant China, Australia, New Zealand, Africa, and America and West Indies stations returned to their former commands. The

Admiralty emphasised that the moves should be made as quietly and unobtrusively as possible [113]. The Admiralty on 11 and 14 July issued messages thanking the personnel involved for their services [114, 115]. The Abyssinian crisis was over.

The end of the crisis did not mean, however, quiet in the Mediterranean. In April 1936 a general strike among the Arab population of the Palestine Mandate expanded into what became a general revolt. This was predominately an army affair but the navy, as it had periodically during past Palestine troubles, provided ships and men to support the authorities. On 18 July, a military revolt began in Spanish Morocco and expanded into a long civil war with international ramifications [116]. The Spanish Civil War would occupy much of the Mediterranean Fleet's attention over the next few years.

27. *Admiralty to Admiral W.W. Fisher*

[ADM 116/3536]   7 August 1935

SECRET.

[Telegram] Addressed Commander-in-Chief, Mediterranean, repeated Commander-in-Chief, Home Fleet. From Admiralty.

IMPORTANT.

Part I.

Important conversations with France take place next week and if a united front to uphold the Covenant were to result, subsequent conversations with Italy may cause a strained situation from about August 17th onwards. This strained period might be the beginning of a time of potential danger to the Mediterranean Fleet in Malta.

Still assuming the united front with France, the meeting of the League Council on September 4th, if no prior agreement had been reached between Great Britain, France and Italy, will be a second and perhaps more marked period of danger.

In view of these facts and the anti-British feeling possibly becoming even more violent than at present in the Italian Press, visits of the Fleet to the Adriatic or Italian ports are undesirable, apart from strategic considerations.

The Fleet Programme therefore must be recast in order that possibility of any precipitate action on the part of Italy should find our Fleet suitably disposed. Your programme dated 8th July is cancelled but no action should be taken yet to inform the British ministers concerned or to promulgate the cancellation. You should telegraph revised movements based on the following general disposition –

1. The Fleet to proceed to Egypt and adjacent harbours to forestall Italian action in Canal zone and to prevent surprise submarine attack.

2. In the hope that the situation may develop peacefully a programme to be proposed from mid-September onwards which could be cancelled if necessary.

If the situation becomes sufficiently strained prior to the League meeting it might be necessary for the Fleet to leave Malta earlier than the 29th August and in that case orders will be issued by the Admiralty to you probably about August 17th, or shortly after. This may involve the Fleet being at 12 hours' notice from that date.

2012.

Part II.

As you will perceive the precautionary action in Part I has been ordered with the dual purpose of ensuring, as a first step, the security of the

Mediterranean Fleet against surprise attack and the command of the Canal Zone.

The course of the proceedings at Geneva in September is at present wholly uncertain, and no useful anticipation can be made. The proceedings may result in stalemate. If, however, they were to result in anything like economic sanctions we must realise that Italy might well retaliate with some action which would bring about a state of war. Also if she anticipated such an outcome at Geneva it is conceivable that she might take similar action earlier.

In the still improbable event of the situation developing to this point Admiralty are fully aware that your present strength would not admit of carrying out general offensive operations. If and when such a course became likely you would receive therefore reinforcements of units of Home Fleet as early as possible, and it must be assumed that full co-operation of the French and other Powers of the League would be forthcoming. Admiralty are considering what effective counter-action could be taken in the event of an attack on Malta and a general appreciation of the situation showing Admiralty intentions, will be forwarded to you at an early date. You should, however, consider whether Navarin Bay would be a suitable base for operations for a striking force.

You will appreciate the importance of the utmost secrecy and in any case no action should be taken which would cause alarm at Malta before the receipt of the further communication which will be made to you after the conversations with France.

Your views on the matters raised in this telegram will be appreciated.

You may communicate such portions of this telegram as you consider necessary to the Governor of Malta.

2105

28. *Vice Admiral Sir C.J.C. Little*[1] *to Fisher*

[ADM 116/3038]   10 August 1935

[Carbon]

I am sending you with this letter the Chiefs of the Staff report to the Government in regard to the 'military' position in the event of the Council of the League deciding to enforce the Covenant, especially Article 16, in

---

[1] VA [later Adm] Sir Charles James Colebrooke Little (1882–1973). Dir of Trade Div, Naval Staff, 1920–22; Capt of the Fleet, Med Station, 1922–24; commanded battleship *Iron Duke*, 1926–27; Dir RN Staff College, 1927–30; RA 2nd BS, 1930–31; RA Submarines, 1931–32; DCNS, 1933–35; C-in-C China, 1935–38; 2nd Sea Lord, 1938–41; Head of British Joint Staff Mission in Washington, April–Aug, 1942; C-in-C Portsmouth, 1942–45.

the case of the Italo–Abyssinian dispute. The policy of our Government is to support the Covenant, assuming that the other Powers, Members of the League, and notably France, also give active support. The preoccupation of the Chiefs of Staff has been to point out to the Government:

(a) the necessity for the active co-operation of France and the other Powers, and

(b) the desirability of postponing any crisis which could possibly precipitate war so as to give a breathing space for the Services to prepare and also to concert plans with the other Powers.

In the meanwhile, of course, there is always the possibility of a man like Mussolini[1] taking some extreme action of which we have no warning whatever and it is on account of this that Admiralty messages 667 and 668 were sent to you. The First Sea Lord expressed great sympathy in your reply numbers 387 and 388[2] and he hoped that Admiralty message 680 would serve to further elucidate Admiralty intentions at the present time.[3] You will observe that the policy of the Government hinges entirely on the result of the meeting of the three Powers in Paris next week. Preliminary to this, Eden[4]

---

[1] Benito Mussolini (1883–1945). Militant socialist and editor of Socialist party paper *Avanti!* before WWI; broke with party, founded his own paper, *Il Popolo d'Italia*, and became ardent interventionist, 1914–15; served in Italian army until wounded in accidental explosion, 1915–17; founded groups of disaffected veterans which subsequently became Fascist Party, 1919; seized power, Oct 1922; dictator of Italy, Oct 1922–July 1943; after removal by the King and subsequent imprisonment, rescued by the Germans and installed as leader of the 'Italian Social Republic' in the German-controlled northern portion of Italy, Sept 1943–Apr 1945; captured and executed by Italian partisans while attempting to reach Switzerland, 28 May 1945.

[2] On 8 August Fisher suggested the Fleet should remain at Malta and an immediate move to Navarin should the situation deteriorate. Any air attack on Malta should be followed by immediate counter-attacks in the form of air attacks against Port Augusta and Catania, severing the railway in Sicily and Italy, and long-range bombardments of Taranto and Messina. Fisher added that the reinforcements of additional asdic destroyers, aircraft carriers, the 2nd Cruiser Squadron and the 2nd Submarine Flotilla were very desirable but the extreme importance of an immediate counter-attack did not justify waiting for them. Moreover, Fisher recommended that only the 3rd Cruiser Squadron should proceed to the Suez Canal. 'History of the Italian-Abyssinian Emergency, August 1935 to July 1936', 20 Dec 1937, p. 1, ADM 116/3476.

[3] The Admiralty concurred in Fisher's views but stated that it was undesirable for the Fleet at its present strength to be placed in a position of being liable to attack. The Admiralty was considering despatching the Home Fleet to arrive at Gibraltar on 29 August in which case the reinforcements Fisher requested, as well as the battle cruisers in addition, would be sent to join the Mediterranean Fleet by about 4 September. The Admiralty considered the possibility of an Italian attack on Malta as doubtful, approved the Mediterranean Fleet remaining there until 29 August, and stressed the importance of the security of the Suez Canal. Ibid., p. 2.

[4] Robert Anthony Eden (1897–1977). Created 1st Earl of Avon, 1961. MP (C) Warwick and Leamington, 1923–57; Parliamentary Private Sec to Sec of State for Foreign Affairs, 1926–29; Parliamentary Under Sec, FO, 1931–33; Lord Privy Seal, 1934–35; Min without portfolio for League of Nations Affairs, 1935; Sec of State for Foreign Affairs, 1935–38, 1940–45; Sec of State for Dominion Affairs, 1939–40; Sec of State for War, 1940; Leader of the House of Commons, 1942–45; Sec of State for Foreign Affairs and Dep PM, 1951–55; PM, Apr 1955–Jan 1957.

and Vansittart[1] are proceeding to Paris on Tuesday in an endeavour to obtain the co-operation of the French Government and to assure themselves that this co-operation if forthcoming will be material as well as moral.

I enclose a letter from Vansittart of yesterday's date which is the best appreciation we could have of Foreign Office views and I feel that you will agree with him.[2] In the meanwhile we are pushing forward with what unobtrusive preparations we can make and practically all the Units of the Home Fleet will be brought forward in time to sail on the return of the Second Watch from leave on August 29th. The other important step which is being taken is to send the anti-submarine boom for the advance base to Malta, and in the event of the situation petering out we hope you will be able to store it at Malta.

After sending you message 680 the First Sea Lord proceeded abroad on his well deserved leave, but, of course, he will return if, as a result of next week's talks, it is decided to proceed with measures for bringing pressure to bear on Italy.

I hope that as a result of these various papers you will now be in full possession of the situation as seen in London and we shall, of course, keep you immediately informed of further developments ...

29. *Fisher to Admiralty*

[ADM 116/3038] 20 August 1935

SECRET.
[Telegram] Addressed Admiralty.

425. Without underrating characteristics of enemy vessels or morale of Italian Navy I feel content that on receiving suitable reinforcements from Home Fleet any situation at sea could be dealt with without active assistance of other Powers – provided that France was friendly.

---

[1] Robert Gilbert Vansittart (1881–1957). Created 1st Baron Vansittart of Denham, 1941. Entered the FO, 1903; private sec to Lord Curzon, 1920–24; head of the American Dept, FO, 1924–28; Asst Under-Sec and Private Sec to PMs Stanley Baldwin and Ramsay MacDonald, 1928–30; Perm Under-Sec, FO, 1930–38; designated 'chief diplomatic advisor to the government', an essentially meaningless position reflecting his loss of influence during the Abyssinian crisis with leading figures such as Eden and Prime Minister Chamberlain, Jan 1938; resigned July 1941. Known for his strong anti-German views.

[2] Vansittart doubted that the French were prepared to take a strong line at Geneva and the League meeting was likely to end in a fiasco leading to an eventual serious deterioration of the situation in Europe, but no immediate crisis. In the less likely prospect of the French taking a strong line, the possibility of a coup on the Italian side without warning became 'a possible rather than a probable danger'. Vansittart to Little, 9 Aug 1935, ADM 116/3038.

Any hostile acts by Italy should in my opinion be met by strongest possible counter offensives within 24 hours if possible. Delay adds to difficulty. If Mediterranean Fleet is in eastern (bases) and Home Fleet at Malta the latter would be available for immediate minor counter offensive but major offensive would have to await junction of the two Fleets. In either case Home Fleet should have my instructions beforehand to reduce delays. Request an experienced Staff Officer from the Home Fleet should be sent to Malta forthwith so as to explain matters to Senior Officer, Home Fleet on arrival of Home Fleet at Malta.

It is probable that H.M.S. *Centurion*[1] might be of greatest value. Request she may be sailed for Malta at the earliest possible moment.

If Staff Officer could bring latest intelligence regarding coast defences including fixed torpedo positions and possible minefields off Sicilian bases and Taranto, Otranto, Brindisi, Bari, Ancona and Naples it would be helpful.

2133.

30. *Fisher to Admiralty*

[ADM 116/3038] 21 August 1935

SECRET.

[Telegram] Addressed Admiralty.

435. It is assumed that if air attack is made on Malta to which of course anti aircraft defences will reply a state of war will immediately exist and counter offensive can be undertaken without further instructions.

2212/21.

---

[1] The battleship *Centurion* had been converted into a radio-controlled target ship. However, under the London Naval Treaty (Annex II, Section III (c)) she could not be reconditioned for warlike purposes and would not at this time be sent to the Mediterranean for subsidiary services possible without reconditioning. Minute by Director of Plans, 21 Aug 1935, ADM 116/3038.

31. Fisher to Admiralty

[ADM 116/3038] 22 August 1935

SECRET.

[Telegram] Addressed Admiralty (Correction rec'd. 2340/22).

IMPORTANT.

436. Your 738. Fleet will be at notice required from tomorrow 22nd as steady preparations have been going on for some days.[1] I cannot judge what chance there is of surprise attack on Malta, but however remote I think it should be guarded against accepting the considerable alarm it must create. At present harbours are entirely open to entry, guns and lights are not kept manned, no patrol outside harbour and ordinary leave to officers and men is being given. In this matter I must rely on Their Lordships' advice. Two or three Italian merchant ships use harbour daily and it is therefore not possible at this stage to put boom across.

2244/21.

32. Admiralty to Fisher

[ADM 116/3536] 22 August 1935

SECRET.

[Telegram]. Addressed C.-in-C. Mediterranean, repeated C.-in-C. Home Fleet. From Admiralty.

My 1230/21. Cabinet have decided not to raise the Arms Embargo for the present. Mediterranean Fleet will remain at Malta until due to commence the cruise on 29th August and further precautions in your 436 (not to C.-in-C. Home Fleet)[2] for the protection of the harbour in the meanwhile should not be taken. You should now cancel the original programme which included Italian ports and promulgate the whole of the new programme as contained in your messages 1548/11 and 430 (not to C.-in-C. Home Fleet). No reason for cancellation of the first cruise need be given in orders as this cruise is now obviously undesirable. When you are questioned, however, you should follow the reasons for change of

---

[1] Following the breakdown of the Paris conversations on 19 August when the Italians had refused to even consider compromise proposals, the Admiralty on 21 August ordered the Fleet to be at 24 hours' notice from the following day. They also suggested ships should be completed as unostentatiously as possible with extra HA ammunition up to the limit of safe stowage and with outfit torpedoes and depth charges. 'History of the Italian–Abyssinian Emergency ...', p. 4, ADM 116/3476.

[2] Doc. No. 31.

programme as set out in Admiralty Message 1230/21. This message is sent in advance and further information of general policy for your guidance will follow in due course.
1618/22.
D.C.N.S.

33. Admiralty to Fisher

[ADM 116/3038] 22 August 1935

SECRET.
[Telegram] Addressed C.-in-C. Mediterranean, repeated C.-in-C. Home Fleet, V.A. Malta, R.A. Gibraltar. From Admiralty.

Request report concerning local seaward defences at Malta and Gibraltar as follows:–
(1) How long would it take to complete all preparatory work that can be undertaken without attracting comment.
(2) After preparatory work is completed how long would it take to lay A/T Baffles, Anti C.M.B. Boom and A/S Boom respectively at Malta.
(3) After preparatory work is completed how long would it take to lay A/T Baffles at Gibraltar.
1907.

34. Rear Admiral Gibraltar[1] to Fisher

[ADM 116/3038] 23 August 1935

SECRET.
[Telegram] Addressed C.-in-C. Mediterranean, repeated Admiralty, C.-in-C. Home Fleet.

With reference to Admiralty message 1907 22nd August no preparatory work is necessary. A/T Baffles could be completely laid with seven repetition seven days working continuously in shifts.
1035/23.

---

[1]RA [later VA Sir] Murray Pipon (1882–1971). Naval attaché in Paris, 1925–28; COS to C-in-C Plymouth, 1929–30; commanded battleship *Royal Sovereign*, 1930–32; Cdre, RN Barracks, Devonport, 1932–34; RA-in-charge and Adm Superintendent, HM Dockyard, Gibraltar, 1935–37; retired list, 1936; recalled to active service and Senior British Naval Officer, Suez Canal, 1940–42; Flag Officer-in-charge, Southampton, 1942–45.

35. *Fisher to Admiralty*

[ADM 116/3040] 23 August 1935

SECRET.

[Telegram] Addressed Admiralty, repeated C.-in-C. Home Fleet.

444. Following are proposals re H.M.S. *Glorious*. On arrival turn over for Refresher Deck Landing Training for 812 and 825 Squadrons to be carried out. If no danger is apprehended to her unescorted H.M.S. *Glorious* to sail for Alexandria on 2nd September leaving 802, 812 and 825 Squadrons at Hal Far and retain on board the half of 823 Squadron only. H.M.S. *Glorious* would thereby avoid any possible danger from air attack at Malta. If events occur entailing movements of Mediterranean Fleet West from Canal Area, in which case Home Fleet would presumably also be in the Mediterranean, H.M.S. *Glorious* would re-embark all her aircraft left behind at Malta and be ready to operate with Fleet or Fleets. In the meanwhile Malta would have two Bomber and one Fighter Squadrons for offence and defence and Mediterranean Fleet on passage west would have air reconnaissance. In absence of R.A.F. reinforcements to Malta I consider it essential to provide this Air Force for defence of Malta in case of attack occurring when Fleet is too far away to take action.
1638/23.

36. *Admiralty to Fisher*

[ADM 116/3040] 24 August 1935

SECRET.

[Telegram] Addressed C.-in-C. Mediterranean, repeated C.-in-C. Home Fleet. From Admiralty.

Your 1638/23. Proposals for deck landing practice and for *Glorious* to proceed to Alexandria are concurred in. Admiralty, however, are strongly adverse to her aircraft being used for the defence of Malta. Important as that base is it is entirely secondary to the value of the Fleet. The F.A.A. is exceedingly weak in numbers and the loss of these aircraft would immobilize the Carrier no reserve personnel being available other than those pilots serving in the Fleet.

Once landed and if hostilities were imminent or commenced the withdrawal of them even if they were intact would be impracticable and might be exceedingly risky in face of ensuing action. It is regretted therefore they are unable to concur in this proposal.

Air defence of Malta cannot be accepted as a naval duty even under these circumstances. One squadron of modern R.A.F. fighters is being sent by Air Ministry to Malta but will not be ready for operations for six weeks.
2032.

37. *Fisher to Admiralty*

[ADM 116/3040] 25 August 1935

SECRET.

[Telegram] Addressed Admiralty, repeated C.-in-C. Home Fleet.

458. Your 770.[1] Respective duties of three services at Malta appeared to be of less concern than immediate counter offensive by bombers and fighters which can be carried out from Malta and continued till Fleet is ready to operate in Central area, when no difficulty is foreseen in flying on H.M.S. *Glorious* aircraft. Protection of Dockyard and Naval Establishments was entirely secondary in my mind to making war on enemy, but I would observe Fleet Air Arm in Mediterranean is very dependent indeed on unguarded Calafranca, which at present is entirely without defences and could therefore be attacked with impunity and continuously until rendered useless. Not having received Their Lordships' views on the counter offensive foreshadowed in Admiralty message 668[2] I considered better use could be made of numbers 802, 812 and 825 Squadrons from Malta (if available) and only whilst Fleet was in Canal area.

Proposal in my 444[3] was based on above considerations, but without knowledge of the major strategy it is intended to follow, I do not press for this as I appreciate all of the difficulties confronting Their Lordships before deciding.
1618.

[1]Doc. No. 36.
[2]Doc. No. 27.
[3]Doc. No. 35.

38. *Vice Admiral Malta[1] to Fisher*

[ADM 116/3038] 25 August 1935

SECRET.
[Telegram] Addressed C.-in-C. Mediterranean, repeated Admiralty.

691. Admiralty message 749.[2]

(I). Preparatory work to lay boom in Grand Harbour with temporary arrangements without hurdles or winches is complete. Hurdles and winches are expected to be in place by 13th September. Moorings for and preparatory work on anti-torpedo baffles in Marsa Musciet not yet commenced. Time required one week.

(II). Time required to lay anti-torpedo and A/B boom in Grand Harbour now 48 hours and for anti-torpedo baffles 7 days. There is no anti-submarine boom available at Malta, winches have not been supplied and winch houses have not been built.

1227.

39. *Admiralty to Vice Admiral Malta*

[ADM 116/3038] 26 August 1935

SECRET.
[Telegram] Addressed Vice Admiral Malta, repeated C.-in-C. Mediterranean. From Admiralty.

Your 1227/25. Preparatory work on baffles for Marsa Musciet should be commenced.

1912/26.

---

[1] VA [later Adm Sir] Wilfred Frankland French (1880–1958). RA in 2nd BS, 1931–32; VA-in-Charge, Malta, 1934–37; Member of Executive Council of Malta, 1936–37; retired list, 1938; British Administrative and Maintenance Representative in Washington, DC, 1941–44.
[2] Doc. No. 33.

40. *Admiralty to Fisher*

[ADM 116/3038] 26 August 1935

SECRET.
[Telegram] Addressed C.-in-C. Mediterranean, repeated Rear Admiral Gibraltar, C.-in-C. Home Fleet. From Admiralty.

Reference R.A. Gibraltar's 1035/23.[1] Period of 7 days for laying A/T Baffle is not acceptable. It is essential that it should be laid if only on a temporary basis in a maximum of 3 repetition 3 days from the order, i.e. the time it would take Home Fleet to reach Gibraltar if necessary in emergency. You should report further on the matter.

2110/26.

41. *Rear Admiral Gibraltar to Fisher*

[ADM 116/3038] 27 August 1935

SECRET.
[Telegram] Addressed C.-in-C. Mediterranean, repeated C.-in-C. Home Fleet, Admiralty.

986. With reference to Admiralty message 2110 August 26th.[2] Outer line of nets at end entrance could be laid in three days. It is hoped that whole could be in place in less than seven days. This latter estimate was given having regard to lack of experience in handling and consequent difficulty of guaranteeing exact time.

1112/27.

42. *Admiralty to Fisher*

[ADM 116/3038] 28 August 1935

SECRET.
[Telegram] Addressed C.-in-C. Mediterranean. From Admiralty.

822. The assumption contained in your message 435 is confirmed by the Foreign Office[3] and you are free to act in accordance with Admiralty

---

[1] Doc. No. 34.
[2] Doc. No. 40.
[3] Doc. No. 30.

Instructions. In the event of your being absent from Malta you would however probably receive notice from the Admiralty.
1926/28.

43. *Backhouse to Captain K.H.L. MacKenzie*[1]

[ADM 116/3039] H.M.S. *Nelson*, at Portsmouth
28 August 1935

[Carbon]
MOST SECRET.
No. 00941/1
MEMORANDUM.
INFORMATION.
The Admiralty have directed that the following units are to be detached to join Commander-in-Chief, Mediterranean, in the Eastern Mediterranean: –
Group 'A'. *Courageous*, 5th Destroyer Flotilla, *Kempenfelt* and 3rd Division.
Group 'B'. *Guardian*.
Group 'C'. *Mackay*, *Lucia* & 2nd Submarine Flotilla. (Group 'C' is to proceed to Malta.)
2. In order that this reinforcement may be effected at an early date and as unobtrusively as practicable, it is intended –
(a) That they proceed at such a speed as to ensure early arrival in the Eastern Mediterranean (Groups 'A' and 'B'), and Malta in the case of Group 'C', without refuelling en route, and with an adequate margin of fuel remaining in case of emergency.
(b) That they pass the following areas in dark hours and avoid being reported anywhere on passage as far as is practicable –
  (i) Gibraltar Straits.
  (ii) Cape Bon–Malta Channel.
(c) They should deviate from the shipping routes as time permits for the purposes of (b).
SAILING.
3. *Guardian* is to sail from Plymouth about 0800 on 31st August 1935, and proceed to the eastward as if to Portland. An hour or two later course is to be altered and you are to proceed in accordance with these orders, being guided on passage by the instructions contained in paragraph 2.

---

[1] Capt MacKenzie was commanding officer of the netlayer HMS *Guardian*. Copies of the order went to the Admiralty and C-in-C Plymouth.

DESTINATION.
4. Alexandria.
INTENTION.
5. It is intended that *Guardian* should pass through the Gibraltar Straits during the night 3rd/4th September, and through the area Cape BON–MALTA Channel during the night 6th/7th September, and arrive Alexandria p.m. 9th September.

It will probably be found necessary to pass Cape BON at about 2100/6th September and proceed thence at 17 knots passing to the southward of Malta in order to be out of sight of the island at daylight. At other times you should adjust speed as necessary.

TIME TABLE.
6. The following approximate time table is given as a guide only:–
  2330/3rd Sept.   Pass Gibraltar.
  2100/6th Sept.   Pass Cape BON.
  0530/7th Sept.   Pass about 20 miles south west of MALTA.
  1300/9th Sept.   Arrive Alexandria.

SPECIAL INSTRUCTIONS.
7. W/T silence is to be maintained except for essential messages; positions on passage are not to be reported in the ordinary course. The Admiralty and Commander-in-Chief, Mediterranean will be informed of your departure and expected time of arrival. Should it prove necessary to put into port en route your destination is not to be disclosed.

No communication should be carried out with merchant ships or shore stations.

The ship's company may be informed of the ship's destination at 1600/31st August.

MOVEMENTS OF GROUPS 'A' AND 'C'.
8. Group 'A' is expected to pass Gibraltar during the night 2nd/3rd September, Malta during the night 4th/5th September and to arrive at Alexandria p.m. 7th September.

Group 'C' less *Lucia* and *L.23* are expected to pass Gibraltar during the night 4th/5th September and arrive Malta on 9th September.

MAILS.
9. Commander-in-Chief, Home Fleet, will take the necessary action with the Admiralty to divert mails.

44. *Fisher to Chatfield.*

[CHT/4/5]   Commander-in-Chief, Mediterranean Station
0630, 29 August [1935]

[Holograph]

Brownrigg[1] arrived last night & I was so glad to get your letter but I have scarcely had time to more than glance through the appreciation & other information he brought. He is staying here till Sep. 3 & then going on in *Glorious*. He must square up many things with French here & be at Alexandria in lots of time to receive his odd collection.

Binney[2] (if fit – in hosp. yesterday with sandfly fever) due to start home via Marseilles, will report to you at once. He will tell you what we have done & our general ideas for he has been at Staff meetings & in case Roger Backhouse or other H.F. Flag Officer has to join without much preparation. Binney comes with outline sketch plans of certain small operations which may or may not be possible. Nothing fixed of course & the ships to undertake them liable to drastic alteration. I see great possibilities in *Centurion* – and wish I could have her at Malta ready to use and lose!

Brownrigg will hoist his flag in *Resource* and later in *Woolwich* perhaps, but I expect that will have to wait till Port X becomes more of a certainty.

The Dockyard have done splendidly under hot severe [?] conditions. Mr. Hunt, Deputy Victualling Store Officer in the absence of his chief Salter has perhaps been the most conspicuous but all excellent now.

There is a certain amount of pro-Italian feeling in the Yard – & possibly a few potential enemies. Trade Unions are holding a meeting very shortly and have printed & circulated a carefully worded & argued statement for the revision of the Cut. They do feel that not all the facts are properly known in London. French tells me that D of D & the Sea Lords feel there is a certain case for redress but the Civil side or

---

[1] VA [later Adm Sir] Henry John S. Brownrigg (1882–1943). Dep Dir Gunnery Div, Naval Staff, 1925; Dir of Gunnery Div, 1926–27; COS to C-in-C Plymouth, 1927–29; commanded aircraft carrier *Courageous*, 1929–30; Dep DNO, 1931; RA commanding 3rd CS, 1933–35; VA (Base), Alexandria, 1935; Adm commanding Reserves, 1936–38; C-in-C The Nore, 1939; commanded Home Guard, Chatham Area, 1940–41; Cdre of Ocean Convoys, 1942–43; lost when SS *Ville de Tamatave* foundered, 20 Jan 1943.

[2] RA [later Adm Sir] Thomas Hugh Binney (1883–1953). Dir, Tactical School, Portsmouth, 1931–32; Flag Capt (*Hood*) and Ch Staff Officer to RA commanding Battlecruiser Sqdn, 1932–33; COS to C-in-C Plymouth, 1933–35; Chair of committee to examine education and training of naval personnel, 1935–36; RA, 1st BS, 1936–38; Chair of committee to examine work and organisation of the naval staff, 1938; Commandant Imperial Defence College, 1939; Flag Officer commanding Orkneys & Shetlands, 1939–42; Flag Officer-in-charge, Cardiff, 1944; Govr of Tasmania, 1945–51.

Treasury intervene. Anyhow French is for restoring the Cut & has said so for some time – and I'm told that were that done now 80 per cent of the anti-British feeling in the Dockyard wd. disappear. You might say that a concession made just now wd be ascribed to funk but I think it might be awarded in recognition of the good work done in putting the Fleet on a war basis in so short a time or something of that sort. But I believe this to be important and the Dockyard men to have a just grievance.

I don't like Malta having no reply to make to air attack. The Air Ministry has failed us – or rather the Cabinet. You know that I especially reported long ago that all was well except Malta. But apart from that I cannot feel it to be right that Italians should have the run of their teeth anywhere and until the Fleet returns to the Central Area Sicilian bombing which is the real formidable bombing can be well checked by *Glorious*'s aircraft who will be no use as far as I can see in Egypt.

I feel sure that I will not have to send officer spies [?] from the Fleet to picnic in Sicily to get the news I want (though I am sending one more tomorrow). It is too amateurish – and there is no one in this Island who has imagination or initiative. I have to support & go into details – otherwise no one moves. (Sounds bad coming from me but it is only the truth). Everyone has his own proper job & does it well but the special value that I attached to Somerville[1] for whom I asked you was that he had a mind which could function outside the ordinary grooves & make war à la 1935 instead of à la 1914.

So a live organization of Intelligence is essential & a Cryptographic Section as well. These 2 things make all the difference between success & failure. One works with the known & not the unknown or guessed at. Italians walk all over Malta, their ships are in harbour everyday & now & then a flying boat settles in Calafranca. They see everything but I remain comparatively blind. I know you will order this to help me & you.

Just off to sea & please forgive such an ill written & ill expressed letter.

---

[1] RA [later AoF Sir] James Fownes Somerville (1882–1949). Dir of Signal Div, Admiralty, 1925–27; commanded cruiser *Norfolk*, 1931–32; Cdre RN Barracks Portsmouth, 1932–34; Dir of Personal Services, Admy, 1934–36; RA (D), Med Fleet, 1936–38; C-in-C East Indies, 1938–39; placed on retired list because of ill health, 1939; commanded Force H, June 1940–Jan 1942; C-in-C Eastern Fleet, 1942–44; Head of British Admy Delegation, Washington, 1944–45; Admiral of the Fleet, 1945.

45. Admiralty to Fisher

[ADM 116/ 3038] 29 August 1935

SECRET.
[Telegram] From Admiralty.

831. My 802, further consideration is being given to the question of Italian opposition to your establishing the base at Port 'X' [Navarin]. This opposition is most likely to take the form of attack from the air and by light surface craft and submarines. Of this the air attack presents the greatest menace as many untried factors arise. An extreme view is held that attack by long range bombers and flying boats from the heel of Italy would render the task of establishing the base impossible and the anchorage untenable. Admiralty are not prepared to agree to such an appreciation in advance and would be glad of your views to enable the question to be explored from every aspect. The question of reserves of A.A. ammunition is a very important one and any forecasts of your requirements would be of considerable help.
1734.

46. Fisher to Admiralty

[ADM 116/3038] 31 August 1935

SECRET.
[Telegram] Addressed Admiralty.

486. Your 831. The extreme view is not held here in view of the long distance to be flown by aircraft carrying a heavy load of bombs, but unquestionably occasional air attacks must be regarded as probable. It is thought that danger can only be reduced by attack on Italian air bases in heel of Italy.

I would propose to leave H.M.S. *Resource* and employ H.M.S. *Woolwich* at Alexandria until more is known of the scale of air attack to be expected but M.N.B.D.O. should be established as soon as possible. If and when it is found too hazardous for fleet to use Port X for refuelling, refuelling will have to be carried out at other places unprotected but further from main source of air attack.

It is evident that Fleet Air Arm cannot be counted on to continue operations indefinitely and may be more urgently required with Fleet at any time. The establishment and equipment of British air base in Greece or Yugoslavia would make whole difference and be worth every effort to obtain. Two squadrons of flying boats operating from Port X would increase safety of fleet at sea and add to the chance of intercepting the enemy.

Foregoing views are subject to Rear Admiral Aircraft Carriers' confirmation and I shall see him immediately upon his arrival in Egypt. A reserve of as many rounds as possible of 4-inch high angle high explosive and 2-pounder Mark VIII ammunition and a further supply of bombs, aircraft flares and depth charges considered desirable.

2008/31.

47. *Fisher to Admiralty*

[ADM 116/3040]   1 September 1935

SECRET.

[Telegram] Addressed Admiralty, V.A. 1st Battle Squadron, H.M.S. *Barham*, R.A. Gibraltar, V.A. Malta, C.-in-C. Plymouth.

492. H.M.S. *Barham* is to proceed to Alexandria calling at Gibraltar to embark mails and refuel and at Malta to disembark ratings, refuel and replace practice ammunition expended.

On approach to and departure from Gibraltar and Malta sub-calibre, full calibre and high angle practices are to be carried out, arrangements being made direct between H.M.S. *Barham* and R.A. Gibraltar and V.A. Malta.

Speed on passage is not repetition not to be below 16 knots nor more than 18 knots.

2048.

48. *Fisher to Admiralty*

[ADM 116/3038]   1 September 1935

SECRET.

[Telegram] Addressed Admiralty.

493. Should an attack on Italian Air Bases be started whether or not Malta has been attacked from air or should these attacks by us wait enemy action by air? I am inclined to think latter.

2243.

49. *Admiralty to Fisher*

[ADM 116/3038]   3 September 1935

SECRET.

[Telegram] From Admiralty.

882. Your 493. The attack on Italian Bases is only approved at present as a retaliatory measure for attack on Malta by air. If attack is of a different type such as bombardment or submarine attack you should advance to take retaliatory action but instructions will be immediately sent by W/T.

1942 [Holograph notation initialled by D.C.N.S.] Copy sent F.O. 4/9.

50. *Memorandum by Chatfield*[1]

[CAB 16/138]   3 September 1935

MOST SECRET.

D.P.R. 15.

THE NAVAL STRATEGICAL POSITION IN THE MEDITERRANEAN.

1. The Naval situation when the meeting of the Council of the League of Nations takes place on September 4th will be that the Mediterranean Fleet will be in the Eastern Mediterranean, in Egypt and adjacent ports, and that it will within a few days be reinforced by one aircraft carrier, 1½ flotillas (see footnote)[2] of destroyers, and a submarine flotilla from the Home Fleet as approved by the Defence Policy and Requirements Committee at their meeting on 23rd August.

The remainder of the Home Fleet will be at Portland ready to sail at short notice for the Mediterranean.

2. Should any decision to impose sanctions lead to a more dangerous military situation, it would be necessary for the Mediterranean Fleet to be further reinforced and for the remainder of the Home Fleet to proceed

---

[1] Memorandum to the Sub-Committee on Defence Policy and Requirements, Committee of Imperial Defence, circulated at the request of the First Lord of the Admiralty.

[2] Footnote in original: 'NOTE: When consideration was given to the actual Flotillas to be despatched to the Mediterranean, the Admiralty found it preferable to retain the 6th Destroyer Flotilla (A/S fitted), which it had originally been intended to send, in Home waters, as they were not in a state of complete readiness at the time and in view of the difficulty of bringing forward another A/S fitted flotilla without publicity. The 2nd Destroyer Flotilla was accordingly selected instead, but as four ships of this flotilla had not completed their leave (having been serving in Irish Waters) and to send them without their leave would cause publicity, only 5 ships of the flotilla were dispatched making, with the 5th Destroyer Flotilla, 1½ flotillas in all.'

to Gibraltar. Full military concert by the French with corresponding action by them would be necessary before effect was given to any decision to impose sanctions. The agreement of Greece to the use of her ports would also be essential.

This change of dispositions could be carried out in about 4 days if it could be made as a peace-time measure, but had such a stage been reached that Italy might take warlike measures to try to prevent the adoption of the new dispositions, about 6 days would be required, owing to our shortage of destroyers in commission making it necessary for the Home Battle fleet to wait at Portland until the operation of reinforcing the Mediterranean Fleet was completed before destroyers would be available to escort that Fleet to Gibraltar.

The remarks which follow on the strategical situation in the Mediterranean assume that the above changes of disposition have been completed. They also assume that a state of war has been reached, since, as reported by the Chiefs of Staff, this is considered to be the inevitable result of the exercise of effective sanctions.

3. A glance at the map is sufficient to shew that on the <u>broad aspect</u> of the matter Italy is in a disadvantageous general strategical position.

She is dependent for 76% of her total imports on seaborne trade, and of this, 62% comes through Gibraltar, 3% via Suez, and the remaining 11% from the Mediterranean and Black Sea countries.

With our forces based at Gibraltar and in Egypt, her main communications can be cut with comparatively little effort to ourselves, whereas to take any steps (excepting by submarine) to counter our action she would have to send her forces far from their bases where they would be brought to action.

This strategical disadvantage is so great that it is unlikely that Italy could make any serious attempt with Naval forces to interfere with our control of the two exits to the Mediterranean except by the action of her submarines, which could not prove decisive.

Further, Italy's object is the prosecution of her Abyssinian war, and the mere closing of the Canal to her by the presence of our Naval forces (whether closing is done in Canal itself or by action outside it) might be decisive within a measurable period.

4. While the broad strategical aspect remains as described above, and the final outcome of a conflict with Italy cannot be a matter of doubt, the problem of putting warlike pressure on her is not so simple as might appear above, even when we assume that we are operating with the other principal Naval Power Members of the League, and have full use of their ports.

The reasons for this are as follows:–

(1) The fact that Italy has been preparing for war over a considerable period, whereas we have not been doing so.
(2) The excellent strategical position occupied by Italy for operations in the <u>Central</u> Mediterranean.
(3) The initially good position occupied by Italy in the Red Sea.
These three points are dealt with in detail below.
Point (1).

5. Italy has been preparing for war with Abyssinia for some time, and in general her policy in respect of armaments has been one of steadily strengthening her armed forces for some years past. All reports go to shew that Italian factories for the production of munitions, aeroplanes and other warlike articles are to-day working day and night, while in this country no corresponding action is being taken.

6. Consequent on war with Italy not being provided against, no action to strengthen our Mediterranean position has been taken. No defended bases exist in the Eastern Mediterranean to correspond, for instance, to the Italian Island of Leros in the Dodecanese on which Italy has expended so much effort in recent years, and the harbours of which the Fleet would have to make use are undefended.

Further, the strength of the British Navy has been reduced, both relatively and absolutely, in recent years compared to that of the Italian Fleet, so that the latter, which could once be almost neglected, has now to be taken seriously into consideration.

So far as capital ships are concerned, no question arises as by the Washington Treaty Italy accepted a low ratio compared to ourselves and has not even built up to that. In other categories, however, the position is very different, and while we have been limited in these classes, i.e. Cruisers and light craft, since 1930 by the London Naval Treaty, Italy has not been so limited, and has now more 8″ cruisers (7) than we, with our world-wide responsibilities, can afford to keep in the Mediterranean and Home Fleets combined (3 at present day, or 4 if the Commonwealth ship *Australia* can be included).

She has also 47 modern leaders and destroyers to our 43 in Home and Mediterranean Waters, and 15 of these are fast, heavily-armed leaders to which we have no counterpart.

Moreover, our reserve destroyers can only be got ready slowly after mobilisation is ordered.

In submarines Italy has 51 modern vessels to our 15 in Home and Mediterranean Waters.

7. The building up of reserves of ammunition has also been delayed for financial reasons. So far as ammunition for low-angle guns is concerned the situation is fairly satisfactory. The position with regard to ammunition

for Anti-Aircraft guns is, however, serious. These stocks, which are almost a new requirement since the war, are in process of being built up on a programme spread over a considerable number of years.

Further, as immediate war with a Mediterranean Naval Power has not been envisaged, the reserve stocks of H.A. aimed at have been based on the requirements of a Fleet when operating at sea from properly defended bases.

The situation now visualised is quite a different one, and is that of a fleet operating in narrow waters from a base open to air attack and unprovided with the necessary scale of A.A. defences by Army guns and Fighter aircraft of the Royal Air Force.

Under these circumstances the Fleet would be dependent on its own anti-aircraft gunfire for defence, and the scale of anti-aircraft ammunition required would bear little relation to that wanted for operations at sea where air attack is limited and where land aircraft can play only a minor part.

The situation today is consequently that the whole reserve of H.A. ammunition of the Fleet may not be enough to enable it to meet sustained attack by the metropolitan air forces of a Mediterranean Power for even so short a period as one week. Whether sustained attack would, in fact, take place must, of course, depend on the seriousness of such attack, which is uncertain, and on the morale of the Italian Air Force after first attacks. Until this is known it must be assumed that attacks might be sustained.

Such small steps as it is possible to take unobtrusively are being taken, but no real progress to improve stocks is possible in the absence of wartime emergency powers to increase production. Even with such powers the position could not be rectified under a period of some months as has been pointed out by the Supply Board in recent papers.

Point (2).

8. Italy occupies territory on both sides of the Central Mediterranean and in close proximity to the narrow channel between Sicily and Cape Bon in N. Africa, through which all traffic from East to West in the Mediterranean must pass. Her position in this area is so strong that after full consideration the Admiralty have come to the conclusion that, at any rate in the early stages of war with her, all through traffic of merchant shipping in the Mediterranean must stop and the ships be diverted round the Cape route. Such action will cause great inconvenience and loss to our shipping interests, as trade is likely to be diverted to the ships of non-members of the League which could continue to run through the Mediterranean, but it is not a vital consideration from the strategical point of view and supplies of essential commodities could be maintained.

9. Italy's position in the central area of the Mediterranean also places her on interior lines with regard to the operations of the British and French

Fleets, which, as forecast by the Chiefs of Staff in their report (Paper No. C.O.S. 392), must (apart from British control of the Gibraltar area) operate mainly in the Eastern and Western Basins of the Mediterranean respectively. This ability to act on interior lines would, in certain respects, offset the superior strength of the combined fleets.

10. Again our only first-class base in the Mediterranean is Malta within 60 miles of Sicily from which the whole force of Italy's metropolitan air force could be operated.

The Chiefs of Staff have in consequence expressed the opinion that the position of Malta is such that it might not be possible for our fleet to make use of it at all in a war with Italy.

Even, however, should the serious situation arise that Malta had to surrender to avoid undue casualties to the civil population, this though serious in its moral effect and in stiffening Italian resistance, would not be a vital consideration from the main strategical point of view.

11. Docking and Repair Facilities.

With Malta not available for the use of the Fleet, we should be faced with a very difficult situation in connection with docking and repair facilities. In the Eastern basin of the Mediterranean there is only one dock (at Alexandria) capable of taking even a small cruiser, while, although in the Western basin there is a certain number in French and other ports, in addition to Gibraltar (which can dock any vessel except a Capital ship), the only ones capable of taking a battleship are at Toulon or Bizerta, both of which places are again close to Italian territory.

There is the further difficulty, that to escort a damaged ship from the Eastern Mediterranean to the Western would be a difficult operation as she might have to proceed at slow speed between Sicily and Cape Bon in the face of possibly heavy scale of attack by submarines and aircraft.

Strong escort would also be required to prevent attack by Italian surface forces.

Repair facilities on any scale in the Eastern Mediterranean are also absent so that a damaged ship, even if she did not require docking, would take a considerable time to repair.

Alternative Bases.

12. In view of the appreciation that Malta would not be available for the fleet, the Admiralty have given consideration to an alternative base.

In the Eastern Mediterranean, Alexandria is the most suitable harbour and it could provide the accommodation required if restriction is placed on commercial traffic.

With the fleets based on Gibraltar and Alexandria Italy's war in Abyssinia would be in jeopardy and Italy herself could in time be forced to capitulate. Quick results will, however, be difficult to achieve.

A far more rapid decision should be attained if we could retain control of the Central Mediterranean. The Admiralty have consequently investigated the question of an alternative base in the Central Mediterranean from which we could exercise such control.

Harbours suitable for defending and large enough for the fleet are few, and in this area the only practicable one is a port in Greece (referred to in future as Port 'X').

This port is some 250 miles from Italy proper and the same distance from Italian Libya and the Italian base of Leros in the Dodecanese.

13. All distances in the Mediterranean are short compared to those experienced in Ocean Warfare, and though Port 'X' is much less exposed to attack by Italian aircraft than is Malta, it cannot be said to be immune from such attack.

While war experience enables the Admiralty to gauge the threat to a Fleet operating at such a base from the submarine and the mine, no such experience is available in connection with the extent of the threat from the air, and it must be remembered that many of our ships are old. As pointed out previously, whether sustained attack on the Fleet would in fact take place must, of course, depend on the morale of the Italian Air Force after the first attacks. Any support that could be given to the fleet by our own land air forces would be invaluable. In this connection the following extract may be quoted from a recent telegram from the Commander-in-Chief, Mediterranean:-

'The establishment and equipment of British air base in Greece or Yugoslavia would make whole difference and be worth every effort to obtain. Two squadrons of flying boats operating from Port "X" would increase safety of fleet at sea and add to the chance of intercepting the enemy.'

Again, the action of French Air Forces against Italy must be a most valuable diversion and it may well be that such action may prevent a serious scale of attack developing on Port 'X'.

Moreover effective air attacks cannot be concentrated on a number of objectives simultaneously.

Experience on all these questions should rapidly become available in war. The policy the Admiralty would adopt, bearing in mind the shortage of anti-aircraft ammunition, would be to prepare Port 'X' as rapidly as possible for use by the fleet, trusting that they could hold it against such an attack as might be brought against it. If this attack proved, contrary to anticipation, to be more serious than could be accepted, Port 'X' would be abandoned. If compelled to abandon Port 'X' and retire to Egyptian bases, we should largely lose control of the Central Mediterranean, even with the co-operation of France, until the time arrived when our reserves of anti-aircraft ammunition had been made good and the general war preparations of all three Services were further advanced.

It may possibly have to be accepted therefore that for several months after the outbreak of hostilities we could not retain control of the Central Mediterranean even with the co-operation of France and the other Mediterranean Powers.[1]

14. The results of such partial and temporary withdrawal from the Central Mediterranean would probably be as follows.

Firstly, the most rapid way of all of obtaining a decision would be to reduce the strength of the Italian Navy, by destroying as many units as possible. While this might be possible with the fleet operating in the Central Mediterranean, it is less likely that opportunities would arise if it were based in Egypt.

Secondly, our partial abandonment of the Central Mediterranean to Italy would inevitably have some moral effect; it would be likely to stiffen Italian resistance and to have repercussions on the opinions and actions of the minor Mediterranean states.

Thirdly, Greece, Turkey and other such states can hardly be expected to co-operate in the measures taken if we have abandoned the Central Mediterranean to Italy and in consequence a lively trade is likely to spring up with Italy and the closing by Turkey of the Dardanelles to shipping from the Black Sea would be more doubtful.

Should this be the case, Italy's resistance could be considerably prolonged, because she has made special efforts in recent years to increase her sources of supply of important commodities from this area.

A fourth, though perhaps minor point, is the question of Malta – while with our base at Port 'X' we should be able to prevent Malta being subjected to any other form of attack than that from the air, with no base nearer than Alexandria it is more unlikely that we could do so, and Malta might be occupied.

Point (3).

15. While British trade in general can be deflected round the Cape and so kept almost entirely clear of risk from war operations, the fleet and land and air forces operating in the Eastern Mediterranean must be kept supplied, and this would require a considerable volume of shipping.

Supply through the Mediterranean would be a very difficult problem as it would involve putting the ships into convoys and providing strong

---

[1] Footnote in original: 'Note to Paragraph 13. At Admiralty request six Army H.A. guns and 12 Searchlights are being sent to the Mediterranean so as to be available for Port "X" if required, but these constitute only a fraction of the scale of defence needed, and in any case they cannot be in place for some 10–14 days after the base was occupied, during which period attack must be expected.'

escorts against both surface and submarine attack. These would be exposed to deliberate day and night attack by Italy, and thereby, apart from other conditions, provide an excellent opportunity for maintaining Italian morale. The major part of our war supplies would consequently have to proceed by the Cape route and up the Red Sea.

Here again, however, Italy is initially well situated for attack on shipping which has to pass some 450 miles of Italian territory (Eritrea) on its passage up the narrow Red Sea. Attack by surface ships would in this case not be strong and could be countered by escorting vessels.

At present Italy has no submarines in the Red Sea [Note added: 'Since the above was written information has been received that Italy has sent two submarines into the Red Sea.'] and unless she has sent some there before war commenced it would be unlikely that she could send any via the Cape. The submarine problem in the Red Sea would consequently only arise if Italy decided and were allowed (and it is not seen how it could be prevented) to send such vessels there before the war stage commenced. She has, however, no submarine bases in the Red Sea and the threat would, in any case, be unlikely to be a sustained one as fresh supplies of munitions would not be available.

With regard to the threat from the air, however, the situation is less satisfactory. Italy has some 300–400 aircraft in Eritrea and Somaliland and can send further forces there by air, though again, supplies would be strictly limited. Our surface vessels available for Red Sea convoy have, on the other hand, not got up-to-date anti-aircraft armaments and the remarks above as to shortage of reserves of anti-aircraft ammunition would affect their operations also. If then Italy could spare air forces from their work in Abyssinia, and decided to concentrate against our line of supply in the Red Sea, and further, if she adopted unrestricted air attack against such shipping, it is probable that she could achieve some success, especially in the early stages before her stock of bombs was exhausted and before our counter measures would have time to develop. To counter such action on her part, counter offensive action by our Air Force working from the Sudan would presumably be undertaken, but the time such counter measures would take to develop was explained in the Report of the Chiefs of Staff (Paper No. C.O.S. 392, paragraphs 50–55).

Italian action in the Red Sea area, however, differs fundamentally from that in the Central Mediterranean area, in that it must in a relatively short space of time peter out through loss of seaborne communications with the home-land.

16. The strategical position inside the Mediterranean has, in the above remarks, been looked at principally from the point of view of the British Navy.

Operating as we should be in conjunction with the French it is necessary to consider what the reactions of our strategy would have on that of France.

In particular a decision as to whether Port 'X' would be used or not would be likely to have considerable influence on French action. The French main bases at Toulon and Bizerta, though not so exposed as Malta, are within raiding distance of Italian Air Forces. The problem in their case is not so serious however, because, firstly, they have probably installed adequate defences and, secondly, the bases are in their own home territory from which their own metropolitan Air Forces could operate in counter action (for example, against the Italian base of Spezia), whereas in our case either Malta or Port 'X' is in an isolated geographical position.

It is not possible to say how the French will look at this problem, but there is little doubt that if we had to relinquish full control of the Central Mediterranean the scale of attack against the French might be considerably increased.

It is most important therefore that before there was any question of sanctions being enforced, discussion between our Service Departments and those of the French Government should be authorised in ample time with a view to the adoption of a general allied policy failing which friction and misunderstandings would be almost certain to occur.

Conclusions.

17. The conclusions to be drawn from the above are:–

(a) Italy is in a fundamentally weak general strategical position as long as we can hold both exits to the Mediterranean which we should not have great difficulty in doing and that by such action the war with Abyssinia would probably have to be brought to a close and Italy herself in time forced to surrender.

(b) On the other hand owing to Italy's very strong strategical position in the Central Mediterranean and to our general unpreparedness for war in narrow waters, especially in connection with H.A. guns and ammunition, unless Italy surrenders at the mere threat of war, we might have to abandon in the early stages efforts to retain full control of the Central Mediterranean, which control is probably necessary if a quick decision is to be expected.

(c) Such abandonment of full control in the Central Mediterranean would be likely to affect the attitude of Greece, Turkey and other small Mediterranean states and open up to Italy supplies from the Black Sea. Should this be the case, Italy's resistance would be prolonged.

It would also be likely to have considerable moral effect on world opinion and to stiffen Italian resistance. Such effect would however probably be short lived as the results on Italy of the stopping of her supplies to Abyssinia and the closing of both ends of the Mediterranean become apparent.

(d) If there is any chance of sanctions being applied it is most important that discussions should be authorised between our Service Departments and those of France in ample time with a view to the adoption of a broad strategical policy.

(e) Before effect is given to any decision to impose sanctions it is of great importance that time shall be available to enable the necessary further reinforcement of the Mediterranean Fleet and the move of the Home Fleet to Gibraltar to take place – in concert with corresponding action by the French.

Admiralty,
3rd September, 1935.

51. *Admiralty to Fisher*

[ADM 116/3040] 3 September 1935

SECRET.

[Telegram] Addressed C.-in-C. Mediterranean, repeated C.-in-C. Home Fleet.

From Admiralty.

Following A.A. ammunition is being despatched to Mediterranean this week in ships which will also carry foreign station reserves of other natures of ammunition for ships detached from Home Fleet to Mediterranean Fleet:

10,000	4 inch.
2,000	4.7 inch.
25,900	2 pdr. Mark VIII.
2,000	depth charges.

This will leave only 4,000 rounds 4 inch and about 1,000 rounds 4.7 inch A.A. ammunition in reserve at home. Reserves of 2 pdr. Mk. VIII and 0.5 inch machine gun ammunition at home are negligible.

Reference your 492[1] [not to C.-in-C. Home Fleet] *Barham* has been instructed not to use H.E.H.A. ammunition for practice firings.

1709.

[1] Doc. No. 47.

52. *Admiralty to Fisher*

[ADM 116/3038] 4 September 1935

SECRET.

[Telegram] Addressed C.-in-C. Mediterranean, repeated C.-in-C. Home Fleet, V.A. Malta. From Admiralty.

My 1709/3, sense of which has been communicated to V.A. Malta, ammunition referred to is being despatched in S.S. *Spanker* and *Ciscar* leaving about 5th September. S.S. *Heminge* and *Hawkinge*, *Spanker* and *Ciscar* should be loaded on arrival with such ammunition, N.A. stores, torpedoes, bombs and depth charges from Malta as you may desire. Estimated that these four vessels could embark the bulk of stocks at Malta including bombardment reserve and ammunition earmarked for Armed Merchant Cruisers. Their Lordships' general view is that at least 50% should be embarked. Whether all or part of each particular category of ammunition is loaded is, however, left to your discretion as also is the distribution between ammunition ships and the subsequent disposal of these ships. Very early instructions should be given to V.A. Malta. Mr Atkins, D.A.S.O. Malta and necessary Staff from R.N.A.D. Malta will be detached for duty with C.-in-C. Mediterranean for charge of ammunition ships and supply of N.A. [Naval Artillery?] Stores with the Mediterranean Fleet. Cabin and office accommodation should be considered.

1949/4.

53. *Fisher to Admiralty*

[ADM 116/3038] 4 September 1935

SECRET.

[Telegram] Addressed Admiralty, repeated C.-in-C. Home Fleet.

507. Your 882.[1] Policy outlined is understood. The Fleet at present in Eastern Mediterranean will move west when ordered or after an act of war. It will probably consist of three battleships, three 8-inch cruisers, four 6-inch cruisers, two aircraft carriers, twenty five or twenty six destroyers and two oilers, to be joined later by two battle cruisers, two 6-inch cruisers and 6th Destroyer Flotilla. H.M.S. *Barham*, H.M.S. *Kempenfelt* and Third Division will be left at Port Said. If Portland Flotilla could be used to cover Home Fleet in Gibraltar approaches it would save

[1] Doc. No. 49.

me detaching 6th Destroyer Flotilla from central area where they are almost indispensable.
2113/4.

54. *Fisher to Admiralty*

[ADM 116/3038] 5 September 1935

SECRET.
[Telegram] Admiralty repeated C.-in-C. Home Fleet from C.-in-C. Mediterranean.

513. Your 822 and my 507.[1] General outline that I have in mind is as follows.

Fleet reserve ammunition leaves Egypt after dark with two oilers in company which will be detached to Skutari 17 miles 007° from Cape Matapan during following night. Main Fleet kept concentrated by day with air reconnaissance. On arrival in central area light forces will be used as most desirable in carrying out inshore sweeps by night between Syracuse and Cape Spartivento covering bombardment of railway near Taormina by 8-inch cruiser and possibly attacking ships in Port Augusta if they are unprepared and accessible. Air striking force to carry out dawn attack on Port Augusta and Catania air base if such retaliatory action is authorised. Battleships and destroyers will provide protection for aircraft carriers. On conclusion of 48 hours operations light forces will proceed to fuel in relief from oiler at Skutari. 3rd Cruiser Squadron and one asdic flotilla will proceed to Alexandria to convoy H.M.S. *Woolwich*, H.M.S. *Guardian* and M.N.B.D.O. to Port X under cover of main forces. Every opportunity will be taken to harass Taranto area when M.N.B.D.O. arrives to keep their surface and air forces occupied. For whole of this period all submarines will be disposed to intercept enemy forces at likely places. I shall hope to get M.N.B.D.O. to Port X about 8 or 9 days after fleet leaves Egypt. All above will be subject to any enemy movements reported.
1928/5.

---

[1]Docs Nos 42 and 53.

55. *Admiralty to C.-in-C. East Indies*[1]

[ADM 116/3038] 11 September 1935

SECRET.

[Telegram] Not by w/t.
Addressed C.-in-C. East Indies at Aden, repeated C.-in-C. Africa at Simonstown and C.-in-C. Mediterranean at Port Said. From Admiralty.

Part I.

* * *

The course of proceedings at Geneva is uncertain. Should the conciliation committee fail and a decision eventually be reached to support the League Covenant in some manner, it is the intention of H.M. Government not repetition not to take any action without ensuring that France and other members of the League fully co-operate. It is visualised that proceedings at Geneva may last an appreciable time. If the decision was only to impose sanctions of a non military nature, the naval situation would be one of suspended action. Should such a limited decision be arrived at it is possible but not probable that Italy would take some action which could bring about a state of war. We must therefore be prepared for war to follow the imposition of any form of sanctions.

Limit of the East Indies Station is hereby temporarily extended from the Straits of Bab el Zandeb to the southern entrance of the Gulf of Suez and the following ships are placed under your command:–

For Red Sea area *Norfolk, Emerald, Colombo*, 5 destroyers from Mediterranean, 3 sloops from Persian Gulf, 2 sloops from Red Sea.

In the event of war two sloops from Africa Station will also be placed under your command for Italian Somali Coast operations.

Part II.

Italian naval units in Red Sea are: Cruisers *Bari, Taranto*; Leaders *Pantera, Tiger*; Destroyers *Palestro, Audace, Impavido*; Submarine Depot Ship *Volta*; Submarines *Settimo, Septembrinni*; Sloops *Oetia, Axio*, Patrol Vessels *C. Berta, P. Corsinni*; Armed transport *Lussin*, 4–4.7″ guns; Armed tug *Ausonia*; one floating dock. At Mogadishu armed transport *Cherso* 4–4.7″ guns.

---

[1] VA [later Sir] Frank Forrester Rose (1878–1955). Commanded destroyer *Laurel* in Heligoland action, Aug 1914; commanded 3rd DF, Atlantic Fleet, 1921; DOD, Admy, 1925–27; commanded RN Barracks, Portsmouth, 1928–29; RAC DFs, Med Fleet, 1931–33; C-in-C East Indies, 1934–36; invalided, 1937.

Between 225 and 350 aircraft are in Eritrea and Somaliland with aerodromes at Asmara, Assab, Massawa, Tesseni (40 miles N.E. of Kassala, Sudan), Mogadishu, Bandar Alaia, Baidoa and some 43 landing grounds, including one at Bandar Kassim.

Latest information regarding Italian Bases is as follows:–

MASSAWA. One battery of 4″ guns mounted to northward of harbour entrance and one battery of 3″ guns.

Khor Zakhiliya and Harkinko Bay are being used as anchorages for merchant ships.

ASSAB. Batteries of 4.5″ and 3″ guns.

MOGADISHU. Two 4.5″ guns.

KISIMAYU. Some 4.5″ guns are being mounted.

The ports in Italian Somaliland being used BENDAR KASSIM, ILLIU, UMBIA, MOUENISHU, SELVA AND ZENEC.

0937.

PART III.

It is realised that if war situation is reached Aden is exposed to bombing attack from the large number of aircraft in Italian colonies. The extent to which these aircraft would or could be diverted from their military campaign for such a purpose is necessarily uncertain. Assuming however that they were so diverted, since Italy's own communications with Colonies would be cut whatever the scale of attack it should be a diminishing one.

Army are increasing garrison at Aden to higher colonial establishment and total of A.A. guns is being increased to 8 fully manned. These arrangements should be complete by 20th September.

One squadron of Fighter aircraft is being sent out and 205 Flying Boat Squadron from Basra will follow later.

As our through shipping in the Mediterranean would be diverted round the Cape, the main line of supply for the Fleet in the Eastern Mediterranean and for Army and Air Force in Egypt will be through the Red Sea and you would be responsible for its protection.

Unless Italy were to adopt unrestricted warfare with aircraft against shipping she would be unlikely to achieve any success so long as convoy was adopted.

If she did adopt unrestricted warfare with aircraft it is realised that your power of action in convoying ships in Red Sea is likely to be governed by the extent of air attack. Request you will consider practicability of retaliatory action against Italian bases and the possibility of exhausting Italian air effort as soon as possible by diversionary operations.

PART IV.

To be ready for any developments you should prepare for the institution of the Contraband Control Base at Aden for the examination of ships proceeding to the westward through the Gulf of Aden with contraband for Eritrea. The organisation of the base should be similar to that which is provided for in nucleus form for the initial stage in your war orders and in the Aden Defence Scheme. A full scale 'B' contraband control base would not be instituted unless it were found to be necessary at a later stage. Directions as to what if any ships were to be examined would depend upon decisions taken at Geneva.

S.O.(I) Colombo should be directed to send books, etc., needed for the base to Aden by first available vessel.

The subject of the employment of H.M.I.S. *Cornwallis* as guardship is being taken up with India Office.

As situation progressed Admiralty would endeavour to reduce the need for examination at Contraband Control Base by the adoption of a Navicert or similar system.

Commander-in-Chief Mediterranean would at the same time set up a Contraband Control organisation in Port Said area.

A blockade of Somali coast would probably be declared and enforced reinforced by two sloops from Africa Station mentioned in Part I.

0951.

56. *Admiralty to Fisher*

[ADM 116/3038]　　　　　　　　　　　　　　　　11 September 1935

SECRET.

[Telegram] From Admiralty.

958. My 956 and 957. Request you will take necessary action in connection with change of station limits and to prepare for the institution at a later date of Contraband Control organisation in Port Said area. Legal questions in this connection are being discussed with Foreign Office and no visible action should be taken at present.

1140.

57. *Fisher to Admiralty*

[ADM 116/3038] 15 September 1935

SECRET.
[Telegram] Addressed Admiralty repeated C.-in-C. Home Fleet.

582. Agree with employment of Battle Cruisers, H.M.S. *Furious* and 6th Destroyer Flotilla in the Western Area and there are capital objectives near Naples, Savona and Genoa.[1]

Will the 2nd Cruiser Squadron join me? H.M.S. *Rodney* is not required.

My proposed operations have been mainly understood by Their Lordships but referring to Admiralty 882 3rd September[2] may it be assumed that air attack on or ship bombardment of enemy vessels in their own ports are not retaliatory in nature and need not therefore wait for prior attack on Malta from the air.

I would again urge the provision of every available high angle gun, naval if necessary, for Malta and despatch of fighters and bombers up to the limit that can be operated there. Interception of and counter attacks by Malta aircraft reinforced by force of Battle Cruisers could I believe render Malta capable of being used as a base and this would obviate many serious difficulties if not dangers.

2033/15.

---

[1] 'On 13 September, the Admiralty informed the Commander-in-Chief that it was by no means impossible that a single-handed war with Italy might be forced upon us before the League Powers had agreed on action. If this occurred it was considered better to use the Battlecruisers actively in the Western Mediterranean with *Furious* and Sixth Destroyer Flotilla to cause anxiety in Italy as to their Northern and Western ports. The Commander-in-Chief was offered *Rodney* instead of Battlecruisers, but this offer was refused.' 'History of the Italian–Abyssinian Emergency …', p. 9, ADM 116/3476.

[2] Doc. No. 49.

58. *Chatfield to Admiral Sir Frederic C. Dreyer*[1]

[CHT/4/4] 16 September 1935

[Carbon]

* * *

Your letter of the 28th July[2] reached me two days ago and it is most interesting and I agree with most of what you say but not all. What you say about the German Colonies I am not so sure about. It was always said before the war that they would be a threat to our Imperial communications and to some extent they were but actually they proved one of the Achilles' heels of the German Navy and were no doubt one of the facts which made her realise what a disaster it was when we came into the war because those Colonies were bound to fall into our hands. As long, therefore, as we keep command of the sea any nation which has Colonies oversea will hesitate to go to war with us. It is for that reason that I personally should have no objection to the Italians being established in Abyssinia. I think it would prove to them a weakness rather than a strength and make them more anxious to be friends with us than the contrary. Actually it is a disaster that our statesmen have got us into this quarrel with Italy who ought to be our best friend in the future as she has been in the past because her position in the Mediterranean is a dominant one. Once we have made her an enemy the whole of our Imperial responsibilities become greater for obvious reasons. However the miserable business of collective security has run away with all our traditional interests and policies and we now have to be prepared as far as I can see to fight any nation in the world at any moment. Everybody will hate everybody else, and suspicion and intrigue will be worse than ever.

Meanwhile, I am in the unpleasant position of preparing for war and for a war which we have always been told could never happen. It is a war which if it takes place, there could be no doubt as to its end; we shall have many losses in ships and men, thereby our world position as a naval power will be weakened. I have no doubt Mussolini will go on with his war despite the League. If war comes then I may have to hang on to Little a bit longer but otherwise you may expect him out on his proper date.

[1] Adm Sir Frederic Charles Dreyer (1878–1956). Flag Capt to Jellicoe in *Iron Duke* (including Battle of Jutland), 1915–16; Cdre and COS to Jellicoe in *New Zealand* on mission to India and the Dominions, 1919–20; commanded battlecruiser *Repulse*, 1922–23; Asst CNS, 1924–27; commanded Battle Cruiser Squadron, 1927–29; DCNS, 1930–33; C-in-C China Station, 1933–36; retired list, 1939; Cdre of Convoys, 1939–40; Inspector of Merchant Navy Gunnery, 1941–42; Chief of Naval Air Services, 1942–43.
[2] Not in the file.

59. *Admiralty to Fisher*

[ADM 116/3536] 18 September 1935

SECRET.

[Telegram] Addressed C.-in-C. Mediterranean, repeated R.A. (D), V.A. Malta, R.A. 3rd Cruiser Squadron, V.A. 1st Battle Squadron, R.A. 1st Cruiser Squadron. From Admiralty.

Your 1403/13 and Admiralty message 1721/17.[1] Further consideration has been given to this matter in view of present situation. Assuming that report of 5-Power Committee will not be accepted by Italy which is highly probable as the proposals appear to contribute little or any advance on what was offered at Paris, subsequent steps at Geneva may be prolonged. It is further probable that nothing more than restricted economic sanctions in some moderate form will be agreed to judged by present disposition of many Members of League. If this is the case Italy is likely to ignore threat of such sanctions and proceed with her war with Abyssinia without attacking any Member of League. Despite these facts Admiralty do not wish to weaken your strategic position by despatching to Greek or other ports any units which would be exposed to submarine attack or whose absence would menace your tactical security should hostilities unexpectedly arise. It is considered therefore that any general carrying out of second summer cruise programme would be unsound especially as our present main dispositions are having the desired effect on the Italian mind but visits of isolated small units to certain Greek harbours such as Piraeus and Crete would have political advantages. Admiralty do not consider it necessary for ships to be at less than 6 hours' notice.

1645.

60. *Admiralty to Fisher and Backhouse*

[ADM 116/3038] 18 September 1935

SECRET.

[Telegram] (BY LAND LINE)
Addressed C.-in-C. Mediterranean, C.-in-C. Home Fleet, repeated V.A. B.C.S. From Admiralty.

Reference Admiralty telegrams 1215/17 and 1514/17 discussion of plans with French Admiralty is to be taken up in Paris.

---

[1]The Admiralty ordered that adherence to the programme of the Second Summer Cruise should be abandoned and requested Fisher to notify the British Minister in Athens of any changes of visits to Greek ports.

THE ABYSSINIAN CRISIS, 1935–1936 111

In the meantime it is considered that if a war situation is reached we must be ready to act single-handed at any rate in the early stages and until it is seen what assistance we shall receive from France and other nations. Present dispositions are based on this.

Chiefs of Staffs' Sub-Committee have prepared appreciation of single-handed war of which copy will be sent C.-in-C. Mediterranean and C.-in-C. Home Fleet.[1]

Their general conclusion is that economic pressure and naval action alone could not be relied on to defeat Italy as in case League war but that Italian army in East Africa would have to surrender before long and this would very probably bring Italy to terms.

Also that control of central Mediterranean would be even more important in single-handed war but for this purpose it would be more than ever important to endeavour to use Port X. Docking and repair facilities would present very serious problem.

With reference to outline of possible operations in event of war contained in C.-in-C. Mediterranean's 1928/5,[2] Admiralty now visualise that concurrently with operation of forces in Eastern basin forces from Gibraltar would carry out sweep in vicinity of Italian West Coast with the object of making minor attacks on military objectives by gun or air intercepting Italian warships if possible and to spread alarm in Italy and cause dispersion of effort. The two fleets might also carry out a simultaneous operation in Malta Channel before returning to their bases. It is thought that massed Italian forces now in Sicily will endeavour to occupy the Malta channel and they would be very vulnerable to a sweeping operation from East–West simultaneously mainly with destroyers and cruisers supported by larger units and carriers at a distance. The seven flotillas that will be available in the two fleets should thus have a good opportunity to bring Italian forces to battle or run them into port. In any case some such operation is envisaged for passing through important convoys when essential.

2155.

---

[1] D.P.R. 21 (also Paper No. COS 397). Memorandum by the Chiefs of Staff Sub-Committee on a single-handed war with Italy, 16 Sept 1935, CAB 16/138.
[2] Doc. No. 54.

61. *Admiralty to Fisher*

[ADM 116/3038] 21 September 1935

SECRET.
[Telegram] BY LAND LINE.
Addressed C.-in-C. Mediterranean, repeated C.-in-C. Home Fleet. From Admiralty.

Admiralty letter M.03317/35 of 24th August.

Consideration has been given to the High Command in the Mediterranean in the event of war, observing that for certain operations that require to be co-ordinated between Home and Mediterranean Fleets it will be necessary for all British forces to be under Command of C.-in-C. Mediterranean.

Again if we have the French fleet operating with us in the Western basin close co-operation between C.-in-C. Home Fleet and French C.-in-C. will be necessary.

It is considered too early to make final decision in case of a League War, but in the event of a single-handed war C.-in-C. Mediterranean would be in general command of all forces operating in the Mediterranean.

In view of this the operations visualised in your 1928/5 and Admiralty 2155/18[1] should be discussed as necessary with C.-in-C. Home Fleet.

1716/21.

62. *Admiralty to Fisher*

[ADM 116/3038] 3 October 1935

MOST SECRET
[Carbon] M. 04321/35.

Copy to:– Commander-in-Chief, Home Fleet.

With reference to your telegrams 1928/5/9 and 2203/20/9 and Admiralty telegrams 2155/18/9 and 1716/21/9, I am to inform you that further consideration has been given to the operations visualised, with special reference to the probability that the Battle Cruiser Squadron, 2nd Cruiser Squadron and the 6th Destroyer Flotilla will not now proceed to the Eastern Mediterranean, and the whole problem has been discussed at a meeting at which Admiral Sir A.D.P.R. Pound, K.C.B., was present.

2. Their Lordships are of the opinion that while the establishment of Port X must clearly be your chief preoccupation at the commencement

---
[1] Docs Nos 54 and 60.

of hostilities, if it were possible to carry out operations in the vicinity of the coasts of Italy, Sicily and in the Malta Channel in the early stages, such operations would have considerable effect, especially on the morale of the Italian Navy and nation.

3. It appears that the form that operations would take would depend largely on whether or not Italy had bombarded Malta by air. If she had done so retaliatory action against shore objectives as already authorised would be available, but if she had not Their Lordships consider that bombardment of shore objectives should not take place at this stage.

4. Their Lordships consider that it must be assumed that Alexandria and Port Said will both be liable to attack in the absence of the Fleet, until our establishment at Port X has made it clear that operations by Italian surface forces to the Eastward of that place would be of a very risky nature. In the early stages there will be numerous supply ships and other merchant ships causing congestion at these ports which will present objectives for attack. The whereabouts of the Italian 8" cruisers at the outbreak of war will consequently be a matter of great importance.

5. The attached two outline plans represent in a very general way Their Lordships' views as to the possible sequence of the operations visualised in your telegrams referred to above, assuming that you would arrange simultaneous operations by the Home Fleet units in the Western basin.

In formulating these outlines Their Lordships do not in any way wish to hamper your freedom of action in planning your operations and they request that when you have considered the matter you will forward by telegram your comments and proposals for the actual operations you consider might be carried out.

[*Attachments*]

OUTLINE PLAN No. 1.

Assumption.

Imposition of sanctions by the League leads sooner or later to hostilities with Italy. All Mediterranean Powers are involved but none are ready for operations but ourselves.

Italians commence bombarding Malta by air.

Operation.

Mediterranean Fleet leave Alexandria after reconnaissance and A/S operations off the approach have been carried out and escort base defence ships etc. to Port X, subsequently proceeding for operations. Possibly on the fourth night after leaving, Mediterranean Fleet operate off the coast of the foot of Italy and off Sicily.

Battlecruisers with *Furious* and 6th Destroyer Flotilla leave Gibraltar so as to arrive off Genoa about 1800 on the day before the Mediterranean Fleet arrive for operations off the Italian Coast.

Aircraft from *Furious* attack Ansaldo's fitting out works and the oil depot at end of mole.

Nelson, Rodney and 2nd Cruiser Squadron with 20th Destroyer Flotilla and/or 21st Destroyer Flotilla leave Gibraltar so as to arrive off Cagliari, Sardinia, about 1700 on the same day and carry out a bombardment of the Naval Base at that place.

Home Fleet destroyers subsequently fuel as necessary at Port Mahon, Majorca or in French African port from oilers sent there or possibly fuel at one of these places before the operation commences.

Subsequent Operation.

Provided that progress in the establishment of Port X permits, it is considered that as soon as possible a further operation should be carried out preferably by the Combined fleets, with the object of clearing the Malta channel of enemy vessels. For this operation it would be necessary for arrangements to be made for the flotillas and possibly other vessels of the Home Fleet to fuel in the Balearic Islands or French North Africa.

The time of this operation would be concerted by Commander-in-Chief, Mediterranean, and it might take place on the 5th or 6th day after leaving Alexandria.

OUTLINE PLAN No. 2.

Assumption.

As in Outline Plan No. 1, but the Italians have not bombarded Malta.

In this event attack on shore objectives would not take place and the operations of the fleet would be confined to those against Italian naval forces and shipping.

Other military objectives ashore should not be attacked till specially ordered.

The two fleets would proceed towards Malta and carry out a concerted movement as planned by Commander-in-Chief, Mediterranean, base defence ships etc. having been first escorted to Port X.

It is not envisaged that battlefleets would themselves operate in the Malta Channel so as to avoid inviting attacks on them.

63. *Admiralty to Admirals Fisher, Backhouse and Rose*

[ADM 116/3487] 4 October 1935

SECRET.

[Telegram] Addressed C.-in-C. Mediterranean, C.-in-C. Home Fleet, C.-in-C. East Indies. From Admiralty.

Question of belligerent rights at sea by Italy is under consideration. Admiralty consider that until a state of war between Italy and Abyssinia is declared she is not entitled to exercise belligerent rights. Should she do so local protest should be made and facts reported to Admiralty with a view to diplomatic action. No military measures should be taken by H.M. Ships. If Italy declares war Admiralty view is that she must be permitted to exercise full belligerent rights unless and until a decision to deprive her of such rights is taken by the League of Nations. You should be guided by this opinion until matter has been fully considered by H.M. Government when further instructions will be sent if necessary.
2034/4.

64. *Fisher to Admiralty*

[ADM 116/3038] 8 October 1935

SECRET.

[Telegram] Addressed Admiralty.

790. At a conference with High Commissioner,[1] G.O.C.[2] and Air Vice-Marshal[3] I advised their not relying on Navy to evacuate garrison and inhabitants of El Sollum by sea <u>at the last moment</u>.

Lack of shelter, proximity of hostile forces on shore, also chance of air and submarine attack during and after evacuation made this advice

---

[1] Sir Miles Wedderburn Lampson (1880–1964). Created 1st Baron Killearn, 1943. 1st Sec in Peking, China, 1916–20; acting High Commissioner in Siberia, Nov 1919–Feb 1920; member of British delegation, Washington Naval Conference, 1921–22; head of Central European Dept, FO, 1922–26; Min in Peking, 1926–33; High Commissioner for Egypt and the Sudan, 1933–36; Amb to Egypt, 1936–46; Special Commissioner in South-East Asia, 1946–48.

[2] Maj Gen [later Gen Sir] George A. Weir (1876–1951). Commanded 3rd Cavalry Brigade, Irish Command, 1920–22; Commandant Equitation School and Inspector of Cavalry, 1922–26; Gen Officer commanding Bombay District, India, 1927–31; commanded 55th (West Lancashire) Division, Territorial Army, 1932–33; GOC British troops in Egypt, 1934–36; GOC-in-Chief, Egypt, 1936–38; retired, 1938; during Second World War, Zone Commander, Home Guard, 1941; retired, 1942.

[3] Air Vice Marshal Cuthbert T. MacLean (1886–1969). AOC Aden Command, 1931; AOC RAF Middle East, 1934–38; AOC No. 2 (Bomber) Group, 1938–40.

obvious. If a dangerous situation can be anticipated four or five days beforehand evacuation can be carried out gradually during dark hours by road or coastguard vessel.

As soon as Italian forces pass to the Eastward of El Sollum they can be effectively attacked from sea and this chance will be taken advantage of if possible.

1558/8.

65. *Backhouse to Admiralty.*

[ADM 116/3038] *Nelson* at Portland, 9 October 1935

MOST SECRET
THE COMMANDER-IN-CHIEF, HOME FLEET.
No. 001325.

With reference to Admiralty Letter M.04321/35 of 3rd October 1935,[1] be pleased to submit the following remarks for the consideration of Their Lordships.

Outline Plan No. 1.

2. I would not feel easy about the operation allocated to the two Battle-cruisers, *Furious* and the 6th Destroyer Flotilla in case, for example, any considerable force of Italian ships were encountered, or one of the ships was unfortunate enough to be mined or torpedoed.

I regard *Furious* as no match for a 6-inch cruiser. Her A/A armament is very weak and she is not fitted with the H.A.C.S., so that no dependence could be placed upon it. Consequently, I do not consider she could be safely left unguarded for a moment.

The area of the operation and the time of day make it probable that the whole force would be subject to aircraft attacks which would be expected to grow in intensity until nightfall.

3. It appears to me, therefore, that, however desirable a nearly simultaneous attack against Cagliari might be, it would be preferable for *Nelson* and *Rodney* to be in close support of the Genoa force and I should prefer them to be accompanied by two of the ships of the 2nd Cruiser Squadron and one of the 20th or 21st Flotillas.

If necessary, a demonstration could be carried out against Cagliari by *Royal Sovereign* and *Ramilles*, with the other two ships of the 2nd Cruiser Squadron and the other flotilla. This force might accomplish little, but it would certainly draw some Italian aircraft and might achieve a useful purpose thus.

[1] Doc. No. 62.

4. These operations would tax severely the destroyers of the 20th and 21st Flotillas, but would be greatly facilitated if these flotillas were able to fuel at say Port Mahon, or Algiers, on both the outward and return passages.

I should have little anxiety about the return passage of the heavy ships to Gibraltar without destroyers, if it came to this, but if it came to necessity the two '*Royal Sovereign*' class battleships and *Furious* could also go to one of the ports mentioned above to refuel on their way back.

5. I beg, further, to remark that it seems to me hardly necessary that the attack on Genoa should be timed for 1800 on the day previous to the Mediterranean Fleet operations. I suggest that it would be better to carry it out early in the day.

Depending on the resultant situation, *Furious* could then be detached, with a suitable escort, or retained for further air operations if conditions were suitable; these would probably consist mainly of her fighters engaging enemy aircraft.

Outline Plan No. 2.

6. In this case, the Home Fleet forces could sweep through as far as Sicily, whence they would be certain to draw Italian aircraft from Sardinia and Sicily, and probably the submarines from the naval bases in Sardinia and Western and Northern Sicily.

This, it is thought, should prove as good a diversion to assist the Mediterranean Fleet as an operation carried out to the southward of Sicily in the direction of Malta, where, incidentally, there would be greater risk of encountering mines.

66. *Fisher to Admiralty*

[ADM 116/3038] 19 October 1935

SECRET.

[Telegram] Addressed Admiralty, repeated C.-in-C. Home Fleet.

894. Admiralty letter M.04321/35 of 3rd October. As regards operations by Mediterranean Fleet I am in agreement with both plans Number 1 and Number 2.

Operations of Western Force cannot be decided in detail by me without personal conference with Flag Officers concerned.

The following comments are submitted, however.

Operations of Gibraltar Force should precede those of Mediterranean Fleet by 12 to 24 hours if possible, otherwise synchronise.

Ansaldo works at Genoa is good objective for air attack and Iron Works, oil tanks and Railway bridges near sea at Savona could be shelled at night by destroyers.

If air attack on Genoa is made in late evening as suggested, landing on would have to be by night, which is undesirable, or aerodromes at San Raphael or Nice made use of.

In general I recommend dawn for air attack as approach of carrier is unobserved beforehand. In any case, effect of possible Italian retaliation on French aerodrome should be considered.

Bombardment of Cagliari by H.M.S. *Nelson* and H.M.S. *Rodney* is not recommended. They would be open to air attack for many hours, and though moral effect would be great, actual damage inflicted without air spotting would not compensate for injuries which must be expected by such slow and large targets.

Second Cruiser Squadron and 21st Destroyer Flotilla could make diversionary bombarding raid on Cagliari at high speed, but that also would be safer and more effective at early dawn.

If, subsequent to Genoa and Cagliari operations, air reconnaissance or intelligence establishes presence of Italian ships in Maddalena, Gibraltar force might concentrate about sixty miles west of Bonifacio and send air striking force in as soon as ready during the day. If no enemy in Maddalena, Gibraltar force proceeds to western end of Malta channel by which time it is hoped Mediterranean Fleet may be approaching Eastern end. If this is possible, any enemy surface force there should be cut off. Whether or no it is possible for concerted arrival of Gibraltar force and Mediterranean ships in Malta Channel it is my intention before leaving that area to carry out an intensified submarine hunt. 21st Destroyer Flotilla would probably need fuel in North African port the night before the attack on Cagliari.

Not considered possible to forecast operations while position of enemy force can only be guessed at. The main principle is, however, agreed, that is to make the fleet felt at widely separated places at the very beginning while keeping important heavy ships outside areas of air concentration until danger from these can be estimated.

Port Said itself will be sufficiently guarded I think by H.M.S. *Barham* and two destroyers; Alexandria by fixed defences now being mounted and by aircraft from Aboukir. Shipping between the two ports must use dark hours as far as possible and rely on air reconnaissance before starting. Further it is my intention to station H.M.S. *Rowena*, H.M.S. *Torrid*, H.M.S. *Thruster* at Alexandria and to employ them on anti-submarine duties in Alexandria–Port Said route. Consideration will be given to establishing a submarine patrol as a deterrent if this route is attacked by surface ships.

1708.

67. *Backhouse to Admiralty*

[ADM 116/3038]

Nelson at Portsmouth
25 October 1935

MOST SECRET
THE COMMANDER-IN-CHIEF, HOME FLEET.

No. 001325. With reference to my letter No. 001325 of 9th October, 1935, I beg to forward for the consideration of Their Lordships the following remarks on the subject of dawn and daytime attacks by carrier borne aircraft.

2. The great advantage of dawn attack lies in surprise.

It is much less probable that the aircraft will meet with opposition from enemy aircraft, while the anti-aircraft gun defences are likely to be less ready than later in the day.

Consequently, the attacking aircraft should be able to make their attacks with considerably greater probability of success, both in the selection and attack of intended targets.

I find that this opinion is held strongly by the Air Officers I have consulted.

Further considerations are that:–

(I). The carrier can fly off aircraft before dawn undisturbed by fear of hostile action while her aircraft are onboard.

(II). Landing-on subsequently will be carried out in daylight.

3. On the other hand, the disadvantages of a dawn attack lies in the fact that the carrier herself may be attacked at any time after daylight by enemy aircraft (and possibly by surface vessels and submarines) until she is out of their range, which may not be for a considerable time.

4. With a late afternoon or dusk attack, the risks attached to the carrier aircraft and to the carrier herself as outlined above are practically reversed.

The aircraft would be unlikely to effect surprise against a defended port or coast and must expect opposition, its strength depending on the extent of the enemy defences. They would, therefore, be less likely to achieve success.

The risk of the carrier would be greater while flying-off, which is not, however, for long, but, assuming the operation to be timed to be completed just before dark, she should be immune from attack by hostile aircraft subsequently and thus have the whole night to get clear.

The above does not take into account the possibility of the carrier and Force accompanying her being sighted and reported by enemy aircraft or vessels during their approach, which, depending upon the air and surface forces the enemy were liable to concentrate in the area, might lead to strong opposition being met, especially from the air, thus prejudicing the whole operation and possibly necessitating its abandonment on account of the safety of the carrier.

5. The conclusions I draw from these considerations is if that if good results against military objectives are to follow aircraft attacks, it is better to plan for dawn.

If, however, accuracy of bombing attacks is not of consequence – being designed primarily to affect enemy public opinion – it is of relatively little consequence at what time of day attacks are made. I suggest, however, that a feeble or ineffective attack would not effect its purpose, while, if our own aircraft suffered many casualties the effect would be the opposite of what was intended.

6. On the whole, therefore, I favour dawn attack as I consider this more likely to achieve its purpose.

It is a necessary corollary that the strength of the surface escorting forces should be considerable in order to cover and protect the carrier.

7. As regards possible objectives for the Western Mediterranean force, I suggest the following:–

GENOA. Ansaldo's fitting out works and the oil tanks on the jetty (as previously proposed). Ansaldo has a number of works at this place and it is thought that some others could be recognised easily such as the explosive works SAMPIER DARENA.

MADDALENA. The naval station appears to be a good objective and it seems probable that some naval vessels would be found there. N.I.D. reports show that it has been used considerably recently by vessels of all classes, including some 8" cruisers. This is a legitimate military objective.

CAGLIARI. The seaplane and airplane station at ELMAS is a suitable objective for air attack and should not be difficult to find. It does not appear that a bombardment by ships on the barracks or oil tanks would have much probability of success at the range at which it would be necessary to fire.

8. In formulating these proposals, it has been kept in mind that indiscriminate bombing or bombardment of towns is definitely to be avoided, military objectives being chosen.

If bombardments are carried out at works or forts situated in or close to a large town, even by cruisers and destroyers at relatively short range, it is thought that it would be practically impossible to avoid some projectiles going astray and hitting buildings for which they were not intended.

68. *Fisher to Chatfield*

[CHT/4/5]   *Queen Elizabeth*, Alexandria
                                                 25 October 1935

[Holograph]

I have felt all along in such close touch with you that I have not written. Every conceivable want of the Fleet seems to have been foreseen and provided for and of course I have noticed that every wish of mine has been met. No C.-in-C. has been so considerately treated.

Cunningham[1] of *Resolution* takes this & whom I am sure will give you the Fleet news.

All is well. *Very* well I think. In spite of retarded leave, a great deal of hard work, poor recreational facilities and fabulous price of beer, there has been no grumbling – on the contrary great keenness.

We keep up our programmes of exercise and firings at sea just as if we were at Malta. Every day & most nights squadrons or flotillas are at sea and so far we have managed to get in and out of this congested harbour without incident. Yesterday I counted 84 ships or vessels under my orders here in Alexandria alone.

Air attacks on the Fleet in harbour & at sea day & night all days except Sat. & Sun.

The H.A. fire is variable. I saw the 1st B.S. fire the other day at a sleeve [target], single ships & then massed firing. No formation could have survived either. Later I saw the 8″ cruisers who had previously been better than battleships and was very disappointed. But a very big target is I feel sure going to be very different from the generally almost invisible sleeve. I am satisfied that we are getting better & that an occasional shoot is not a bad thing if only to avoid complacency. I felt the A/S flotillas might be losing their cunning and so shall keep 1 or 2 s/ms permanently round here for them to hunt.

I get more glad every day that Dudley Pound came out. We can discuss things so openly together, even [?] never disagree! And his orderly logical mind is a great example to the staff. I think he is happy in his job. We get out together between 3–5 p.m. for either a swim or a game of tennis most

---

[1]Capt [later AoF Sir] John Henry Dacres Cunningham (1885–1962). Commanded minelayer *Adventure*, Atlantic Fleet, 1928–29; Dep DP, Admy, 1929–30; DP, Admy, 1930–32; commanded battleship *Resolution*, Med Fleet, 1933–35; Asst CNS, 1936–37; Asst CNS Air (subsequently designated as 5th Sea Lord), 1937–38; VAC 1st CS, 1938–41; 4th Sea Lord, 1941–42; C-in-C Levant, 1943; C-in-C Med & Allied Naval Cdr Med, 1943–46; 1st Sea Lord, 1946–48; Chmn Iraq Petroleum Co., 1948–58.

days. Raikes[1] is also happy & his job gains in importance as he gathers the staff to compete [?] with it. The Army & R.A.F. want to make him Fortress Commander & I shall agree for when the Fleet leaves for [Port] X there must be a King in Alex^d & Raikes will wear the crown.

Marriott[2] is busy in the land, Lyon competing with gun defences at Haifa (and how troublesome the Army is over this) – & a mine dept. whilst here. I have had the most signal help from Charles Forbes[3] who has taken a mass of Fleet Administrative details off my shoulders. He is absolutely splendid. R.A.(D)[4] perpetually smiling but more broadly than ever now he has his little *Galatea*, and then there is Gerry Wells[5] ashore who has slaved to meet all our demands for more berths, buoys, quays, lighters, tugs. He is a jewel and if anyone else had been in his place things might have been so different.

I have just sent you a W/T re docking programme. Alexandria is well known as the worst place in the world for collecting coral growths.

Cutty Sark is here with the Duke.[6] He wanted to go to Sollum but I said he couldn't – & now to Mersa Matruh, which I also advised against. His presence on that coast would start a crop of rumours & his yacht wd certainly be taken as Q ship disguised destroyer.[7] I send you two photographs to show you the crowds attracted by our last march through the town.

---

[1] RA [later Adm Sir] Robert Henry Taunton Raikes (1885–1953). Commanded cruiser *Sussex*, 1929; commanding officer Boy's Training Establishment HMS *St Vincent*, 1930–31; Dir of RN Staff College, Greenwich, 1932–34; Cdre and COS, Med Fleet, 1934–35; Admiral-in-charge, Alexandria, 1936; RA (Submarines), Gosport, 1936–38; VAC Reserve Fleet Destroyer Flotillas, 1939; C-in-C Northern Patrol, 1940; C-in-C South Atlantic, 1940–41; Flag Officer-in-charge, Aberdeen, 1942–44.

[2] Capt (Retd) J.P.R. Marriott, Senior British Naval Officer, Suez Canal Area.

[3] VA [later AoF] Sir Charles Morton Forbes (1880–1960). DNO, Admy, 1925–28; RA (D), Med Fleet, 1930–31; 3rd Sea Lord, 1932–34; VAC 1st BS and 2nd-in-command, Med Fleet, 1934–36; C-in-C Home Fleet, Apr 1938–Dec 1940; C-in-C Plymouth, 1941–43.

[4] RA [later AoF Sir] Andrew Browne Cunningham (1883–1963). Created Viscount, 1946. RAC DFs, Med Fleet, 1934–36; VAC Battle Cruiser Sqdn and 2nd-in-command, Med Fleet, 1937–38; Dep CNS, 1938–39; C-in-C Med, June 1939–April 1942; Head of British Admy Delegation in Washington, 1942; Naval C-in-C, Expeditionary Force, North Africa, 1942; C-in-C Mediterranean, 1943; 1st Sea Lord, 1943–46.

[5] RA [later VA Sir] Gerald Aylmer Wells (d. 1943). British Naval Attaché, The Netherlands and Poland, 1922–25; commanded battleship *Royal Sovereign*, 1927–29; retired list, 1930; Dir Gen Ports and Lighthouses Administration of Egypt, 1932–43.

[6] Hugh Richard Arthur Grosvenor (1879–1953). Succeeded grandfather as 2nd Duke of Westminster, 1899. Generally known by his nickname Bendor. Lord Lieut of Cheshire, 1906–20; Major, Cheshire Yeomanry; led motorised column 120 miles into the desert to rescue prisoners held by the Senussi in Libya, March 1916; Asst to the Controller, Mechanical Warfare Dept, Ministry of Munitions, 1917.

[7] Westminster's yacht *Cutty Sark* had been converted from materials intended for a destroyer. The yacht's origin is clear from photographs.

69. *Pound to Chatfield.*

[CHT/4/10]

[Holograph]

H.M.S. *Queen Elizabeth*
26 [October 1935][1]

This is a great life and I am enjoying every minute of it. W.W. [Fisher] has been kindness itself and is delightful to work with. At first I think his staff were a bit nervous as to what it would be like having a full Admiral as C.O.S. but they are no longer shy. They (his staff) are an excellent lot but were looking a bit jaded about the time I arrived as they had a very strenuous 2 months – things are easing up a bit now and they are full of ginger. I am afraid the replies to the various Admiralty letters I brought out have been slow in reaching you but I had to spend a day at Port Said to get au fait with conditions [at] the Canal and it allowed me to go straight to *Q.E.* [*Queen Elizabeth*] as there was no accommodation in Alex. Also, at this time all the various orders for [Port] 'X' were just reaching finality and until they were completed there was not much chance of tackling the other big questions. I hope from now on you will get any replies you require much quicker. During my stay at Port Said I motored down to Ismailia with Marriott who is thoroughly in the saddle.

We had a bit of excitement the other day when the *Ausonia* caught fire just to the windward of 1st B.S.[2] We didn't quite know whether her oil tanks would blow up – she had about 900 lives onboard. The Italians were pretty futile – about as futile as our fire engines. As she started to heel over she had to be beached and as she couldn't slip her cable we had to cut her cable [with?] our arco-acetylene plant – it took about 20 minutes.

She is still burning – 3 days.

Jerry Wells (D.G. Port) did very well but it was rather amusing when he brought the Captain of *Ausonia* onboard to thank C.-in-C. to see the Captain of *A*'s white hardly creased [uniform] whilst Wells, like all our people was wet through and black as your hat.

I don't think the Fascist regime has made the I.T.'s into Merchant seamen whatever it may have done to other branches.

---

[1] The dating of this letter poses problems. It bears the notation 'Bet. October 1935 & Feb. 1936'. The Italian liner *Ausonia* burned out in Alexandria harbour on 18 October 1935 and Pound's mention of her still burning after three days would imply the letter was written on the 21st. However, he dates it 'Sunday, 26th'. The problem with this is there is no Sunday the 26th until January. Pound may have erred as to the date (not likely given Pound's reputation) or the letter was written over a period of days.

[2] The Lloyd Triestino liner *Ausonia* (12,955 tons) caught fire and burned out in Alexandria harbour on 18 October. Although sabotage was sometimes rumoured, the cause of the fire remained unknown. In 1936 the hulk was towed to Trieste and broken up.

What a tragedy it was about Neville Fisher.[1] W.W. was of course absolutely knocked out but hid it in a marvellous way. We kept him hard at work, which was about all we could do to help.

I do hope you will not feel that I am itching for the change of command to take place. I am leaving a tremendous lot and am absolutely happy and if as I hope the period of my command will not be reduced because I start late I am getting in some active [?] time at sea which is a great satisfaction.

I see you had an unsatisfactory first round with Air Ministry but feeling as you do about it there can only be one result.

Marvellous weather here and with leisure time a week and a long swim on other days we keep very fit.

All good luck.

70. *Admiralty to Fisher and French*

[ADM 1/8804] 26 October 1935

SECRET.
[Telegram] Addressed C.-in-C. Mediterranean and Vice Admiral, Malta. From Admiralty.

Reference Vice Admiral's Signal 2155/21, Government policy is (1) to encourage families now at Malta to leave voluntarily and (2) to discourage fresh families from going out from United Kingdom until after the emergency period. In view of (1) offer of a return indulgence passage after the emergency contained in my 2115/16 October applies to all families who have left or will leave Malta voluntarily owing to the emergency. Vice Admiral Malta should so far as practicable render a nominal list of all such persons on the lines laid down in my 1842/19 October to Vice Admiral Malta only. It is suggested that particulars of families who have left at their own expense in ordinary steamships should be obtained from local Shipping Agents. Their Lordships do not wish to issue any general Fleet Order on the subject. The offer of an indulgence passage will of course not apply to families who have taken or take passage to Malta contrary to Admiralty advice.

As regards (2) measures for discouraging families from proceeding from the United Kingdom were promulgated in my 2158/11 October. Copies of the letters that have been sent to families known to have booked passages were forwarded to you with Admiralty Letter P.M.3027/35 of 22nd October. Wives who persist in their intention to proceed to Malta forthwith have been advised to consult their husbands by telegram before

[1] Fisher's son Neville had joined the RAF and was killed in a flying accident.

coming to a final decision. Some wives have produced telegrams from the husband advising immediate sailing. Their Lordships therefore suggest that officers and ratings concerned should be advised to discourage their wives from sailing until the emergency is over unless there are exceptional circumstances. The objections to families proceeding to Malta apply also to Alexandria although to a lesser degree.

War Office are discouraging wives of Army personnel from proceeding to Alexandria.

Reference my 2158 of 11th October, letters have also been sent to wives and relatives of naval personnel who are known to have booked passages in S.S. *Chitral*, *Rajputana* and *Carthage*.
1650/26.

71. *Fisher to Admiralty*

[ADM 116/3050] 29 October 1935

SECRET.

[Telegram] Addressed Admiralty.

999. Your 1825 of 25th October, Base Defences Mediterranean is the only organisation containing Royal Marines on special duty and has assumed responsibility for shore defences of Alexandria and Port 'X'.

In round numbers 250 officers and men have been allocated for Alexandria and 920 for Port 'X'.

No officers and men have been allocated for the Canal terminal ports should these be required later.

I have personally examined requirements both at Alexandria and Port 'X' with the following results.

The 250 officers and men at Alexandria are barely sufficient for manning two 6-inch batteries and control stations but I have directed that as soon as four-inch battery is ready the 250 officers and men are to be divided between the six-inch and four-inch batteries so that all the guns are manned, any loss of efficiency and readiness due to under manning being accepted.

As regards Port 'X' important matters of principle arise. Up to the present time Base Defences organisation has been maintained in the same state of instant readiness as the Fleet and with total of 920 officers and men defences could be rapidly installed and efficiently manned when complete.

If men are now taken away to be replaced on mobilisation it seems certain they will not be at Port 'X' when convoy arrives and that base must be established with number available when convoy leaves Alexandria.

If delay in establishing base and delay in its reaching proper efficiency which I have hitherto held to be inadmissible can be accepted then a total of 150 men can be withdrawn.

A withdrawal of any number in excess of 150 would not permit of the installation of the Base.

Full consideration has been given to the help that will be available from battleships and other vessels that will be at Port 'X'.

2358/29.

72. *Admiralty to Fisher*

[ADM 116/3487] 1 November 1935

SECRET.

[Telegram] From Admiralty.

815. Your 2238/31/10.[1] Fact that members of League are applying sanctions against Italy under Article Sixteen of the Covenant is conclusive that they regard a state of war as existing but Italy has not herself declared that a state of war exists or taken any other action preliminary to the exercise [of] belligerent rights such as issue of list of contraband and Admiralty consider there is no reason to suppose she intends doing so. In circumstances Admiralty are strongly of opinion that no repetition no action should be taken to raise this very delicate and dangerous question at any rate for present. For your own information Foreign Office view is that as a matter of law Italy is not in the present circumstances entitled to exercise such rights though we are under no obligation to deny them. Policy of H.M. Government is that no repetition no forcible measures should be taken to withhold them except in pursuance of a policy unanimously adopted by members of the League. Consequently instructions in my 2034 of 4th October[2] remain in force viz. that if Italy does exercise such rights local protest should be made and facts reported to Admiralty but no repetition no military measures should be taken by H.M. Ships.

2007/1.

---

[1] Fisher enquired whether past Admiralty messages could be read as acknowledging a state of war existed between Italy and Abyssinia and that Italy was entitled to exercise belligerent rights. Fisher to Admiralty, Telegram No. 14, 31 Oct 1935, ADM 116/3487.
[2] Doc. No. 63.

73. *Admiralty to Fisher*

[ADM 116/3038] 2 November 1935

SECRET.

[Telegram] Addressed C.-in-C. Mediterranean, repeated C.-in-C. Home Fleet, C.-in-C. East Indies. From Admiralty.

Reference Admiralty telegram 1840/24/10 and your 1708/19[1] and 1648/17.

Assurances asked for from France were received but negotiations with Italy for a 'detente' are at present at a standstill and there seems at the moment little hope of progress.

As a result, however, of assurances from France conversations have been commenced between Admiralty and French Naval Staff with a view to concerting action in emergency. The first point that has arisen in the conversations is that French forces are in such a state of unreadiness that instead of France being able to give any active help to us in the early stages she would require assistance from us until her mobilisation and particularly the transport of troops from Africa were completed.

The mere fact that France was mobilising would however have the main effect which the operation of Home Fleet off western coast of Italy was intended to produce, i.e. of containing Italian surface and air forces in that area and consequently Admiralty consider that operations of Mediterranean Fleet already foreshadowed are unaffected. Operations in Western basin would depend on discussions with French but our forces in that area would preferably be required in the early stages to assist in ensuring the safety of French Colonial troop movements.

Responsibility for control of Mediterranean in war is being discussed with French on the basis of Mediterranean Fleet being mainly responsible for Eastern Basin and French Fleet mainly for Western basin for Cap de Gata to Cap Bon Sicily line. British Gibraltar Force to be free to co-operate with the British Eastern Forces or the French Fleet as required.

1544/2.

[1] Doc. No. 66.

74. *Admiralty to Fisher*

[ADM 116/3050] 2 November 1935

SECRET.

[Telegram] From Admiralty.

832. Your 2358/29/10.[1] Concur that delay in establishing base and delay in reaching proper efficiency are inadmissible.

1920/2.

75. *Admiralty to Fisher*

[ADM 116/3038] 2 November 1935

SECRET.

[Telegram] Addressed C.-in-C. Mediterranean, repeated C.-in-C. Home Fleet. From Admiralty.

My 1544/2.[2] Remarks which follow refer to points previously under discussion when assistance of French was assumed to be unlikely.

In the view of present situation of assumed use of French ports and French mobilisation it is considered that if any operation by Home Fleet is carried out in Western basin it should be by a single combined force.

Instructions in Admiralty telegram 1316/28/8 would apply to above operations.[3] Admiralty concur that if intelligence shows that enemy ships are at Maddalena attack by air striking forces would be suitable operation to consider.

With reference to the best time for making air attack Admiralty consider that dawn attack though more likely to achieve its purpose is also more likely to lead to prolonged counter attack on carrier. On the other hand risk to aircraft is likely to be greater in dusk attack.

It is considered that decision as to which to adopt must depend on circumstances and the strength of A.A. defences of escorting vessels.

2143/2.

---

[1] Doc. No. 71.
[2] Doc. No. 73.
[3] Any Italian success would counterbalance the advantage gained by retaliatory operations which involved the Fleet in approach to waters under enemy control. Consequently retaliatory action against shore objectives should not be undertaken if it would result in serious loss to the Fleet. 'History of the Italian–Abyssinian Emergency …', p. 6, ADM 116/3476.

76. *Admiralty to Fisher and Backhouse*

[ADM 116/3050]　　　　　　　　　　　　　　　5 November 1935

<u>Secret</u>.
[Carbon]　　　　　　　　　　　　　　　　　　　M.04386/35.

I am to transmit for your information copy of correspondence with the General Post Office and Foreign Office concerning the maintenance of cable and wireless communication between the United Kingdom and Egypt in the event of hostilities with Italy.

2. I am to observe that the cables in the Mediterranean and the Red Sea follow generally the shortest distances between Gibraltar–Malta–Alexandria, and Suez–Port Sudan–Aden. They avoid to a great extent the proximity of coasts (except at termini) and are laid approximately in mid channels between ports.

3. In view of the considerable depths in which these cables are for the most part laid, it would seem easier for them to be cut near their termini, and this would tend to simplify interception of the vessel carrying out this operation.

4. There are, however, areas in which, both by reason of the depth of water and of proximity to Italian territory, interference with these cables would present the least difficulty to the enemy. Such areas exist between Sicily and Cape Bon, and between Jebel Zukur and the Eritrean coast. Their Lordships are of opinion that it would not be practicable in either area continuously to prevent interference with the cables.

5. It has been ascertained that cable ships are at present in the following vicinities:–
Dominia (9273 tons) – Cadiz. Telegraph Construction and Maintenance Co.
Lady Dennison Pender (1984 tons). Left Aden 25/9. Cable and Wireless Co.
Levant (283 tons) – Malta.　　　　　　　　"　　"　　　　"　　"
Retriever (674 tons) – Left Malta for Levant 22/9.　"　　"　　　"　　"

In addition there are a few vessels in Home Waters, normally maintaining cables in the Atlantic, Channel and North Sea, etc.

6. Their Lordships consider that protection would be necessary to a cable ship while effecting repairs, the composition of the escort depending on forces available and as determined by the Senior Naval Officer concerned.

7. Italy possesses 3 Government owned cable ships which might be employed for the purpose of cable cutting, but as far as can be ascertained no privately owned ones exist. Such particulars as are available in addition to those given in C.B. 1815 are as follows:–

Città di Milano (ex German *Baron von Oldenburg*). Continuously reported during the last 10 days in the vicinity of Sicily and recently as working on Italian cable 1891 (C.B. 1865 Chart T.10) near Marittimo Island.
Capable of working in great depths.
Glasone. Left Port Said 12.9.35 for Italian ports, present location unknown. Capable of working in depths up to 1,400 fathoms.
Città di Siracusa. At Massawa. Capabilities believed to be similar to those of *Glasone*.

8. Having regard to all the circumstances, Their Lordships have reached the conclusion that, on the assumption that the secrecy of British cable routes in the Mediterranean and Red Sea has been preserved, the cutting of these cables would be a difficult operation. The risk of interruption of our cable services does, however, undoubtedly exist, and in view of the numerous demands likely to be made on our naval forces to provide convoy escort, local defence and patrols, in addition to any major operations contemplated, it appears doubtful whether adequate protection against enemy action in this respect can be guaranteed.

77. Fisher to Chatfield.

[CHT/4/5] Commander-in-Chief, Mediterranean Station
Queen Elizabeth, Alexandria
8 November [1935]

[Holograph]

We have had our first severe blow – squalls measured from 60 to 100 m.p.h. – and have learnt a good deal. Sailing Directions & local witnesses describe it as unprecedented both as regards force & time of year. They always do.

Devonshire & *Australia* dragged badly & brought with them some of our auxiliary vessels – but in this result no man of war was damaged & other injuries have now been made good by *Resource* & *Woolwich*. I experienced one day of moderate hard blow & then sailed for Haifa thinking it was all over. I was wrong and we had the full force of it & heavy sea & *Durban* went ashore in the breakwater but got off very soon and without damage. It was a good warning & I don't look for further risk. Am having two tugs from Malta and one from Gibraltar sent here & with a slight rearrangement & thinning out of berths we ought to be alright.

I went over long [?] part of the ground at Haifa with Lyon and the result you will have seen in the telegrams sent. Harbour Authorities at Haifa are quite helpful. Lyon is King there & well placed.

I was very glad to get approval for getting on with the docking. Here at Alexandria we dock a new destroyer every 24 hours. One leaves the dock at 6.30 a.m. every morning & her successor was in at 7 a.m. Cost per ship under £100 [and?] 2 coats of antifouling paint put on.

The Sailors Club here is a great success & booming. It is in Claridge's Hotel, 2000 men in it on the opening night. Bar is 5d a pint, [*illegible*] 1/3. A team of ladies do most of the waiting in the restaurant (which is a huge place) & the sailors act accordingly.

I haven't gone beyond 6 hours notice for steam[1] which means no night leave for either officers or men. No relaxation is necessary yet but if we have to stay here very long I may have to consider it later but think I can arrange for a system of small quotas only & with machinery for recall.

To remind you of old days I enclose this week's programme of exercises just completing. You will see some new items but I send it really lest it might be thought that we [are] all bottled up inside & getting rusty. The movement of ships in & out is almost incessant and our firings can be heard in the town & the ships seen from that long Esplanade. Some of the 12,000 Italians in the town no doubt keep Rome informed & I hope the effect is not lost.

If & when we get to Port X I shall probably transfer to *Ajax* so as to control the operations at sea & leave Forbes with the Battleships inside. When the defences on shore are really established it may be advisable to withdraw some or all of them to Alexandria again out of harm's way. The fewer targets presented the better provided that the shore defences can guard the oilers & repair ships, etc. Experience alone can decide this. In the early stages they are necessary for the A.A. protection of the convoy and the installation & defence of the Base.

I am going out this afternoon with Alec Ramsay[2] to see his Desert Bombing range where the F.A.A. practice and tonight have 46 to dinner on the Quarter Deck – all the masters of auxiliaries & merchant ships in the harbour. There's virtue for you!

---

[1] Underlined by Chatfield with marginal note: 'DCNS: I feel this ought to be eased.'
[2] VA [later Adm] The Hon. Sir Alexander Robert Maude Ramsay (1881–1972), 3rd son of 13th Earl of Dalhousie. Naval Attaché, Paris, 1919–22; commanded cruiser *Dunedin*, 1922–24; Flag Capt and COS to C-in-C North American Station, 1924–26; commanded aircraft carrier *Furious*, 1928–29; Cdre RN Barracks, Portsmouth, 1929–31; RA Aircraft Carriers, 1933–36; C-in-C East Indies, 1936–38; 5th Sea Lord and Chf of Naval Air Services, 1938–39.

78. *Admiralty to Fisher*

[ADM 116/3046] 18 November 1935

SECRET.

[Telegram] From Admiralty.

56. Progress of negotiations with Italy for detente in Mediterranean is proceeding very slowly and withdrawal of Battle Cruisers from Gibraltar may not repetition not take place as soon as at one time anticipated if at all. This does not repetition not imply greater tension but is due to unwillingness of Italy to take such a limited step only. A comprehensive plan for giving leave, recommissioning certain units and refitting Home Fleet ships at home and abroad has been prepared. This plan provides for relieving Home Fleet ships serving in Mediterranean with other Home Fleet units some of whom will commence leave shortly in order to set the scheme in operation. Full details will be sent to you as soon as the whole plan is approved by H.M. Government.[1]

1829/18.

79. *Fisher to Admiralty*

[ADM 116/3046] 3 December 1935

SECRET.

[Telegram] Addressed Admiralty, repeated C.-in-C. Home Fleet.

213. Your 1646 30th November. The general intention expressed in para. 3 of Admiralty Letter[2] is noted and in view of Admiralty message 2106 29th November (not to C.-in-C. Home Fleet) this, it is considered, should not be departed from. As long as the emergency exists in present acute form when Italy is making every possible warlike disposition and preparation the fleet should be kept at full strength and instantly ready, and consideration of leave should be disregarded. When emergency is definitely over objection to all Home Fleet ships giving leave at the same time is not apparent. I am in full agreement however with arrangements Z and Zero + 6 and sequels to them but those proposed for H.M.S. *Courageous*, 5th Destroyer Flotilla, H.M.S. *Mackay* and part of 19th Destroyer Flotilla on Zero + 25 and Zero + 27 days and their sequels

---

[1] A table with the proposed movements was submitted for Fisher's comments in Admiralty to Fisher, M.05292/35, 20 Nov 1935; copy in ADM 116/3046.

[2] 'The general intention is, as far as possible, to maintain the present strength of the Fleet in the Mediterranean.' Ibid.

constitute such a weakening of forces on Eastern Mediterranean as to be unacceptable unless situation greatly improves. All cruisers are now highly trained and have specific tasks if war breaks out. If situation deteriorates I consider that H.M.S. *Furious*, H.M.S. *Cairo*, 20th and 21st Flotillas should leave United Kingdom for Gibraltar immediately and that H.M.S. *Renown*, 2nd Cruiser Squadron and 6th Destroyer Flotilla should join me in Eastern Mediterranean. H.M.S. *Hood* to await arrival at Gibraltar of units from United Kingdom and then to operate as previously arranged if hostilities have commenced. It should be noted that apart from intervals when there will be only one carrier in Eastern Mediterranean as provided in the table it is estimated that it will be necessary for H.M.S. *Furious* to have at least one month's intense working up before being used as it is proposed to use H.M.S. *Courageous* and H.M.S. *Glorious*.

1618/3.

80. *Fisher to Admiralty*

[ADM 116/3398] 6 December 1935

SECRET.

[Telegram] Addressed Admiralty, repeated V.A. Malta.

239. Unless information is already available at Admiralty is there any objection to a destroyer being sent now from Malta to Bizerta to make the following enquiries as regards that port.

(i). Amount of fuel oil available for British Medn. Fleet, description and quality.

(ii). Rate and manner of supply. Whether connections are suitable or if adaptors need be made. In latter case necessary sketches are required.

(iii). Number of (a) destroyers, (b) cruisers that can be fuelled simultaneously.

(iv). Regulations concerning entry by day and night in time of war.

(v). Swept channels and minesweeping forces employed and sufficiency of latter.

(vi). High angle defence.

(vii). What is maximum draught permissible to enter canal and are there any restrictions as regards length.

(viii). What tugs are available and what is their H.P.

(ix). Would provision of a British man of war as base and wireless guardship be desirable.

2338.

81. Admiralty to Fisher

[ADM 116/3046] 7 December 1935

SECRET.
[Telegram] Addressed C.-in-C. Mediterranean, repeated C.-in-C. Home Fleet. From Admiralty.

Your 1618/3.[1] Your proposed amendments to programme of reliefs are noted. No weakening of forces in Eastern Medn. is contemplated unless situation improves. As stated in covering letter programme has been drawn up as a general guide and each movement will be governed by the situation at the moment. Commanders-in-Chief Home Fleet and Mediterranean Fleet will, if possible, be notified well in advance of any contemplated changes in their forces. The arrangements to give leave to Home Fleet ships are being made not only in order to avoid being faced with the problem of giving leave to all Home Fleet ships at the same time when the emergency terminates but also in order to make provision for re-commissionings and resumption of foreign service drafting which will become an important commitment and so prevent the simultaneous immobilization of a large part of the Navy. The several points in connection with your operations contained in your 1618/3 are being considered and will be the subject of a further telegram.

1502.

82. Admiralty to Admirals Fisher, Rose and Bailey[2]

[ADM 116/3046] 18 December 1935

SECRET.
[Telegram] Addressed C.-in-C. Mediterranean, C.-in-C. East Indies, Rear Admiral Commanding, Battle Cruiser Squadron, repeated C.-in-C. Home Fleet. From Admiralty.

The Admiralty fully appreciate that disappointment may be felt by men in certain units of the Home Fleet and other men who would normally be on Home Service at their enforced absence from home this Christmas.

In the national interest it is essential that a considerable proportion of the Fleet should remain away from England for the present, but as soon

---
[1] Doc. No. 79.
[2] RA [later Adm Sir] Sidney Robert Bailey (1882–1942). Flag Cdr to Adm Sir David Beatty, C-in-C Grand Fleet, 1916–19; Capt (D), 9th DF, Atlantic Fleet, 1923–25; Naval Asst to 1st Sea Lord, 1925–27; COS to C-in-C Med, 1931–32; Asst CNS, 1933–34; RAC Battle Cruiser Sqdn, 1934–36; Pres RN College, Greenwich, 1937–38.

as circumstances admit, it is the intention of the Admiralty to arrange for units to be withdrawn or relieved in order that the men concerned may be given the leave which it is at present necessary to defer.

You are at liberty if you consider it desirable to have this message communicated verbally to Ships' Companies. If this is done, opportunity could be taken to inform them as follows – that the situation is still menacing, and it is the realization of the complete readiness of our forces in the Mediterranean and Red Sea that is the greatest, if not the only, guarantee that Peace will be preserved.

2001.

83. *Chatfield to Fisher*

[CAB 53/5] 20 December 1935

SECRET.

[Telegram] From C.N.S.

Personal.

Following for communication to G.O.C. and A.O.C. in C. (begins):–

PART I.

1. The Chiefs of Staff Sub-Committee have considered the most secret memorandum by the Commanders-in-Chief of the three Services, dated 30th November (Med.00200/1), and generally approve your proposals,[1] including the following:–

(a) the abandonment in the earlier stages of hostilities of the proposal to establish a base at Port 'X';

(b) the utilisation for the defence of the base at Alexandria of the anti-aircraft guns and searchlights of the mobile Naval base organisation; subject to the requirements that they shall be re-embarked if the occupation of Port 'X' shall subsequently be decided upon.

2. The military reinforcements you ask for and the extension of the railway to Mersa Matruh having been approved by His Majesty's

---

[1] The memorandum consisted of three parts: Part I consisted of a combined appreciation; Part II contained comments on the paper prepared by the Chiefs of Staff Sub-Committee of the Committee of Imperial Defence concerning the defence of Egypt and the Sudan; and Part III, a paper by the C-in-C Mediterranean giving the naval considerations. Fisher 'showed that the advantage of having a base situated in such a central position as PORT X did not justify the risk of trying to establish it in the face of the Italian air threat, the strength of which it was not possible to estimate'. Fisher therefore proposed 'that the establishment of an Advanced Base at PORT X immediately on the outbreak of war should not be attempted, but that ALEXANDRIA should be considered as the Main Fleet Base and that certain re-arrangement of ships in the Mediterranean should take place in order to facilitate the execution of what was considered to be the function of the Fleet in the event of war'. 'History of the Italian–Abyssinian Emergency …', p. 15, ADM 116/3476.

Government, the Commanders-in-Chief should consider seizing the initiative in the event of hostilities, and, by means of a combined operation, driving the Italians back from their frontier to deprive them of advanced air bases and secure early success. You should investigate and report on the possibilities of this operation.

3. The question of sending an additional Army Co-operation Squadron is being considered and enquiries are also being made as to the possibility of placing at the disposal of the Air Officer Commanding-in-Chief on the outbreak of war an additional bomber squadron from Iraq., and relieving it by a squadron from India.

4. In the event of French military co-operation after the outbreak of hostilities, French and/or British air forces will be sent to Tunis to operate against Italian objectives in Southern Sardinia and Sicily.

5. Consideration is being given to the improvement of the existing courier service by Imperial Airways, and also to the provision of an alternative service on the outbreak of hostilities.

6. Questions of detail will be taken up by the Service Departments concerned with their respective Commanders-in-Chief.

PART II.

7. For your information it has been ascertained by direct enquiry that other Mediterranean Powers have made no preparations to date for active co-operation in the present emergency, and it is unlikely that they will be induced to make such preparations. Nothing can be done in any of these countries without mobilisation, and you must assume that they can give no active military co-operation in the earlier stages of war. It therefore appears that our own forces will have to sustain the war for a not-inconsiderable period. France and Greece, however, have promised full use of their ports, and Turkey is willing to co-operate with her limited air forces.

8. The situation as regards the military co-operation of France is at the present time profoundly unsatisfactory. She has made no preparations for a war with Italy, and it is very unlikely that any precautionary measures will be taken before an emergency arises, as this involves mobilisation, which, under the existing political situation in France, is not feasible.

9. You are probably following the general course of events at Geneva in connection with the peace proposals. The course taken there in the immediate future will more clearly confirm the accuracy or otherwise of the situation portrayed above. (ends).

1615.

84. *Admiralty to Fisher*

[ADM 116/3398] 21 December 1935

SECRET.

[Telegram] Addressed C.-in-C. Mediterranean, repeated V.A. Malta. From Admiralty.

IMMEDIATE.

No objection to your 2338/6.[1]

Destroyer should arrive Bizerta p.m. Sunday 22nd December so as to have Monday 23rd for collecting required information. Object of visit is to be kept as secret as possible and knowledge thereof limited to officers directly concerned. At Bizerta only Prefet Maritime and his chief of staff are in possession of the facts. Any tour of inspection will probably be made in plain clothes. Question Number VI repeat VI is being dealt with elsewhere and should not repeat not therefore be asked.

Owing to the number of Italians resident in Tunis it is essential that anything in the nature of an 'incident' should be avoided.

1144.

85. *Fisher to Chatfield*

[CAB 53/5] 23 December 1935

SECRET.

[Telegram] Addressed Chief of Naval Staff.

PERSONAL.

Contents of your 1615 20th December were communicated to General Officer Commanding and Air Officer Commanding-in-Chief at a meeting in Cairo yesterday 22nd. It was agreed that a more forward policy than that referred to in para. 24 of Mediterranean 00202/1 was not possible at the present time but when the reinforcements referred to in that paragraph are ready to move forward it was considered possible to base a mechanised force at El Sollum and from there to operate in Italian territory. This is not expected to be before the first week in February. The details above which would require maintenance of this force entirely by sea is being worked out by the three services and will be communicated later.

From every point of view forward policy appears to be the correct one and it would be to our advantage to avoid any action which might commit us before we were ready to follow it. General Officer Commanding thinks

---
[1] Doc. No. 80.

that if Italians increase their numbers in Libya it would prove more an embarrassment to them than a danger to himself.
2148/23.

86. *Commander Louis Mountbatten*[1] *to Vice Admiral Malta*

[ADM 116/3398]   H.M.S. *Wishart*, Malta
24 December, 1935

MOST SECRET.

I have the honour to report that, after having embarked Lieutenant-Commander E. Neville, H.M.S. *Wishart* left Malta at 2100 on 21st December, 1935, in accordance with your sailing orders No. 99 of the same date and shaped course for Pantellaria Island, which was reached shortly after daybreak on 22nd.

2. In accordance with your verbal instructions *Wishart* passed the Island to northward on a normal course at a distance of not less than one mile. A very good lookout was kept but nothing could be seen of any guns, emplacements, harbour defences or obstructions, barbed wire, searchlights or observation huts. Cape Spadillo light was observed to have been reduced in strength and altered in characteristic. A separate report on this point has been made.

3. H.M.S. *Wishart* then proceeded to Bizerta which was reached at 1400 on Sunday, 22nd December. The ship was met by a launch containing the Préfet Maritime's flag lieutenant (Enseigne de Vaisseau H.L.R. Cazenave), and a pilot. In accordance to their instructions I secured to number two buoy at the entrance to Baie Ponty opposite the Admiralty House.

4. The Flag Lieutenant impressed on me that besides himself only the Préfet Maritime (Vice Admiral J.J.J.N. De Laborde[2]) and his Chief of

[1] Cdr [later AoF] Louis Mountbatten (1900–79). Created Viscount, 1946, Earl, 1947. Younger son of Prince Louis of Battenberg (after 1917 Marquis of Milford Haven). Fleet Wireless Officer, Med Fleet, 1931–33; commanded *Daring*, 1934; *Wishart*, 1935; *Kelly* and 5th DF, 1939–41; *Illustrious*, 1941; Cdre Combined Operations, 1941–42; Chief of Combined Operations, 1942–43; Supreme Allied Cdr, Southeast Asia, 1943–46; Viceroy of India, Mar–Aug 1947; Gov-Gen of India, Aug 1947–June 1948; commanded 1st CS, Med Fleet, 1948–49; 4th Sea Lord, 1950–52; C-in-C Med, 1952–54; C-in-C Allied Forces, Med, 1953–54; 1st Sea Lord, 1955–59; Chief of UK Defence Staff and Chmn of Chiefs of Staff Cttee, 1959–65; assassinated by IRA bomb, 1979.
[2] Vice Amiral Jean-Joseph, Comte de Laborde (1878–1977). Developed interest in aviation in the years before the war and received brevet de pilot, 1914; commanded naval aviation centre, Dunkirk, and subsequently air patrols in zone of northern armies, 1917; commanded maritime aviation centre, Saint-Raphaël, 1919; Chef, Service Centrale de l' Aéronautique Maritime, 1924–26; commanded first French aircraft carrier *Béarn*, 1926–28; commanded 2e escadre, 1930; Vice Amiral and Préfet Maritime Bizerte, 1932–36; commanded Escadre

Staff (Captain V.P.E. Trucy) were aware of the nature of my mission, and that they regarded it as of the highest importance that it should be kept a close secret. He accordingly requested me to have no communication with the shore except for my official call on the Admiral, which had been arranged for 0930 on the following morning.

5. I at once agreed to do whatever the Admiral wished but requested permission to land the Ward Room Messman and the Postman. He seemed in such evident distress over this small request that I withdrew it. I pointed out that in order to preserve the secrecy of my mission I had been instructed to report my arrival by L/T and since he objected to the Postman landing I asked whether he could allow me to land with Lieutenant-Commander Neville, in plain clothes, to send the telegram myself. We undertook to try and find some clothes that did not look too obviously English and to speak only French.

6. The Flag Lieutenant regretted that only the Admiral could decide such an important matter and that as he was playing polo he would not be accessible until later in the afternoon. I then dictated to the Flag Lieutenant a list of the questions I proposed to ask on the following day so that the answers could be prepared in advance.

7. Finally at 1600 I prevailed on the Flag Lieutenant to come ashore with me to ask the Admiral permission for us to land. To this he rather reluctantly agreed and I waited in his office while he saw the Admiral. After half-an-hour permission was granted.

8. I asked whether it were possible to send a telegram from the local suburb, La Pecherie, to which he replied 'Not on Sunday.' I asked if we could get a taxi. He made some enquiries and then replied in the negative. Finally he agreed to meet us at the Officers' Club if we found our own way there after changing into plain clothes. He then drove off in his car and left us to do the two and a half mile journey by boat and on foot.

9. When we met him he hurried us out of the Club and took us to the telegraph office which he correctly warned us would be shut. I asked to be taken to an hotel to see if they could send a telegram for us but this he seemed loath to do. When eventually he left us we found an hotel for ourselves and had no difficulty in sending our telegram.

10. The Flag Lieutenant was personally friendly and spoke a little English, but he was obviously frightened of the possibility of incidents

de l'Atlantique, 1936–38; Inspector-General of Maritime Forces, 1938–39; C-in-C Western Maritime Forces (Brest), 1939–40; C-in-C High Seas Forces at Toulon in unoccupied zone of France, Sept 1940–Nov 1942; on German invasion of the unoccupied zone scuttled French fleet at Toulon, 27 Nov 1942; after the war condemned by High Court and sentenced to death, Mar 1947; sentence commuted to 15 years' imprisonment and set free Sept 1951; amnestied Sept 1959.

with Italians, who, according to him, abounded in Bizerta. He did, however, offer to give us dinner ashore, but this I refused on the grounds that no one else on board was being given leave and so I felt we must return.

11. At 0930 on Monday, the 23rd, I landed with Lieutenant-Commander Neville and my Engineer Officer, Mr. Mardon, Warrant Engineer, in No. 3 dress, having been told by the Flag Lieutenant that the Admiral would expect us in this dress. The Flag Lieutenant met us in a Monkey Jacket and conducted us to the Admiral's office, where the Admiral and his Chief of Staff held a meeting with us.

12. On greeting me the Admiral at once said that he hoped that I would agree to the negotiations being carried on in French as the mission was far too secret to permit of an interpreter being employed. To this I agreed.

13. The Admiral opened the meeting by expressing his surprise and annoyance at a British destroyer being sent to Bizerta at the present time with insufficient warning for him to be able to protest. He added that he had had his first intimation on Friday afternoon and had spent most of Saturday trying to get the Ministry of Marine in Paris to cancel the visit.

14. He stated that most of the information required could have been obtained from Paris and that any supplementary information could well have been obtained by one or two officers, who could have come from Malta by the Mail Steamer in plain clothes, without arousing suspicions.

15. The Admiral was at pains to point out that his indignation at *Wishart*'s arrival was in no way directed against the British Navy and that he would be only too glad to supply us with the necessary information. He explained that Tunisia was only a French Protectorate and that the French had not the same powers here as in Algeria and other colonies. He stated that a very high percentage of the European population was Italian or of Italian extraction, whose sympathies remained whole-heartedly with their land of origin, even in cases where they had taken out naturalisation papers in order to obtain work in the Dockyard.

16. Up to now his policy had been to give the civil population the impression that whatever other bases might be placed at the British Fleet's disposal there could be no object in their using Bizerta, since it would only take Italian Bombers half-an-hour longer to reach Bizerta than Malta. His object in doing so, he explained, was to avoid stirring up trouble amongst the Italian element, which would most probably lead them to commit acts of sabotage at the Naval Base, Dockyard and aerodromes.

17. He did not mean to suggest that Bizerta should not be used if war came but he did wish to preserve his facilities intact until such time as they might be needed. He had hoped that if it were used the British ships would arrive without the local population having their suspicions aroused beforehand.

18. However, in his view, the cat was as good as out of the bag. Did I realize that our Consul-General at Tunis had been sufficiently indiscreet to inform him over the public telephone lines that *Wishart* was arriving on a highly secret mission? After that he regarded the secret as a 'Secret de Polichinelle' [open secret].

19. He then proceeded to read his instructions from the Ministry of Marine, telling him what information he was to give me. He complained that as this letter had arrived after working hours on Friday there had only been Saturday forenoon in which he and his Chief of Staff could obtain the answers to the more technical questions without arousing suspicion. He pointed out that neither he nor his Chief of Staff were technicians and now that a British ship had actually arrived it was no longer possible to ask any of his staff questions, without clearly indicating that he was doing so in order to supply us with the answers.

20. I asked whether he would allow us to visit the Dockyard at Sidi Abdallah to view the oiling arrangements and docks ourselves. To this he at first objected altogether, and then proposed to let my Engineer Officer go alone in plain clothes. I pointed out he spoke no French and finally after much persuasion I obtained permission for all three of us to visit the Dockyard provided we did so in the Dockyard Workmen's dinner hour.

21. The Admiral then proceeded to deal with the questions I had given his Flag Lieutenant. Nothing would induce him to reveal the position of his proposed swept channels or booms[1] but he undertook to have a fourteen knot examination vessel permanently on duty at the entrance to the swept channel to lead British ships into the harbour, as explained in Enclosure Number two.[2] He also supplied me with a chart of Lake Bizerta, showing approaches to the Dockyard (which can be purchased in Paris for 20 francs a copy) and two secret plans, one of the Naval Base at Bizerta and the other of the Basin at Sidi Abdallah Dockyard. These three are forwarded with Enclosure Number two.

22. He welcomed the idea of sending a British man-of-war as base and wireless guard ship and considered it almost essential if the base were used by British ships. He informed me that he had been extremely surprised to hear that the post of British Vice-Consul at Bizerta had been abolished within the last six months, as this removed our easiest channel of confidential communication with him. The Consul-General at Tunis

---

[1] Mountbatten subsequently wrote: 'The impression received was that no preparation for war or any form of war plan was in existence. The way in which the Vice-Admiral constantly brushed aside questions concerning the swept channels gave much more the impression that he had no concrete plans for swept channels than that he had any he wished to keep secret.' Enclosure No. 2 of *Wishart* Report of Proceedings of 25 Dec 1935, ADM 116/3398.

[2] Not reproduced.

could only communicate over the public telephone or by letter and he urged strongly that the British Vice-Consulate be re-opened forthwith.

23. I asked the Admiral whether, in view of my short stay, he considered I should call on the civil and military authorities. He first said that it would rouse suspicion if I did not visit them, but later he changed his mind and said that he would prefer me not to call and that he would explain matters to them.

24. After this the meeting broke up. I informed the Admiral that I had been instructed to convey my Commander-in-Chief's compliments to him which appeared to gratify him greatly and he requested me to convey his warmest thanks and reciprocal messages.

25. After changing into plain clothes, we were driven by the Chief of Staff to Sidi Abdallah, which is about sixteen miles from Bizerta, with which it is connected by a very good road. The Chief of Staff became much more affable and helpful once he was on his own and reiterated his desire to help.

26. On arrival at the Dockyard we were at once taken to see the Admiral Superintendent (Rear Admiral E.H.D. Muselier[1]). The Admiral Superintendent had not been in the secret and had only received a guarded telephone message when we started. He gave us a very real welcome and said that he had served on the Naval Inter-Allied Commission of Control in 1920. He would have been obviously delighted to do everything for us personally but the Vice Admiral had apparently decided that he should not identify himself with us publicly. He even went to the length of fetching some refreshments and handing them round himself, rather than allow any of his naval servants to see us together.

27. He answered all our questions very frankly and fully but as he had had no warning he had to get a staff officer to look up some of the details. This officer was kept in an adjoining room. When a question had to be referred to him, we all kept silence while the Admiral Superintendent slipped through the door and presently returned with the desired details. What his servants and staff officers must have thought of their chief's movements can only be conjectured, but it is doubtful whether these tactics are more likely to have allayed than aroused suspicion.

28. The Admiral Superintendent agreed that it would be essential to have adapters to enable connection to be made between their oil fuel

---

[1]Contre Amiral [later Vice Amiral] Émile-Henri-Désiré Muselier (1882–1965). Commanded yard at Sidi Abdallah, 1934–36; naval commander in Tunisia, 1936–37; commanded 2nd cruiser division, 1937–38; commanded Marseille defence sector, 1938–39; placed on inactive list, Oct 1939; joined Free French in London and first C-in-C Free French Naval Forces, 1940–42; noted for stormy relationship with Gen de Gaulle which curtailed his career.

connections and ours. He allowed my Engineer Officer to make a sketch (which is forwarded with Enclosure Number two) and finally yielded to my suggestion to exchange adapters so as to enable both navies to make suitable ones themselves.

29. I accordingly supplied him with *Wishart's* 3½ inch (No. 4 size) double female adapter for oiling ship Pattern No. 1703 and received in return the corresponding French adapter, which is now on board *Wishart*. Instructions are requested for the disposal of this article.

30. At my suggestion the Admiral Superintendent, backed by the Chief of Staff, undertook to make twenty-four male screwed ends to fit into our 3½ inch adapters, as soon as war broke out or became obviously imminent. As French men-of-war carry their own fuelling hoses these are not normally supplied at French oiling jetties or in French oilers, but the Admiral Superintendent said he would prepare a few 120 foot lengths of flexible 4¾ inch hose with ends suitable for our connections.

31. The Chief of Staff supplied me with the technical specification of their oil fuel which I have translated and added to Enclosure Number two, but as their methods of arriving at the figures they give and the details of their testing apparatus are not known, I considered it essential to obtain a sample of the actual fuel and brought an empty bottle ashore with me. This bottle was filled with the oil fuel Grade A and is now in *Wishart*. Instructions are requested for the disposal of this bottle.

32. Finally the Chief of Staff took us on a tour of inspection of the Dockyard. We visited every jetty and dock and obtained particulars in each case, which are given in Enclosure Number two.

33. When we asked to be allowed to inspect the various machine shops and foundries the Chief of Staff seemed impatient to leave and so we gave way. I do not, however, regard his attitude in this as suspicious since it was by then 1400 and he had not yet had any lunch.

34. On return on board we hurriedly compared notes, and prepared a list of final questions to put to the Chief of Staff who had announced his intention of returning my call on behalf of the Admiral at 1600.

35. At 1600 the Chief of Staff arrived off in frock coat and epaulettes in a high speed launch, which stopped its way by breaking my gangway. We succeeded in hoisting the Chief of Staff aboard, after which his boat shoved off and came round and broke off the bits of the ladder that were still holding together and sank them.

36. The Chief of Staff answered those of the final questions which were not too technical for him. I said that we required at least another day to make sure that the information we had been hurriedly collecting was throughly understood, but he held up his hands in horror and said that every extra hour we stayed was one more hour of worry and anxiety for

the Admiral. I accordingly slipped and proceeded at 1700, five minutes after the Chief of Staff had left the ship.

37. I arrived at Malta at 0730 on Tuesday, 24th December.

38. I should like to record my appreciation of the assistance I received from Lieutenant-Commander Neville and Mr. Mardon in making notes and preparing the enclosed reports. Lieutenant-Commander Neville's assistance was of particular value on account of his knowledge of French. He and I took down independent notes in English while the answers were being given in French and the fact that our notes were in close agreement makes me feel confident that we have taken down correctly what we were told.

39. I am not equally confident that we were always correctly informed, chiefly on account of the Admiral's intense desire for secrecy preventing him from verifying his facts.[1] For instance he volunteered the fact that the largest dock was 200 metres long, whereas I know from information signalled by the Admiralty that the length was supposed to be 250 metres. When I taxed him with this he appealed to the Chief of Staff who supported him. It was only when I questioned the Admiral Superintendent that I got the truth and the Chief of Staff then excused himself and the Admiral by saying that it was not really their job to know all these details and they had not had time to verify them.

87. *Summary of Revised Mediterranean War Plan.*[2]

[ADM 116/3476] [January 1936]

SECRET.

* * *

79. With the temporary abandonment of the establishment of PORT X the Plan underwent certain modifications, though fundamentally it remained the same. The outline was as follows:–

The HAIFA Force would proceed to ALEXANDRIA if time permitted, and the whole Fleet would leave ALEXANDRIA in company on the outbreak of war or as soon after as possible and proceed to take control of the area to the Westward of CRETE, as far as the MALTA Channel, with the object of cutting off supplies going in to Italy and from Italy to her Army in Libya.

---

[1]This was subsequently found to be true in the question of oil fuel hose adapters. Mountbatten to Vice Admiral Malta, 8 Jan 1936, ADM 116/3398.

[2]Excerpts from the Commander-in-Chief's Narrative of Events in 'History of the Italian–Abyssinian Emergency …', pp. 18–19, ADM 116/3476.

80. On the way to the central area Third Cruiser Squadron and First Battle Squadron would carry out a bombardment of TOBRUK, which would be synchronised with an air attack. The bombardment would start half an hour after sunrise on Zero plus 1 day (Zero being the day the Fleet left ALEXANDRIA). The First Cruiser Squadron and Aircraft Carrier would not take part in the TOBRUK operations, but would proceed on the course for the central area. On completion of the bombardment the First Battle Squadron would return to ALEXANDRIA escorted by the Local Defence Flotilla and the remainder of the Force would rejoin First Cruiser Squadron.

81. The defence of MALTA during the first 24 hours would devolve on the First and Nineteenth Flotillas and the First Submarine Flotilla, but the two destroyer flotillas would join the Commander-in-Chief at about 1200 on Zero plus 2 day.

82. On arrival in the central area the whole Fleet would carry out offensive operations against the Italian and Sicilian coasts for 48 hours, after which it would occupy the area for about 6 days before returning to ALEXANDRIA for fuel and rest. First Flotilla would return to MALTA and 19th Flotilla to ALEXANDRIA to assist in the local defence and in convoying supplies from PORT SAID.

83. One Battleship and two Destroyers would remain in the Suez Canal, which would be denied to Italian ships, and these would be reinforced by two of 19th Flotilla when it arrived at ALEXANDRIA.

84. The question of maintaining control in the central area was investigated and the conclusion come to that it would not be feasible to occupy the area continuously in sufficient force to deal with any opposition likely to be encountered. The governing factor was the necessity for the Flotillas to boiler clean after 21 days steaming. This entailed a regular cycle of Flotillas cleaning boilers and consequently a permanent reduction of one Flotilla from the force available. The remaining Flotillas can only spend from 4–6 days at sea, which further reduces the destroyer strength.

85. It was therefore decided to occupy the area for about a week and then the whole force would return to the Base for fuel. ALEXANDRIA would be considered as the Fleet Base, but it was hoped that it would be possible to use MALTA and the Greek harbours (see paragraph 86a) refuelling both Cruisers and Destroyers, and possibly for boiler cleaning Destroyers also, which would simplify the problem of maintaining the necessary control.

* * *

86a. It was decided, partly to give the ships' companies a change of scene and partly to obtain some information about certain harbours, to send the Fifth Flotilla to Greece. The Flotilla left ALEXANDRIA on 7th

January and visited the PIRAEUS, sub-divisions visiting POROS and NAUPLIA in addition. The Flotilla returned on 21st January.

* * *

88. *Fisher to Chatfield*

[CHT/4/5] Commander-in-Chief, Mediterranean Station
Queen Elizabeth, Alexandria
11 January 1936

[Holograph]

Just a <u>glance</u> only at these photographs may interest you and then perhaps Controller or D.N.O. may store them in the Archives. The men who man these batteries are enjoying 'the war' like anything.

Now I have got 12 more H.A. guns & 12 more lights for A.A. defence at Alexandria & have helped the Army at Mersa Matruh in this respect too, all on the understanding that they come back to me in 48 hours if & when our base moves forwards. <u>Water</u> is the great problem for both M. Matruh & Sollum. We have gone into it on the spot. Situation <u>now</u> precarious at M.M. as the [?] Roman wells are failing & *Bacchus* has to go on distilling there until I can get hold of a water carrier or fit out a special distilling ship.

Sollum was very interesting. I explored every yard & had a good look at the Italian fort & mechanised force on the other side of the wire. The Egyptian company at Sollum now that one company of 11th Hussars are there have their tails well up whereas before they were windy when a whole Egyptn battalion was there. You will hear more about Sollum and M.M. by signal before you get this as Military opinion is still very fluid though we are trying to get ideas crystallized.

We gave Brownrigg a good run for his money the other night – as a sort of swan song. Flying boats, 2 Blockships, C.M.B.s (R.A.F. motor boats), Q Ship, sabotage parties inside the harbour. All attacked the Fleet between 2 & 3 a.m. It was a great success for <u>both</u> sides! – & we shall do it often – in whole or in part. He will tell you about it if you can spare him 5 minutes. I know what you are going through with the Japs. Its an unfinished business these Disarmament Conferences and you have had more than your fill of them.

We had a very cheerful party onboard on New Years Day. The Masters, Chief Officers & Chief Engineers of the R.F.A.s & all British merchant ships in port – about 240 in all and it wd have done Benito Mussolini good to see & hear them. Fifth Flotilla are I gather doing useful service in Athens so that rather dubious spot I hope to get tapped [?]. Almost as good as Port X if the Greeks play up? Enclosed signal will show that there is a lighter side to our life.[1]

89. *Admiralty to Fisher*

[CAB 53/5] [13 January 1936]

[SECRET.]

[Copy of Telegram]

Your 1558/7 of 7/1/36[2] and General Officer Commanding Egypt's 5646 dated 6/1/36. Without combined appreciation from Commanders-in-Chief, difficult to assess problem completely, but on known information Chiefs of Staff consider that the preparation of an advanced base at Sollum on the scale suggested, is inadvisable. The position of Sollum on Italian frontier would enable Italians to make all preparations for attack with aircraft and artillery fire on the base organisation. In view of the fact that if war breaks out it will be started by Italy, such attack could be carried out without warning and might form first act of aggression. Moreover, on political grounds, such extensive preparations in close proximity to frontier, are at present undesirable.

In the event of war, we consider that first action by mobile force should be from Sidi Barrani with object of driving Italian forces back from frontier and clearing Sollum area. Base at Sollum should then be established and further advance made when preparations are complete.

Please forward early combined Report asked for in paragraph 2, Part I of C.N.S. wire to Commander-in-Chief, Mediterranean, 20/12/35 preceded by telegraphic summary. On receipt of this Chiefs of Staff will make further communication.[3]

Meanwhile telegraph any modification to requirements given in (a) to (h) of your telegram No. 1558/7 to accord with above considerations stating your immediate needs.

---

[1] Not found. Chatfield noted on the letter: 'Not to be answered for there is nothing to answer.'
[2] On 7 January the three Commanders-in-Chief reported to the Chiefs of Staff that, in order to implement the policy of initiative and defence, Sollum had first to be prepared and held as a base from which a mobile force could operate. It would require at least four weeks to prepare and stock the base. 'History of the Italian–Abyssinian Emergency …', p. 19, ADM 116/3476.
[3] On 18 January the three Commanders-in-Chief forwarded their appreciation emphasising the advantages of holding Sollum but the Chiefs of Staff in London were unconvinced and still considered the proposals inadvisable for the moment. On 29 January they directed that preparations for operations from Sidi Barrani only were to be made. 'History of the Italian–Abyssinian Emergency …', pp. 21–2, ADM 116/3476.

90. *Admiralty to Chatfield and Backhouse*

[ADM 116/3038] 15 January 1936

SECRET.

[Telegram] Addressed C.-in-C. Mediterranean, repeated C.-in-C. Home Fleet. From Admiralty.

Reference Admiralty telegram 1704/20 December[1] not to C.-in-C. Home Fleet. Short appreciation of operations of Gibraltar Force in early stages of hostilities is being forwarded to you under cover of Admiralty Letter M/P.D. 05494/36 of the 11th January. It is clear from the appreciation that if hostilities break out in near future Second Cruiser Squadron might not be able to join you till some seven days after outbreak of hostilities. If your intentions have been read correctly requirement for exchange of Second and Third Cruiser Squadrons was based on maintaining continuous control of Central Mediterranean Area by Squadrons operating as reliefs. In view of policy of whole force operating together put forward in paragraph 2 of Admiralty telegram mentioned[2] request confirmation that this delay in Second Cruiser Squadron joining you in early stages can be accepted. The use of one or two battleships for the operation mentioned in your 1558/7 January (not to C.-in-C. Home Fleet) is considered desirable in itself to counter inevitable criticism of their inactivity in other operations.

1639/15.

91. *Admiralty to Admirals Fisher, Backhouse and Rose*

[ADM 1/9946] 17 January 1936

SECRET.

[Telegram] Addressed C.-in-C. Mediterranean, C.-in-C. Home Fleet, C.-in-C. East Indies. From Admiralty.

The subject of objectives in the event of hostilities with Italy has been considered by His Majesty's Government. Instructions have been framed so as to be common to all three services and are as follows:–

(i) As a general principle we must avoid taking the initiative in any naval or air bombardment directed against the civilian population in order that Italy may not be provided with a case for stating that we had

---

[1]Doc. No. 83.
[2]The recommendation to seize the initiative with a combined attack on Italian territory.

broken humanitarian practice and that, consequently, she was entitled to take similar retaliatory action;

(ii) That the above should be interpreted as meaning that in the event of an act of war against any legitimate British naval, military or air objective by Italian ships or aircraft, only counter attacks of a corresponding nature are authorised;

(iii) That even if Italian action takes the form of attacks on non-military objectives, no retaliatory action of a similar nature is authorised unless specific approval is given from Home for such action. In any case, a report should be rendered immediately by the Commander chiefly concerned and instructions asked for;

(iv) Consequently, in the event of attacks being made upon any objectives which lie in parts of the British Empire or countries in occupation by military forces, the Commanders of the military, naval and air forces are authorised to take action against legitimate military targets without further reference to Home authorities, but they are not authorised to undertake operations directly against the civil population as such, even if such action is taken by the Italians, without the specific approval from Home referred to in (iii) above;

(v) On the outbreak of war the above instructions do not preclude the immediate commencement of offensive operations against military objectives nor do they preclude the necessary defensive action on the part of the anti-aircraft defences or fighter aircraft, whether or not Italian aircraft actually drop bombs on their objectives.

Legitimate naval, military and air objectives are to be considered as governed by the following rules:–

Naval Objectives – Hague Convention No. IX of 1907.
Military   "   – Hague Convention No. IV of 1907.
Air   "   – Articles 21 to 25 of draft rules of Aerial Warfare drawn up by Commission of Jurists at the Hague in 1923.

As regards Hague Convention No. IX, the rule in Article 6 which provides (subject to military exigencies) for prior warning before the bombardment of a defended port is inapplicable in modern conditions, and in any bombardments of defended ports carried out in accordance with the foregoing instructions it may be assumed that military exigencies render notice impracticable. In the case of a bombardment of military objectives in an undefended town in accordance with Article 2 of the Convention, it may also be assumed that military reasons render a prior summons to destroy those objectives impracticable, and that the last paragraph of that Article will apply. All possible steps should be taken, as there provided, to ensure that the town suffers as little harm as possible.

A case under Article 3 is most unlikely to arise in modern conditions, but if it did the notice provided for must be given.

Further instructions will follow as to the interpretation of the Hague Rules of Aerial Warfare. You should report whether copies are available on the station.

Proposals contained in C.-in-C. Mediterranean's 2033/8/8 to Admiralty only for attack on air bases at Port Augusta and possibly Catania, and also the railway cutting operations you suggest are all considered to fall within the category of legitimate naval, military and air objectives. Portions of my 1316/38/8 (not to C.-in-C. East Indies) which deal with above questions are cancelled.

1430/17.

92. *Backhouse to Fisher*

[CHT/4/5] *Faulknor*
5 February 1936

[Holograph]

First I want to thank you most warmly for doing so much for me while I was with you. I was treated most hospitably & royally & I am very grateful. It was good of you also to take in Kinloch (F.L.)[1] who I did not intend to take with me but had not the heart to leave behind.

[*Added in margin*] If you thought me odd landing in plainclothes, I had thought it out & decided that it wd be best in view of Admy intention [instruction ?] to be as unobtrusive as possible.

I was v. glad to see you again & to find you so well & in such good form. I hope the lumbago is now written off for a long time.

After I had gone, I began to wonder if you had gotten any good out of my visit. I'm afraid it was v. little. The only thing I can do for you is to give you ships & as you have already got a good many of them I may not have seemed v. generous. However I expect you'll get what you want without difficulty, so that part of the business will be all right. I have felt all the time that the War area must have first claim – but as I told you, I do not feel able to say I can do the little I have to do with nothing except 2 battleships & *Hood*, plus one flotilla of old Destroyers (mostly overage).

---

[1] Lt [later Cdr] David Charles Kinloch (1906–69). Flag Lt to C-in-C Home Fleet, 1935–38; commanded destroyers *Middleton*, 1941–42; *Obedient*, 1942–43; commanded cruiser *Dauntless*, 1943–44; Staff Officer (Operations) on staff of Cdre (D), Eastern Fleet, 1945; retired, 1946. Kinloch married Backhouse's daughter in April 1938.

Perhaps, if War comes, it will be found that these 3 big ships can do some good in conjunction with your party. I hope so.

I do not know if I am to be allowed to see the French C.-in-C., but I much doubt it. If I did, I doubt if he'd commit himself far. I thought Mountbatten's remarks on this subject interesting.[1]

We ran into the dickens of a gale on Monday night & I am sorry I did not turn back at midnight. We had a hard day on Tues. A v. nasty steep, confused sea which shook us up a lot & we had to heave to to avoid continuous bumping at each end. Blowing v. hard too. Good experience of course & that is not a small matter. The noise in my cabin under these conditions is marvellous, principally caused by oil crashing about in the tanks underneath.

This morning when it seemed a little better we got to C. Bon but there met the sea again & we decided to go & shelter [*added in margin*: 'Kelibia Nord']. There are 6 or 7 large tramps in the same bay. I have to get on tonight. I did think of going on to Bizerta but decided against it, altho' I shd not have appeared in any form.

I suggest it might be worth [?] your while getting the minelaying rails [?] belonging to the two '*E*'s [destroyers] sent to Haifa. They are no use in England. It wd give you 2 more potential minelayers.

I have a sense that if it is war, we shall have to have a go at those Sicilian aerodromes & possibly the harbour at Augusta. Just to shake the Italians up nearer home. There is not a great choice of objectives, but I think some minelaying off their Western coast shd be advantageous. Perhaps the French wd like to capture Sardinia if they have any troops to spare or it could be a joint operation. If there is war, it will be well to go for it & the sooner & more heavily the pressure is applied, the quicker it will all be over. But I have no idea what our Govt. contemplates in the way of forming an army or armies – after mobilisation. If, as usual, we start slowly, we shall probably regret it. Our prestige would depend a great deal on making a quick & decisive job of it.

Thank you again very much for receiving me so kindly. All good wishes to you & the M.F. [Mediterranean Fleet].

[*Added in margin*] I am sure you must be feeling the strain of the great responsibility you have carried these last 5 months, & still carry. I would not like you to think that I did not realise this – mine has been nothing in comparison.

---

[1] See Doc. No. 86.

93. *Fisher to Admiralty*

[ADM 116/3038]   7 February 1936

SECRET.

[Telegram] Addressed Admiralty, repeated C.-in-C. Home Fleet.

658. C.-in-C. Home Fleet have discussed your 1639 15th January. If hostilities broke out in near future the delay in 2nd Cruiser Squadron joining the Eastern Force can and indeed must be accepted. I am of the opinion that sooner 6th Destroyer Flotilla joins the Eastern Force the better whether situation remains as it is at present or not in order that they may become practised in their new duties.

2. If political considerations admit it would also be better for 2nd Cruiser Squadron to join Eastern Force as soon as they are collected at Gibraltar so that they may work with Fleet. I would not however be prepared on that account to surrender 3rd Cruiser Squadron until after first 48 hours of operations as they have undergone special training for these and could not be replaced without prejudice to operations.

3. Re these two proposals C.-in-C. Home Fleet makes following representations. He considers a Cruiser Squadron is necessary at Gibraltar to carry out duties referred to in Admiralty Letter M/P.D.05494/36 of 11th January 1936 and that at least two flotillas are essential for same purpose. Also if all modern destroyers are removed from Home Fleet, possible operations of Home Fleet ships will be greatly restricted pending arrival of 'G' Class.

4. My observations to this are as follows. I agree that a cruiser squadron is desirable and this will be provided after initial operations. The duties of fleet are enumerated in paragraph 5 of Chiefs of Staffs' instructions to C.-in-C. dated 22nd January 1936 and ability fully to carry out these duties depends on reinforcement of Eastern Forces by units named and I consider that result of war will be decided in water east of Cape Bon and that some reduced degree of naval control in Gibraltar area must be accepted if it conduced to successful carrying out of main object.

If there was a harbour in Eastern Mediterranean which could accommodate H.M.S. *Hood* I would ask for that ship as well for at present I find difficulty in suggesting for her and also H.M.S *Nelson* and H.M.S. *Rodney* an objective worthy of them other than supporting of French Forces in Northern Area if active French co-operation materialises.

C.-in-C. Home Fleet is in general agreement with me but is anxious to make it clear that full discharge of his obligations in Straits of Gibraltar as laid down in paragraph No. 5 of Chiefs of Staffs' instructions to Commander-in-Chief dated 22nd January is not feasible if his force is reduced as proposed in paragraphs one and two above.

He considers also that whatever situation may be at outbreak of hostilities great demands will be made on western forces later for convoying reinforcements and supplies to Eastern Mediterranean. Although forces for Gibraltar are expected to be sent from Home after mobilisation these would necessarily be untrained.

0023/7.

94. *Admiralty to Fisher*

[ADM 116/3058] 19 February 1936

SECRET.

[Telegram] Addressed C.-in-C. Mediterranean, repeated C.-in-C. Home Fleet. From Admiralty.

Your 0023/7 and Admiralty Letter P.D.05494/36. Following important duty for Gibraltar Force should be added to those detailed in paragraph 5 of Admiralty letter mentioned. Begins. To occupy the Western basin with adequate force immediately before outbreak of war, so as to give cover to our merchant shipping and that of League States against attack and to prevent sporadic action in western basin by Italian forces as first act of war. Ends. In view of this and the other duties of the force it is essential that sufficient cruisers shall at all times form part of Gibraltar Force.

Admiralty would consequently not be prepared for Second Cruiser Squadron to join you without Third Cruiser Squadron being surrendered simultaneously. *Sydney* will however be sent to join you shortly, for administrative reasons this is also advantageous.

In view of above and observing that you will then have four modern 6-inch cruisers you may no longer consider exchange of Second and Third Cruiser Squadrons necessary. If exchange is to be made it is desirable that it should be made now as an exchange shortly before or after outbreak of war is definitely undesirable.

On the other hand at this particular moment when situation is somewhat easier there are administrative advantages in retaining cruiser squadrons under the command of their own Commanders-in-Chief.[1]

In view of above considerations request your remarks on this exchange.

Leander's leave is not up till end of March but *Ajax* could remain temporarily in Eastern Mediterranean if exchange proceeded.

Achilles should be available to join Gibraltar Force if necessary about beginning of April.

---

[1] The 2nd Cruiser Squadron and 6th Destroyer Flotilla were part of the Home Fleet although included in Fisher's operational plan.

Sixth Destroyer Flotilla will leave England about 4th March and will work up at Gibraltar. When working up is complete consideration must be given to exchanging Sixth and Fifth or Second Destroyer Flotillas so that one of the latter may give leave but this must depend on political situation. Request your views and those of Commander-in-Chief, Home Fleet as to which flotilla it would be preferable to relieve. When '*G*' Class destroyers become available at end of April further reliefs of destroyer flotillas may be possible subject to maintaining 2 destroyer flotillas with Gibraltar Force.

Admiralty appreciate necessity for convoy duties by Gibraltar Force in early months of a war but demand for this is not likely to be immediate and mobilisation should provide sufficient forces though they would inevitably be untrained.

1850.

95. *Fisher to Chatfield.*

[CHT/4/5]

Commander-in-Chief, Mediterranean Station.
Alexandria, 25 February [1936]

[Holograph]

I know how busy you have been and are. This is merely to say again that whatever you settle about ships, transfers, reliefs, leave, etc. I accept implicitly as I know that you have first & foremost in your mind meeting every <u>reasonable</u> wish of mine.

So if I even do not get my way in everything I know it is for a good reason.

You will I'm sure see Ramsay on his return in *Courageous*. He expressed a fear to Dudley Pound that he had not been a success! I think I successfully disabused him of that idea which from anything I have said or felt was quite unfounded – even fantastic.

He has done most exceptionally well. The personal energy & leadership he has shown have been unbounded, also the courage of his opinion & the risk he has taken. The Carriers are ready for war now. War of every kind against ships or shore, day or night. He & his flagship are great losses.

We have started our Queen Bee[1] firings. Going slow with the attack on them so as to preserve their lives, but despite firing one gun at a time with powder filled shell, we have shot down 3 out of the four at 8000'.

---

[1] De Havilland 'Queen Bee', a radio-controlled version of the well-known trainer the 'Tiger Moth'. It first flew in January 1935 and a sea-plane version could be catapulted from warships for gunnery exercises by the Fleet. By the time production ended in July 1944 a total of 380 had been built. Owen Thetford, *British Naval Aircraft since 1912*, 4th edn (London, 1978), p. 387.

Coventry has not got going properly yet & *Curlew* has only just arrived, but both ships are pretty keen.[1]

Lampson is very anxious that my name shd be given as one of his Technical Advisers as well as that of the G.O.C. and A.O.C.-in-C. So I have agreed on condition that Raikes can deputise for me whenever I cannot get away from the Fleet. I mean that Raikes shall do the work & both are pleased with this arrangement. Our first meeting is to be on March 2 & adjourn till March 9.

Don't be uneasy at my choice of *Ajax* as Fleet Flagship in operations. I intend to lead the 8″ cruiser line but leave to Horton[2] selection of target fire distribution, etc. which he can do better after all the practice he has from his ships. I can't help feeling it will always be a case of a 'chase' or 'close interception' and *Renown* will in that case have to do her best from whatever position she can reach. *Sydney* will be rear ship of the line. *Ajax* is very good from wireless and V/S Command point of view & we have arranged for the adequate accommodation of all essential staff. She is also (but less so when *Sydney* arrives) conspicuous – different from all other ships so that the position of C.-in-C. can readily be seen which wd not be the case if I was in an 8″ ship where there are already 2 Admirals. The *Ajax* I also feel justified in moving without an A/S escort which I'm not sure I should in a '*London*' class.

Although nothing startling or novel emerged from meeting Rose & Roger Backhouse, it was good to talk things over face [to face] & for Rose a real relief to get away from Aden for a spell. I enclose letters from them to show this.[3]

We have made great strides I think in the guarding against surprise attack here and I feel comfortable that a bolt from the blue whether by S.81s[4] or C.M.B.s or sabotage inside the harbour can be met. I have not yet relaxed leave arrangements or notice for steam. I have felt that immediate readiness helped to justify our all being here whereas something less than that might raise questioning as to the necessity of our being here at all.

I think we have got a pretty fine edge on most things or ought to have with such wonderful opportunities of practice.

---

[1] HMS *Coventry* and HMS *Curlew* had recently been converted into AA cruisers.
[2] RA [later Adm Sir] Max Kennedy Horton (1883–1951). Successful submarine cdr in the North Sea and Baltic during the 1914–18 War winning the DSO and 2 bars; commanded 1st CS, 1935–36; commanded Reserve Fleet, 1937–39; C-in-C Northern Patrol, Sept 1939–Jan 1940; Flag Officer Submarines, 1940–42; C-in-C Western Approaches, 1942–45; retired, 1945.
[3] For Backhouse's letter, see Doc. No. 92.
[4] Savoia-Marchetti SM.81. A large three-engined aircraft and in 1935 in terms of range and bomb load probably the most effective bomber in the Italian Air Force.

May I assume that *Penelope* will come to 3rd C.S.?
Marriott is here and we have been reviewing the Canal situation. There are no difficulties.

96. *Fisher to Chatfield*

[CHT/4/5]

Commander-in-Chief, Mediterranean Station
H.M.S. *Queen Elizabeth*, Alexandria
6 March [1936]

[Holograph]

Thank you for your last long & interesting letter – the best part of which by far was that you were going to remain at the helm for another year or two. I hope the latter & even longer for no one can replace you.

It was very kind of the 1st Lord to offer me Portsmouth. I so highly appreciate the compliment but I had genuine hesitation in accepting because there are so few places to go round and I feel I have had a great deal more than my share. Cork[1] & his Countess would I know do Portsmouth splendidly. So would Dreyer. This last command of mine satisfies me to the brim. I should jolly well think so! And before my F.S. [Foreign Service] leave is expired almost I drop straight into another job. I hope the Service will forgive me. It looks so grasping but I couldn't find a way of saying no without making everyone uncomfortable and being thought a humbug. But I would have understood if you had said, 'Look here William. You have been employed almost without a break for 9 years (I think!) And though you happen to be the Senior we think that Bill Boyle or F.C.D. [Dreyer] should have a turn now.'

We have had some splendid clear days with the Queen Bees and some with crossing clouds. I have seen most of them myself and believe it is not an exaggeration to say that given an uninterrupted view and the Q.B. steering a steady course the majority of ships if they use H.E. will hit it in the first run. But when bursts are close the Q.B. is made to jink. This saves her if only one ship is firing. Neither she nor a formation could bomb accurately under those circumstances. [*Added in margin*] A big formation of S.81s approaching to bomb the Fleet and keeping together, as they should do, is now a pleasant dream, certainly not a nightmare.

---

[1] Adm Sir William Boyle, the Earl of Cork and Orrery.

97. Fisher to Admiralty

[ADM 116/3468]   H.M.S. *Queen Elizabeth*, at Alexandria
19 March 1936

SECRET.

M.0122/00200.
Be pleased to submit to Their Lordships condensed reports on certain subjects connected with the recent emergency in the Eastern Mediterranean which may be of value if only for record. A schedule of the enclosures is attached.[1]

2. As Their Lordships are aware the initial distribution of the Mediterranean Fleet at the end of August, 1935, and which remained practically unchanged throughout, had for its objects (a) the removal of the most important units outside the range of large scale air attack from Sicily and Italy whilst leaving a sufficiency of destroyers of low endurance and submarines at Malta to repel a seaborne attack of any description on that island and (b) to occupy a strategic position near the Suez Canal and to afford additional security to Egypt as a whole – both internally and externally.

3. Had it been possible to accommodate all the remainder of the Fleet at Alexandria this would have presented great advantages, in planning operations, starting them from a common departure point and concentrating all measures and material for defence. As it was Haifa had to be used as an overflow from Alexandria and the Canal ports and it became the base for the Third Cruiser Squadron, one Destroyer Flotilla and the Minelaying Organisation and thus some dispersion of resources was necessary.

4. The strategic plan in outline was to obtain and maintain control of the Central area between Italy, Tripoli and Greece attacking the enemy whenever he might be found at sea and when feasible in harbour. By such control it was anticipated that Malta would be safeguarded from sea attack and that Libya would be cut off from Italy and the military, air and naval forces in that colony would succumb in the same manner as those in Italian East Africa.

5. Without detriment to the achievement of the main objective support to military and air operations from Egypt against Italian forces in Libya was arranged for and operations were concerted for the reduction of the Italian main base at Tobruk by bombardment and for supporting the

---

[1] Not reproduced. The enclosures on diverse subjects are too numerous and voluminous to reproduce.

advance of the Army by gunfire, conveyance of supplies by sea and preparation of distilling apparatus where required on the coast.

In connection with these matters meetings between the General Officer Commanding, the Air Officer Commanding-in-Chief and myself were frequent and plans elaborated in complete agreement.

6. Meanwhile at Alexandria, Malta and Haifa arrangements had to be made for the continuation and development of the war training – under the special conditions foreshadowed – of the units stationed there and the necessary provisions to guard against a surprise attack.

7. Attention was of course specially directed to meeting the more probable forms of offence, i.e. air attack, submarine attack and W.M.B. [War Motor Boat] attack without neglect of low angle firings by day and night and bombardment practice. A typical weekly programme of exercises from Alexandria is attached and as a result of many weeks practice I think it can be said that efficiency was so advanced as to ensure an enemy flying formation approaching the Fleet in clear weather being broken up and partially destroyed and a submarine attack being detected and dealt with in time to avoid danger.

8. The surprise attack was a more difficult problem. The actual putting into force of the precautionary measures decided on had to depend on the appreciation of the general situation and should this have taken a grave turn a variety of measures were capable of being put into operation at once. It would have been possible for instance for a large scale air attack to have been launched from Libya and the Dodecanese Islands on the Fleet at Alexandria. W.M.B.'s might have been brought in special ships within striking distance of Alexandria. The Italian Fleet itself could have sailed and not have been reported and could have found our ships closely packed in harbour presenting an easy target for a bombardment from the sea at decisive ranges at dawn.

9. These possibilities so fruitful for an enterprising enemy were continually in our minds and I think adequately provided for. For these reasons, amongst others, no all night leave was granted to officers or men and the Fleet was never at more than six hours' notice for steam, emergency ships being ready at immediate notice. (N.B. – Leave arrangements were slightly relaxed at the beginning of March allowing a few officers and men from certain ships to sleep ashore.)

10. By degrees the gun and searchlight defences of Alexandria against ship and air attack were developed until they may now be described as formidable and, after many night exercises, efficient, and the same may be said on a lesser scale of Haifa.

11. There remained the final risk of a treacherous attack by night or sabotage from within the harbour. This question was fully investigated

and from time to time certain ships or squadrons were directed to organise such attempts and many ingenious schemes and ruses resulted which keyed up the vigilance of the Fleet.

12. The local population of Alexandria soon got used to the many A/A batteries mounted in the town and outside it, to the presence of gun detachments and search light crews, to the illumination of aircraft over the town which was almost a nightly exercise, to the constant sound of heavy and light firing at sea and eventually to the blackout of Alexandria.

13. In all these activities and in services connected therewith officers and men evinced the greatest keenness and this applies also to store ships, ammunition ships, oilers, depot ships, repair ships, Base Defence ships and tugs, the personnel of which worked with unflagging zeal.

* * *

In Admiralty message 398 of 18th December, 1935, Their Lordships informed me that the situation was still menacing and that it was the realisation of the complete readiness of our forces in the Mediterranean and Red Sea that was the greatest if not the only guarantee that peace would be preserved. Thanks to the devoted zeal of Flag Officers and Captains and the really wonderful spirit and discipline of all those they commanded I believe that the fleet was completely ready and hope that this fact was indeed a contribution to the preservation of the peace.

98. *Pound to Chatfield.*

[CHT/4/10] Commander-in-Chief, Mediterranean Station
31 March [1936]

[Holograph]

You may have wondered why I have only written you one letter since I left the Admiralty.[1] The reason and one which I am sure you will appreciate is that I soon made up my mind that it was impossible for a C. of S. to write about anything of interest without poaching on the preserve of his C.-in-C. When I came out W.W. very kindly said, 'We'll run this show together like two C.-in-C.s' but I don't think he imagined it was possible even more that I did. It was just I think to ease what he in his kindness thought might be a difficult position for me – which it never was. I enjoyed every moment of the time and I hope you have not felt that I was 'champing at the bit' to take over because I certainly should not have recommended it any sooner had I been in your place.

[1] Doc. No. 69.

We had completed the Queen Bee firings just before I took over and as this was a kind of peak of an intensive exercise programme lasting over several months I have given the Fleet a fortnight off these weekly programmes before commencing them again. The first of these weeks was a harbour week for all ships except the newly commissioned 6th D.F. and except for the latter the week will be a harbour week [with] the exception of 24 hours in which I took the whole Fleet to sea for two exercises – we completed them this morning and everything went very well and I think we shall learn a considerable amount from both of them at the discussion we are having on Friday.

The First Exercise was to test out:

(a) Whole Fleet getting quickly to sea.

(b) Our 'Central Mediterranean' covering diagram.

(c) Whether *Coventry* and *Curlew* could protect *Glorious* against air attack (provided by No. 4 F.B. Wing) whilst she was flying on and off.

(d) Some special fighting instructions to deal with the British–Italian Fleets in Medn.

(e) New signal book which was used throughout the exercise and which proved admirable.

(f) *Ajax* as a Fleet Flagship for which she proved most satisfactory.

In these exercises I had Lyon commanding a skeleton force, representing Italians, composed of *Exeter* (7– 8″ Cruisers), *Arethusa* (6– 6″ Cruisers), *Torrid*, *Rowena*, *Thruster*, each representing Italian D.F.

The situation was that the Italians had been intercepted but didn't want to fight – this resulted in a chase which brought home a point I had realised all along; that the '*D*' Class Cruisers were out of place in the Eastern Medn Force because:

(a) Owing to their low endurance they would hardly ever be available in the Central Area.

(b) If they were present they would never get into action with their low speed.

This was the only point during my time as C.O.S. that W.W. and I held contrary opinions. I only mention this so you will not think I have suddenly changed my mind. He was a great believer in Lyon and so am I and I would sooner have him commanding the 6″ Cruisers than Meyrick[1] good as the latter no doubt is. But there is no question that Meyrick with 2nd C.S. would be invaluable whereas Lyon with 3rd C.S. would really mean only Lyon with *Arethusa*. In the exercise yesterday *Sydney* even,

---

[1]RA [later Adm Sir] Sidney Julius Meyrick (1879–1973). Dir of Training and Staff Duties Div, Admy, 1926–27; Flag Capt and Capt of the Fleet, HMS *Nelson*, Atlantic Fleet, 1927–29; Capt RN College, Dartmouth, 1929–32; Naval Sec to 1st Lord, 1932–34; commanded 2nd CS, Home Fleet, 1934–36; C-in-C America and West Indies Station, 1937–40.

never got into the picture, as neither did *Renown*. No doubt *Ajax*, *Galatea* and *Arethusa* would take on 6 I.T. 6" cruisers but I do not think it is a situation which should be accepted if it can be avoided. I am sending a letter asking that if the situation deteriorates that 2nd C.S. should immediately exchange with 3rd C.S.[1]

I do hope you did not think I was tiresome about *Douglas* and *Wolsey* relieving 2nd D.F. at Aden but 3 Flotillas of high endurance are really the absolute minimum and unless one resists all demands they soon get reduced in strength. So long as the Force here can go into the Central Area, then everyone understands:

(a) The concentration here.

(b) Keeping at highest state of efficiency.

(c) Restrictions on leave.

but once the force here was reduced below that required for the Central Area the necessity for (b) and (c) would not be so readily understood.

W.W. as you can readily imagine left the Fleet in magnificent condition.

The Second Part of our 24 hour Fleet exercises depicted a surprise situation in which everyone opened envelopes and had to issue their orders rapidly.

It was the old problem of preventing a junction between *Valiant* coming from Port Said and *Barham* from the Westward (I was in *Ajax*). I gave Cunningham 1st C.S. and 3 Flotilla to tackle *Barham* whilst *Glorious* and 3rd C.S. (*Arethusa*, *Sydney*, *Coventry* and *Curlew*) tackled *Valiant*. *Glorious* for the first time I believe in an exercise shadowed *Valiant* by night with her aircraft until 3rd C.S. got to grips.

The Battle Fleet – *Ajax* & *Renown* attacked *Valiant* and then *Barham* – night action with indirect illumination.

I think everyone had come to the conclusion that this was a comparatively easy problem as a result of W.W.'s attack on Boyle, but in this case Boyle, with a convoy, was tied to a steady course.[2] When the enemy Battle fleet is free to manoeuvre it becomes much more [difficult] to reach the correct position before ordering the shadowing vessels to switch on. If one makes a mistake one's cruisers get blown out of the water for nothing. This will be an interesting problem to tackle. The Destroyers found the Battle Fleet by directional bearings of the shadowers but neither this nor the plot is I think sufficiently accurate for a battle fleet engaging the enemy battle fleet.

---

[1]Underlined by DCNS [James] who added the comment: 'This has been settled by telegram for the present.'

[2]Probably a reference to the Combined Exercises of 1934 when Boyle was C-in-C Home Fleet.

I do hope you will manage to get some leave this summer as I know how much you must need it however fit you feel.

Please remember me to Lady Chatfield and with the best of wishes for all your important work.

99. *Minutes by James[1] and Chatfield*

[ADM 116/3042] [22–7 April 1936]

Register No. M.02618/36.

During this week the international situation has undergone still one more metamorphosis.

On the previous occasion when the Council of Thirteen met at Geneva to explore the possibilities of making peace between the belligerents, the proceedings, despite M. Flandin's[2] appeal, commenced with the usual threats, which, as far as I can ascertain, most people regard as futile, but, after the threats had been uttered and raised the usual storm in Italy, the Council did do some work towards bringing the belligerents together.

This second meeting of the Council, which has just concluded, presents certain new features. It is true that once again the proceedings started with something that was bound to irritate Italy, but this time we have been compelled to eschew Sanctionist threats and bow to French determination to leave any more punitive measures against Italy alone until after the French elections in May.

I particularly mention the threats uttered by our representative because it is the subsequent inflammation in Italy that causes a periodical renewal of thoughts about unilateral war.

---

[1] VA [later Adm] Sir William Milbourne James (1881–1973). Grandson of the artist John Millais, portrait of him as a child was used as an advertisement for Pears Soap. Consequently, the nickname 'Bubbles' followed him throughout his career. Flag Capt and COS to C-in-C China Station, 1921–22; Dep Dir, RN Staff College, Greenwich, 1923–25; Dir, RN Staff College, Greenwich, 1925–26; Flag Capt and Chief Staff Officer to RA, 1st BS, Med Fleet, 1926–27; Naval Asst to 1st Sea Lord, 1927–29; COS, Atlantic Fleet, 1929–30; COS, Med Fleet, 1930–31; Flag Officer commanding Battle Cruiser Sqdn, 1932–34; Dep Chf of Naval Staff, 1935–38; C-in-C Portsmouth, 1939–42; Chf of Naval Information, Admy, 1943–44; MP (Unionist) for North Portsmouth, 1943–45. Prolific author.

[2] Pierre-Étienne Flandin (1889–1958). Member and subsequently head of centrist Alliance démocratique group; deputy from Yonne, 1914–40; Min of Commerce, 1924, 1929–30; Min of Finance, 1931–32; Premier, Nov 1934–May 1935; Min of Foreign Affairs, Jan–June 1936; Foreign Min in Vichy Government, Dec 1940–Feb 1941; tried for collaboration and sentenced to 5 years' 'national indignity', but reprieved, 1946. Anxious for alliance with Italy, although inclined to oppose German reoccupation of the Rhineland, he subsequently became a strong advocate of the policy of appeasement, especially during the Munich crisis of 1938.

It is by no means impossible that the Italians will be in Addis Ababa before the Council of Eighteen meet again, and, though the Council are warned for 11th May, it now appears that the French Parliament will not meet until June 1st, and will have formal business to transact before they turn their attention to political problems. It, therefore, appears probable that the Council will only talk and take no action, if any action is indeed possible, until June is well advanced.

Ecuador has lifted Sanctions; there are indications that Yugo-Slavia is yearning to do likewise as she is hard hit; France would lift Sanctions if she could present to the world a sufficient reason for so doing. The Sanctionist front is cracking. And our Fleet remains at its 'war stations' and there is nothing happening at the moment to indicate that it will not be still at its 'war stations' in 6 months' time.

The Cabinet are well aware of the dispositions of the Fleet and of the disadvantages inherent to those dispositions. They also know how extremely anxious we are to relieve the situation and to take some first steps towards a return to normal. But I feel that a Cabinet blown hither and thither by winds beyond their control will not initiate anything with regard to the Fleet, and indeed are probably giving no thought to the Fleet.

Some months ago there appeared to be a chance that Italy, incensed by our attitude and misunderstanding our motives, and disappointed with the course of the campaign in Abyssinia, might risk all in an attack on our Fleet and the consequent conflagration in Europe. But today, with Addis Ababa almost in sight, with French policy openly favourable, with Hitler's adventitious entry into the Rhineland, there does not seem to be one single reason why Italy should turn from her successful campaign in order to commit some act which would bring the forces of Europe against her.

It is admitted on all sides that we must recover our relations with Italy, if such a thing is possible; that Italy must come back into the League, if the League still exists; that Italy must, for the stability of Europe, be persuaded to come back into military alliance with France. Should we not be taking steps now to implement this <u>recovery</u> policy? The decision, of course, rests with H.M. Government, but the signs are not wanting that it is beginning to attract attention.[1]

If that is so, there is a measure that might assist, and that is to modify our naval dispositions so that we no longer <u>appear to the world</u> to be ready at a moment's notice to grip Italy with armed force. I underline <u>appear</u>,

---

[1] James added a footnote calling attention to a recent speech by Prime Minister Baldwin; reproduced extracts from a Foreign Office telegram to the British ambassador in Rome on 21 April about the possibility of better Anglo–Italian relations; and a recent letter in *The Times* from the former Governor of New Zealand.

as the mobility of naval forces is so high that it is possible to reconcentrate at will in a very short time.

With every public declaration by Ministers that military sanctions will not be considered, with every Italian success, military and diplomatic, it becomes more and more difficult to explain away the policy behind the rigid retention of our Fleet at its 'war stations'. We went to them for a very good reason, and it seems as though we do not know how and when to quit them.

My conclusion is that it might help our Government and help to recover the situation if, <u>without actually diminishing our power to concentrate in certain eventualities</u>, we relaxed the present concentration to some extent, and thus brought our military preparations more into conformity with the <u>real situation</u> and in conformity with proceedings at Geneva.

The execution would be simple. It would mean that units of the Mediterranean Fleet could proceed on cruise, that we could withdraw ships as necessary from Gibraltar for leave or refit, and that the ships of the East Indies Station could cruise within a certain radius. I do not suggest that the base organisations should be broken up. [*added in holograph*: 'i.e. the defence guns, booms, etc. at Haifa & Alexandria'.] They should remain, and the Fleet should be ready to reconcentrate if necessary.

I feel that there are possibilities that a gesture such as I have outlined might be of some aid towards solving the difficulties of our Government, and it is obvious to everyone that those difficulties are extremely great at the present moment.

There is another aspect to this question, and one which more nearly concerns the Board of Admiralty. I refer to the efficiency and well-being of the Fleet. The personnel have never faltered under the particularly trying circumstances which arise when a fleet is kept tuned up for a war, which becomes daily more nebulous. But now the hot weather is approaching; Aden is becoming more trying as the season advances; with the exception of a few officers whose wives have joined them, the personnel at Haifa and Alexandria must be intensely weary of their surroundings. It seems to me that to continue in a state of being highly keyed up for war throughout the summer is only justifiable if there is a real and urgent political necessity for so doing. It is not today clear that the necessity exists.

As I remarked before, I do not think we can trust the Government, with its heavy preoccupations abroad and at home, to initiate action about the Fleet; they do not really realise all the implications of keeping the Fleet in the state it has been since last August; their foreign policy has been rudely shaken up and I prophesy is going to be subjected to greater shocks ere the summer is through; and so they might even welcome suggestions from the Services which might go some way towards getting them out of their difficulties.

I have discussed this with Second Sea Lord, and he wished me to add more concrete proposals for movements of ships, but I feel that the principle must first be accepted.

W.M.J. [William M. James]
22nd April, 1936.

First Lord.[1]

This is the minute by D.C.N.S. that I spoke to you about this morning. I feel the time has come when the Foreign Office should be again compelled to pay attention to the serious consequences to the Navy of our present policy in the Mediterranean and its demands on that Service which have never been eased since they were inaugurated last August.

The Fleet have now been at Alexandria for 8 months and kept on an absolute war footing, reinforced by a large part of the Home Fleet and ships from China, North America and East Indies, all of which Stations have been depleted during that period. When the Admiralty took the action in August which saved the League it was never understood that it would last so long as it has and I feel that excepting for Admiralty pressure the Foreign Office will complacently leave the Services in the position they are now for an almost indefinite period.

The question to be faced at the moment, therefore, is this – Is this immediate readiness for war in the Mediterranean any longer necessary with all the administrative difficulties which it involves and the serious consequences that are growing daily as to the readiness of the Fleet for use elsewhere? The situation is further complicated by the expansion in our naval personnel that is now taking place. Large numbers of boys and young seamen are ready to go to sea and their disposition is a matter of growing difficulty.

It is, of course, a fact that the Admiralty have taken certain action to ease the naval situation without definite Cabinet approval. Ships have been brought home to give leave and relaxation and for repair, others are still at Alexandria with men entitled to Home Service who have had neither Christmas nor Easter Leave but who cannot be spared so long as this state of strain is kept up. I understand that considerable numbers of men from the R.A.F. Squadrons in Egypt have been brought home and I

---

[1] Bolton Meredith Eyres-Monsell (1881–1969). Created 1st Viscount Monsell of Evesham, 1935. MP (C) South Worcestershire, 1910–35; Civil Lord of the Admy, 1921; Financial Sec to the Admy, 1922–23; Parliamentary Sec to the Treasury, 1923–24, 1924–29, 1931; 1st Lord of the Admiralty, Feb 1931–Jun 1936; Chf Conservative Whip, 1923–31; Regional Commissioner for Civil Defence, South Eastern Region, 1941–45.

am not sure whether similar action had not been taken as regards our military forces. Surely Italy now has no intention of attacking us when she has got all she wanted without doing so! In her moment of victory is she likely to hazard her existence by attacking us?

I suggest, therefore, that the Cabinet should be asked to definitely ease the naval situation and the strain on our Admirals, Officers and Men that they have borne uncomplainingly during the past 8 months. Such action might be somewhat as follows:–

(1) Fleet to be officially [*added in holograph*: 'but secretly'.] informed by the Admiralty that the immediate danger of war in the Mediterranean can now be discounted, that it remains a possibility, but an unlikely one; that the Commanders-in-Chief are therefore authorised to consider themselves at a certain period of notice for hostilities, that is to say, that hostilities could not commence unexpectedly and that a fortnight or 3 weeks' notice can be relied on.

(2) Advantage to be taken of this notice to ease the strain on the personnel in any way considered desirable, including the return of the Mediterranean Fleet to Malta and the Home Fleet to England, leaving in the Alexandria area and Gibraltar a minimum number of ships as a representative force. Similarly at Aden the Commander-in-Chief to be authorised to take similar action. Time is important as one of the ships it is most desired to bring home is the battlecruiser *Renown* from Alexandria which is not justifiable so long as the instant readiness for war attitude is maintained. It is desired to bring the *Renown* home early in May.

E.C. [Ernle Chatfield]
27th April, 1936.

SECRET

First Lord.

With reference to the paper I sent you yesterday about the Mediterranean situation (M.02618/36), may I suggest that you should make use of the following additional point, namely the great mobility of the British Navy.

When the alarm occurred last August and we were expecting the possibility of hostilities at a moment's notice the Fleet was moved to its new strategic position in a week. The Air Force and the Army reinforcements, on the contrary, took many weeks, if not months, in taking up their new positions. If therefore these slow moving forces can be brought back to England, as I understand is being done, how much more justifiable is it to move our ships and not to keep our very mobile Fleet fixed at Alexandria, Aden and Gibraltar.

E.C. [Ernle Chatfield]
28th April, 1936.

[Holograph]
The following is an extract from Conclusion 9 of Cabinet 31 (3b) (29/iv/36):

'(b) That, pending the Meetings at Geneva arranged to begin on the 11th May, no major re-distribution of the Fleet should take place, but that this decision should not preclude some relaxation of the present state of instant readiness of the Fleet at Alexandria nor unostentatious movements of ships such as the Admiralty had lately been carrying out, provided that these were not on such a scale as to reflect on our foreign policy.

(c) That the question of the position of the Fleet in the Eastern Mediterranean should be considered by the Cabinet immediately after the conclusion of the Geneva meetings.'

C.B. Coxwell
Private Office
30/iv/36.

Consequent on this decision certain action as to Fleet dispositions have been taken on M.02437/36.[1]

E.C. [Chatfield]
30th April, 1936.

100. *Pound to Chatfield*

[CHT/4/10]   Commander-in-Chief, Mediterranean Station
30 April [1936]

[Holograph]

I like to write you every now and then to let you know how we are getting on. I do <u>not</u> expect an answer as I know how more than fully occupied your time is.

Anderson (U.S. Attaché).[2] Appears to have been very pleased with his tour as I received two letters from him, one after leaving here, and one after leaving Haifa. Full of gratitude for what had been done for him. He is quite useful at tennis and we had a good four one afternoon.

---

[1] Decisions summarised in Doc. No. 111.
[2] Capt [later VA] Walter Stratton Anderson, USN (1881–1981). Asst COS and Operations Officer to C-in-C US Fleet, 1927–29; commanded battleship *West Virginia*, 1932–33; Naval Attaché in London, 1934–37; member of US delegation to London Naval Conference, 1935–36; commanded Cruiser Division 4, Scouting Force, 1937–39; Director of Office of Naval Intelligence, 1939–40; commanded Battleships, Pacific Fleet, 1941; Pres of Board of Inspection and Survey, Navy Dept, 1942–44; Cdr Gulf Sea Frontier and Commandant 7th Naval District, 1944–45; retired, 1946.

Visit to Haifa and Port Said. As soon as King Fouad's[1] condition became serious I cancelled my visit as it was expected that the funeral would be within 24 hrs. of his death and I could not have got back for the funeral. Shall hope to visit these places in about a fortnight's time if the international situation permits.

Training. We have got a large number of recently commissioned ASDIC trawlers as well as 6th D.F. so I am getting two more S/Ms from Malta as we cannot go on working *Pandora* & *Proteus* (standing by to go to Aden if required) as hard as if they were only here for a fortnight.

Minesweeping Trawlers. Arrived yesterday and I went round them this morning – a pretty queer crowd a good many of whom have apparently never followed the sea life before. Cowall, a R.N.R. Lt. from P. & O. who is in command is a good hand I think and will gradually whip them into shape. He has had to get rid of a considerable number of their crews and I expect a few more will have to go.

R.A.(D).[2] Is settling in well and will I think be a great success though he has a difficult man to follow in Cunningham.

Boyd (Capt. D. 4).[3] Was I think a very good selection.

Danckwerts[4] and 6th Flotilla are doing well and will soon be on the top line.

Fleet Exercise. I am taking the Fleet to sea next week and shall do so about once a month as it is just as necessary to keep them wired up in this as in G. & T. practices.

Whites. We went into whites about a fortnight ago so that will give you some idea of the weather we are experiencing. I have never experienced such a marvellous 6 months of weather as we have had since I came out.

Contentment of the Personnel. There is no doubt but that the great majority of the men would like to get back to Malta but they are quite happy and contented. We give a very limited amount of night leave but as far as the men are concerned there is very little demand for actually sleeping on shore but they like to feel that they are not working to the clock and can come off in a shore boat. I am allowing a very few officers

---

[1]Fouad I (1868–1936). Sultan of Egypt, 1917–22; King of Egypt, 1922–36.
[2]Rear Admiral Somerville.
[3]Capt [later Adm Sir] Denis William Boyd (1891–1965). Commanded destroyer *Valentine*, 1932; Dir of Tactical Div, Admy, 1934–35; commanded destroyer *Hardy* and Capt (D), 4th DF, 1936–37; commanded HMS *Vernon* (Torpedo School, Portsmouth), 1938–39; commanded aircraft carrier *Illustrious*, 1940–41; RAC Med aircraft carriers, 1941; RA Eastern Fleet aircraft carriers, 1942; 5th Sea Lord and Chf of Naval Air, 1943–45; Adm (Air), HMS *Daedalus* RN Air Station, Lee-on-Solent, 1945–46; C-in-C British Pacific Fleet, 1946–48; C-in-C Far East Station, 1948–49; retired, 1949.
[4]Capt [later VA] Victor Hilary Danckwerts (1890–1944). Commanded cruiser *Caradoc*, 1930–32; Asst DP, Admy, 1932–34; Capt (D) 6th DF, 1936–38; DP, Admy, 1938–40; Dep C-in-C Eastern Fleet, 1942–44.

and men from each ship at one time to go to Cairo so long as this does not [affect] readiness of the ships to proceed to sea and fight.

Sports and Regatta. The sports are coming off in 10 days and then if it looks as if we shall be here for some months yet we shall probably get on with the Regatta.

One of our great problems is bathing because the harbour is so full of oil that it is impossible to bathe from ships, except those in a few berths. However we are making arrangements for various beaches.

Funeral of King Fouad. I am going up to Cairo tonight with 20 officers – all specially selected and the same height so if they are allowed to be together and not mixed up with the soldiers and airmen they should be very impressive.

Hood. The decision regarding *Hood* coming east or not will probably have been taken before this reaches you but I hope it may have been found possible to let her join me not only because it is undesirable to have reinforcements going up at the last minute but also it would save V.A.1 from a possible double shift.

21st Flotilla. I hope it is the intention to send them to the Medn as otherwise I shall be one flotilla down as the result of the general shift round which has taken place.

To work the Central Area from Alexandria 4 Flotillas with the Fleet are really necessary and then with one Flotilla at Malta the minimum is really 5.

Sunday 3rd. I kept this open to tell you about the funeral. Very well organised considering the short time they had to organise it. We had a couple of hours walk. We were all right in white but some of the diplomats and ex-Prime Ministers in their blue things couldn't stay the course. I had a long talk with H.C. [High Commissioner] who thinks they are going to have a very difficult fortnight over the Regency as everyone is intriguing to get their nominees in and some of them are not too reputable.

H.C. arranged for a meeting of A.O.C., G.O.C. and self. I think he was a little uneasy that the impression was getting about in Army & Air Force that everything was over. He read out some F.O. telegrams which certainly dispelled any such idea.

Relaxation of Present Conditions. Your telegram has enabled me to give the usual night leave until the 12th May and to curtail [?] the notice for steam a bit which will help matters. We go back automatically to 'emergency' conditions on the 12th. I think it would be a pity to extend the relaxation beyond the 12th as if we frequently alter the leave conditions they tend to become an international barometer.

Visit of *Australia* and *Sydney* to Killia Liman. Very much appreciated by these ships and went off very well.

<u>Sand Storm</u>. About a fortnight ago we had the worst desert storm for 20 years. Visibility only 2 cables in the harbour and every scuttle in the ship had to be closed. Cannot complain over a trifle like this when we get such glorious weather as we do. I only wish you could have a little of it – the glorious weather, <u>not</u> the desert storm.

<u>Committee on yearly Fleet routine</u>. I have asked Charles Forbes over a committee to go into this in readiness for return to normal times as something will have to be altered if we are to keep weapons training at a high pitch of efficiency and compete with individual training at the same time. This emergency has given very good proof of the necessity for a high standard of weapons training – the balloon might have gone up early in September without any time for special preparation.

<u>20th and 4th Flotillas</u>. I think I shall probably change them round and send 4th to Malta and get 20th here. The 4th can then do their docking at Malta and 20th complete their working up under R.A.(D).

I do hope you will manage to get some proper leave this summer as however fit you may feel you must be in need of it after all these months of work and anxiety.

Please remember me to Lady Chatfield.

[P.S.] Admiralty decision about *Hood* just arrived. Both V.A.1 and I both understand.[1]

101. *Admiralty to Backhouse, Pound et al.*

[ADM 116/3042]   1 May 1936

SECRET.

[Telegram] Addressed C.-in-C. Home Fleet, C.-in-C. Mediterranean, V.A.C. B.C.S., C.-in-C. Portsmouth, C.-in-C. Plymouth, C.-in-C. The Nore, repeated V.A.C. 1st B.S., R.A. 2nd B.S., R.A.(D), V.A.C R.F., Commodore (D). From Admiralty.

My 1446/22/4 (to C.-in-C. Home Fleet and C.-in-C. Mediterranean) and C.-in-C. Home Fleet's 0957/25/4 to Admiralty only. Pending the outcome of the meetings of the League of Nations commencing May 11th, the situation does not admit of any material changes in Fleet dispositions, but the following decisions have been taken.

Rodney is to leave Devonport for Gibraltar on 7th May. Admiralty would prefer that on her arrival at Gibraltar, C.-in-C. Home Fleet should shift his Flag to her and send *Nelson* home to give leave. He should report

---

[1] The battlecruiser *Hood* did not join Pound in the Eastern Mediterranean.

programme for *Nelson*'s return and leave. *Nelson* to be at 48 hours notice during leave.

R.A.(2) will transfer his Flag before *Rodney* sails to *Ramilles* or *Royal Sovereign*, reporting which.

Reference C.-in-C. Mediterranean's 1708/10/4 and 2118/10/4 to Admiralty and Home Fleet only,[1] after full consideration Admiralty have decided that the best arrangement would be for *Renown* to sail from Alexandria in time to arrive at Gibraltar on the 19th May. Flag of V.A.C. 1st B.S. to be transferred to *Valiant*, V.A.C.B.C.S. to shift his Flag to *Renown* at Gibraltar. Provided the situation admits, *Renown* with Flag will sail for England on May 20th. Flag will remain in *Renown* until ship pays off.[2] *Renown* on arrival is to give leave being at 48 hours notice during leave, and subsequently will pay off for large repairs at Portsmouth. Programme for her return, leave and paying off to be submitted by V.A.C.B.C.S.

On return from Las Palmas *Hood* is to remain at Gibraltar, ready to proceed to Alexandria if necessary.

Queen Elizabeth will sail for the Mediterranean May 19th.

Repulse will sail on 8th June for Alexandria and if situation permits *Hood* will then return, pay off and recommission for service in the Mediterranean hoisting flag of V.A.C.B.C.S.

Diomede will join Reserve Fleet on completion of refit.

Reference to C.-in-C. Home Fleet 1359/24 to Admiralty and Commodore (D) only, approved for *Cairo* with Commodore (D) to return to Chatham to give leave and refit.

Further review of the destroyer position and the measures required to facilitate the recommissioning of the *Hood*, has made it necessary to vary the decision conveyed in my 2037/21/4 (not to R.A. 2nd B.S., V.A.C. 1 B.S., V.A.C.B.C.S., V.A.C. R.F.). When the 20th Flotilla arrives Malta about May 19th, one division of the 1st Flotilla is to proceed to Gibraltar. On its arrival the three remaining vessels of 21st Flotilla (*Campbell*, *Windsor* and *Westminster*) are to return home and reduce to reserve. 21st Flotilla will then cease to exist. At the same time *Mackay* and *Wessex* will reduce to reserve. Decision as to the return of the 1st Flotilla to U.K. will be taken after the Geneva Meeting.

2000/1.

[1] Concerning the future movements of the battlecruisers *Hood* and *Renown*.
[2] Based on the considerations given in a private letter from Backhouse to the First Sea Lord, the Admiralty on further consideration approved the Flag of the VABCS remaining in the *Hood* and *Renown* returning to England as a private ship. Should, however, the development of the political situation make it desirable to send *Hood* to Alexandria, the VABCS would return to England from Gibraltar. Admiralty to C-in-Cs Home Fleet and Mediterranean, 13 May 1936, ADM 116/3042.

102. *Admiralty to Backhouse, Pound and Rose*

[ADM 116/3042] 1 May 1936

SECRET.

[Telegram] Addressed C.-in-C. Home Fleet, C.-in-C. Mediterranean, C.-in-C. East Indies. From Admiralty.

Until the conversations at Geneva commencing on May 11th develop, it is not possible to forecast the political situation with any accuracy. In the meantime however you can assume that the situation is sufficiently eased to enable you to take any opportunity which you may think fit to afford relaxation in Fleet routine.

2025/1.

103. *Admiralty to Backhouse and Pound*

[ADM 116/3042] 14 May 1936

SECRET.

[Telegram] Addressed C.-in-C. Home Fleet, C.-in-C. Mediterranean, repeated C.-in-C. East Indies, Rear Admiral (D), Commodore (D). From Admiralty.

My 2000/1/5/36 (not to C.-in-C. East Indies).[1]

Meeting of Council at Geneva on May 11th has not resulted in any marked change in the military situation, and it appears probable that there will be no change until the next meeting in June.

It has therefore been necessary again to review the destroyer situation.

Four destroyers of 2nd Destroyer Flotilla are now at forty-eight hours notice and could join Commander-in-Chief, Home Fleet at Gibraltar at short notice should the necessity arise. Hence it is no longer essential for one division of First Destroyer Flotilla to join the Gibraltar Force when the remaining three of 21st Destroyer Flotilla return to United Kingdom. The latter three vessels should accordingly leave Gibraltar as convenient about twentieth May.

If the emergency arose, the present strength of the destroyer force at Aden is inadequate, since it consists of four destroyers against three leaders and five destroyers at Italian ports. It would therefore be essential to reinforce Aden as soon as possible. The retention of the whole of the 1st Destroyer Flotilla at Malta will it is considered enable the Commander-in-Chief, Mediterranean, to reinforce the destroyer force at Aden by a

---

[1] Doc. No. 101.

division of one of the Mediterranean flotillas without unduly reducing flotilla strength at Alexandria or Malta.

Request C.-in-C. Mediterranean will report his proposals to effect this reinforcement should the necessity arise.

This telegram should not be taken as indicating any change in the situation as shown in my 2025/1/5/36.[1]

1232/14.

First Sea Lord.

104. *Admiralty to C.-in-C.s Mediterranean, Home Fleet and East Indies*

[ADM 116/3042] 21 May 1936

SECRET.

[Telegram] Addressed C.-in-C. Mediterranean, C.-in-C. Home Fleet, C.-in-C. East Indies. From Admiralty.

During the present period of eased diplomatic tension, fleets can be at 72 hours' notice in order to give increased opportunities for self-maintenance and reduction of strain on personnel. Precautionary state of war readiness to be resumed by 15th June. Notice at which fleets are should continue to be confidential.

1634/21.

105. *Pound to Chatfield.*

[CHT/4/10] Commander-in-Chief, Mediterranean Station
30 May [1936]

[Holograph]

Very many thanks for your letter with so much of interest in it. I hope the League Council will really get done with Abyssinian business in June and not postpone it until September. I think it would be undignified to go on with sanctions but I heard privately that the F.O. are divided 50–50 about whether to continue them or not.

However we are quite happy to stay at Alex. for as long as we are required. We have just completed a very successful series of Squadron and Fleet sports and now all hands are busy practicing for the Regatta – though when we shall be able to hold it remains to be seen.

[1] Doc. No. 102.

We have now got three good <u>bathing places</u> for the men – thanks to the generous way the Municipality has treated us.

I am giving the usual <u>all night leave</u> now and though many men do not take it – they of course appreciate being able to stay on shore if they want to.

We are taking advantage of the Admiralty telegram regarding <u>72 hrs. notice</u> until 15th June and getting in some good self maintenance work. I took advantage of this extended notice to send the 4 'longest out of dock' 8″ cruisers to Malta to dock and the Squadron as a whole will have a good speed on the 15th.

I feel that the League Meeting (if it comes off) must result either in a better or worse position and that it is unlikely that things will go on as they are. I have therefore been spending 10 days working out plans for a 'return to the normal – with evacuation of Alex., Haifa & c.' which will be quite a big business and now during the next 10 days we are going to overhaul all our operation orders so we shall be on the top line whatever happens. Your letter did give me the impression that you thought there might be an 'intermediate phase' in which the 'Bases' would be maintained as such but that the whole of the Fleet would be able to cruise away from its base. I cannot at the moment envisage, with the information I have, what kind of an international situation would require this unless it is that we are going to remain 'war based' so long as there are large concentrations of Italian troops in Africa.

I hope my telegram about *Sydney*, *Australia* and *Exeter* did not strike a note of alarm – it certainly was not my intention to do so but I thought the stage had arrived when the facts should be put before you. I am very grateful for the action you have taken as regards *Sydney* and *Achilles*. Whilst the 72 hours notice has been in force I have had the *Sydney* at Cyprus putting a lot of their men under canvas and making all the remainder sleep on the upper deck whilst the mess decks are being fumigated to try and get rid of their mumps which have been hanging round for nearly 3 months – just one or two at a time.

I have also taken advantage of the quiet time before 15th June to visit Cyprus, Haifa and Port Said – the two former places to get to know the Governor & High Commissioner and the latter place to entertain the Canal officers & c. We are just finishing our visit to Port Said as we have tonight and what with 2 large dinners and a lunch with Governor I think I have consulted with all the important people. Marriott has been a great help as he knows exactly who should be asked. Today I went down to Ismailia in a 'felucca' and lunched with the Canal authorities and worked in two bathes as we had a very pleasant day.

At Cyprus I motored up to Nicosia from Famagusta and lunched with the Governor[1] who was most friendly. Cyprus gave one the impression of being not only very flourishing in a small way but also one of the only really peaceful spots in Europe & c. at the moment. They do however want a decent harbour. The Governor has got a really nice new Government Ho. but I gather there has been rather a fuss over it as, in spite of the cheapness of building and labour in Cyprus, it cost £50,000.

I had two days at Haifa – one with the Haifa force going round all Lyon's ships which I found in very good order. They are not only a very efficient lot but also very happy and contented party there. The second day I flew up to Jerusalem and lunched with the H.C.[2] – very friendly and taking things very calmly. He was very grateful for what we had done for him at Haifa. Outwardly the country appears quite normal except for police about everywhere. I gather a pretty serious search for arms is going on the procedure being to surround a village and carry out a thorough search in which the whole place is turned upside down and everyone made more than uncomfortable. If it is pretty certain there are arms in the village and the search doesn't discover any, then the Headman of the village is lined up and given the 'bastinado' which has a miraculous effect.

I hope I did not embarrass the Admiralty by using *Capetown* for the Negus[3] but it seemed such a waste of fuel to send one of the Mediterranean ships.

I gathered from the H.C. that it was unlikely at the moment that any leaders would be deported but I have always got a ship ready.

I do hope you are making headway with all your committees and the very best of luck with the 'air battle' when it comes along.

I only sent you the telegram about Badoglio[4] in case anything got into the Press.

Please remember me to Lady Chatfield.

---

[1] Sir Herbert Richmond Palmer (1877–1958). Lt-Govr, Northern Provinces, Nigeria, 1925–30; Govr and C-in-C, Gambia, 1930–33; Govr of Cyprus, 1933–39.
[2] Maj Gen Sir Arthur Grenfell Wauchope (1874–1947). British representative on Military Inter-Allied Commission of Control in Germany after World War I; High Commissioner of Palestine, 1931–38.
[3] Title for the sovereign of Abyssinia. The Royal Navy had conveyed Emperor Haile Selassie into exile: first to Palestine in the cruiser *Enterprise* and subsequently from Palestine to Gibraltar in the cruiser *Capetown*.
[4] Marshal Pietro Badoglio (1871–1956). Chf of Gen Staff of the Army, 1919–21; Capo di Stato Maggiore Generale [Supreme General Staff], 1925–40; Govr of Tripolitania and Cyrenaica, 1928–33; C-in-C Italian forces in Ethiopia, Nov 1935–May 1936; created Duke of Addis Ababa by the King, July 1936; named by the King to replace Mussolini, 25 July 1943; fled along with the King to Southern Italy when Germans occupied the peninsula following armistice with Allies, Sept 1943; leader of 'co-belligerent' government on side of Allies; retired June 1944.

[P.S.] Poor Hemsted's death[1] has been a great blow to me – he was not only a first class secretary but a most loyal friend.

106. *Admiralty to Pound et al.*

[ADM 116/3042] 19 June 1936

SECRET.

[Telegram] Addressed C.'s-in-C. Mediterranean, Home Fleet, China, East Indies, America & West Indies, Africa, repeated C.'s in C. Portsmouth, Plymouth, The Nore, V.A.C.R.F., Navy Office Melbourne and Wellington. From Admiralty.

Should a change in the international situation make it no longer necessary to maintain the present state of readiness in the Mediterranean and Red Sea, the Admiralty's general intentions, on which your observations are requested, are as follows:–

(1) The *Australia, Sydney, Achilles, Ajax, Berwick, Adventure, Penzance* and Submarines and Destroyers of the China Fleet to return to their normal station as soon as possible.

(2) All ships of the Home Fleet, including those detached for service under Commander-in-Chief Mediterranean, to proceed as ordered by C.-in-C. Home Fleet.

(3) Mediterranean Fleet to return to Malta or as ordered by C.-in-C. Mediterranean, and proposals for Autumn cruise to be forwarded in due course, observing that a display of strength in Eastern Mediterranean Ports is considered desirable this autumn.

(4) *Hood* to return to Portsmouth in order to refit, give leave and subsequently pay off and recommission for service in the Mediterranean.

(5) *Exeter* to return to U.K. to pay off.

(6) Second Anti-Submarine Flotilla and Fifth Minesweeping Flotilla to remain for present in Mediterranean and Second Minesweeping Flotilla to be reduced to reserve at Malta.

(7) First Destroyer Flotilla to return to United Kingdom to give leave and pay off.

(8) Fourth Destroyer Flotilla to return to U.K. in the near future to refit and recommission. Date will be ordered.

---

[1]Paymaster Captain James Rustat Hemsted died suddenly on the tennis courts at Alexandria. He had been Pound's Secretary ever since the latter became ACNS in 1927. On the relationship see Robin Brodhurst, *Churchill's Anchor: Admiral of the Fleet Sir Dudley Pound* (London, 2000), pp. 59–62.

(9) The various base organisations to be demobilised and personnel returned to the United Kingdom as soon as possible.

(10) East Indies Squadron to proceed as ordered by C.-in-C. East Indies, and base organisations at Aden reduced and personnel returned to United Kingdom as soon as possible.

(11) It will probably be necessary to retain a capital ship at Alexandria whilst the Egyptian Conversations are proceeding.[1]

Instructions with regard to disposal of material, armament stores, etc., would follow.

1759/19.

107. *Pound to Admiralty*

[ADM 116/3042] 22 June 1936

<u>SECRET</u>.

[Telegram] Addressed Admiralty, repeated C.-in-C. Home Fleet, C.-in-C. China, C.-in-C. East Indies.

486. Your 1759 19th June 1936.

Paragraph One. Intend holding Mediterranean Fleet Regatta week commencing 5th July and, to avoid disappointment after training which has been done, would propose retaining cruisers from other stations for this even if earlier release were possible.

Before decision as to return of cruisers to East Indies Station is reached, request consideration may be given to the point that if H.M.S. *Shropshire*, H.M.S. *Sussex* and H.M.S. *Devonshire* are taken in hand for mounting anti-aircraft guns as directed by in Admiralty Letter G.01683/36 of 4th June 1936 effective strength of squadron may be reduced to H.M.S. *London* only for a considerable period and the present position in the Eastern Mediterranean during the Autumn cruise would be an unbalanced one.

China submarines at Alexandria would return to Malta forthwith to collect gear and all would sail thence for China as soon as possible heavy spare gear following by freight or H.M.S. *Cumberland*. H.M.S. *Adventure* will proceed to China as soon as possible after completion of trials ordered

---

[1] The negotiations ended with the formal signature in London on 26 August 1936 of a treaty between Great Britain and Egypt. Among the provisions, the British agreed to withdraw their forces except for a 10,000-man garrison in the Suez Canal Zone. This number could be increased in time of war. The British would maintain a naval base at Alexandria for not more than 8 years, Egypt would become a member of the League of Nations and a treaty of alliance between Great Britain and Egypt would be in force for 20 years.

by Admiralty Letter T.048/36 of 20th February 1936 in which it is also intended to employ H.M.S. *Narwhal*.

Propose that H.M.S. *Ajax* calls at Malta for quick docking before returning to the America & West Indies [Station]. This should result in economy.

Request information of intentions regarding H.M. Ships *Coventry*, *Curlew*, 1st Minesweeping Flotilla, *Kate Lewis*, *Garry*, *Liffey* and *Dee*.

Part II follows.

1408/22.

487. Following is Part II of my 1408 22nd June.

Paragraph (2). 6th Destroyer Flotilla would like to remain until after Regatta. H.M.S. *Guardian* is ready to sail. Suggest transfer of flag to R.A. Aircraft Carriers and 820 Squadron should take place at Malta if my 2043 21st June is approved.

Paragraph (3). Intend that Mediterranean Fleet should return to Malta with following exceptions: (a) H.M.S. *Valiant*, wearing flag of R.A.C. 1st B.S. to remain at Alexandria in accordance with paragraph (11), H.M.S. *Valiant* being relieved by H.M.S. *Barham* after latter has docked.; (b) probably necessary to retain 3rd C.S. and destroyers in Palestine. Red Sea Sloops to remain at Port Said as wireless link while demobilization of Canal Area is taking place.

Instructions regarding autumn cruise are noted.

Paragraph (4). Can any indication be given of the probable dates of arrival of H.M.S. *Hood*?

Paragraph (5). Assume R.A. Evans[1] should return to England in H.M.S. *Exeter*.

Paragraph (6). 2nd A/S Flotilla and 5th Minesweeping Flotilla return to Malta. concur regarding reduction of 2nd M.S. Flotilla to reserve.

Paragraph (7). Concur.

Paragraph (8). Concur. It is hoped that interval between paying off the 4th Flotilla and arrival of 'H' Class Destroyers will be as short as possible.

Paragraph (9). About 3 weeks will be required to demobilise the whole base defence organisation at Alexandria.

Paragraph (10). Concur with proposals contained in C.-in-C. East Indies 1350 20th June. Question of station limits to be revised at later date.

Paragraph (11). See Paragraph (3).

1530/22.

---

[1] RA [later VA Sir] Alfred Englefield Evans (1884–1944). Dep DNI, 1929–30; Capt of the Fleet, Home Fleet, 1930–33; Cdre South America and commanding officer cruiser *Exeter*, 1933–35; RA 2nd-in-command, 1st CS, 1935–36; RA-in-Charge and Superintendent of Dockyard, Gibraltar, 1937–39; retired list, 1939; Cdre of convoys, 1939–40; Head of Naval Technical Mission in Ottawa, 1940–44.

108. *Memorandum by Admiral Sir William W. Fisher*[1]

[CAB 16/147] n.d. [29 June 1936]

SECRET.

My Flag has been flying with one brief interlude for the last five years in the Mediterranean and I have, therefore, been in a position to watch the progress made in the H.A. fire of the Fleet since this was first seriously tackled by my predecessor in Chief Command, the present First Sea Lord.

2. In August of last year a period of danger supervened and since that date A.A. training has been intensive in all units of the Fleet resulting in the doubling of the rate of fire of the 4-inch gun (now 16 to 20 rounds per minute) and an increase in effective accuracy that is difficult to measure but very obvious to see.

3. Being well aware that the principal menace to the Mediterranean Fleet lay with submarines and aircraft, it was natural that every important exercise set to combat these should have been witnessed by me. In the last six months I saw over 100 runs by 'Queen Bees' and many hundreds of runs with the sleeve target. The defence of the Fleet in an open anchorage has been frequently practiced and those factors making for combined A.A. defence by ships and shore batteries have been exercised and developed at Malta, Haifa and Alexandria daily and nightly, excepting Saturdays and Sundays. The appointment of an Air Defence Officer in each ship and the inclusion of a Senior Officer on my Staff as Fleet Air Defence Officer has led to an all-round improvement in control, communications, lookouts and loading.

4. As regards the Fleet's main base – Alexandria – we had to guard against long-range bomber attack from Libya and attack by flying boats from the Dodecanese. When the situation warranted two arcs of sea patrols at radii of about 75 and 150 miles were arranged for (and exercised) in addition to the military shore look-outs in the Western desert. I counted on getting sufficient warning by telephone or W/T to man all guns, and lights, both afloat and ashore, and to have fighters in the air to intercept. I mention this as being a typical state of affairs applicable to almost any harbour that the Fleet might be using. With all the Fleet in Alexandria I calculated that from 40 to 80 H.A. guns (4-inch and above) could be brought to bear on an enemy formation depending on the direction from which it approached.

5. The defence of the Fleet at sea depends on the suitability of the 'Cruising order' which was such as to give the greatest chance of early

---

[1] The memorandum was circulated on 29 June 1936 as V.C.S.20 to the Sub-Committee (of the CID) on the Vulnerability of Capital Ships to Air Attack.

warning of the approach of enemy aircraft and the development of maximum gunfire as soon as they came in range.

6. The result of the intensive training was that instead of being extremely apprehensive of what might happen were the Fleet attacked in harbour or at sea, I became tolerably sure that the enemy, and not we, would suffer most. No Fleet could, however, be expected to resist repeated attacks such as might have been possible if a base closer than Alexandria to Italian shore-based aircraft had been used or if the Fleet had continuously cruised in waters within easy striking distance of them.

7. The small 'Queen Bee' target provided our most realistic test. I expected a single well-trained ship firing two gun salvoes to be on the target in under 30 seconds after her first shell had burst. With a larger target to observe on (such as an S.81) and still more with a formation, effective fire should occur earlier and if instead of one ship a squadron is firing I am sure a formation would be broken up and receive many casualties.

8. High speed, good manoeuvrability and the presentation of as small a target as possible are, of course, factors making for the increased safety of the Capital Ship as are also her H.A. gun power and armour protection. Our present battleships do not possess yet the above factors in sufficient degree to warrant their use within effective range of large concentrations of enemy aircraft unless the objective in view is of special importance, the carrying out of which justifies a liberal use of auxiliary vessels in a purely protective capacity. In the recent emergency I had in mind operations that would and would not have called for the use of battleships within effective range of Italian aircraft.

9. I can see no escape from the necessity of building Capital Ships, their existence in the Navies of other Great Powers as well as the necessity of dominating foreign 8-inch Cruisers makes them essential to us. The nature of them will take into consideration the opponents they may have to face and the dangers against which they have to be guarded.

109. *Sir Samuel Hoare[1] to Rt Hon. Anthony Eden[2]*

[FO 371/20382] Admiralty, Whitehall
3 July 1936

SECRET.

My attention has been drawn to your telegram (No.65) of the 26th June from Geneva, giving an account of your interview with M. Pouritch,[3] in the course of which you stated that it was the intention of His Majesty's Government 'to maintain forces in the Mediterranean in future which would be in excess of those which were at our disposal in that Sea when this dispute had broken out'.

So far as the Naval forces are concerned, this statement is literally correct, because the Mediterranean Fleet at the end of last summer was not up to the strength which had been announced in the First Lord's Estimates Statement of March last year, but I think I ought to make it clear to you in case your statement should be called in question, that any increase in the actual strength of the Mediterranean Fleet or in its standard strength as announced in the Estimates Statement of 1935, cannot be achieved at the present time or in the near future. In 1939, however, when the ships of the New Construction begin to come into service and the process of modernising the old capital ships will have been almost completed, an appreciable increase in the numerical strength and power of the Mediterranean Fleet will take place.

As things stand at present, the actual strength of the Fleet in terms of large vessels is greater than it was in July 1935 by 1 battleship, 1 battle cruiser, 1 aircraft carrier and 2 cruisers.

---

[1] Samuel John Gurney Hoare (1880–1959). Created 1st Viscount Templewood, 1944. MP (C) Chelsea, 1910–44; Sec of State for Air, 1922–24; Nov 1924–June 1929; Sec of State for India, 1931–35; Sec of State for Foreign Affairs, 1935; 1st Lord of the Admy, June 1936–May 1937; Sec of State for Home Affairs, 1937–39; Lord Privy Seal, 1939–40; Sec of State for Air, 1940; Amb to Spain, 1940–44.

[2] Robert Anthony Eden (1897–1977). Created 1st Earl of Avon, 1961. MP (C) Warwick and Leamington, 1923–57; Parliamentary Private Sec to Sec of State for Foreign Affairs (Austen Chamberlain), 1926–29; Parliamentary Under-Sec, FO, 1931–33; Lord Privy Seal, 1934–35; Min without portfolio for League of Nations Affairs, 1935; Sec of State for Foreign Affairs, 1935–38, 1940–45; Sec of State for Dominion Affairs, 1939–40; Sec of State for War, 1940; Leader of the House of Commons, 1942–45; Sec of State for Foreign Affairs and Dep PM, 1951–55; PM, Apr 1955–Jan 1957.

[3] Bojidar Pouritch, Yugoslav Delegate at the League of Nations.

110. *Chatfield to Vansittart*

[ADM 116/3042] 6 July 1936

[Carbon]

SECRET

The First Lord has asked me to get into touch with you at once about the raising of Sanctions and the redistribution of the Fleet. He said that he did not think that any Cabinet decision was required and it is a matter that can be perfectly well arranged between the Admiralty and the Foreign Office. If you agree to this, as I hope you do, perhaps you will give me the word over the telephone as early as you can, as I do not want to keep the Fleet in what they may feel is rather a false position a moment longer than I need.

As far as I know the only immediate moves that would take place would be – the First Flotilla would come home from Malta and, possibly, one or two other ships. The remainder of the Mediterranean Fleet is having their Annual Regatta at Alexandria, which will not finish until the end of this week, but the important thing is for them to get the abandonment of Alexandria as a base started, which includes such things as dismounting guns and getting away our store ships, oilers, ammunition ships, etc. At Gibraltar no <u>immediate</u> move is essential, and I should direct the Commander-in-Chief to make proposals for return to England. As regards the Red Sea forces, most of these are now on the East Coast of Africa, as you will remember, and I want to get the destroyers from there back to China at once.

What, therefore, we need here is to be able to send out a general message confidentially that the state of war readiness is ended, ordering a few important movements to take place which will not cause undue comment, and then inviting Commanders-in-Chief to make proposals for the dispersal of their Fleets and for the early return of them to their normal dispositions and stations.

[*Attached note*]
D.C.N.S.

Sir Robert Vansittart telephoned a reply to the attached letter of 6th July in the following sense:–

'The Secretary of State agrees to the Admiralty proposals so long as any withdrawals are not so large as to nullify his undertaking in the House that we would keep larger forces in the Mediterranean than previously. This seems especially important at this moment in view of reports abroad that we intend to abandon the Mediterranean.'

R.C. Jerram,[1]
Secy to 1st Sea Lord.

[Holograph]
Proposed telegram does not nullify the undertaking, which is a declaration of future policy. For a few months the actual figures will not show an increase over 1935 Fleet, but *Hood*, *Courageous* and new '*H*' flotilla will soon be out there.

W.M.J. [James]
8.7.36.

111. *Admiralty to Pound*

[ADM 116/3043] 6 July 1936

SECRET.

[Telegram] Addressed C.-in-C. Mediterranean, repeated V.A.C. 1st B.S., R.A. Alexandria, V.A. (B), R.A.C. 3rd C.S., S.N.O. Haifa, Defence Commander Base Defences Med., R.A. Gibraltar, R.A. Malta. From Admiralty.

Your 1533/27/6 (not to R.A. Gibraltar):–
Following action is to be taken on termination of emergency conditions:–

(i) All naval fixed defences at Haifa are to be evacuated. Instructions regarding Alexandria will be issued later.

(ii) All Booms and Underwater Defences laid at Alexandria and Haifa are to be recovered. It is not desired to store Booms or baffles at Alexandria or Haifa. If possible these defences, together with Indicator Net in S.S. *Benarty*, should be stored at Malta or Gibraltar; if this cannot be arranged they are to be sent back to United Kingdom.

(iii) Subject to decision regarding defences at Alexandria, all naval equipments ex Malta are to be returned to Malta, and M.N.B.D.O. equipment to United Kingdom.

(iv) Anti gas stores for Base Defences and ships attached and for Bases other than Malta are to be returned to United Kingdom. Anti gas stores at Malta are to be retained at Malta. Ships from other stations attached to Mediterranean Fleet are to retain their Anti gas stores on board.

---

[1]Paymaster Cdr [later Paymaster RA Sir] Rowland C. Jerram (1890–1981). Sec to Chatfield during the latter's appointments as 4th Sea Lord, 3rd Sea Lord, C-in-C Home Fleet, C-in-C Med, 1st Sea Lord, and Minister for Co-ordination of Defence, 1919–40; served in cruiser *Cleopatra*, 1942–43; Sec to Combined Operations HQ, 1943; Comptroller, Headquarters, Southeast Asia Command, 1943–45; head of Admy Mission to the Med, 1945; retired, 1945.

(v) My 1220/1 (to C.-in-C. Mediterranean). V.A. Malta and R.A. Alexandria refers.

1652/6.

112. *Admiralty to Pound et al.*

[ADM 116/3042] 8 July 1936

SECRET.

[Telegram] Addressed C.'s-in-C. Home Fleet, Mediterranean, China, East Indies, America & West Indies, Africa, repeated C.O.C.O.S., V.A.C.R.F., C.'s-in-C. Portsmouth, Plymouth, The Nore, R.A.(S), Navy Board Melbourne, Navy Board Wellington, V.A. Malta. From Admiralty.

Part I.

Their Lordships direct the following steps are now to be taken to revert to normal conditions:–

1. C.-in-C. Home Fleet to signal proposals for the return to the U.K. of all units of the Home Fleet abroad. The general policy outlined in C.-in-C. Home Fleet's 0014/22 to Admiralty, Home Ports, Mediterranean and 2nd Battle Squadron only is approved.

2. C.-in-C. Mediterranean to signal proposals for all other ships in his command not belonging to Home Fleet, which are commissioned with home service complements, to return to the U.K., including *Coventry* and *Curlew* whose future service is now under discussion. 2nd A/S Flotilla and 5th M/S Flotilla are to remain for the present in Mediterranean and as directed by C.-in-C. Mediterranean. Vessels of 2nd M/S Flotilla are to reduce to reserve at Malta. *Seahorse* and *Starfish* are to return to U.K. forthwith. Decisions as regards *Garry*, *Liffey* and *Dee* will be given later.

3. Mediterranean Fleet to be returned to Malta or as ordered by C.-in-C. Mediterranean. Proposals for Autumn cruise to be forwarded in due course. At least one Capital Ship is to remain at Alexandria while Egyptian Conversations are in progress. C.-in-C. Mediterranean's intentions contained in his 1408 22nd June (to Admiralty, Home Fleet, China and East Indies only)[1] concerning the proposed regatta are noted.

4. 1st Destroyer Flotilla to return to U.K. to give leave and to pay off. The 4th Destroyer Flotilla will return to U.K. to recommission for service in the Home Fleet.

5. *Australia* and *Sydney* to proceed to Australia Station, date of sailing to be reported.

[1] Doc. No. 107.

6. *Achilles* to proceed to New Zealand Station via Panama Canal. Date of sailing to be reported.

7. *Exeter* with Rear Admiral Evans to return to U.K. to refit and recommission.

8. *Ajax* after docking and making good minor defects at Malta to proceed to Bermuda to rejoin America and West Indies command.

9. *Berwick* and Submarines of China Fleet to return to China, except *Proteus* which will first install her new battery. *Adventure* will return to China after carrying out trials with *Narwhal* as ordered by Admiralty letter T.048/36 of 20th February.

10. *Penzance* to proceed to Africa Station as ordered by C.-in-C. East Indies.

11. *Ormonde* to be sailed for the U.K. at the earliest opportunity.

12. C.-in-C. Mediterranean to consider and report whether *Kate Lewis* can be steamed home if all easily removed top hamper such as derricks, mine bogies, etc. are made efficient for a sea voyage.

1605.

Part II.

13. The base organisations specially established for the late emergency at Alexandria, Haifa, Mersa Matruh, Malta and Suez Canal to be demobilised and personnel returned to U.K. with as little delay as possible, together with all personnel dispatched from home for service in 2nd M/S Flotilla, repair ships and similar duties, e.g., *Shamrock*, *St. Day*, *Wolsey*, *Douglas*, etc.

14. Naval A.A. and Coast Defences installed at Haifa and Mersa Matruh are to be returned to pre-emergency storage. Instructions regarding A.A. and Coast Defences at Alexandria will be sent separately.

15. All Booms and Underwater Defences laid at Alexandria and Haifa are to be recovered. It is not desired to store booms or baffles at Alexandria or Haifa. If possible these defences together with Indicator Net in S.S. *Benarty* should be stored at Malta or Gibraltar, if this cannot be arranged they are to be sent back to United Kingdom

S.S. *Nuddea*, *Rio Azul* and *Benarty* are to return to Rosyth to unload, calling as necessary at Malta or Gibraltar.

S.S. *Atreus* is to return to Plymouth to unload Indicator Loop and then to proceed to Portsmouth.

16. Mine depot at Haifa is to be closed down.

17. Instructions regarding disposal of Armament Stores (including mines) in Mediterranean will be sent later.

18. Vessels of East Indies Squadron are to proceed as ordered by C.-in-C. East Indies. 15th Destroyer Division to return to China.

19. C.-in-C. East Indies is requested to signal proposals for reduction of base organisation at Aden and return of personnel to U.K. as soon as possible. Instructions as regards disposal of Armament Stores will be sent later.

20. R.N.R. and R.N.V.R. officers should return to England as convenient locally in time to take leave due to them before the date when their temporary engagements terminate.

1606.

113. *Admiralty to Pound et al.*

[ADM 116/3042] 8 July 1936

SECRET.

[Telegram] Addressed C.-in-C. Mediterranean, [and same distribution as preceding message]. From Admiralty.

IMPORTANT.

Admiralty Telegrams 1605 and 1606/8. Movements should be made as quietly and unobtrusively as practicable.

1630/8.

114. *Admiralty Message Home and Abroad*

[ADM 116/3043] 11 July 1936

[Telegram] From Admiralty.

157A. Their Lordships wish to express Their high appreciation of the services rendered by Officers and Men of H.M. Ships which have been stationed in Mediterranean and Red Sea Waters for the past ten months, a duty which has been arduous and has involved much hardship.

1156/11.

115. *Admiralty Message Home and Abroad*

[ADM 116/3043] 14 July 1936

[Telegram] From Admiralty.

158A. Admiralty message 157A is to be regarded as applying also to Officers and Men of the M.N.B.D.O. and other Naval and Marine personnel employed on shore. The Board also desire to express appreciation to Civil officers and work people of Shore Establishments

and the officers, crews and other staff employed in R.F.A. Supply Ships, Oilers and Chartered Vessels employed in the Mediterranean and Red Sea since August, 1935. Their Lordships highly appreciate the manner in which all concerned have performed their duties, frequently under conditions of difficulty and urgency.
1600/14.

116. *Pound to Chatfield.*

[CHT/4/10] Commander-in-Chief, Mediterranean Station
23 July [1936]

[Holograph]

Things have moved rather fast since I wrote the other day. Our entry into the Grand Harbour was, as far as numbers are concerned, rather pathetic – *Q.E.*, *Barham* and two ships *Curlew* and *Coventry* which don't belong to me.

The autumn cruise looks as if it would devolve into a collection of independent cruises as I shall have to give everyone a turn at Malta before proceeding on the cruise.

I have *Q.E.* and *Galatea* (R.A.(D)'s flagship) yet available for stunts.

I think it is a good thing that V.A.A. is returning to England and *Courageous* as he has v. little to do with *Glorious* and there are a lot of things want cleaning up, particularly the question of those wireless shacks.

I have sent a telegram today saying I would rather keep *Grafton* at Cannes for the whole time. Firth[1] is admirably fitted for the job and on second consideration I came to the conclusion that I could not provide anyone of the same calibre for the second period and I don't want H.M. to say, 'Why on earth couldn't they have left *Grafton* there for the whole period.' I had the impression that a few years ago H.M. was not too keen on the Navy and though I think this feeling has probably changed it is all important that he should have no possible grounds for criticism. *Grafton* was in good order before but I think she will look really beautiful by the time H.M. sees her.[2]

---

[1] Cdr [later Capt] Charles Leslie Firth (1900–71). Flag Lieutenant to RA (D), Med Fleet, 1931–33; commanded destroyers *Grafton*, 1936–38; *Imogene*, 1940; *Jackal*, 1940–41; Dep Dir Signals Dept, Admy, 1941–43; commanded flotilla leader *Troubridge* and Captain (D), 24th DF, 1943–44; Dep Dir Signals Div (V/S and W/T), Admy, 1944–45; CO HMS *Mercury*, Signal School, 1946–47; Flag Capt to C-in-C America and West Indies Station, 1948–49; retired, 1950.

[2] The destroyer *Grafton* served as escort for the yacht *Nahlin* which had been chartered by King Edward VIII, under the pseudonym 'Duke of Lancaster', for a cruise to the Eastern Mediterranean. The *Nahlin*'s cruise became notorious because of the presence of Wallis Simpson who the King would subsequently marry, abdicating the throne in order to do so.

I was amused by the way Charles Forbes hustled his Gordons off – a very good evolution for which I am sure he was responsible.[1] I gather from the papers that the Military Clauses of the Egyptian Treaty have been settled or perhaps some commitments will expire shortly. I think the Governor (Bonham-Carter)[2] will be extremely easy to work with – they are both most friendly.

I am glad that we were on the scene at Barcelona before the Italians.[3] I have not gone to the Western Mediterranean myself as I thought C.S.1 and R.A. Gib. each looking after an area with the Admiralty giving general directions as they must with the general situation to consider was a good arrangement. Should the situation at Tangier deteriorate it would be a different matter – in any international business we should naturally run the show.

I think the evacuation of the Eastern Mediterranean is getting on well and it ought not to be long before we have returned to normal there, except for the 'war' in Palestine.

---

[1] The battlecruiser *Repulse* arrived at Gibraltar from England on 12 July, embarked the Gordon Highlanders for service in Egypt, and left for Malta the same day. The *Repulse* arrived at Malta early on the 15th and the Gordons were immediately transferred to the cruisers *Exeter* and *Shropshire* who had recently completed docking. The Gordons were disembarked at Alexandria on the forenoon of the 17th. 'History of the Italian–Abyssinian Emergency ...', p. 27, ADM 116/3476.

[2] Lt Gen [later Gen] Sir Charles Bonham-Carter (1876–1955). Dir of Staff Duties, WO, 1927–31; commanded 4th Div, 1931–33; Dir-Gen of Territorial Army, 1933–36; Gov and C-in-C, Malta, 1936–40.

[3] A military revolt began 18 July in Spanish Morocco. It quickly spread to the mainland of Spain, touching off violent disorders and what proved to be a long and bloody civil war.

PART III

THE SPANISH CIVIL WAR AND THE NYON AGREEMENTS, 1936–1937

The Mediterranean Fleet was in the process of returning to normal after the Abyssinian crisis when the Spanish Civil War began on 17 July in Melilla on the north-eastern coast of Spanish Morocco as a military rebellion against the Popular Front Government.[1] The insurgents quickly gained control of Spanish Morocco and a number of points in mainland Spain. However, the revolts were crushed in many of the major Spanish cities such as Madrid and Valencia, ending hopes for a quick insurgent victory. In the Spanish Navy many crews mutinied and killed their officers, but the majority of the fleet remained loyal to the Republic. This posed a dilemma for the insurgents. How could they move forces across the waters to help the rebellion on the mainland before it could be crushed? The solution was provided by the Germans and Italians, who transported troops by air to the mainland. The uprising was not crushed and the Republic had sufficient inherent strength to resist the Nationalist forces on land. The struggle evolved into a long stalemate that would last the better part of three years. The Germans and Italians provided aid and 'volunteers' to the Nationalists led by Francisco Franco, while the Republic received substantial aid from the Soviet Union. The most effective foreign assistance was probably active intervention by German and Italian aircraft, wearing Spanish insurgent colours. The insurgents gained control of the port of Cadiz and in February 1937 captured Malaga on the southern coast as well. However, the Government retained control of the eastern or Mediterranean coast of Spain, including the ports of Barcelona, Cartagena, Valencia and Almeria, for the majority of the war. In the North the government also held on to the northern coast on the Bay of Biscay, notably Bilbao, until 1937. The Nationalists succeeded in taking control of Majorca and Ibiza in the Balearic Islands. Minorca remained in Republican hands but an attempt by a largely Catalan force

---

[1] Among the numerous histories of the subject, a valuable work remains the classic by Hugh Thomas, *The Spanish Civil War* (New York, 1961), which has gone through a number of editions. A useful survey is George Esenwein and Adrian Shubert's *Spain at War: The Spanish Civil War in Context, 1931–1939* (London, 1995). The literature on the war and its various aspects is huge and constantly changing, often with sharply divergent interpretations. This is only to be expected given the passions the struggle aroused among contemporaries.

in August 1936 to recapture Majorca for the Government was defeated. The Nationalist position in the Balearics as well as their control of Spanish possessions in North Africa put them in an advantageous position on the flank of the Mediterranean lines of communication to the Republican ports. They were also able to commission two powerful ships, the 10,000 ton 8-inch gun cruisers *Canarias* and *Baleares*. In addition, possibly their most important asset at sea was the clandestine support of Italian warships and submarines. This was particularly true for the better part of 1937.

The question of how much German and Italian support for the Nationalists was due to ideological affinity and how much to European power politics may be debated. The French, despite the ideological sympathy of many on the political Right for the Nationalists, also looked with alarm on Italian intervention in Spain as a move to outflank them and threaten their communications in the Mediterranean. The revolt began when a Popular Front government was in power in France so there was a natural sympathy for the Spanish Republic. Foreign intervention in Spain was a threat to European peace, a factor certainly recognised by the British. The British and French quickly proclaimed their policy of non-intervention and took the lead in associating other powers with this policy. By September 1936, after much diplomatic wrangling, some 27 countries had ostensibly reached an agreement not to send war material to Spain. The so-called non-intervention agreements were not, however, legally binding. Although a Non-Intervention Committee was established in London to monitor the situation, it was in reality powerless with no means to enforce its authority. The Italians and Germans evaded the restrictions and continued to supply the Nationalists, and the Soviet Union replied by continuing material support for the Republicans. The Non-Intervention agreements worked to severely restrict the Republic's source of arms and made Russia, for reasons of its own, the major material support of the Republic with all this implied for strengthening Communist influence in Spain.

From the beginning of the Spanish Civil War and throughout the year 1937, the situation in Spain was the major concern of the Mediterranean Fleet. It was not its only concern or activity. The Arab revolt in Palestine required the presence of ships at Haifa in support of the civil and military authorities. In November 1936 the Turkish Fleet paid its long delayed visit to Malta.[1] During the first half of March 1937 the combined exercises of the Mediterranean and Home Fleets took place in the western Mediterranean, the ships under orders not to approach closer than 12 miles to the Spanish coast or Balearics.[2] In the spring of 1937 the majority of

---

[1] Pound, 'Visit of the Turkish Fleet to Malta', 10 Dec 1936, ADM 116/3536.
[2] Combined Fleet Exercises 1937, ADM 116/3873.

the battleships and cruisers of the Mediterranean Fleet returned to home waters for the 20 May Coronation Review at Spithead where their light grey hulls and upper works contrasted sharply with the dark grey of the Home Fleet. In August Pound conducted Exercise 'S.Z.' to the east of Malta, the exercise simulating conditions which might exist in a war with Italy.[1] But it was the Spanish Civil War that occupied the most attention, particularly for the flotilla craft which rotated in tours of duty involving tedious and occasionally dangerous presence in the war zone off the Spanish coast.

The beginning of the war was accompanied by considerable violence, seen by many observers as a class war. The immediate concern of the British Government was the safety of British subjects caught in the war zone. They included tourists in Madrid and in towns on the Mediterranean coast of Spain as well as businessmen and employees associated with industrial concerns owned by British companies. Warships of the Mediterranean Fleet were ordered to proceed at high speed to the major Spanish ports and the evacuation of British subjects was soon expanded to include subjects of other nations which did not have warships on the scene. The Home Fleet performed similar duties on the northern coast of Spain.[2] Those ships of the Mediterranean Fleet engaged in duties on the Spanish coast had a busy and lively time evacuating refugees [118, 125]. The Admiralty, still mindful of the negative publicity the Navy had received as a result of the Invergordon Mutiny a few years before, sensed the possibility of favourable publicity and asked Pound to compile accounts of the Navy's activities, with an eye to human interest stories [124]. The reports were duly created, collected and forwarded and would fill an entire volume.[3] In the period from mid-July to mid-August, 31 British warships evacuated approximately 6,000 refugees of 55 nationalities, only about one-third of whom were British. The system that developed was for the large County-class cruisers or repair ships *Woolwich* and *Resource* to act as depot ships and clearing stations for refugees who arrived overland or were picked up at various points along the coast by destroyers. They were then ferried to Marseilles by destroyers, or when circumstances demanded, a larger warship.[4]

British warships were authorised to resist any attacks, whether Government or insurgent, outside of Spanish territorial waters which were defined as extending three miles from shore [117]. British ships were

---

[1] Exercise 'S.Z.' – 24–26 Aug 1937, ADM 116/3872.
[2] The role of the Navy is covered in Roskill, *Naval Policy between the Wars*, Vol. II, chap. 12.
[3] Only a few can be included in this volume. They may be found in ADM 116/3051.
[4] 'The Mediterranean Fleet in Spanish Waters', pp. 1–2, Enclosure No. 1 in Battle Cruiser Squadron Letter No. BCS 237/S.17, 24 Oct 1936, ADM 116/3051.

warned to stay out of Spanish territorial waters. However, British merchant ships engaged in normal trade with Spain involving non-military material were not to be interfered with on the high seas by either side. To enforce this, to cite a notable example, the battlecruiser *Repulse* supported the destroyer leader *Codrington* in shielding from a Government cruiser a small steamer of the Gibraltar-based Bland Line trading with Spanish insurgent-held Melilla in North Africa [120]. The British Government did prohibit British merchant ships from carrying war materials to Spain and, while remaining strictly neutral in the conflict, declined to grant belligerent rights to either side. This meant British merchant shipping could not be interfered with on the high seas except to establish identity. Chatfield disagreed with the Government's refusal to grant traditional belligerent rights, especially after the Nationalists appeared to gain the upper hand at sea. In future wars the British might depend on those rights themselves.[1] Similar sentiments were expressed by Admiral James, Deputy Chief of Naval Staff, who would later characterise the government as 'weak and bewildered'.[2] In December 1936 the Admiralty provided Admiral Pound with a lengthy memorandum on the duties of British warships towards British merchant ships during the war [126]. The British Government emphasised its neutrality but it is evident from reports emanating from the Mediterranean Fleet that many naval officers could not forget the circumstances surrounding the beginning of the war and the murder of naval officers.[3] Mutiny and murder of officers were not actions likely to arouse sympathy among others in the same profession, nor was the apparent lack of discipline and unseamanlike condition of Republican warships. The often turbulent behaviour of workers' militias or radical organisations in the ports, the apparent bullying of people with whom at least some officers might have felt a class identification, aroused open distaste. On the other hand, Nationalist atrocities were more likely to take place outside of the Navy's sight. They were not necessarily unknown, as the chilling report of the British Vice Consul in Palma after the failure of the Republican attempt to recapture Majorca demonstrated. He reported 'at least fifteen communists have been shot every night, but this cannot last long as there are not many left to shoot'.[4] Ironically, despite the sentiments of many officers, the great majority of 'incidents' in which

[1] Chatfield, *It Might Happen Again*, p.92.
[2] Adm Sir William James, *The Sky was Always Blue* (London, 1951), p. 187.
[3] Roskill also draws attention to the frequent pro-Franco tone of reports. Roskill, *Naval Policy between the Wars*, Vol. II, p. 374.
[4] Alan Hillgarth, 'Memorandum No. 4 on the Situation in Majorca for the Information of S.N.O. Barcelona, Consul-General Barcelona and the Secretary of State', 25 Aug 1936, copy in FO 371/20537. Hillgarth was a retired naval officer in regular contact with British warships calling at Palma.

British warships or merchant ships became embroiled involved the Nationalists or, in the case of Italian submarines, those acting in their behalf.

The Non-Intervention Committee after months of wrangling finally agreed on a scheme of naval control to take effect at the end of April, 1937. There would be naval patrols in specific zones assigned to the British, French, German and Italian navies and recruitment of 550 'sea observing officers'. Ships bound for Spanish ports were obliged to stop at specific ports or anchorages to take on the observers. They were: Dover, the Downs, Cherbourg, Lisbon, Gibraltar, Sète, Oran, Marseilles, Palermo and Madeira. The multinational observers, often retired naval officers, would ensure that ships in which they were embarked neither carried nor unloaded prohibited material at Spanish ports.[1] Ships carrying observers flew a special non-intervention flag, a white banner with two black balls. On land there would also be observers on the French and Portuguese frontiers with Spain. The scheme was to be administered by an International Board chaired by a Dutch Vice Admiral and one member from Great Britain, France, Germany, Italy and the Soviet Union. Warships of the participating nations would assure that vessels passing through their zone and bound for Spanish ports had submitted to this preliminary control and when employed on this duty would fly the international fisheries patrol flag, a triangle with blue and yellow stripes. However, the scheme applied only to ships belonging to countries who had signed the non-intervention agreements and warships would have no power to actually search a ship, but merely to verify its identity [128, 129]. Apparent infractions would be signalled to the governments concerned. The scheme was denounced as a 'mockery' as Germany and Italy, two of the leading violators in arming the Nationalists, were given a role in enforcing it and German ships carrying arms to Spain reportedly flew the Panamanian flag. Moreover, the Germans and Italians withdrew their ships from the patrol in June after the German armoured cruiser *Deutschland* had been bombed in error by Republican aircraft and a mistaken belief the cruiser *Leipzig* had been attacked by a submarine.[2]

British warships in the war zone ran certain risks even when well outside of Spanish territorial waters. On 13 February the destroyers *Havock* and *Gipsy* were bombed by an aircraft wearing Nationalist

---

[1] A very informative typescript diary by one of the sea observing officers is at the Imperial War Museum, London. Commander V.F. Smyth Mss., 7683 74/89/1.

[2] Thomas, *The Spanish Civil War*, pp. 394–5; Gathorne-Hardy, *Short History of International Affairs*, pp. 439–40; Gerald Howson, *Arms for Spain* (New York, 1998), pp. 230–31; René Sabatier de Lachadenède, *La Marine française et la guerre civile d'Espagne, 1936–1939* (Vincennes, 1993), pp. 25–6, 105–7.

markings [127]. The ships were undamaged, the bombs fell wide of their mark and the attack was almost certainly a mistake made because British destroyers and Spanish Government destroyers were similar in outward appearance. On 6 April the destroyer *Gallant* was attacked twice during the day by Nationalist aircraft, again without effect [130]. Pound realised that the British distinguishing markings might not be visible to aircraft and proposed that if they notified the Nationalist naval authorities at Palma in advance of their movements, there should be no more attacks on them [131, 132]. The Nationalist naval authorities at Palma, with whom the British were usually on good terms, naturally insisted the attacks were all a mistake. Spanish Government destroyers were at sea in the vicinity that day and in the future information on British movements would minimise the possibility of mistakes. There was also a hint that the pilots involved were not Spanish, probably Italian, and difficult for the Spanish authorities to control [133, 134]. The Admiralty approved advanced notification of British movements, at least until the naval observation scheme entered into force later in April. The reason was obvious, after the scheme was in operation advance notice might enable the Spanish authorities to arrange for ships carrying arms to evade observation thereby rendering the scheme even more ineffective than it was already likely to be [135].

Relations with the Nationalists were not always cordial. The destroyer *Arrow* had a decidedly chilly reception at Malaga at the beginning of the system of control, and also had to manoeuvre to stay out of the line of fire when Government destroyers attacked the port [138]. Moreover, new Nationalist artillery emplacements on the mainland near Gibraltar seemed to threaten the use of the Gibraltar dockyard in the event of hostilities. The Admiralty comments on this development were philosophical, the Director of Plans pointing out that ever since the invention of long-range cannon the British position in the Strait of Gibraltar was dependent on a weak or friendly Spain and it was unfortunate they were antagonising the side likely to make Spain strong (i.e. the Nationalists). Chatfield delivered the final word on the importance of a friendly Spain for the whole Mediterranean position with the words: 'When we return to the Foreign Policy of "keeping our friends" we shall be on safer ground. The greatest danger would be for Spain to be allied to another Power' [136].

The Mediterranean Fleet suffered its most serious loss during the Spanish Civil War when the destroyer *Hunter* was mined while on patrol off the Republican port Almeria on 13 May 1937. Immediate assistance was given by Spanish Government vessels and the ship was towed into the harbour [139–41]. Eight men of the ship's company were killed and fourteen wounded. At first it was not clear what had caused the explosion, subsequent investigation concluded it was probably a mine [142, 143,

147, 149, 153]. The Nationalists denied responsibility although it was eventually revealed that the Nationalists had laid a minefield off the port. The damaged ship was brought to Gibraltar for temporary repair and the Admiralty eventually prepared a bill for £127,054 as compensation for the damages. It was a futile exercise; the Spanish Nationalists never paid.[1]

Following the *Hunter* incident, Admiral Pipon, Vice Admiral Gibraltar, suggested that non-intervention patrol duties might be carried out by less valuable ships than destroyers [152]. Pound supported the idea, at least for the southern Spanish ports, and suggested ships from the minesweeping flotilla. He still considered destroyers more suitable for the north-eastern ports and Balearic Islands because of their speed and the 'prestige' factor. He also wanted two destroyers at Gibraltar to deal with incidents at sea, such as the interference with British merchant ships. The release of some destroyers from the non-intervention patrols would also have the advantage of allowing them to pursue training in flotilla work with the fleet [155]. The Captain commanding the Third Destroyer Flotilla at the end of the flotilla's commission pointed out the disadvantages of rarely having the whole flotilla together, especially as he believed a massed torpedo attack might still take place in a future naval battle [137]. The Admiralty was willing to temporarily provide eight trawlers to replace destroyers in non-intervention work provided the additional ratings required came from the Mediterranean Fleet. Pound was unable to meet this condition [156, 157]. Pound remained concerned over the wear and tear of machinery and personnel in the destroyers as a result of long periods on the Spanish coast and in June again suggested that at least temporarily the 1st Minesweeping Flotilla should be used on the southern coast. This time the Admiralty agreed for the minesweepers to be so employed for a period of 30 days [160, 161].

One of the paradoxes of the Spanish Civil War is that, despite the sometimes ill-concealed support of the Nationalists by the Italians and Germans, the Royal Navy's relations with the German and Italian navies participating in the non-intervention patrols remained generally cordial, an example of professionals working together. The Italians, who had been on the verge of war with Britain the preceding year, seemed anxious to be back in the good graces of the British [164]. This certainly could not be said of French–Italian relations which were anything but friendly. Anglo–German relations through 1937 were also cordial [159], especially as this was before the crises of the following year heightened tension in Europe. After the German armoured ship *Deutschland* had been bombed in May 1937 by Republican aircraft, the authorities at Gibraltar arranged

---

[1] Roskill, *Naval Policy between the Wars*, Vol. II, p. 382.

for the wounded to be treated in hospital and organised a funeral with full military honours for the German dead, although Hitler later ordered the bodies exhumed and brought back to Germany for an even bigger ceremonial re-interment.[1] The destroyer *Hereward* witnessed the heavy handed German retaliation, the bombardment of Almeria by the 28cm guns of the *Admiral Scheer* [158].

The alleged submarine attack on the *Leipzig* led Pound to consider the possibility of a similar 'foul attack' on a British warship. He accordingly issued instructions for the contingency including the authorisation for destroyers and minesweepers to hunt and destroy the submarine if possible [162]. In August of 1937 attacks on merchant ships in the Mediterranean by unknown submarines became common. Pound prepared for the eventuality the attacks would include British ships. In case a British merchant ship was attacked by a submarine without warning British warships were authorised to counter-attack [165]. The so-called 'pirate submarines' were actually Italian, a fact widely suspected by the press and public and known for certain by the Royal Navy. The Admiralty, having penetrated the Italian naval cypher, had warned Pound a submarine campaign was imminent.[2] The attacks were directed at merchant ships carrying supplies to the Spanish Republic and took place not only in the Western Mediterranean but also in the Sicilian Narrows and Aegean.[3] In the Aegean they were directed at ships coming out of the Dardanelles from Russian ports. Italian surface ships also took part in the attacks, perhaps an unwise action if secrecy was to be maintained. When the Spanish tanker *Campeador* was torpedoed during the night of 12 August off Cape Bon by the Italian destroyer *Saetta*, the glow of the flames enabled the Italian destroyer which had shadowed the tanker during daylight hours to be identified and named in a Spanish protest to the League of Nations.[4]

The question of what exactly constituted a 'British' ship was also contentious. As the British enforced the rights of their ships to trade and

---

[1] The incident is treated in detail with numerous photos in Markus Titsch, 'Panzerschiff *Deutschland*: The Attack off Ibiza Island', *Warship International*, vol. 45, no. 3 (2008), pp. 229–48.

[2] Roskill, *Naval Policy between the Wars*, vol. II, p. 383. See also Christopher Andrew, *Her Majesty's Secret Service: The Making of the British Intelligence Community* (New York, 1986), pp. 401–2.

[3] The Italian campaign, once very secret and a sensitive topic, is covered in depth in a publication of the historical office of the Italian Navy: see Franco Bargoni, *L'impegno navale italiano durante la guerra civile spagnola (1936–1939)* (Rome, 1992). There was actually an earlier Italian submarine campaign in late 1936, notable for the torpedoing of the Republican cruiser *Miguel de Cervantes* by the Italian submarine *Torricelli* off Cartagena on 22 November. Operations against merchant ships, however, were carried on under restrictive rules and without success, although the Italians were nevertheless embarrassed by the discovery of an unexploded torpedo with Italian markings. Ibid., pp. 138–44.

[4] Ibid., pp. 297, 303, 306.

had a strong navy to back up their claims, there was a movement by shipowners to transfer their ships to the British flag, creating a situation where there was little that was 'British' about some ships except the flag [164]. The case of the destroyer *Hostile*'s attempt to go to the assistance of the British-flagged *Noemijulia* which had been bombed by Nationalist aircraft and the almost comic difficulties in communication is an example [170]. The misuse of the British flag could easily lead to charges that the British were conniving with foreign shipowners to bring contraband to Spain and Pound recommended the practice of transferring registries of ships for the purpose of trading on the Spanish coast be stopped [171]. The Admiralty replied that instructions had been issued by the Board of Trade to British Consuls in Europe and the Mediterranean to refer applications for provisional certificates of registration to London [172].

British merchant ships were also attacked outside of Spanish territorial waters by Nationalist aircraft operating from the Balearics. The planes had Nationalist markings but could be flown by Spanish, Italian or German pilots. Rear Admiral Somerville, then Rear Admiral (D), was ready to take a hard line on the principle that according to International Law merchant vessels could not be attacked without their identity being established and the safety of passengers and crew assured. Somerville wanted to warn the Nationalists that if the attacks did not cease he would be compelled to take offensive action against Nationalist aircraft operating on the trade routes followed by British ships [166]. The proposal appears to have alarmed Pound and he pointed out problems with its practical application for it implied either a declaration of war against the Nationalists or employing resources the British did not possess [167, 169]. Nevertheless the Admiralty strongly supported the principle that aircraft had to follow the same rules of International Law in regard to merchant vessels as warships [168]. Admiral Bastarreche, the Nationalist naval commander in the Balearics, proposed that the British investigate the bona fides of any ship designated by the Nationalists that was approaching the coast of Spain and suspected of illegally flying the British flag. Pound, however, wondered if this was practicable with the forces at their disposal, given the potentially large number of ships and wide area involved [173].

The Mediterranean Fleet nearly suffered another loss on the night of 31 August when an unknown submarine attacked and narrowly missed with a torpedo the destroyer *Havock* off Cape San Antonio. *Havock* counter-attacked with depth charges but eventually lost contact. Other British destroyers in the area joined in the search along with Somerville in the light cruiser *Galatea*. The cruiser was equipped with a float plane that could be employed during daylight hours. The British may have regained contact with a submarine, although it is not certain the contact was actually a submarine or, if it was a submarine, the same one which had attacked the *Havock*. The Admiralty to the great disappointment of those involved halted the search, worried that an 'innocent' submarine might be attacked [174–7]. The submarine in question was the Italian *Iride* under the command of the then Lieutenant Junio Valerio Borghese who would later achieve fame during the war for his leadership in the use of the special underwater weapons such as torpedoes guided by frogmen. The amount of damage, if any, to the submarine or even how close to destruction it was is not clear, but given the international uproar over the incident *Iride*'s escape undetected was a great relief to the Italians.[1] Borghese had mistaken *Havock* for a Republican destroyer. When the British Government, ordered the 11th Destroyer Division (4 ships) of the Home Fleet to reinforce the Mediterranean, Somerville recommended any destroyer operating in the area be repainted the light grey of the Mediterranean Fleet to avoid the similarity between the darker colours of the Home Fleet and the Republican warships [179].

On 2 September the British-flagged and newly renamed tanker *Woodford* was torpedoed and sunk off Cape Oropesa by a submarine operating on the surface. The submarine was actually the Italian *Diaspro*. No one aboard the *Woodford* except the Observing Officer was British. Nearly all of the crew were Greek.[2] Pound weighed the alternatives for protecting British ships and reluctantly concluded convoy would be the only way of protecting them although a 'poor substitute for clearing seas of piratical submarines' [181]. Somerville agreed, but considering the 'dubious character' of many ships in the Spanish trade and its 'highly remunerative nature' did not think the convoys need be too frequent [182]. The jaundiced eye with which many officers viewed the ships trading with Spain is evident in many of the contemporary reports.

The crisis created by the submarine attacks led to the convening of an international conference in Switzerland, complicated by the refusal of the Italians to attend following Russian accusations of their complicity. The Admiralty had initially hoped for an agreement whereby no power would have submarines at sea in the Mediterranean except in certain specially designated submarine exercise areas. Consequently, submarines encountered outside of these areas could be sunk on sight. In the absence of the Italians the British had to settle for the next best solution. The intention was to create hunting groups of warships stationed in zones assigned to the naval powers. These hunting groups would be authorised

---

[1] Ibid., pp. 309–10. On 12 August another Italian submarine, the *Jalea*, had torpedoed the Republican destroyer *Churruca* off Valencia, putting the ship out of action for the remainder of the war.

[2] Paul Heaton, *Spanish Civil War Blockade Runners* (Abergavenny, 2006), p. 73.

to pursue submarines that had attacked merchant ships of any nationality [183, 184, 188]. Pound was unenthusiastic about this proposal, which he dubbed Scheme 'Z', because it would not allow every submarine sighted to be attacked whatever it was doing. Consequently, it was too dependent on luck. Nevertheless, he outlined methods in which it might be put into effect and also, in response to an Admiralty request, estimated the forces necessary for the introduction of convoys [186, 187]. Pound also regarded a demonstration of efficiency in hunting submarines as a potential deterrent to war and therefore recommended a maximum effort during the initial period of any scheme. This would involve the whole of the available Mediterranean destroyer force operating on the coast of Spain with half at sea at a time [189].

The participating powers reached an agreement at Nyon on 11 September to take effect on the 20th. Despite this it seemed at first that British warships exercising off Alexandria had been attacked by a submarine. The 'attack' was subsequently attributed to natural phenomena [216]. Under the Nyon agreement the bulk of the obligations would fall on the French and British navies. Greek and Turkish participation in the Aegean was apt to be minimal although both had agreed to make their ports and anchorages available. The Mediterranean was divided into zones with an area reserved for Italy whose eventual participation it was hoped to obtain. Chatfield naturally kept Pound informed of the major decisions and final agreement [191–3]. This initial agreement was followed by a supplementary agreement on 17 September covering attacks by surface ships and aircraft [215]. Pound obviously intended to concentrate the majority of the fleet in the western Mediterranean but this left the problem of the Aegean. Ships from Russian ports and bound for Spain had been attacked after exiting from the Dardanelles. Pound's solution was to use capital ships, initially the battleship *Malaya* and battlecruiser *Repulse*, attended by a small number of destroyers. The big ships were chosen because they could each carry four float planes, invaluable for patrolling a wide area [194]. The initial thought of using Greek and Turkish destroyers outside of their territorial waters was abandoned, apparently at the desire of those powers. The British did intend to take the opportunity to ask for the use of Greek and Turkish harbours, notably Skyros and Lemnos Islands from the former and Eritra Bay from the latter [200, 203–5]. The Nyon agreements were supplemented by an Anglo–French agreement on the methods of enforcement and Pound was expected to meet with the French admiral to settle the details [195, 196, 198].

Chatfield hoped that after an initial period of maximum effort the patrols could be reduced if there was no pirate submarine activity thereby easing the strain on the destroyers [199]. He and Pound had good reason to believe this might be the case for cryptographic intelligence had revealed that in the Aegean the Italian submarines had been ordered to make no attacks [202]. The Italian government had been alarmed over the international repercussions of the attack on the *Havock* and had called off the submarine campaign. British intelligence knew most of this and as a result the Nyon arrangements had therefore largely succeeded before they went into effect.[1]

Pound proceeded to Oran in his flagship, the battleship *Barham*, to meet with Admiral Esteva, his French counterpart, on 17 September. The meeting went very well and Anglo–French relations in September 1937 were much warmer than the chilly reception given to Mountbatten during the Abyssinian crisis two years before. The French were willing to adopt the British proposals and allowed an RAF flying boat squadron to be based at Arzeu in Algeria [206, 207]. The Mediterranean Fleet provided the repair ship *Cyclops* to serve as an accommodation ship for the RAF personnel [197, 201]. Pound, in a private letter to the First Sea Lord, described flying boats as a 'necessary part of a Fleet' and thought there should be a flying boat squadron with depot ship attached to the fleet permanently [214]. He also appeared sceptical about the extent the Italians had called off their submarine activity, despite the intelligence he had received, and worried about attacks in the unpatrolled area west of Cape Matapan [208, 210]. He also assured the Nationalist naval commander in the Balearics that British aircraft on the Nyon patrols would not fly closer than ten miles to Spanish territory and the Spanish admiral promised to warn him if he would send Nationalist submarines south of a given line [212]. The impending use of aircraft in conjunction with the Nyon patrols also resulted in what the British naval attaché in Berlin termed a 'friendly communication' from the German Admiralty warning aircraft against flying over German warships as the latter had orders to fire on approaching aircraft [213]. The German attitude after the bombing of the *Deutschland* was understandable.

The first day that the Nyon agreements went into force came after a week during which there had been no submarine activity. Pound seemed to be holding his breath and wondering what the Italian reaction would be. He had, however, changed his mind about Scheme 'Z' and was glad they had put it into effect instead of the convoy system he had originally recommended. This was because it would be far easier to ease up on the present scheme than on a convoy system once it had started [214]. Pound need not have worried about the Italians, they reversed their position and joined the Nyon arrangements, although some delicate negotiations were

---

[1] Andrew, *Her Majesty's Secret Service*, p. 402.

involved before agreement was reached on 30 September in Paris. The Italians insisted on equality in the size and division of zones and there was some complex and intricate redrawing of lines on the map. Nevertheless, the new agreement had the advantage of eliminating the theoretically unpatrolled areas of the first agreement [217, 218]. The new agreement also necessitated a meeting at Oran on 30 October between Pound, Esteva and Admiral Bernotti, the admiral designated in command of Italian patrols. Despite potential prickliness between the French and Italians, Pound described the meeting as 'very satisfactory and pleasant' [237]. He was, however, privately rather cynical and wrote Chatfield that it looked 'as if the Italians were more concerned in getting areas which would enable them to report shipping than to hunt submarines which they know well will probably not be there' [245]. The phrase 'poachers turned gamekeepers' has often been applied to Italian participation in the Nyon agreements and it seemed the height of hypocrisy for them to join in policing a situation their own actions had caused. Nevertheless, the 'pirate' submarine attacks ended, at least for a while.

Merchant ships might have been spared but on 4 October it seemed as if submarine attacks on British warships had resumed. The destroyer *Basilisk* on patrol in company with the destroyer *Boreas* reported an apparent attack by a submerged submarine. *Basilisk* counter-attacked with depth charges, without any result. *Boreas* never confirmed the contact. The incident received widespread coverage in the press and this disturbed Chatfield for, as in the case of the *Havock*, a submarine had attacked a British destroyer and apparently escaped. This would give ammunition to those who doubted the Admiralty's confidence in its anti-submarine detection apparatus, notably asdic. To avoid a repetition of the *Havock* incident in which the submarine had also gotten away, Chatfield favoured grouping destroyers so that a submarine detected and attacked would have no chance of escape [219]. The 'confidence' factor was all important, for this was really the first test of the new apparatus and tactics under war conditions.[1] Pound doubted that there had actually been a submarine attack, but admitted that they were 'in a cleft stick' for if a submarine had attacked *Basilisk* and been depth-charged it ought to have been destroyed and if there had been no submarine they could not say so without acknowledging the unreliability of asdics [220]. Pound favoured handling the problem of prestige by what he called the 'indirect' method, that is reducing the hunting units from three to two destroyers so as to keep as many units as possible on patrol in the British zone, thereby making it so

---

[1] Willem Hackmann, *Seek and Strike: Sonar, Anti-submarine Warfare and the Royal Navy, 1914–54* (London, 1984), pp. 133–4.

hot for submarines that they would prefer to transfer their activity to the non-British zone. The investigation of the *Basilisk* incident confirmed Pound's suspicions by concluding there had been no submarine attack.[1] The incident, however, demonstrated the need for more flotilla training in anti-submarine work and to achieve this Pound intended to keep the numbers of destroyers on patrol low enough to allow systematic anti-submarine training [221–4]. The Admiralty approved this reduction in the strain on the flotillas provided submarine activity did not reoccur and Pound arranged for one division to be on patrol at a time [229, 230].

Once the Italians had halted their secret submarine campaign the major danger to British shipping came from the activities of Nationalist surface ships and especially Nationalist aircraft, or rather aircraft acting under Nationalist markings whoever the pilots were. The British protested the bombing of the British steamer *Cervantes* when 12 miles off the coast on 8 October. This led Pound to consider reinforcing the western basin of the Mediterranean by recalling the battlecruiser *Repulse* from the Aegean to Gibraltar 'to show that we mean business' [225]. The Admiralty, however, did not consider such reinforcement necessary at the time [231].

Incidents continued to occur. On 30 October the British-flagged steamer *Jan Weems* carrying a cargo of food stuffs and two Non-Intervention Control Officers was sunk by Nationalist aircraft leading to another strong British protest [238, 239]. Nevertheless, despite the problems with Nationalist aircraft the apparent absence of submarine activity allowed Pound and the Admiralty to plan further reductions in destroyers on patrol in the Mediterranean and the return of Home Fleet destroyers that had been detached to the Mediterranean [240–43]. The flying boat squadrons at Arzeu were also withdrawn in December, although one squadron was held in readiness for a quick return to the Mediterranean and equipment and spares retained at Gibraltar for this purpose [251].

In the major Government-held ports there were so-called 'safety anchorages' where foreign warships on the scene to protect their nationals would not be attacked. The Nationalists claimed that merchant ships were taking advantage of these zones and as a result their immunity would no longer be respected. This threat was joined with an apparent attempt to declare a blockade of the remainder of the Government-held coast. The British Government warned once again that they did not grant belligerent

---

[1] In 1968 Borghese, in a book published in Spain, claimed without many details that he had attacked the *Basilisk* while in command of the *Iride*, then operating under the name *Gonzales Lopez* (but with Italian crew) and based on Soller, Mallorca. He also claimed the subsequent depth-charge attack left two dead and four wounded in his crew. The Italian Navy has denied this; there are apparently no supporting documents in the Ufficio Storico and Borgese's motives in making the claim are uncertain. Giorgio Giorgerini, *Uomini sul fondo: Storia del sommergibilismo italiano dalle origini a oggi*, 2nd edn (Milan, 2002), pp. 200–202.

rights to either side and the Nationalists accordingly had no right to declare a blockade. British ships would continue to be protected by British warships [247–50, 252]. The potential for incidents involving the Navy remained.

In the Aegean the *Repulse* ran into Greek objections over landing armed men and the laying of nets and obstructions in the anchorages on the island of Skyros. The Greeks were afraid that other powers – meaning Italy – would request the same privileges. Obtaining Turkish permission to use anchorages and bays also proved a delicate diplomatic task and with the end of 'piratical submarine activity' there is a distinct impression that a large and powerful ship like *Repulse* was wasted on duties such as the interception of British merchant ships suspected of carrying contraband war material to Spain from Russia [226, 244]. The Aegean patrols had less and less to do but on 23 October the destroyer *Ilex* made contact with a submerged object and the senior officer of the flotilla ordered depth charges to be dropped. No submarine was present but the use of depth charges alarmed both Pound and the Admiralty for even if a Spanish submarine had been resting on the bottom she would have been acting legitimately and a major international incident might have been created [233–6]. Before the end of the year Pound withdrew all the large ships from the Aegean. There was really nothing for them to do and the winter weather greatly hampered use of their float planes, a major reason for sending them there [253].

The Spanish Civil War may have occupied a good deal of the Mediterranean Fleet's attention during 1937 but there were other questions as well. The defence of Egypt's western frontier from a potential Italian attack from Libya was certainly one. In September 1937 Italian forces in Libya had been reinforced by two army corps and the local British commanders in Egypt became alarmed. The Royal Navy would have had the responsibility of escorting convoys with men and material through the Mediterranean to reinforce Egypt in the event of an emergency. Pound and the local commanders in Egypt were ready to implement preliminary measures including those involving ordering the aircraft carrier *Glorious* to Alexandria [227]. The Chiefs of Staff in London were not, however, alarmed at the apparent Italian build-up which they had foreseen, and Pound and the military and air commanders in Egypt were informed the proposed measures were not necessary [228, 232]. At that time Prime Minister Neville Chamberlain was intent on reaching an understanding of some sort with Mussolini and the Italians.[1]

---

[1] The subject of the Italian build-up in Libya in the autumn of 1937 and the response of the British government is examined in Morewood, *The British Defence of Egypt*, pp. 103–5.

In late November of 1937 Chatfield made it clear to Pound that the threat from a resurgent Germany was the primary concern of both the Chiefs of Staff Committee and the Committee of Imperial Defence, and until the Royal Air Force and the Air Defence of Great Britain had been sufficiently strengthened the Mediterranean would take second place. This explained the obvious Government attempt to placate Italy for the Mediterranean position was only one of three anxieties the British had; Germany and Japan were the others. The British Government was still 'hopelessly weak' to meet the responsibilities of all three armed services and since they could not come to terms with either of their two chief opponents it was better to reach agreement with Italy with whom the British had no real enmity [246]. Chatfield's letter made Pound realise that the German menace must be much greater than he had been aware in the Mediterranean but, somewhat unrealistically, he argued that priority to everything anti-German should only be taken into account where actual 'fighting equipment' was concerned and not to questions of finance such as the excavation of submarine shelters at Malta which would not retard anti-German measures [254].

Pound had also been designated to lead a reinforced Mediterranean Fleet to the Far East in the event relations with Japan continued to deteriorate. However, as Chatfield recognised at the end of 1937, any British margin of superiority over the Japanese Navy would have been questionable, although he apparently had some hope of potential cooperation with the US Navy [255].

The year 1938 would see a temporary repeat of submarine attacks on shipping involving British ships and the Spanish Civil War dragged on into the spring of 1939 before the Nationalist forces under Franco achieved their final victory. However, 1938 was the year of major European crises with the threat of a general war and Spain would become more and more of a sideshow.

117. *Pound to British Naval Forces in the Western Mediterranean*

[ADM 116/3534] 23 July 1936

SECRET.

[Telegram]. Addressed R.A.C. First Cruiser Squadron, V.A. Gibraltar, H.M.S. *Devonshire*, Captain (D) 4 [4th Destroyer Flotilla], Captain (D) 20 [20th Destroyer Flotilla], H.M.S. *Repulse*, repeated R.A. (D), Admiralty.

<u>IMPORTANT.</u>

728. Any definite hostile attacks by aircraft on a British Ship whether by bomb or other weapon are to be resisted by force without previous warning should this take place outside Spanish territorial waters. Similar acts by insurgent men-of-war outside Spanish territorial waters should be resisted if persisted in after remonstrance. Any such action is to be reported immediately. Spanish territorial waters are to be taken as extending three miles from shore. Attention is drawn to K.R. and A.I. 943 to 957. British ships are to be warned to keep outside Spanish territorial waters and British men of war are to avoid as far as possible being berthed or anchoring near Spanish men of war.

2254./22.

118. *Captain J.H. Godfrey[1] to Rear Admiral commanding 1st Cruiser Squadron[2]*

[ADM 116/3051] H.M.S. *Repulse*, at Marseilles
31 July 1936

[Carbon]

With reference to King's Regulations and Admiralty Instructions, article 1132, I have the honour to forward the following letter of proceedings of His Majesty's ship *Repulse* under my command for the period Tuesday, 21st July, to Friday, 31st July, 1936.

2. In accordance with the Vice Admiral Commanding First Battle Squadron's message timed 0059 of Tuesday, 21st July, the 2nd Battalion of the Gordon Highlanders were re-embarked p.m., that day, and *Repulse*

---

[1] Capt [later Adm] John Henry Godfrey (1885–1971). Dep Dir, RN Staff College, Greenwich, 1929–31; commanded cruisers *Kent*, 1931; *Suffolk*, 1931–33; Dep DP, Admy, 1933–35; commanded battlecruiser *Repulse*, 1936–39; DNI, Admy, 1939–42; Flag Officer commanding Royal Indian Navy, 1943–46.
[2] RA Max K. Horton.

weighed and left Alexandria at 2145. When clear of the Great Pass speed was set at 22 knots and course shaped for Gibraltar so as to arrive at 0700 on Saturday, 25th July.

3. The voyage was without incident except for the sighting of several Spanish warships during the early hours of Saturday, 25th July, the witnessing from a distance (between 0600 and 0700) the bombing of Algeciras by three Spanish Government flying boats, the engagement between three Spanish men of war, including *Libertad* and *Jaime I*, and the forts at Ceuta.

4. On arrival at 0715 on Saturday, 25th July, the ship secured alongside the South mole at Gibraltar, and the Gordon Highlanders were disembarked during the course of the forenoon.

5. *Repulse* had meantime been placed under the orders of the Rear Admiral Commanding First Cruiser Squadron, by the Commander-in-Chief, Mediterranean's message timed 2149 of 24th July. In accordance with the former's message timed 1137 and the Rear Admiral, Gibraltar's message timed 2318, both of 25th July, *Repulse* sailed at 1000 on Sunday, 26th July, for Palma, Mallorca, conveying £2150 and certain stores and provisions for *Devonshire*.

6. <u>Monday, 27th July.</u>

At 1205 *Repulse* arrived and anchored off Palma and at 1430 the Captain of the *Devonshire* came onboard accompanied by the Acting Vice Consul, Mr. Sayward.

At 1520 the Italian destroyer *Maestrale* arrived. Calls were exchanged and the situation in Spain explained to the Italian Captain who appeared to have very little idea of the state of affairs and had received no orders from his Government or superior authority.

At 1530 *Repulse* took over the shore signal station which had been established at the British Association club.

At 1700 *Devonshire* sailed for Valencia.

At 1745 one flying boat from Minorca flew over the town and dropped pamphlets. The machine gun fire and occasional shots of larger calibre directed against the aircraft, which was at a height of at least 5000 feet, were quite ineffective and the few shell bursts that were observed appeared not more than 500 feet high.

7. <u>Tuesday, 28th July.</u>

S.S. *Derbyshire* arrived at 0650 and was boarded by an Officer of the guard. She sailed at 1020 with about sixty British subjects onboard.

During the forenoon the Italian Ward Room officers called on *Repulse*.

At 1145 two flying boats flew over the town and dropped twelve bombs resulting in four deaths and a certain amount of panic. The Vice

Consul[1] was onboard at the time and on going ashore about noon he found the Consulate besieged by British subjects demanding protection. He returned onboard, and after a detailed discussion of the situation my signal timed 1606 ..., was despatched.

At 1800 six refugees, Captain and Mrs. Dear and four children, near whose home a shrapnel shell had burst during the forenoon, embarked.

At 1730 one aircraft flew near the town but no bombs were dropped.

During the afternoon, as reported in my 2021, the American Consul arrived by air, and was brought onboard by the British Vice Consul. He informed me that arrangements were being made for an American ship to call at Palma on Friday, 31st July, to evacuate the entire American colony.

8. <u>Wednesday, 29th July.</u>

During the night the Rear Admiral Commanding First Cruiser Squadron's message timed 2106/28 was received giving instructions to evacuate to Marseilles as many as could be induced to go of the British and foreign communities in Mallorca.

I accordingly landed at 0830 and met the French, German and Columbian Consuls at the British Consulate, and arrangements were made to embark refugees of all nationalities the next day. It was estimated that about three hundred persons altogether would be willing to leave, but as it turned out, due partly to the short aerial bombardment which takes place daily at about 1100 and 1600, a much larger number availed themselves of the offer.

The French Consul undertook to get in touch with the Swiss, Dutch, Norwegians and Swedes, to inform them of the arrangements made, and to confirm that their respective nationals should be evacuated in *Repulse* if they so desired. The Italian Senior Naval Officer was informed of the action being taken, and said that he had not yet received any definite instructions with regard to the Italian community.

Later I met the American Consul who asked me to evacuate thirty American subjects as accommodation in the ship ordered in for Saturday was insufficient.

9. <u>Thursday, 30th July.</u>

The evacuation of the foreign communities was commenced at 0900 and completed at 1700, a total of 503 men, women and children of all nationalities being embarked.

---

[1] Lt Cdr [later acting Capt] Alan Hugh Hillgarth (1899–1978). Seriously wounded as a Midshipman at the Dardanelles, 1915; retired Dec 1927; became a novelist and travelled extensively including a period prospecting for gold in Bolivia; British Vice Consul, Palma Majorca, 1932–37; Consul, 1937–39; Asst Naval Attaché, Madrid, 1939–40; Naval Attaché, 1940–43; Chf of Intelligence Staff, Eastern Fleet, 1943–44; Chf of British Naval Intelligence, Eastern Theatre, 1944–46; unofficial informant to Churchill on intelligence and defence subjects, 1948–51.

* * *

U.S.S. *Quincy* arrived at 0800 and calls were exchanged. Captain William F. Amsden was most anxious to help in any way and I was obliged to him for the use of the *Quincy*'s boats. As the *Quincy* was returning to the Spanish coast we agreed not to modify the arrangements by which the American subjects were evacuated partly by *Repulse* and partly by cargo vessel.

At 1730 H.M.S. *Repulse* weighed and proceeded, course as requisite for Marseilles. Speed was set at twenty-four knots in order to get as much time in hand as possible in view of the mistral reported blowing in the gulf of Lyons.

As reported in my signal timed 1735/30 two air raids were carried out during the course of the day. The second one occurred most opportunely, just as *Repulse* was getting under way, inasmuch as there had been a tendency on the part of some of the refugees to take the attitude that the Consular authorities were being unnecessarily nervous in ordering the evacuation. Most of the refugees who witnessed the bombing from the safety of *Repulse*, a mile from shore must have realised the wisdom of getting away while the going was good.

10. <u>Friday, 31st July.</u>

By midnight the wind had freshened sufficiently to necessitate a reduction of speed to twenty, then sixteen and by 0030 to twelve knots, on account of the amount of water coming aboard. At 0115 the ship was hove to while the quarterdeck awning was furled. Passengers were cleared off the upper deck and hatchways were battened down. Then the ship proceeded at the highest speed possible, twelve knots. At this speed the catapult deck, quarter deck and forecastle were continuously awash; wind blowing north west force 8 and sea force 6.

Repulse anchored in L'Estaque roads at 1350 G.M.T. and a boat was sent in to bring off the Consul General, foreign Consuls and French Police and passport authorities.

A salute of twenty-one guns was fired on anchoring which was returned by the Pharo battery.

The tender *L'Obstine*, which had been arranged by the Consul General, came alongside at 1500 and the disembarkation commenced. Two trips by the tender were necessary, the second one leaving the ship at 1850.

After embarking general mess stores and mails *Repulse* weighed at 2200 and shaped course as requisite for Valencia, speed being set so as to arrive at 1600 the following afternoon.

119. *Pound to Chatfield*

[CHT/4/10]   Commander-in-Chief, Mediterranean Station
Saturday [8 August 1936]

[Holograph]

* * *

I was very glad that I went up to Barcelona and got in touch with the people at Cartagena, Alicante and Valencia on the way as I was able to get a good idea of what was really required. I had the French and Italian Admirals to dinner, both very pleasant and forthcoming and working closely with C.S. 1. The latter is doing very well but I hope my proposal to relieve him and *London* on 21st by Binney and *Shropshire* will be approved as the *London* will have been quite long enough secured to the wall at Barcelona (which is essential).

I am not sure that the Admiralty order that our ships are to fly a white ensign forward is really sound as whatever special mark a ship shows there are occasions when it is not visible and you put the other side in the position of saying, 'We saw no special mark as you say you carry so thought it couldn't be a British ship.' If I find it disadvantageous I will represent the fact.

Both the Italians and Germans are very unpopular on the East Coast of Spain, being fascist countries and if as is reported in the Press this morning the Germans land a force in Barcelona the balloon is pretty certain to go up.

* * *

Galatea made a very fair passage from Malta to Gib. – 34 hrs. steaming which meant an average speed of just under 29 knots with none too clean a bottom. Practically no vibration. I hope things will clear up so that I can meet the First Lord at Malta when he first goes there – if I cannot he will I am sure understand and I would then either see him at Malta on his return from Cyprus or at Gib.

* * *

Officers and men have been splendid over the evacuation of refugees and are thoroughly enjoying having a definite job to do. Some of the destroyer officers have not slept in their cabins for a week on end but the present arrangements will avoid the necessity of this except very occasionally.

I am afraid things do not look too good for your getting your much needed leave.

120. *Godfrey to Rear Admiral commanding 1st Battle Squadron*

[ADM 116/3051] H.M.S. *Repulse*, At Malta
29 August 1936

[Carbon]

No.191/2158.

With reference to King's Regulations and Admiralty Instructions, Article 1132, I have the honour to forward the following letter of proceedings of His Majesty's Ship *Repulse* under my command for the period Thursday, 20th August, to Saturday, 29th August, 1936.

Thursday, 20th August.

2. Having turned over the duties of Senior Naval Officer, Valencia to H.M.S. *Resource* who arrived and anchored in the bay at 0900, *Repulse* sailed for Gibraltar at 1200.

* * *

Saturday, 22nd August.

8. *Repulse* weighed at 0630 and carried out a patrol off the Straits until 1030 in accordance with Med.097A/1 of 10th August, 1936, Patrol of the Strait of Gibraltar. There were no incidents to report except the sighting of three French destroyers of *La Fortune* class, westward bound.

9. The ship entered harbour at 1038 and secured alongside the South Mole.

Sunday, 23rd August.

10. At 1259 an 'Immediate' signal was received from the Commander-in-Chief, Mediterranean Station …, ordering *Repulse* to proceed to Melilla at full speed to act in support of *Codrington* and *Wolsey* who had previously been ordered to sea by the Rear Admiral, Gibraltar.

11. This action followed the receipt of a report that a British steamer, S.S. *Gibel Zerjon* (Bland Line) had been stopped and boarded off Cape Tres Forcas by the Spanish Government cruiser *Miguel Cervantes*, and ordered to return to Gibraltar.

12. On receipt of the Commander-in-Chief, Mediterranean's 1259, *Repulse* was at four hours' notice for steam, and 175 libertymen were ashore. The co-operation of the 2nd Battalion Gordon Highlanders was enlisted in getting the libertymen back to the ship. Officers in private cars, a regimental lorry and despatch riders at short notice set off round the town passing the word and rounding up *Repulse*'s men, and it was largely due to their efforts that only three libertymen failed to return to *Repulse* before sailing. At 1500 the ship cast off, and having left the harbour by

the southern entrance, went ahead at 1530 and by 1600 a speed of 28 knots was obtained.

13. On receipt of Rear Admiral, Gibraltar's 1536 …, while continuing in the general direction of Melilla, course was altered at 1625 so as to keep a mid-course between Malaga and Melilla.

14. At 1645, the Captain, 3rd Destroyer Flotilla's 1641 … was received, when course was altered to intercept the *Miguel Cervantes* and *Gibel Zerjon*.

Codrington, Miguel Cervantes and *Gibel Zerjon* were sighted at 1731 by which time *Gibel Zerjon* was steaming in a southerly direction by order of the Captain, Third Destroyer Flotilla, who had just returned to his ship after boarding *Miguel Cervantes*, informing the Captain of the non-recognition by the British Government of a six mile territorial limit, and pointing out that the interference with British shipping outside the three mile limit would not be tolerated and was to cease.

15. It was ascertained that the Captain of the *Girbel Zerjon* desired to return to Gibraltar for further orders, his W/T transmitter being defective. *Codrington* was accordingly ordered to accompany her thither.

16. *Miguel Cervantes* after lingering in the vicinity for a while disappeared in the direction of Malaga, while *Repulse* remained in company with *Codrington* and *Gibel Zerjon* until dark, and then increased speed to return to Gibraltar.

17. On receipt of the Commander-in-Chief, Mediterranean Station's 2018 at 2200 course was altered and speed adjusted so as to arrive off Cape Tres Forcas at daylight the following morning.

Monday, 24th August.

18. The day was spent patrolling off Melilla and Cape Tres Forcas and was without incident except for the sighting of the Spanish submarine *B.5*, who also appeared to be on patrol.

19. H.M.S. *Repulse* hauled off from the land during the night.

Tuesday, 25th August.

20. At daylight *Repulse* resumed patrolling. Submarine *B.5* was still in the vicinity.

21. On receipt of the Commander-in-Chief, Mediterranean Station's 1201 …, course was shaped and speed set to arrive at Gibraltar at 1900. *Repulse* anchored in Gibraltar Bay at 1855 and completed with oil fuel the same evening.

Wednesday, 26th August.

22. Having embarked stores and provisions H.M.S. *Repulse* sailed at 1020 in accordance with sailing orders issued by the Rear Admiral, Gibraltar, so as to be in the vicinity of Melilla by 1800 en route for Malta. S.S. *Gibel Zerjon* was expected there about 1900. *Repulse* was ordered

to see that the three mile limit was respected and that S.S. *Gibel Zerjon* was not interfered with outside that limit.

23. This service was carried out without incident. S.S. *Gibel Zerjon* with *Wolsey* near, but not escorting her, was overhauled about 1700 and she entered Melilla harbour at 1955 unmolested.

24. *Repulse* then shaped course and adjusted speed so as to reach Malta at 1030 on Saturday, 29th August.

* * *

121. *Pound to Somerville*

[FO 371/20537] 2 September 1936

SECRET.

[Telegram] Addressed R.A.(D), repeated Admiralty, R.A. 1st B.S.

IMMEDIATE.

70. My appreciation of the situation in Majorca is as follows:

There appear to be 4 situations we may have to deal with.

Conclude the insurgents will be referred to as the Whites and the so called Government forces as the Reds.

Situation A. The Whites either prevent the Reds from advancing further or throw them out of Majorca. In this case Italians will probably gain credit and influence with the Whites for the assistance Italians have rendered latter but there would not appear to be any necessity or likelihood of the Whites demanding that a protectorate should be assumed by the Italians or any other power.

No reason is seen why Italian armed parties should be landed or Italian flag hoisted and consequently the situation is not one which the British Senior Naval Officer present is likely to be faced with.

Situation B.

The Reds advance from their present position and it appears likely that the whole of Majorca will fall into their hands.

In this case Whites might ask the Italians or another power to assume protectorate over the island. This might result in the Italian flag being hoisted over Palma or elsewhere and once the Italian flag had been hoisted the landing of Italian armed parties and action by Italian men-of-war to assist the Whites might follow.

Under these circumstances landing of British armed parties could not be contemplated.

If on the other hand Italians express the intention of landing armed parties to protect their nationals it would be the duty of the British Senior

Naval Officer strenuously to endeavour to deter them from this course of action, but if they insisted on doing so then we should also land an armed party on the plea of protecting our nationals.

Part 2 follows.

1733/2.

IMMEDIATE.

73. Part 2 of my 1732.

Situation C.

Should Whites beat Reds on mainland and this be considered due to assistance given by Italians there might be intention locally of giving Italy protectorate over Majorca in return for their help.

There does not appear to be any action which local British S.N.O. can take.

Situation D.

The Reds beat Whites on mainland and [*corrupt group*] is that Reds intend attacking Majorca with large force.

In this case, in order to stave off attack, the Whites might offer a protectorate to some foreign power. Under these circumstances it is considered that the local British S.N.O. should insist that the protectorate is assumed by Great Britain, France and Italy, and if necessary, Germany until negotiations to determine the future status of Majorca.

I should be glad of your remarks on the above, particularly as to whether there is any other action you consider local British S.N.O. can take in each case.

Would the continued presence of a French man-of-war at Palma be a good thing?

1828/2.

122. *Somerville to Pound*

[FO 371/20537] 3 September 1936

SECRET.

[Telegram] Addressed Commander-in-Chief Mediterranean; repeated Admiralty, Rear Admiral 1st Battle Squadron.

IMMEDIATE.

90. Your 1753 and your 1828 2nd September.

Situation A. Concur.

Situation B. I believe military and political situation at Majorca is such that landing of relatively small armed party from Italian ships would exercise no material or additional moral effect and is consequently unlikely. If situation at Majorca is critical intervention to be effective would entail substantial re-enforcement of land and air forces with possible co-operation of warships. In these circumstances minor issue of protection [of] nationals unlikely to arise. Consider Senior Naval Officer should be furnished with properly worded demand on the lines referred to in Situation D to be handed to Italian Commander if full scale intervention is attempted. Possibility that Rossi[1] may purposely refrain from promoting efficiency of defence in order to secure ultimate Italian intervention is perhaps remote but cannot be entirely excluded.

Situation C. Undoubtedly feeling in Majorca that France has not quite played the game by them and strongly of opinion they would resist most strongly any suggestion of being bartered [for] Italian assistance on mainland. Am informed reliably no suggestion of separatist movement in Balearic Islands has ever been made and will only occur if Majorca finds herself in extremis.

Situation D. In this as in Situation B presume that justification for intervention would be sought on grounds that geographically and politically Balearic Islands entitled to some degree of self determination. Consider action by Senior Naval Officer as in B is necessary. In view of hostility felt for French owing to their attitude in present situation consider presence of French ship will serve no useful purpose more especially as practically all French nationals have left. Presence of British ship desirable as this must undoubtedly act as brake on sub rosa intervention by Italians.

0217/3.

123.   *Pound to Chatfield*

[CHT/4/10]   Commander-in-Chief, Mediterranean Station
18 September [1936]

[Holograph]

As it turned out there was no need for me to refer to the Admiralty over the evacuation of the Civil Governor from Malaga because on getting [in] touch with R.A.1 after my arrival at Malaga this morning I discovered that the F.O. had replied in the affirmative to the Consul General's (Barcelona) question as to whether the Catalan Government could be

---

[1] Arconovaldo Bonacorsi, known to the Spanish as 'Conde Rossi', probably because of his red beard. Console [Colonel] in the MVSN [Fascist Militia]. Described as a fanatical Fascist, he was sent to Majorca to assist the Nationalists resist attempts by Republican forces to recapture the island.

evacuated if necessary. Unfortunately this reply was not repeated to me – though R.A.1's original telegram was – hence I thought there had been no reply.

The history of the Malaga business was this. Before I left Gib. I got *Worcester*'s signal saying that the Civil Governor had asked our Consul whether as a last resort one of our ships could evacuate him. I took the line that we had evacuated Crowned Heads to save them from revolutionaries and as the Civil Governor belonged to the recognised government and friendly power there was no reason why he should not be rescued. So I told *Worcester* to try and get him to evacuate in a Spanish man of war but if this was not practicable to evacuate him in *Worcester* as a last resort. We then left Gib. and were on our way to have a look at Melilla on our way to Malta when I got a message from First Lord [Hoare] at 2 A.M. saying that he was inclined against acceding to the Governor's request but that he left the decision to me. This left me in rather a difficult position as I am not quite certain what the First Lord's position of responsibility is on these yachting trips but I thought well here is an ex Foreign Secretary of great experience in international reactions and he will not have sent his message unless he felt rather strongly about it. I decided that the best thing to do was to go to Malaga myself so I told R.A.1 not to do anything about evacuating the Civil Governor pending my arrival. At the same time I sent a message asking First Lord if he would give me his reasons for being against acceding to the Governor's request because it would help me in coming to a decision. I must say I was rather dumbfounded when I got his reply saying that his only reason was the same as I gave in my instructions to R.A.1 – 'that it was undesirable to get mixed up in evacuating government officials' – and then he went on to say that he quite agreed with me that if it was the only way to save the man's life he should be evacuated.

The situation was somewhat burlesque because we subsequently learnt that the Governor had been replaced during the night – our friend having gone to Madrid. I am only telling you all this so that you will be au fait with the incident and not for one moment with the idea that the First Lord should be made aware of the somewhat difficult position he put me in because I am sure his message was only sent with the idea of being helpful and he is so very correct in his attitude and I like him so very much. I am getting A.C.1 to come over by destroyer from Gib. to see me before I leave for Malta so that he could be put wise about the whole business because I think the Malaga situation may blow up at anytime.

The incident such as the above makes me wonder whether I am right in personally going on the cruise[1] and leaving the Western basin to anyone

---

[1] To the Aegean.

else. However, unless the Admiralty tell me that they would like me to remain in the Western basin I shall stick to my proposal, which you approved to go on the cruise¹ for the following reasons. Most of the work is purely routine work connected with evacuating refugees and though incidents such as the Malaga one do occur from time to time and loom rather large at the moment they are really very trivial and all the Flag Officers under me are entirely capable of dealing with them. If two admirals only are kept on the Spanish coast, and I think it would be a pity to keep more, than the one not locked up in Barcelona must be available for being 'on guard' watching the foreigners at such places as Tangier and Palma and I do not think it would enhance our prestige for the C.-in-C. to do this. There may occur at any time some reason unknown to me why my presence in the Western basin may be desirable and I know T.L.'s would not hesitate to tell me. A.C.1 will also let me know if I should come west. I could not loaf at Malta whilst half the Fleet is cruising and the other part manning the coast of Spain.

I saw the First Lord at Gib. and there is no doubt that he has thoroughly enjoyed his time out here.² He seemed a little apologetic about going to see the H.F. so soon but I tried hard to make him realise that the oftener the Fleet saw him the better they would be pleased. I am so glad you have got your leave after all because I know what a heavy autumn you have in front of you.

Binney is sick today and I am afraid it looks like para-typhoid.

124. *Pound to Mediterranean Fleet*

[ADM 116/3051]   23 September 1936

[Telegram] Addressed: Mediterranean Fleet, repeated R.A. Gibraltar, Admiralty.

237. Admiralty have called for a report of work done by the fleet in evacuating refugees from the coast of Spain and any other incidents of interest to the public in connection with the Spanish crisis. The report should include following:

(a) Mileage steamed including passage to and from Malta.

(b) Relations with local officials and any difficulties experienced and how overcome.

---

¹Underlined by Chatfield with marginal notation: 'D.C.N.S. Please return. I propose to concur ..., E.C., 24/9.' There is a marginal note by the DCNS that a telegram approving the cruise had been sent.

²Hoare visited Malta in late August and also spent a day with part of the Fleet at sea, observing exercises.

(c) Co-operation with foreign navies.
(d) Outline of organisation on board H.M. Ships to deal with refugees.
(e) Any stories of human interest.

All ships now at Malta, including Hospital Ship *Maine*, are to forward reports to C.-in-C. Mediterranean by noon, Monday, September 28th. These reports will be forwarded to V.A.C.B.C.S. by C.-in-C. Mediterranean. R.A. 1st C.S. and R.A. Gibraltar are requested to collect reports from ships now in their areas and forward them to V.A.C.B.C.S.

V.A.C.B.C.S. is requested to co-ordinate all reports into one comprehensive story and forward it to Admiralty sending a copy to C.-in-C. Mediterranean. The story should particularly deal with the human element.

1237.

125. *Captain F.H. Goolden[1] to Vice Admiral Geoffrey Blake*

[ADM 116/3051]

H.M.S. *London*, at Barcelona
10 October 1936

No. 213

In accordance with Commander-in-Chief's signal 1237 of 23rd September 1936, the following report from H.M.S. *London* is submitted.

2.– NARRATIVE.

The ship arrived off Barcelona at 1100 on 23rd July, 1936. A local pilot came on board from whom it was learnt that there was no de facto Port Authority. The motor boat equipped with W/T was therefore sent inside the harbour to the Seaplane Station, and permission for the ship to enter was given through the Commanding Officer of the establishment. *London* entered the harbour at 1300 and berthed with two anchors down and stern to the south east corner of the Muelle de Barcelona. A moveable pontoon projects at right angles to this mole, and this was secured alongside the ship, thus providing an ideal means of access to the ship. *London* was the first foreign man-of-war to arrive at Barcelona.

3.– The ship remained in this position until Saturday, 22nd August, when, having been relieved by H.M.S. *Shropshire*, she proceeded to Malta, arriving there on the 24th. During this period (31 days) a total of

---

[1]Capt [later RA] Francis Hugh Goolden (1885–1950). Commanded cruiser HMAS *Sydney* and chief staff officer to Cdre commanding Australian Squadron, 1926–27; Dep DOD, Admy, 1932–33; commanded cruiser *London*, Flag Capt and Chief Staff Officer to VAC 1st CS, Med Fleet, 1934–37; retired, 1937; commanded training establishment HMS *Ganges*, Shotley, 1939–40; Flag Officer-in-charge, Harwich, 1941–42; Naval Asst to 1st Sea Lord, 1942–43; Naval Asst to 4th Sea Lord, 1943.

2300 refugees were evacuated through the ship. 1081 of these were British subjects, and the remainder comprised the nationals of 53 other countries.

4.– While at Malta every opportunity was taken to give leave to the ship's company, no leave having been given, and very little form of recreation being possible since the ship left Alexandria on 18th July. The ship left Malta on Monday, 14th and arrived again at Barcelona on 16th September. During this second period (up-to-date) the number of evacuations have been 678, of which 47 have been British subjects. The total for the whole period up to date of evacuations at Barcelona through H.M.S. *London* is 2978, of which 1128 are British.

5.– <u>ORGANISATION</u>.

… Speaking generally the first days were the busiest, when some hundreds of refugees were accommodated for a night, or longer. At this time the ship was open for embarkation all day, and the refugees streamed on board as they were able to effect their passage through the Custom's barrier.

As many as possible were then transferred to destroyers which lay alongside and taken to Marseilles as opportunity offered. After the first few days destroyers were sailed at a routine time – 0500, so as to arrive at Marseilles before the Customs closed. After the initial comb out of British subjects by destroyers from various places on the coast (and by *Devonshire* and *Repulse* from Valencia and Palma) which involved direct passages to Marseilles, all destroyers thereafter called at Barcelona and remained for the night before proceeding to Marseilles, the refugees being accommodated for the night in *London*, who thus became a clearing ship for all refugees in the Northern area. As time went on, and refugees became fewer, it was possible to confine the time of embarkation from the shore to certain hours of the afternoon and evening, and later, to certain days. The reduction in the number of refugees, and the introduction of intervals between sailings of destroyers, which became possible, made conditions on board easier, especially during the second period, when the ship has sometimes been clear of refugees for two days at a time, and it has been possible to organise recreational games to a considerable extent.

* * *

7.– <u>MILEAGE</u>.

The only steaming which *London* can be said to have done in connection with the evacuation has been the passage to Malta and back, to which may be added the original passage from Malta to Barcelona – a total of 2010 miles.

8.– CO-OPERATION WITH FOREIGN NAVIES.

During the time *London* has been at Barcelona, men-of-war of the following nations have been met:– United States, Germany, France, Italy and Portugal. Of these, the German, French and Italian Navies have retained ships constantly anchored off the port but, apart from a French cruiser in the early days of the civil war and later on an aircraft carrier[1] for a short period, none has been berthed inside the harbour. The U.S.S. *Quincy* has paid periodical visits to Barcelona and the Portuguese destroyer *Douro* stayed for 48 hours. The latter was sent from manoeuvres off Portugal to the East Coast of Spain without warning or preparation and after two days' visit was supplied by *London* with oil fuel, bread and charts to enable her to return.

9.– Relations with all foreign navies have been cordial and all have shown themselves most willing to co-operate in cases where a combined action was required. This was most gratifying in the early days when the British Admiral was not the Senior Officer present.

10.– Information concerning movements of ships has been freely exchanged and facilities for the despatch of mails given. Transfer of refugees to ships of their own country, or of those preferring to proceed to Genoa rather than Marseilles, has been another form of co-operation. In this connection, great care has been necessary to ensure that such transfers are voluntary, since it must be appreciated that many of the Germans and Italians in Spain are Anti-Nazis or Anti-Fascists who, being refugees from their own countries, naturally prefer to avail themselves of British facilities for evacuation.

11.– With regard to the Americans, it has naturally been impossible for one cruiser to cope with the evacuation of United States citizens on the whole of the coast and more than 300 have been taken to Marseilles in British ships. Moreover, our destroyers have conveyed American consular officials and correspondence up and down the coast as a matter of routine and our lines of communication have been constantly utilised in keeping the American cruiser in touch with United States Consular Authorities in the various ports.

That this work has been appreciated is evident, for thanks to the courtesy of the Americans engaged in the film business, a full programme of the latest films is provided nightly on board the *London*, a much appreciated recreation for the ship's company, destroyers' crews and refugees alike.

12.– As regards the social side of the relations with foreign navies, calls have been exchanged with all foreign warships and in spite of the inherent

---

[1] The seaplane carrier *Commandant Teste*.

difficulties of boat communication with ships outside the harbour, such calls have resulted in a good deal of unofficial entertaining, more especially as regards the German Navy, whose officers have been most friendly.

Invitations to dinner and the cinema have been freely exchanged and deck hockey matches arranged.

It has been interesting to note that whereas all the foreign navies have been on most friendly terms with our own, they are not equally so with each other. Ill-concealed coolness on at least one occasion was evident between German and Italian officers in the Ward Room. The former quite openly asked how the British could be so friendly with the Italians 'after Alexandria'. In this connection, when Italian officers first met the British off the Spanish coast it was painfully obvious that they were extremely nervous as to the type of reception they would be accorded and their gratification on perceiving that the British were fully prepared to be friendly was evident.

13.– **RELATIONS WITH LOCAL OFFICIALS.**

On arrival the Rear Admiral Commanding, First Cruiser Squadron paid official calls on His Excellency Don Luis Companys, President of the Generalitat of Catalonia, and on Don Senor Espana, Councillor for Internal Affairs. These calls were returned on the following day by Senor Ventura Gossol, Councillor of Culture. Calls were also exchanged with General Aranguren, Head of the Guardia Civile.

The Admiral was cordially received by all these authorities who readily agreed to any assistance required.

A visit was also received from the President of the Seccion de Alimentation, Senor Sanchez, and four of his Committee. Though belonging to the Anarchist Party, and wearing no collars, they were disposed to be friendly and to help as regards supplies.

14. At the outset the government still maintained considerable authority over the city's destiny and were most helpful. As an example of this two cars with chauffeurs – armed members of the militia – were placed at the disposal of the ship to facilitate communication with the British Consulate-General, which is situated a considerable distance from the harbour. These cars are still retained and have proved invaluable. The drivers are unpaid by us but are given meals from time to time. They are armed with automatics of which they are inordinately proud and they were most reluctant to leave them behind on the gangway when coming on board.

The Catalan Government were also helpful in providing guards for the Consulate and the Muelle de Barcelona at the end of which *London* was berthed, and placed no obstacles in the path of evacuating refugees.

15.– As time went on, the initial short sighted policy of arming the workers indiscriminately resulted in a gradual transference of power from the government to the Workers Unions, the C.N.T.[1] and the F.A.I.[2] This at first took the form of coercing the government into issuing all manner of communist decrees but our associations remained friendly. By the end of September, however, a new Government of the Left had been formed and so many rival organisations existed that it became extremely difficult to know who held any power at all in the city. For example, in order to be allowed through the barrier to proceed on board *London*, it was necessary to have a pass stamped by the British Consulate General, the Catalan Government, the C.N.T., the F.A.I., the U.G.T.[3] and the Anti-Fascist Militia. All the last four had representatives at the barrier who would not allow even a prominent British citizen to go by if their own particular stamp was missing.

16.– HARBOUR AUTHORITIES.

When *London* first arrived information was received that the Captain of the Port had either been imprisoned or killed. He has not been replaced. All berthing arrangements are made by the Pilots. As almost everywhere else in the world, these have been found most reliable, helpful and obliging, whenever anything has been required for H.M. Ships. Their task has been made no easier for them by the fact that they have to take orders, they say, from 'anybody with a gun'.

17.– COMMUNICATIONS.

The communication problem at Barcelona was not confined to ordinary service lines of communication. Information has been exchanged with foreign navies by visual signalling. *London* used to make messages in English and very often used to receive replies in three different languages. At first this international signalling was slow but as time wore on it increased its speed until there was a certain amount of competition about who could make the fastest signals in the transmitting ship's language.

A great deal of international signalling has been done by wireless. Every nation has used us to pass messages for them to various ports in Spain where they had not got a man-of-war. We have also used other nations in the same way but to a lesser extent.

18.– In the early days of the revolution the only communication it was possible to have with the British Consulate was by wireless. As the affairs of the city got more stable a telephone was placed on board. This telephone was thought to be tapped and so it was never used for any

---

[1]Confederación Nacional del Trabajo, the anarcho-syndicalist trade union.
[2]Federación Anarquista Ibérica, the most radical and revolutionary faction of the Anarchist movement.
[3]Unión General de Trabajadores, the well-organised Socialist trade union organisation.

confidential conversation. A portable wireless transmitter was erected in the Consulate and this was moved to the Vice-Consul's house at night. In the early days the crew consisted of three ratings but this was later reduced to two.

19.– Many messages were received and sent on the Consulate wireless set for foreign consulates and foreign men-of-war. The German Consulate also had a portable wireless transmitter and when they had no men-of-war in harbour we used to act as a link for them. Although communication with the German Consulate was satisfactorily carried out by one of our destroyers considerable difficulty was experienced in passing messages on to their men-of-war. Our own lines of communication were very full and during the short periods we were able to devote to trying to get hold of German ships we were unsuccessful. Messages were however cleared via H.M. Ships who were in visual touch with the Germans.

* * *

126. *Admiralty Memorandum*

[ADM 116/3917] [22] December 1936

SECRET.

M.05465/36.

DUTIES AND RESPONSIBILITIES OF H.M. SHIPS IN CONNECTION WITH BRITISH MERCHANT SHIPS DURING HOSTILITIES IN SPAIN.

Spanish Government and Insurgent Warships or Aircraft.

(1) These instructions are to be regarded as applying equally to dealings with both Spanish Government and Insurgent warships or aircraft.

Policy of H.M. Government.

(2) H.M. Government have decided that, at present, they will not grant belligerent rights at sea to the contending parties in Spain; their policy is to continue to prohibit the supply to Spain of war material as hereafter defined, and to this end they have made it illegal for British ships to carry war material consigned to or destined for any port or place in Spanish territory or Spanish Morocco. A copy of the Act is appended.[1]

(3) The above policy involves:–

    (a) The protection of <u>all</u> British shipping, whether or not war material destined for Spain is on board, against attempts by Spanish warships

---

[1] Not reproduced.

to search them, or to interfere in any way with their movements, outside Spanish territorial waters.

(b) The arrest under the legislation mentioned above of any British ship (other than a Dominion ship as defined in paragraph 10) intercepted on the high seas with war material on board consigned to or destined for any port in Spain. For this purpose, British ships include ships licensed under the local law of any British Colony or Protectorate (e.g. ships on the local Cyprus register) and ships registered in any territory administered under mandate by H.M. Government in the United Kingdom, (e.g. Palestine ships).

Notification sent to Spanish Authorities.

(4) Both the Spanish Government and Insurgent authorities have been informed that H.M. Ships have instructions to take any British ship known or suspected to be carrying war material to Spain to a British port and they have been asked to bring this to the notice of all Spanish warships.

Powers of H.M. Ships.

(5) The following powers may be exercised by any Naval Officer on full pay towards any British ship (other than a Dominions ship as defined in paragraph 10) which is suspected of contravening or of having contravened the Act.

(a) He may go on board the ship and for that purpose may detain the ship or require it to stop or to proceed to some convenient place.

(b) He may require the master to produce any documents relating to any cargo which is being carried or has been carried on the ship.

(c) He may search the ship and examine the cargo and require the master or any member of the crew to open any package or parcel which he suspects to contain any articles to which this act applies.

(d) He may make any other examination or inquiry which he deems necessary to ascertain whether this Act is being or has been contravened.

(e) If it appears to him that the Act is being or has been contravened, he may, without summons, warrant or other process, take the ship and her cargo and her master and crew to the nearest or most convenient port in a country to which the Act extends, in order that the alleged contravention may be adjudicated upon by a competent court.

War Material.

(6) A list of articles included in the term 'War Material' is enclosed as an appendix to this memorandum.

British Ships carrying War Material.

(7) As far as possible, advance information will be obtained of any shipment of war material to Spain in British ships. On the receipt of such

information, steps will be taken by the Admiralty to inform Commanders-in-Chief and Senior Officers of the probable movements of any ships carrying war material, in order that appropriate action may be set on foot to intercept such ships.

Special patrols for the interception of all British ships on the chance that they may be carrying war material consigned to or destined for Spain need not be instituted without further orders.

Any British ship (other than a Dominion ship) discovered to be carrying war material as herein defined consigned to or believed on good evidence to be destined for Spain should be sent in to the nearest British port under armed guard. As a general rule it is intended that proceedings will be taken at Malta or Gibraltar, or in the United Kingdom.

War Material sent to Spain through Portugal or other Territory contiguous to Spanish Territory.

(8) Attention is drawn to the fact that the Act covers war material intended to be sent to Spain indirectly by way of Portugal or other foreign countries, whether adjacent to Spain or not, where it can be shown that the goods are consigned to or destined for Spain. You should not however arrest a ship containing war material intended to be forwarded from a foreign port to Spain without obtaining instructions to this effect from the Admiralty.

If you have information of such a transaction, a full report should be made at once to the Admiralty, repeated to the Commander-in-Chief and other authorities concerned, by telegraph, giving full information of the evidence that the ultimate destination is Spain.

Advance information of movements of British Ships.

(9) It is anticipated that advance information of the movements of British merchant ships with legitimate cargoes destined for Spain will be available for communication to H.M. Ships concerned. Such information will be communicated by the Admiralty to Commanders-in-Chief and Senior Officers, who will communicate it to H.M. Ships as necessary, for assistance in carrying out these instructions.

Dominion Ships.

(10) The act applies to all ships entitled to fly the British flag, except those registered in the Dominions of Canada, Australia, New Zealand, South Africa and the Irish Free State, and in any territory administered by the aforesaid Dominions.

It is very improbable that any ships registered in the Dominions will attempt to carry war material to Spain. If any information is received which suggests that such an attempt may be made, an immediate report should be made to the Admiralty.

It would require the concurrence of the Dominions to apply the provisions of the Act to Dominion ships or to withdraw British protection from them.

Hostile Act by Spanish Aircraft or Warship outside Spanish Territorial Waters.

(11) Outside Spanish territorial waters, any definitely hostile attacks by aircraft on a British ship, whether Man-of-War or merchant ship, by bomb or other weapon, are to be resisted by force without previous warning.

(12) Outside Spanish territorial waters, any definitely hostile attacks by Spanish Government or Insurgent Men-of-War on a British ship are to be resisted by force if persisted in after remonstrance.

Remonstrance is to take the following forms:–

The international signal O.F.A.–G.F.A. Cease Fire, is to be hoisted and a round of blank (if carried) fired to call attention to the signal. If this is not obeyed two single shotted rounds are to be fired across the bows of the Spanish Man-of-War and if the latter does not then cease firing she is to be engaged until her fire has been silenced.

Interference by Spanish Warships with British Ships outside Spanish Territorial Waters.

(13) Outside Spanish territorial waters, a Spanish Man-of-War has the right to ask a British merchant ship where she is bound but H.M. Government do not recognise that either of the contending parties in Spain has the right of visit and search. A Spanish Man-of-War is not entitled to order a British ship to stop outside Spanish territorial waters whether for search or other purpose.

(14) If any ship flying the British flag is stopped or captured by a Spanish warship outside territorial waters and one of H.M. Ships is present at the time, the latter should signal to the Spanish warship and ask her to desist, explaining that the British warship will board the merchant ship and that if she proves to be a British ship carrying war material to Spain will send her into a British port to be dealt with according to British law: also that, if she proves to be a foreign ship disguised, the British warship will, after confiscating the British flag, take no steps to prevent the Spanish warship proceeding with the search or resuming her prize. (See paragraph 26 as to use of British flag).

(15) If one of H.M. Ships is summoned to assist a British merchant ship which has been intercepted by a Spanish warship she is, whenever practicable, to proceed to the scene with all despatch and act in accordance with paragraph (14).

(16) If weather conditions make it impossible for boarding to be carried out, the merchant ship flying the British flag should if possible be escorted

to a British port. It should be explained to the Spanish ship that if she is a British ship, and is found to be carrying war material to Spain, she will be dealt with according to British law. It should also be made clear to the Spanish warship that the latter is not expected to enter the British port.

(17) If the Spanish warship refuses to acquiesce in the procedure in paragraph (16) on the ground that the merchant ship is a foreign ship disguised, or if a British port is too far distant for the procedure in paragraph (16) to be carried out, the Spanish warship may be allowed to proceed in company to some convenient sheltered anchorage, or to a port in the hands of the party to which the Spanish warship belongs.

In no case however, should a merchant ship flying the British flag be taken into Spanish territorial waters unless the Spanish warship will give an undertaking not to arrest her if she proves to be a British ship.

(18) If an incident, similar to that envisaged in paragraph (14) and (16) occurs with a British merchant ship which, though not carrying war material, is carrying supplies of other kinds which an intercepting Spanish Man-of-War wishes to prevent reaching the other side, the following action is to be taken. The Spanish Man-of-War is to be informed that H.M. Government cannot permit action being taken against her. The British vessel is to be warned as in paragraph (19). If the Spanish Man-of-War follows the British vessel the British Man-of-War is to follow her also and see that she is not interfered with as long as she remains outside Spanish territorial waters.

Entry of British Ships into Spanish Territorial Waters.

(19) Masters of British merchant vessels wishing to proceed to ports in Spain with cargoes other than war materials should, on application to H.M. Ships, be given the following warning and advice:–

(a) They cannot be escorted by a British Man-of-War.

(b) They must appreciate that there is risk that ships carrying cargoes to Spain other than war materials may be interfered with and that it may not be possible for any of H.M. Ships to assist them, especially if the merchant ship is not fitted with W/T.

(c) If arrested on the high seas they should appeal by W/T to one of H.M. Ships if in waters off the north of Spain or to the Rear-Admiral, Gibraltar, or the Senior Naval Officer at Palma or Barcelona if in waters off the south or east coasts of Spain. They are also to be furnished with the appropriate naval call signs.

Inside Spanish Territorial Waters.

(20) Once a British merchant ship is inside Spanish territorial waters, any action taken by a Spanish warship to stop her is not to be resisted by any of H.M. Ships, whether or not the British merchant ship has

previously been warned by one of H.M. Ships in accordance with paragraph (19). This applies even if the British ship has appealed to one of H.M. Ships for help but has been taken into Spanish territorial waters before the arrival of H.M. Ship, but in such a case the facts are to be reported immediately.

(21) If one of H.M. Ships is present when a British merchant ship, whether carrying war material or not, is brought into a Spanish port, the Commanding Officer of H.M. Ship, after satisfying himself as to the case, is to remonstrate with the local authorities pointing out that the ship has been illegally arrested and that unless she is released with her cargo intact the facts will be reported to H.M. Government, also that if she is carrying war material to Spain she will be dealt with in a British port according to British law; but apart from this and making a full and immediate report of the facts, no other action should be taken.

(22) If, in the circumstances of an arrest described in paragraph (21), no British warship is present at the port to which the British merchant ship has been sent, a warship is NOT to be sent there in connection with the illegal detention of the vessel.

This, however, is not to preclude a British warship being sent to the same or other Spanish port on other business such as for the protection of His Majesty's Consuls or the evacuation of refugees, but in that case no action should be taken with the local authorities respecting the release of the merchant ship detained, unless under instructions from the Admiralty.

Report to Admiralty.

(23) A full report is to be made to the Admiralty, repeated to the Commander-in-Chief and other authorities concerned, of any instance either of interference with British shipping by Spanish warships, or of carriage or attempted carriage of war material by British ships to Spain.

Extent of Territorial Waters.

(24) Territorial waters are limited to those within three miles of the coast of any country.

As regards the extent of territorial waters in bays attention is drawn to O.U. 5316, Naval Prize Manual, Article 2, as amended by Addendum No. 2.

Although H.M. Government do not recognise as territorial a bay exceeding 6 miles in width at the entrance, for the present purposes bays which otherwise come within the definition i.e. as regards configuration, and which do not exceed 10 miles in width at the entrance, should be regarded as territorial.

Preparation for Action.

(25) Before getting involved with Spanish warships in any of the situations described above, H.M. Ships are to be at action stations and ready to open fire, but guns should not be trained on a Spanish warship before it is necessary to do so.

Foreign Ship Flying British Flag outside Spanish Territorial Waters.

(26) If a neutral ship is discovered to be flying the British flag as a 'ruse de guerre', the British flag should be confiscated if this is feasible, but the ship is not to be arrested in pursuance of section 69 of the Merchant Shipping Act.

If a belligerent (i.e. Spanish) ship is so discovered, no action is to be taken. H.M. Government consider the use of a neutral flag by a belligerent as a ruse de guerre to be legitimate, and for this purpose Spanish ships are to be regarded as belligerent, despite the fact that belligerent rights have not been granted.

Secrecy.

(27) The instructions in this memorandum are to be kept secret except when dealing with any actual incidents which may arise, since it is desired to avoid giving the impression that we expect Spanish warships to interfere with British shipping.

Personal Contact.

(28) There is no objection to the establishment of personal contact, exchange of informal calls, etc. between naval officers and the Spanish Insurgent authorities so long as the word 'official' is not used in this connection.

* * *

Appendix No. 2.

List of articles included in the term 'War Material':–

(i) Cannon and other ordnance and component parts thereof;
(ii) Carriages and mountings and accessories for mountings for cannon and other ordnance and component parts thereof;
(iii) Cartridges, charges of all kinds, and component parts thereof;
(iv) Explosives of every description;
(v) Firearms of every description and component parts thereof;
(vi) Grenades and component parts thereof;
(vii) Machine Guns, interrupter gears, mountings for machine guns and component parts thereof;
(viii) Projectiles of all kinds (except air gun pellets) and component parts thereof;
(ix) Mines, land or sea, and component parts thereof;

SPANISH CIVIL WAR AND NYON AGREEMENTS, 1936–1937   231

(x)   Depth charges, apparatus for the discharge of depth charges, and component parts thereof;
(xi)   Bombs, bombing apparatus, and component parts thereof;
(xii)   Flame-throwers and component parts thereof;
(xiii) Fuses and component parts thereof;
(xiv)   Torpedoes and component parts thereof;
(xv)   Torpedo tubes, or other apparatus for discharging torpedoes.
(xvi)   Fire-control and gun-sighting apparatus and component parts thereof;
(xvii) Appliances for use with arms and apparatus exclusively designed and intended for land, sea or aerial warfare;
(xviii) Bayonets, swords and lances, and component parts thereof;
(xix)   Tanks and armoured cars and component parts thereof;
(xx)   Aircraft, assembled or dismantled, and aircraft engines.

127. *Lieutenant Commander R.E. Courage to Pound*

[ADM 116/3534]   H.M.S. *Havock*, at Malta
15 February 1937

SECRET

No. A/4

I have the honour to submit the following report on the bombing [13 February] of H.M. Ships *Havock* and *Gipsy* while on passage between GIBRALTAR and MALTA.

2.– At 1615 H.M.S. *Havock* and *Gipsy* had completed picking up lifebuoys and were proceeding on a course of 083° *Gipsy* at 14 knots *Havock* at 17 knots to take up position for T.D. 55.[1]

3.– At 1637 an aeroplane was sighted bearing 150° and as *Havock* was in position 37° 00′ N 01° 10′ E, it was assumed to be a French Mail or Military plane.

4.– The plane passed slightly ahead of *Havock* on a course of 350° and at 1640 two large bombs assumed to be 250 lbs. fell on the port beam of *Havock* abreast 'B' Gun and about 50–70 yards away.

5.– Action Stations were immediately sounded and ammunition provided at all guns.

6.– H.E. ammunition had been struck down earlier as ships were well clear of Spanish Waters and bad weather was expected.[2]

---

[1] *Havock* was taking up a position 10 miles ahead of *Gipsy* for a night encounter exercise. Report of H.M.S. *Gipsy*, No.002L, 14 Feb 1937, ADM 116/3534.

[2] The position was more than 100 miles from the nearest Spanish territory and 65 miles from the direct route between Melilla and Majorca. It was only 30 miles from the coast of French Algeria, hence the initial assumption the aircraft was French. Ibid.

7.– At 1640 H.M.S. *Gipsy* reported that she was ready to open fire and I ordered her to carry on, though her Commanding Officer[1] had already exercised his initiative and commenced with his machine guns. At 1642 he fired two rounds of S.A.P. which was the only ammunition he had on deck to indicate that he was being engaged. At the same time both ships hoisted White Ensigns at the masthead and spread Union Flags on the Torpedo Tubes.

8.– At 1643 the aeroplane circled round to port to 180° and steered to pass astern of *Gipsy*.

9.– At 1645 when astern of *Gipsy* the aeroplane altered to the same course obviously preparing another attack. Both ships altered course to Starboard and increased to full speed.

10.– Fire was opened with 0.5 machine guns. The aeroplane altered course to counteract *Gipsy*'s manoeuvre and about 1646½ four bombs were seen to leave the machine.

10a.– *Gipsy* promptly altered course further to Starboard and *Havock* to Port and the bombs fell between the two ships about 100 yards from *Gipsy* and 300 from *Havock*.

11.– As the bombs fell the aeroplane altered course to the North and steered towards Majorca.

12.– The markings on the under wing of the aeroplane appeared to be light and dark bands running fore and aft which were assumed to be the Insurgent Markings as confirmed by *Gipsy* who has seen these before.

13.– The aeroplane was a four-engined Junker Type monoplane[2] and was proceeding about 90 m.p.h. at a height of 6000 ft. throughout the attack.

14.– The bombs dropped in the second attack were at first thought to be two 400–500 lb. bombs and two light bombs, but it is possible that all were of 250 lbs. and that the two 'heavy ones' burst further under water and gave an impression of greater weight, while the two 'light ones' failed to explode.

15.– No damage was sustained by either ship, nor did the fire of our guns appear to effect the machine.

---

[1] Lt Cdr P.C. Ransome.

[2] Although the captains of both destroyers described the aircraft as having four engines, the description is puzzling because neither side appears to have employed four-engine aircraft during the war. The most likely aircraft to be involved were the three-engine Junkers Ju. 52, Savoia Marchetti S.79 or Savoia Marchetti S.81. See Gerald Howson, *Aircraft of the Spanish Civil War, 1936–39* (Washington, DC, 1990). A possible explanation derived from films of a Ju 52 in flight is that from certain angles the front of the fixed undercarriage appears to an observer below to protrude from the leading edge of the wing thereby giving the impression of another pair of engines.

16.– H.M.S. *Gipsy*'s report is enclosed and also her diagram which is concurred in.[1]

128. *Admiralty to Commanders-in-Chief Home Fleet, Mediterranean and Vice Admiral Gibraltar*

[ADM 116/9170] 20 February 1937

SECRET.

[Telegram]

1. As a result of deliberations of Non-Intervention Committee it must now be envisaged that a system of supervising both the land and sea frontiers of Spain will be established in the near future with the object of detecting the entry of volunteers and war material. Full details will be communicated by letter but the following brief summary is being telegraphed now in order to enable naval commitments involved to be examined.

2. Supervision of land frontiers will be carried out inside territories adjacent to Spain. Although Portugal has raised objections to a scheme on shore within her territory, it may be assumed for the present that these objections will be overcome, and that there will be no naval supervision of Portuguese coast.

3. The system envisaged for coasts of Spain is based on the idea of watching the unloading of cargoes and landing of passengers in Spanish ports by means of supervisors carried in all ships belonging to Non-Intervention powers proceeding to Spain. These supervisors will be embarked at suitable ports such as The Downs, Brest, Gibraltar and Palermo, so disposed as to cover the approaches to Spain and will return in the ship to a convenient port of disembarkation outside of Spain.

4. The role of the naval forces will be to watch the coasts of Spain with the object of identifying the names of any vessels belonging to the Non-Intervention powers which enter Spanish ports without supervisors on board. For this purpose Naval Authorities will be informed of names of all ships proceeding to Spain which have embarked supervisors. Naval vessels will not repeat not have authority to search merchant ships or prevent them proceeding to Spanish ports.

5. In order however, to enable identification to be carried out in cases of doubt, authority will be given by legislation to board all vessels belonging to Non-Intervention powers and to examine the certificate of registry and clearance papers.

[1] Not reproduced.

6. The coastline of Spain and Spanish possessions will be divided into zones for each of which one of the participating powers will be responsible. The participating powers will probably be Great Britain, France, Germany, Italy, Russia and Portugal and the present proposals are for the zones to be allocated as follows:–

> Great Britain: North coast of Spain from Cape Busto (subject to the allocation of a portion of this zone to Portugal, see below). South coast of Spain from the Portuguese frontier to Cape de Gata. Canary Islands.
>
> France: West coast of Spain from the Portuguese frontier to Sisargas Island. Spanish Morocco. Iviza and Majorca.
>
> Germany: South East coast of Spain from Cape de Gata to Cape Oropesa.
>
> Italy: East coast of Spain from Cape Oropesa to the French frontier. The Island of Minorca.
>
> Russia: North west coast of Spain from Sisargas Island to Cape Busto.
>
> Portugal: A portion (to be agreed later) of the north coast of Spain between Cape Busto and the French Frontier.

End of Part I. Part II follows.

1800.

Following is Part II. of my 1800.

7. The present intention of the Committee is that this scheme should be brought into force on the night of 6th to 7th March, but it seems probable that this date will have to be postponed.

8. Their Lordships nevertheless consider it desirable to examine fully the strength and disposition of forces which would be necessary to carry out the scheme as explained above in the zones allocated to Great Britain. C.-in-C. Home Fleet, is therefore requested to make proposals for the zone on the north coast of Spain, and C.-in-C. Mediterranean, for the zone on the south coast of Spain.

Although the Canary Islands are on the Home Fleet Station, it would probably be more convenient for them to be covered by the Mediterranean Fleet based on Gibraltar, and if C.-in-C. Home Fleet and C.-in-C. Mediterranean agree, C.-in-C. Mediterranean should therefore submit proposals for this area also.

9. It is not yet possible to say how large a zone Portugal will be able to take over and for the time being this factor should be disregarded.

10. Their Lordships recognise that if the scheme is put into force it may, on account of the number of ships required, no longer be possible for us to maintain ships for humanitarian or other purposes as in the past in zones outside those allocated to this country. The possibility of sending ships for these purposes to other zones would however remain, and in case of

urgent need it would be necessary to send ships at the expense of the efficiency of the scheme of supervision.

11. Their Lordships also recognise that it may not be possible to take such effective steps as in the past for the protection of British shipping and the enforcement of the Act prohibiting the carriage of war material to Spain.

12. The whole scheme is drawn up on the basis that Spanish concurrence should not be required and this has not been obtained.

13. You will see from the allocation of zones that except for ourselves warships of participating powers will be operating off what may be described as 'enemy' coasts. In consequence their warships will not be able to go into Spanish ports and will have to carry out their supervision outside Spanish territorial waters.

14. It has been the Admiralty's hope throughout the preparation of the scheme that our ships will on account of our 'neutral' reputation be able to carry on their duties by remaining in Spanish ports. You should assume that this procedure will be unobjectionable to both Spanish parties, and act accordingly from the commencement of the scheme without asking permission or giving any formal intimation of your intentions.

15. It will be realised that the above scheme involves legislation by all the non-Intervention powers conferring rights on all participating warships to board merchant ships and verify their identity and destination. It is obviously desirable in the interests of our own shipping to limit this right as much as possible and your remarks are requested as to whether a proposal to limit the exercise of the right to merchant ships within 10 miles of the Spanish coast would suffice. Through traffic could be advised by Admiralty to keep outside ten miles except at Gibraltar, which is within our zone.

16. Their Lordships much regret that it is necessary to consider imposing this new task upon the fleet. You will however appreciate that it was impossible for H.M. Government to refuse to participate in an agreed international scheme such as this. Every endeavour has been made to restrict the zones to be supervised to the minimum possible, having regard to the fact that we are the only power whose impartiality can be fully trusted, and that we have therefore to take a portion of the coast line of both parties together with an overlap wherever the dividing line between the two parties reaches the sea.

2126/20.

129. *Pound to Admiralty*

[ADM 116/9170] 21 February 1937

SECRET

[Telegram] Addressed Admiralty, repeated C.-in-C. Home Fleet, V.A. Gibraltar.

IMPORTANT

852. Admiralty Message 2126 February 20th.

The role of the Naval Forces given in para. 4 in practice will necessitate the interception outside ports of every ship not known to be carrying a supervisor.

2. So long as we are able to carry out our non-intervention duties by one of our ships stationed inside ports, or in the case of Malaga and Almeria from the outer anchorage, any restriction as regards position in which a merchant vessel can be boarded at sea is immaterial, from point of view of carrying out these duties.

3. Whether or not our non-intervention duties can be carried out from inside ports will depend not only on whether port is liable to bombardment but also on Spanish Authorities.

If latter are not in favour of the non-intervention scheme, it will be easy for them to prevent our ships carrying out their duties by posting sentries round the merchant vessels, or other obstructive measures.

4. It is considered therefore that we must be prepared at any moment to revert to interception at sea in which case we shall be faced with the same problem as remaining non-intervention countries.

5. Any restriction placed on area in which a ship can be boarded will undoubtedly hamper operations carried out by men of war. In particular, the introduction of the ten miles rule would considerably assist a merchant vessel to reach a given port, because she could either:–

(a) approach Spanish coast at some distance from port she is making for and then keep inside territorial waters, or

(b) endeavour to slip through zone between ten miles and three miles limits at night.

It appears quite likely therefore that if other non-intervention powers are really out to make this scheme a success, they will object to the ten mile rule. On the other hand they may welcome the ten mile rule because it will make control in all areas less effective.

6. If we are forced to carry out our task at sea it is considered that the only feasible plan would be to watch each port as traffic passing round Cape St. Vincent and through the Straits of Gibraltar is so great that

control in those areas could not be made effective without using a prohibitively large number of ships.

7. Hence from point of view of carrying out our task in our Portuguese frontier to Cape de Gata area, I see no objection to ten miles rule other than those given in para. 5.

End of Part I. Part 2 follows.

2148/21.

853. My 2148 21st February. Part 2.

8. In the following paragraphs, in which the number of ships required and the disposition of forces is given, allowance is made for evacuation of refugees on the north-east coast of Spain and the provision of a cruiser for Haifa.

9. There is no reason why the Mediterranean Fleet should not take over Canary Islands but if more than one cruiser is required there disappointment will be caused to any additional ships which have to remain behind during review period.[1]

As Home Fleet have been looking after Canary Islands I should be glad if C.-in-C. Home Fleet would comment on my proposal to send only one cruiser there.

10. <u>Case A</u>. Complete control exercised by our ships stationed in Spanish ports.

(a) One destroyer lying inside each of the following ports – Huelva, Seville, Cadiz and outside Malaga and Almeria on account of the danger of bombardment.

(b) Smaller ports in south Spain to be visited when the destroyers in (a) are relieved.

(c) Control of Algeciras to be exercised by tug(s) or boat(s) from Gibraltar.

(d) One cruiser in Gibraltar area for Tangier and general duties.

(e) One cruiser in Canary Islands.

(f) Either H.M.S. *Penelope* while docking and self refitting, or H.M.S. *Delhi* whilst working up, to standby at Malta for Haifa.

(g) Hospital Ship *Maine* and one destroyer and either H.M.S. *Cyclops*, *Woolwich* or *Resource* for evacuation duty north-east coast of Spain and keeping in touch with the Chargé d'Affaires, Valencia, Consul General at Barcelona and Vice-Consul Palma. The destroyer would also carry out mail duties.

---

[1] The majority of the Mediterranean Fleet was to return to home waters for the Coronation Review in May 1937.

(h) Relief destroyer in (a) and (g) would require an additional five destroyers to those stipulated above.

(i) The ships which it is approved to leave behind for review period, with additional two destroyers from First Destroyer Flotilla, would meet above requirements.

11. <u>Case B</u>. Reasonable instead of complete control as in Case A.

(a) One destroyer permanently at Cadiz and two others to provide relief for Cadiz destroyer and visit Huelva and Seville at intervals.

(b) One destroyer permanently stationed at Malaga and one at Almeria, with two additional destroyers as reliefs.

(c) Other dispositions as in Case A.

(d) It would not be necessary to leave any of the 1st Destroyer Flotilla behind during the review period.

(e) It appears unlikely that other non-intervention powers would exercise more control than we should do under this scheme.

12. <u>Case C</u>. Reasonable control by patrol at sea (complete control could not be guaranteed under any circumstances).

(a) One destroyer off Huelva and two off Cadiz–Seville area. Owing to arduous nature of this patrol in winter weather, destroyers should work in three reliefs and a total of nine would be required.

(b) One destroyer off Malaga and one off Almeria working in three reliefs – total of six would be required.

(c) Allowing for other requirements, as in Case A, a total of 17 destroyers would be required which would necessitate the whole of the First Destroyer Flotilla remaining in Mediterranean.

13. <u>Case D.</u> To maintain best patrol possible using only 2nd Destroyer Flotilla and H.M.S. *Vanoc*.

2323/21.

130. *Lieutenant Commander R.C. Beckett[1] to Vice Admiral Gibraltar[2]*

[ADM 116/3534]
H.M.S. *Gallant*, at Gibraltar
10 April 1937

CONFIDENTIAL

No. L.P./2

I have the honour to submit the following report of the bombing attacks by Spanish Insurgent Aircraft on His Majesty's Ship under my command whilst on passage from Valencia to Alicante on Tuesday, 6th April, 1937.

2. FIRST ATTACK.

At 1330, H.M.S. *Gallant* was in position 337 degrees, 15 miles from Cape San Antonio, course 144 degrees, speed 16 knots. One boiler was connected, another at 15 minutes notice, and the third had been flashed up to test a safety valve in view of the forthcoming full power trial. The sea was calm, weather fine with slight cirrus cloud, wind west force 1. Ten rounds of H.E. shell were in position at 'B' and 'X' guns. The T.S.D.S. was being prepared for running.

3. At 1331 four bombs fell in the sea in a line approximately parallel to *Gallant*'s course, about 400 yards to port. Only the third and fourth appeared to explode. A multi-engined monoplane was then sighted flying away, bearing Red 45 at about 10,000 ft. The patches of cirrus cloud made it difficult to pick up the aircraft. The splashes made by the bombs were consistent with 250 lb bombs.

4. The hands were ordered to Action Stations, and both engine telegraphs put to 'Full Speed Ahead'. The second and third boilers were connected in 3 and 4 minutes respectively; at 1334 revolutions for 22 knots had been reached; at 1335, 25 knots; and at 1337, 29 knots.

5. The aircraft, which was above maximum elevation (40 degrees) of the 4.7" guns, circled to port and approached from Red 90. When it was steady, I altered course about 45 degrees to port. The aircraft immediately re-adjusted its bombing course, and a bomb was seen to be released. *Gallant* altered course about 90 degrees to starboard. A bomb fell and exploded on a bearing Red 85, about 600 yards away, and a second or two later a sixth bomb fell Red 135, about 100 yards, but failed to explode.

6. The aircraft withdrew astern and re-approached from the starboard beam. In doing so it came below maximum gun elevation, and three single

---

[1] Lt Cdr [later Cdr] Roger Caton Beckett (1901–73). Commanded destroyers *Sardonyx*, 1931–32; *Wren*, 1933–34; *Defender*, 1934–35; *Grenade*, 1935–36; *Gallant*, 1936–38; *Brazen*, 1939–40; *Garland*, 1940; served in HMS *St Angelo*, Royal Naval Base, Malta, 1940–44; Executive Officer, Combined Operations Base, Southampton, 1944–45; CO, Combined Operations Base, Auchengate, 1946.

[2] VA (Retired) Pipon.

rounds of H.E. were fired from 'B' Gun in director firing and primary control before maximum elevation was again reached. At this time the aircraft gained height and withdrew in the direction of Majorca ...

7. SECOND ATTACK.

At 1643, *Gallant* was in position 075 degrees, 18 miles from Cape Huertas, course 232 degrees, speed 16 knots, with two boilers connected. An aircraft, similar to that which had attacked previously, was sighted approaching high on the port beam, and a few seconds later a single large bomb was seen to fall and explode, bearing Red 90, about 500 yards.

8. The hands were ordered to Action Stations and both engine telegraphs put to 'Full Speed Ahead'. At 1647 revolutions for 26 knots were obtained; at 1654 the third boiler was connected and at 1700 revolutions for 30 knots were obtained.

9. At 1645 I altered course 45 degrees to port. The aircraft, still flying high, circled astern and was joined by a second machine of the same type which had come up from the port quarter. The two aircraft then attacked in formation (starboard quarter line about 200 yards apart) from slightly abaft the port beam at a height of at least 8,000 ft. When I judged them steady on their bombing course, I altered course 45 degrees to port. They immediately adjusted their course to port, and as they released their bombs, I ordered the wheel hard over to alter 100 degrees to starboard. The Coxswain failed to hear the order at first, so the wheel went over 10 seconds late. Halfway through this turn the bombs fell, two at Red 150, about 80 and 100 yards, which did not explode, and two at Green 300, about 300 yards, which exploded. In my signal 2220/7, all these bombs were reported as exploding. It is now considered that the concussion of those closest to the ship would have been felt had they done so.

10. At about 1652, the two aircraft, which had drawn ahead while *Gallant* was taking avoiding action, took a wide sweep to port and came below maximum elevation. I steadied the ship on 260 degrees and ordered fire to be opened. The leading aircraft was engaged with H.E. from 'B' and 'X' guns. The Officer of Quarters, After Group, later ordered H.E. to be supplied to 'Y' Gun on his own initiative, when he saw the final attack developing.

11. One aircraft again approached from 160 degrees (the other machine disappeared to the southward as soon as the H.E. shell bursts were seen), and was engaged with three gun salvoes until over maximum elevation. At 1656 I altered course to 220 degrees. The pilot again reacted and one bomb was seen to be released, at which I altered course 90 degrees to starboard under full rudder. This bomb fell and exploded on a bearing of Red 90, about 300 yards, followed by 3 or 4 small bombs which fell together at Red 90, about 400 yards. Only one of the latter appeared to explode.

12. The aircraft withdrew astern and made off to the north eastward at 1700. Further fire was withheld …

13. At 1725 another alarm was caused by two 'Government' aircraft from Alicante being sighted to starboard heading north-east. I gather later that *Gallant*'s engagement had been reported from Benidorme. These aircraft returned to Alicante half an hour later.

14. *Gallant* expended 21 rounds of H.E. in the last engagement. The allowance carried is 10 rounds per gun. As ships on duty in Spanish waters are liable to similar attacks, it is submitted for consideration that the allowance of H.E. shell may be temporarily doubled whilst so employed.

131. *Pound to Blake*

[ADM 116/3534] 6 April 1937

SECRET.

[Message] Addressed V.A.C., B.C.S. repeated Admiralty, V.A. Gibraltar, V.A.C. 1st C.S., *Shropshire*, *Cyclops*.

IMMEDIATE.

9. As you are aware H.M.S. *Gallant* has been attacked by aircraft twice today and in view of greater activity of Government naval forces a single British Destroyer proceeding along Spanish coast in the vicinity of Majorca will always be liable to attack however much we protest and sooner or later will be hit. There appears to be 2 alternatives, A or B.

A. To either maintain at or send a ship to Palma with orders to convey verbally to local authorities accurate information regarding movements of our destroyers.

Unless given verbally there would be no certainty that information had been received.

B. To use ships other than destroyers for duties now carried out by destroyers in the northern area.

In either case another ship will be required in the northern area and I am ordering H.M.S. *Shropshire* to remain in the northern area.

Request you will deal with situation on lines of A or B and issue necessary instructions.

If you consider second ship is necessary in the Gibraltar area you should retain H.M.S. *Devonshire* until relieved by H.M.S. *London*.

1928/6.

132. *Pound to Blake*

[ADM 116/3534] 6 April 1937

SECRET.

[Message] Addressed V.A.C., B.C.S., repeated Admiralty, V.A. Gibraltar, H.M. Ships *Shropshire*, *Cyclops* and *Garland*.

IMPORTANT.

11. On arrival at Palma Commanding Officer of H.M.S. *Shropshire* should inform local authorities that H.M.S. *Gallant* was twice attacked. I do not consider however that it would be of any use demanding that it should not occur again for the following reasons.

A. War is being waged in the area in question and

B. Our distinguishing marks are not visible at heights of 8,000 to 10,000 feet.

C. It is unreasonable to expect that aircraft will come low enough to make out distinguishing marks before attacking. On the other hand fact of H.M.S. *Gallant* having been bombed should be made use of to insist that if we let them know movements of our destroyers we must insist that there will be no more attacks on them.

2313/6.

133. *Captain W.E.C. Tait[1] to Blake*

[ADM 116/3534] 8 April 1937

SECRET.

[Message] Addressed V.A.C., B.C.S., repeated C.-in-C. Mediterranean, V.A. Gibraltar, Admiralty, H.M.S. *Cyclops*, R.A.C. 1st C.S.

Accompanied by Vice Consul, I have visited the Port Admiral and Colonel Franco[2] who commands the Air force in Majorca. They state that movements of their aircraft at the time of H.M.S. *Gallant*'s bombing were

---

[1]Capt [later Adm Sir] William Eric Campbell Tait (1886–1946). Commanded cruisers *Dragon*, 1928; *Capetown*, 1929; *Delhi*, 1929–31; Dep DNI, Admy, 1932–33; commanded cruiser *Shropshire*, 1934–37; Cdre RN Barracks, Portsmouth, 1937–40; Dir of Personal Services, Admy, 1940–41; C-in-C Africa Station, 1941; C-in-C South Atlantic Station, 1942–44; retired, 1944; Govr of Southern Rhodesia, 1945–46.

[2]Colonel Ramón Franco y Bahamonde (1896–1938). Younger brother of Nationalist leader Francisco Franco. A noted aviator, he became a popular hero after a pioneering flight in a flying boat across the South Atlantic and down the coast of Brazil and Uruguay to Buenos Aires, Argentina, January 1926; although associated with left-wing politics and a Freemason, he left his post as Spanish Air Attaché in the US to join the Nationalists; killed in air crash, October, 1938.

as follows. A seaplane on a recently established patrol off the Spanish coast reported some Government destroyers proceeding south from Valencia. Immediately three bombers were sent out and later another three. These on returning reported having attacked the destroyers which opened fire on them first. Time and position agrees with H.M.S. *Gallant*'s report and it seems a possible explanation that there actually were Government destroyers not many miles distant from H.M.S. *Gallant* and that some aircraft attacked H.M.S. *Gallant* by mistake although nothing has been said yet about meeting a destroyer by itself. I am calling upon the Military Governor tomorrow, Thursday, and shall doubtless hear the result of the enquiry he is making but in the meantime he has sent a message expressing great regret if investigation should prove that Nationalist aircraft are to blame. As a result of my call on Colonel Franco, he is issuing positive orders that no destroyer is to be attacked until aircraft has sighted and recorded letter and figure on the destroyer's bows, these being easier to distinguish than the ensign or marking. He admitted that seaplane pilots being all Spanish Naval Officers know very well what a British destroyer looks like, his land plane pilots do not, and although he would instruct them to the best of his ability it would minimise the possibility of mistakes being made if he were furnished with probable movements of our destroyers in his area and he would greatly appreciate such a courtesy. His area is the whole of the Mediterranean east of a line joining Cape de Gata to Africa and extending to Sardinia. I therefore propose that this should be done and that Destroyer Leaders in this area have a letter and figure painted on their bows.

0001/8.

134. *Tait to Blake*

[ADM 116/3534] 8 April 1937

SECRET.

[Telegram] Addressed Flag Officer, Second-in-Command, Mediterranean, repeated C.-in-C. Mediterranean, V.A. Gibraltar, Admiralty, R.A.C. 1st C.S., H.M.S. *Woolwich*.

Accompanied by Vice Consul who had informed him yesterday of the substance of H.M.S. *Gallant*'s report, I have called upon Military Governor, Majorca. He informs me that as a result of his enquiries it is established that Nationalist bombers attacked four separate destroyers believed to be Government vessels in the area and about the time reported by H.M.S. *Gallant*. Without positively admitting these bombers were responsible for attack on H.M.S. *Gallant*, the Governor agreed they probably were and therefore tenders his sincerest regrets and deepest apologies that such a thing should have occurred. While affirming orders already exist that no vessel shall be attacked without previous identification he confirmed that new instructions were being issued that previous to hostile action against a destroyer the number must be read and recorded. He further guaranteed to the best of his power that such an incident should not recur.

This being so I said that I was prepared to propose that intended movements of British destroyers should be communicated to him verbally if that would assist him but added that such communications could in no way guarantee that no emergency movements would take place outside this programme and I was certain H.M. Government would expect aircraft to take all precautions whether this programme was given him or not. In thanking me for the suggestion he said he would much appreciate the courtesy and for the rest quite understood and agreed.

The Vice Consul then hinted that perhaps foreigners in Air Force might not be so zealous to avoid incidents as were Spaniards. In reply the Governor emphasised that in such case discipline would be enforced on them as well as any others but added for our private information that he hoped we realized what a difficult orchestra he had to conduct.

I hesitated to suggest that in return for destroyer programme we should insist that bombing British destroyers must cease. As a result of my conversation I can see that Spanish Authorities might treat it as a bargain but that a destroyer found off her supposed course or late in her anticipated movements or going in any direction not on communicated programme might well be target for any prowling undisciplined Italian bomber who would then not trouble to investigate further.

1430/8.

135. *Admiralty to Blake*

[ADM 116/3534] 10 April 1937

SECRET.
[Message] Addressed V.A.C., B.C.S., repeated C.-in-C. Mediterranean, Vice Admiral Gibraltar, *Shropshire*, *Cyclops*, R.A.C. 1st Cruiser Squadron.

Shropshire's 0001/8 and 1430/8. No objection to communicating movements of destroyers in area to the eastwards of Cape de Gata to Nationalist Authorities as proposed.

It should however be borne in mind that, when the Naval Observation Scheme comes into force, prospective movements of vessels engaged in connection with Scheme within our zones must be regarded as confidential, since advance information would assist Spanish authorities in arranging for vessels to evade observation.

1200/10.

136. *Minute by Vice Admiral Pipon*

[ADM 116/3917]   Gibraltar, 11 April 1937

SECRET

No. 6084.

Director of Naval Intelligence, Admiralty, Commander-in-Chief, Mediterranean-N.131/6084, Commander-in-Chief, Home Fleet, Vice Admiral, Second-in-Command, Mediterranean, Rear Admiral Commanding, First Cruiser Squadron, Staff Officer (I), Malta.

Forwarded.[1]

2.– It is significant that the 12-inch guns being mounted near Mount Bujeo are howitzers which are not ordinarily designed for use against warships. Although their exact topographical position cannot be given it is quite possible that they may be so placed as to be invulnerable from the 9.2-inch guns on the Rock and at the same time to be able to throw their projectiles into the Dockyard. It is of interest that the main road to Tarifa, which until recently was open to all, is now closed beyond the village of Pelayo (about 4 miles from Algeciras) except to holders of special military passes which are not easily obtainable.

3.– Although the erection of this particular battery constitutes a definite menace to the value of Gibraltar as a repairing arsenal and port of refuge for the Fleet, a moveable battery of howitzers of a smaller calibre placed behind any of the hills around San Roque would be almost as effective.

4.– In view of the large expenditure about to be embarked on in connection with the widening of No.1 Dock this matter is considered to be most important and emphasises that Gibraltar can only be of use to Great Britain if the Government in power in Spain is friendly.

[Admiralty Minutes]

* * *

Noted.

2. The question of the vulnerability of Gibraltar to a hostile Spain or to a hostile power in Morocco was reported on by the Chiefs of Staff in

---

[1] A copy of Staff Officer (I), Gibraltar, 'Coast Defences, Spain, South Coast. Straits of Gibraltar. Tarifa – Algeciras', 6 April 1937.

C.O.S. 509 (24/8/36) and C.O.S. 565 (5/3/37) and has been taken note of by the Cabinet on 10th March, 1937.

3. The remarks in paragraph 4 of V.A., Gibraltar's letter are fully concurred in, but no action to improve the situation would appear to be possible at the present time. With a hostile Spain a mobile battery of howitzers, as V.A. Gibraltar has suggested, could be almost as great a menace to the Dockyard as fixed batteries installed in peace.

4. It is, of course, quite true that since the invention of the long range gun our whole position in the Straits of Gibraltar and consequently the Mediterranean has been dependent on a weak or friendly Spain. It is consequently unfortunate from the strategical point of view that our present policy appears likely to antagonize that side in the Spanish Civil War which it seems may eventually make Spain strong.

T.S.V. Phillips[1]
D. of P.
28.5.37.

[Holograph] C.N.S. & First Lord might like to see this.

Re X [concluding sentence in above] I suggest that whichever way the civil war ends, the winning side (if there is one) Spain will be weak for a generation.

W.M.J. [James]
31.5.37.

[Holograph] A friendly Spain is of the greatest importance to this country in view not only of Gibraltar but the whole Medit^n question. When we return to the Foreign Policy of 'keeping our friends' we shall be on safer ground. The greatest danger would be for Spain to be allied to another Power. I concur with Para. 3 of D. of P.

E.C. [Chatfield]
10.6.37.

---

[1] Capt [later Adm Sir] Thomas Spencer Vaughn Phillips (1888–1941). Capt (D), 6th Destroyer Flotilla, 1928–29; Asst DP, Admy, 1930–32; Flag Capt (HMS *Hawkins*) and COS to C-in-C, East Indies, 1932–35; DP, Admy, 1935–38; Cdre/RAC Home Fleet Destroyer Flotillas, 1938–39; Dep CNS, 1939–41; C-in-C Eastern Fleet, 1941; lost with his flagship *Prince of Wales*, 10 Dec 1941.

137. *Captain G.J.A. Miles[1] to Somerville*

[ADM 116/3681]   H.M.S. *Codrington*, at Plymouth
28 April 1937

SECRET.

No. 23/28

I have the honour to submit the following records and observations on conclusion of the Third Commission of the *Acasta* class destroyers in the Third Destroyer Flotilla.

2. The Flotilla recommissioned at their Home Ports on 23rd July, 1935, having recently completed a refit and attended the Jubilee Review.

3. After carrying out some torpedo discharge trials at Portland the Flotilla arrived at Gibraltar on 3rd August, 1935, for a three weeks working up period. *Achates* and *Acheron* were absent all this time in England; and *Anthony* developed condenser trouble on arrival and was alongside the Dockyard at Gibraltar for the whole period.

4. On arrival at Malta on 26th August, 1935, the Italo–Abyssinian crisis was starting and after 48 hours there completing with war stores the Flotilla proceeded to the Eastern Mediterranean and remained there until 18th July, 1936. During this period the Flotilla was scattered between Haifa, Alexandria and the Canal Ports, and only came together as a whole on rare occasions …

5. On conclusion of the Italian crisis the Flotilla returned to Malta on 21st July, 1936, when it was hoped that they would remain together; but this was frustrated by the outbreak of the Spanish Civil War. Again, except for passages to and from Spanish Ports, the Flotilla remained scattered for the rest of the commission.

6. A study of Enclosure No. 1 [schematic of ship movements] shows that:–

(a) The whole Flotilla met for the first time after they had been in commission for eight months, and then only for two hours.

(b) In the next three months they met twice for periods of two and three days respectively.

(c) At the end of the first year of the commission the Flotilla had been together for a total of only 25 days.

---

[1] Capt [later Adm Sir] Geoffrey John Audley Miles (1890–1986). Commanded minesweeper *Pangbourne* and SNO 1st Minesweeping Flotilla, 1931–32; Dep Dir, RN Staff College, Greenwich, 1933–35; commanded flotilla leader *Codrington* and Capt (D), 3rd DF, Med Fleet, 1935–37; Asst Dir, Tactical School, Portsmouth, 1937–38; Dir, Tactical School, 1938–39; commanded battleship *Nelson* and Flag Capt Home Fleet, 1939–41; Head of Military Mission in Moscow, 1941–43; Naval Force Cdr, Eastern Expeditionary Force (Bombay, India), 1943; Dep Naval Cdr, Southeast Asia Command, 1943–44; Flag Officer, Western Med, 1944–45; Flag Officer Commanding, Royal Indian Navy, 1946–47.

(d) For the whole commission of 20 months, the Flotilla was together for a total of only 38 days when not refitting or docking.

(e) The majority of the Flotilla spent only four months out of 20 at Malta. Certain destroyers were considerably less.

7. From the training point of view this dispersion has been most disappointing, but I feel that individual commanding officers have had certain compensations, especially on the Spanish Coast where they have been confronted with unusual situations which had to be dealt with without any precedent to guide them. I think one can say they have done this with great success.

* * *

TACTICS.

43. The tendency during the past two years has been to minimise the value and even to discredit the massed attack on a battle fleet. This is perhaps natural under the stimulus of the Italian crisis, but I wish to put on record that it is my opinion that the massed attack is by no means dead.

44. The argument usually put forward that a battlefleet of the future will be too small and too agile to offer a suitable target I think is exaggerated. A battlefleet when under fire and engaging other heavy ships tends to lose a good deal of its agility, but even if this is not so, surely we have a good counter in the shape of the Air Striking Force?

45. I would suggest that our torpedo tactics in the future should be directed to synchronising the Flotilla and Air Striking Force attacks.

This has been tried in the past but I believe without much success, usually on account of bad timing by the Air Striking Force. Communications have however improved greatly and where there is a will there is a way.

46. The chances of a fleet action are few and far between in a war, but the results to be obtained by a real victory are so immense that it would seem that no effort to achieve that victory should be spared. The Navy's reputation and the public's confidence in us would probably be irretrievably lost if another indecisive action such as Jutland were again to be fought.

* * *

55. In conclusion it is to be hoped that the international situation will soon ease up and once more permit a Flotilla to be kept together as much as possible. So much of the Flotilla work depends on a common doctrine permeating through it so that Commanding Officers will know instinctively what to do in certain situations without waiting for signals. This can only be achieved by continuous working together and by the Captain (D)

holding frequent meetings of the Commanding Officers at which these controversial points can be discussed and his ruling given.

138. *Commander G.R. Deverell to Vice Admiral Gibraltar*

[ADM 116/3518] H.M.S. *Arrow*, at Gibraltar
30 April 1937

CONFIDENTIAL

I have the honour to submit the following letter of proceedings of H.M.S. *Arrow* between the dates 19th–30th April 1937.

2. In accordance with your orders I sailed from Gibraltar at 2030 19th April 1937 and adjusted course and speed so as to arrive at Malaga at 1200 20th April.

3. The Fishery Pendant was hoisted A.M. on 20th April on crossing the 10 mile line from the Spanish coast.

4. On arrival off Malaga I anchored in the usual berth used by British Destroyers, that is, about 2 cables south of the breakwater, and then immediately landed to call on Mr. Clissold the Acting British Consul.

5. It was arranged that Mr. Clissold, Mr. Mifsud (Interpreter) and myself should pay a round of official calls on the Military and Civil Governors and on the Naval Commander during the afternoon.

6. Both of the Governors received us most cordially. I arranged with them the details of landing the 2 tons of foodstuffs, the property of the International Relief Committee, which I brought up from Gibraltar.

I also obtained their willing permission to land liberty men during daylight hours, and for Officers to land in plain clothes. The only restriction was 'No cameras'.

7. Our reception by the Naval Commander was very different. This officer was obviously very hostile to us. The brusqueness of our reception was only exceeded by that of our congé, and the traditional Spanish politeness was not at all in evidence.

8. The gist of our conversation was as follows:–

In answer to the blunt query 'Why are you here?' I explained that we were here as the servants of the International Non-intervention Committee. As he said that his Government had given him no information or instructions on the subject, I explained the scheme to him as tactfully as possible, and emphasised the point that I had no intention or authority to examine ships cargoes or to spy on them. My duties were almost entirely concerned with checking and reporting their names.

9. He was in doubt as to whether he should request me to leave at once or await instructions from his Government. I stated that if he wished I

would leave at once, but that of course the patrol would be carried on outside the three mile limit.

10. It was finally decided that I should remain where I was until instructions had been received from Salamanca.

11. Two Italian Esploratores were in harbour on arrival. *Nicoloso da Recco* and *Falco*. The former wore the flag of Rear Admiral Jachino.[1] The usual Officer of the Guard visits were exchanged, though Admiral Jachino said that he was sailing that day and did not wish to see me. His ship eventually sailed at about 2300.

12. At 1930 (dusk) I shifted berth to a night anchorage off the village of Palo to the eastward of the town.

13. At 0600 on the next morning 21st April, I returned to the day anchorage.

14. During the forenoon I received calls, first from the Captain of the Italian *Falco*, who was friendly and expressed rather the Naval Officer's than the politician's view of the general situation, and secondly from the Military and Civil Governors, their A.D.C.s and the Captain of the Port representing the Naval Commander.

15. The latter party arrived in a body, and the Captain of the Port immediately opened fire with information that his Government objected to us interfering in any way with Merchant Ships off or in Malaga or indeed in Spanish territorial waters at all.

16. The engagement soon became general with the Military and Civil Governors joining in. It was conducted in the most friendly spirit however. The Captain of the Port, who had been smuggled out of Malaga by H.M.S. *Gypsy* during the communist regime, was particularly friendly and understanding of our point of view. This was in marked contrast from the attitude of his superior.

17. However he was as adamant as he was charming, and intimated, as stated in my 1302/21 ... that even signalling to merchant ships would be regarded as 'interference'.

18. I promised not to 'interfere' and it was agreed that the status quo ante should be maintained until either side received fresh instructions.

19. To prevent misunderstanding, I requested the wishes of his Government in writing.

---

[1] Contrammiraglio [later Ammiraglio di squadra designato d'Armata] Angelo Iachino (1889–1976). His surname often cited as Jachino. Commanded gunboat *Ermanno Carlotto* in Chinese waters and Naval Attaché in China, 1923–28; Naval Attaché in London, 1931–34; commanded Italian naval forces on east coast of Spain, 1936–37; Commandant, Italian Naval Academy at Livorno, 1939–40; commanded Seconda Squadra Navale (heavy cruisers), July–Dec 1940; C-in-C Italian Fleet, Dec 1940–Apr 1943.

SPANISH CIVIL WAR AND NYON AGREEMENTS, 1936–1937   251

20. Leave was given during the dog watches and then the usual move to the night anchorage was made.

21. Before the visit of the Spanish authorities the German ship *Sexta* was boarded as she was picking up her pilot. She was found to be in possession of Certificate issued at Gibraltar. A notification of this fact was received from the Nonintervention Board Gibraltar some ten hours later.

22. 22nd April. H.M.S. *Gipsy* arrived with mail A.M. and sailed at once for Almeria.

23. I returned the call of Captain of R.N. *Falco*.

24. At about 1830 a Spanish Naval Officer boarded the ship and requested that in future I would anchor further from the harbour.

25. I replied that I would anchor wherever requested, and that the only reason that I liked being reasonably close to the entrance was because of boat work. I had no sinister intentions.

26. As he had no suggestions I suggested that I should anchor in future half a mile south of the breakwater. He protested politely that this was unnecessarily far.

27. However on return to the day anchorage on the following day 23rd April I anchored in this position.

28. At 1130 23rd April the same officer again boarded me and conveyed a formal and verbal request from the 'Salamanca Government' that I should proceed outside 'Spanish Jurisdictional Waters'.

29. He volunteered the opinion that the chief reason for this request was as stated in my 1254/23 ... that the inhabitants were muttering because a British ship was sitting outside their harbour 'controlling' 'their' shipping.

30. The use of this word 'control' which both the Italians and Spanish use where we use 'non-intervention', is unfortunate, as it appears to give an impression to the uninitiated of a more active interference than is contemplated by the Non-intervention Committee.

31. At 1230 23rd April I left the anchorage and proceeded to a position approximately 4 miles 155 degrees from Malaga. I have maintained this approximate position day and night since, moving as necessary to intercept approaching ships, but keeping outside the three mile limit.

32. The ship was able to remain stopped, or using one screw at slow speed for long periods.

33. The details of all ships investigated are given in Enclosure No. 1.[1]

34. At about 1230 on Saturday 24th April, the Spanish Government destroyers *Lepanto* and *Sanchey Barcaiztequi* were sighted about 6 miles to seaward, steering to the westward at high speed.

[1] Not reproduced.

35. When abreast the town *Lepanto* turned directly for the harbour and hoisted a very large silk battle ensign at the gaff.

36. I immediately made off to seaward on a parallel and opposite course to her at 17 knots. (Steam on 1 boiler).

37. After I had passed *Lepanto* I turned to the eastward (about 090°) in order to get as far off the line of fire as possible. By this time *S. Barcaiztequi* had also turned and was closing Malaga at high speed.

38. Contrary to my expectations fire was not at first opened on *Lepanto*, and she closed to within 2 to 3 miles of the harbour in a most dashing manner before opening her A arcs to port and opening apparently director controlled fire on the west portion of the town – possibly on the Campsa Oil Tanks.

39. It was not until about three minutes later that the shore defences opened fire. Judging by the splashes there appeared to be perhaps two 12 pounders and one six inch gun firing.

40. The 6-inch whose rate of fire was very slow was mounted on the hillside to the eastward of the village Palo. While the smaller shell possibly came from 2 field guns which are now stationed on the end of the breakwater.

41. *Lepanto* now shifted target to the 6-inch battery and steaming at a very high speed appeared to make some accurate shooting.

42. In the meantime *S. Barcaiztequi* had come in more cautiously and had opened fire on (apparently) the aerodrome situated between Torremolinos and Malaga. However when a 6-inch shell fell alongside her she turned tail and retired firing her stern guns at some target unknown.

43. *Lepanto*'s retiring Southeasterly course and high speed were again bringing the line of fire close to the fleeing *Arrow* now proceeding at full speed on one boiler. (The E.R. Department reached the creditable speed of 21½ knots).

44. The range was now large and rapidly increasing. A few rounds were mistakenly fired in *Arrow*'s direction. I accordingly turned 16 points and broke back to the westward.

45. After a few more rounds fired at random from the shore the battle ended with *Lepanto* and *S. Barcaiztequi* disappearing to the South Eastward.

46. Although the 6-inch gun had some near misses no hits were observed on the destroyers, and the damage done on shore is unknown.

47. The patrol off Malaga was continued without further incident until 1300 <u>30th April</u>, when the ship was relieved by *Grafton*.

48. After turning over duties to her *Arrow* returned to Gibraltar.

49. During most of the time the ship was on patrol the weather has been bad with rain, low visibility and heavy swell. Owing to the low visibility

it was necessary to lay out a dan buoy[1] for 48 hours. This buoy was lighted at night and provided the only mark by which the ship could maintain her position. Watch keeping with one Officer short has been arduous.

50. The following points on the working of the Non-intervention patrol are noted:–

(a) All ships investigated were amenable to the scheme. Relations with ships boarded were markedly friendly.

(b) Almost all ships were 'out of order'. That is they either had not called at a port of embarkation, or if they had, notification had not been made. On one occasion notification arrived over a day and a half after the ship had arrived at Malaga.

(c) One Master of a German ship was doubtful as to how long his certificate would be valid. He proposed to load and unload cargo in several Spanish Ports.

(d) The position of foreign ships trading between Spanish Ports is not clear, e.g. is it incumbent on a German ship which has loaded a cargo at Seville for Malaga to call at Gibraltar for an Observer en route?

(e) It would simplify identification if certificates were numbered and dealt with in the same way as Observers. The Boarding Officer reported that one certificate issued at Gibraltar was not even dated.

(f) The scheme will not begin to be efficient until Observers are embarked in the majority of ships.

139. *Vice Admiral Gibraltar to Admiralty*

[ADM 116/3521] 13 May 1937

SECRET.

[Telegram] Addressed Admiralty, repeated C.-in-C. Mediterranean, R.A.C. 3rd C.S., H.M.S. *Resource*.

IMMEDIATE.

298. Reliable wireless report from two sources indicates H.M.S. *Hunter* on patrol off Almeria is in sinking condition due to explosion on waterline port side. Cause unknown. Spanish Government destroyer *Lazaga* is standing by. H.M.S. *Hyperion* is proceeding from Malaga and Captain (D)2 in H.M.S. *Hardy* from Gibraltar.

1826/13.

---

[1] Buoy generally used in minesweeping to mark a channel searched or cleared.

140. Vice Admiral Gibraltar to Pound

[ADM 116/3521] 13 May 1937

SECRET.

[Telegram] Addressed C.-in-C. Mediterranean, Admiralty, R.A.C. 3rd C.S., H.M.S. *Resource*.

IMMEDIATE.

300. My 1826/13. H.M.S. *Hunter* reported at 1810 through Span[ish] Battleship *Jaime I* ship floating and being towed alongside. Ministry of Marine, Valencia, reports at 1824 through H.M.S. *Resource* H.M.S. *Hunter* then half a mile off Almeria port with heavy list to starboard and down by the head. Explosion occurred four to five miles off shore, believed due to external cause.

At 1930 same source reports H.M.S. *Hunter* alongside coaling wharf at Almeria. Casualties confirmed three dead and twelve believed wounded taken on board *Jaime I*. Spanish Authorities rendering all possible assistance. H.M.S. *Hyperion* expects to arrive at Almeria at 2045. Capt. (D) 2 in H.M.S. *Hardy* with one Surgeon and four sick berth ratings from *Maine* and Chief Constructor representative to report on damage, will arrive about 0200 14th May.[1]

1933.

141. H.M.S. Hunter to Admiralty

[ADM 116/3521] 13 May 1937

[Telegram] Addressed: Admiralty, repeated Comdr. R.N.B. Devonport, C.-in-C. Mediterranean, V.A. Gibraltar, Capt. D.2.

IMMEDIATE.

At 1415 13th May an explosion occurred on board H.M.S. *Hunter*, cause unknown. Number 1 boiler burst. Ship was towed into Almeria by Spanish Authorities with a heavy list. Number 1 boiler room, W/T Office, ship's galley and two after mess decks entirely destroyed. Unable to raise steam.[2]

* * *

2317/13.

---

[1] Pipon subsequently ordered the hospital ship *Maine* to proceed to Almeria as soon as possible.
[2] A list of casualties (8 killed, 14 wounded) is given, with names and addresses of next of kin.

142. *Admiralty to Captain (D), 2nd Destroyer Flotilla*

[ADM 116/3521] 14 May 1937

SECRET.

[Telegram] Addressed Captain (D) 2, repeated C.-in-C. Mediterranean, V.A. Gibraltar, R.A.C. Third Cruiser Squadron from Admiralty.

IMMEDIATE.

Admiralty are anxious to learn as soon as possible fullest details which may establish cause of explosion. In what depth of water was *Hunter* when explosion occurred? What are the details of damage? Was a column of water thrown up and if so how high? And any other information you may think useful.

1105/14.

143. *Rear Admiral L.V. Wells[1] to Admiralty*

[ADM 116/3521] 14 May 1937

SECRET.

[Telegram] Addressed Admiralty, repeated C.-in-C. Mediterranean, V.A. Gibraltar, Rear Admiral (D).

IMMEDIATE.

48. Following is finding of enquiry held on board H.M.S. *Hardy* into damage to H.M.S. *Hunter*. Begins.

No evidence of initial explosion of internal origin. Evidence points to following rapid sequence of events: (a) External detonation port side; (b) Explosion of number one boiler.

No conclusive evidence whether cause was torpedo or mine. Nothing seen to denote presence of submarines or torpedo track. One witness heard tapping on hull starboard side before explosion which might have been mine float. Presumptive evidence therefore suggests that ship struck floating mine. Ends.

1154/14.

---

[1] RA [later Adm Sir] Lionel Victor Wells (1884–1965). Commanded cruiser *Diomede*, 1927–30; commanded aircraft carrier *Eagle*, 1933–35; Dir Tactical School, Portsmouth, 1935–37; commanded 3rd CS, 1937–39; VA Aircraft Carriers, 1939–40; Flag Officer commanding Orkneys & Shetlands, 1942–44; Naval Officer-in-Charge, Aberdeen, 1944–45.

144. *Admiralty to Pound*

[ADM 116/3521]　　　　　　　　　　　　　　　　14 May 1937

[Telegram] Addressed C.-in-C. Mediterranean, repeated Captain D.2, R.A.C. Third Cruiser Squadron from Board of Admiralty.

Admiralty deeply regret to hear of the serious accident to H.M.S. *Hunter* whilst performing her humane service under the Non Intervention Scheme and offer their sympathy to you and your command. Request you will convey to Captain of *Hunter*[1] Board of Admiralty's deep sympathy with the ship's company.

1159/14.

145. *Rear Admiral Wells to Vice Admiral Gibraltar*

[ADM 116/3521]　　　　　　　　　　　　　　　　14 May 1937

SECRET.
[Telegram] Addressed V.A. Gibraltar, repeated Admiralty, C.-in-C. Mediterranean, Captain D.2.

IMPORTANT.
Your 1104. H.M.S. *Arethusa* will tow H.M.S. *Hunter* with H.M.S. *Hardy* standing by. Tug *St. Omer* is not required at present.

1237.

146. *Admiralty to Pound*

[ADM 116/3521]　　　　　　　　　　　　　　　　14 May 1937

SECRET.
[Telegram] Addressed C.-in-C. Mediterranean, repeated Captain D.2, R.A.C. Third Cruiser Squadron, R.A. Second Battle Squadron from Admiralty.

In view of accident to *Hunter* which may have been caused by a mine do you consider any further precautions to safeguard our ships operating off Spanish Coast are practicable?

1258/14.

---

[1] Lt Cdr B.G. Scurfield.

147. *Rear Admiral Wells to Admiralty*

[ADM 116/3521] 14 May 1937

SECRET.

[Telegram] Addressed Admiralty, repeated C.-in-C. Mediterranean, V.A. Gibraltar, R.A.(D).

IMMEDIATE.

49. H.M.S. *Hunter* was stopped just outside 100 fathom line. My 1154 nothing can be added until vessel is docked. Column of water and oil covered side of bridge to height of foremast. Details of damage to follow.

1311/14.

148. *Pound to Ships and Authorities in Western Mediterranean*

[ADM 116/3521] 15 May 1937

SECRET.

[Telegram] Addressed Ships and Authorities in Western Mediterranean, repeated C.-in-C. Home Fleet, Admiralty.

IMPORTANT.

201. Until situation regarding damage to H.M.S. *Hunter* has been cleared up the following instructions are to be brought into force forthwith:

A. Ships are to proceed at not less than 12 knots except in case of fog.

B. If necessary to stop ships are to be kept pointed head to wind.

C. Proposals contained in Admiralty Telegram 2009 of 13th May[1] to V.A. Gibraltar and R.A.C. 3rd C.S. only are not to be brought into force for the present.

D. Ships operating off Almeria are to be kept one mile outside the 100 fathom line.

1155/15.

---

[1] Notation on messsage: 'One of H.M. Ships appearing periodically off Carthagena, Alicante, etc.'

149. *Vice Admiral Gibraltar to Pound*

[ADM 116/3521] 15 May 1937

SECRET.

[Telegram] Addressed C.-in-C. Mediterranean, repeated Admiralty, R.A.C. 3rd C.S., Capt. (D)2.

315. After consulting R.A.C. 3rd C.S. and Capt.(D)2 I am of opinion that risk of an accident similar to that of H.M.S. *Hunter* has always existed and still exists. It is considered Almeria is least important of places to be watched in non-intervention scheme, its facilities for landing war material being very bad and there has been practically no traffic into port except by Spanish coasting vessels since scheme came into force. In view of this and my available destroyers being reduced by one it is proposed to discontinue Almeria beat for the present and to visit it thoroughly by vessel patrolling at Malaga on her relief.

2123/15.

150. *Pound to Admiralty*

[ADM 116/3521] 16 May 1937

SECRET.

[Telegram] Addressed Admiralty, repeated C.-in-C. Home Fleet, V.A. Gibraltar, R.A.C. 3 C.S., Capt. (D)2.

IMPORTANT.

203. Taking into consideration the evidence available to date as to cause of damage to H.M.S. *Hunter* I am of opinion:

A. As ship was on 100 fathom line and explosion apparently occurred on water line it is unlikely it was caused by moored mine.

B. As explosion occurred on water line a torpedo which caused it must have been running on the surface and track would probably have been seen.

C. Taking into consideration direction of wind and ship's head damage from a floating mine would almost certainly have occurred on water line and explosion would have taken place on port side of ship, as it did.

Further evidence may be obtained where ship is docked, and a more complete deduction may then be possible, particularly as to damage at 136 station which at present is unexplained.[1]

---

[1] Pipon subsequently informed Pound on 20 May that the Commanding Officer of HMS *Hunter* reported the damage aft was caused by the same explosion and several officers who were aft at the same time felt a distinct whip. V.A. Gibraltar to Pound, 1700/20.

For the time being I am assuming damage was caused by a floating mine and have accordingly issued instructions contained in my 1155 of 15th May.

0013/16.

151. *Vice Admiral Gibraltar to Pound*

[ADM 116/3521] 17 May 1937

SECRET.

[Telegram] Addressed C.-in-C. Mediterranean, repeated Admiralty, R.A.(D).

IMPORTANT.

329. It seems most probable that no mines have been laid off Nationalist ports and therefore it is submitted that precautions as laid down in your 1155 15th May need only be carried out when working off Red territory.

2051/17.

152. *Vice Admiral Gibraltar to Admiralty*

[ADM 116/3521] 17 May 1937

SECRET.

[Telegram] Addressed Admiralty, repeated C.-in-C. Mediterranean, R.A.(D), R.A.C. 3rd C.S., Capt.(D)2.

331. It is submitted that accident to H.M.S. *Hunter* involved loss of several lives, the disablement of a valuable ship and consequent high cost of repair. Strongly suggest destroyers should be replaced by some less valuable vessels such as armed trawlers who would do the work equally efficiently and release the destroyers for their legitimate service.[1]

2221/17.

---

[1] The Admiralty replied: 'Your suggestion will be given full consideration'. 1604/18.

153. *Vice Admiral Gibraltar to Admiralty*

[ADM 116/3521] 17 May 1937

SECRET.

[Telegram] Addressed Admiralty, repeated C.-in-C. Mediterranean, R.A.(D), R.A.C. 3rd C.S., Capt.(D)2.

332. Court of Enquiry as to explosion in H.M.S. *Hunter* sat this morning, Tuesday, and reported that damage was almost certainly caused by either floating or moored mine. Portion of distorted metal ⅛ inch thick with eye bolt welded on by acetylene and stamped with monogram W.S. or H.S. number 12682 was found in damaged part of ship.

H.M.S. *Hunter* was in position Almeria Light 168 degrees 6 miles, depth 100 fathoms, ship stopped, Wind force two south by west, ship was drifting slowly to port.

2359/17.

154. *Pound to Vice Admiral Gibraltar*

[ADM 116/3521] 18 May 1937

SECRET.

[Telegram] Addressed V.A. Gibraltar, repeated Admiralty, R.A.(D).

211. Your 2051 17th May. Whilst agreeing probability that mines are not laid off Nationalist ports is correct, possibility that they may be and risk from floating mines which have broken adrift from Nationalist field cannot be ignored. The instructions contained in my 1155 May 15th are therefore to be adhered to for the present.

2258/18.

155. *Pound to Admiralty*

[ADM 116/3521] 22 May 1937

SECRET.

[Telegram] Addressed Admiralty, repeated V.A. Gibraltar, R.A.(D), R.A.C. Third Cruiser Squadron, H.M.S. *Emerald*.

227. Reference message 2221 of 17th May from V.A. Gibraltar.

A. It is agreed that Destroyers carrying out non-intervention duties should be replaced by less valuable ships. For this purpose Armed

Trawlers are adequate but are not less liable to suffer damage from mines than destroyers.

B. The most suitable ships for this duty are vessels of First M.S.F. H.M. Ships *Kingfisher, Mallard, Puffin*, which should be immune from all but drifting mines.

C. For the four patrols off Huelva, Cadiz, Malaga and Almeria at least eight vessels are required.

D. In addition it is considered two destroyers should be available at Gibraltar for dealing with incidents at sea such as interference with Merchant Ships.

E. Although the coast under Spanish Government control is more likely to be mined than elsewhere, the duties of H.M. Ships outside British non-Intervention area are such that ships on passage can keep well clear of coastal waters and can proceed to anchorage guaranteed clear of mines.

F. Owing to their speed and necessity for maintaining prestige, destroyers are considered most suitable for north eastern area and Balearic Islands. A total of two destroyers and one receiving ship is normally sufficient for this area.

G. If vessels other than destroyers can be provided for non Intervention patrol, all Destroyer commitments in Spanish waters can be met by one Division.

H. During present period when only two Flotillas are available on Station, the release of one Division would enable essential training to be carried out, both in Flotilla and the Fleet.

I. H.M.S. *Vanoc* is not considered suitable for D or F unless brought up to full complement, but there are many duties on which she can be employed in southern area with her present complement.

1433/22.

156. *Admiralty to Pound*

[ADM 116/3521]   25 May 1937

SECRET.

[Telegram] Addressed C.-in-C. Mediterranean, repeated V.A. Gibraltar, R.A.(D), R.A.C. 3rd C.S., C.-in-C. Plymouth, C.-in-C. Portsmouth, C.O.C.O.S., V.A.C.R.F., S.O.R.F. Rosyth, S.O.R.F. Devonport from Admiralty.

Their Lordships have approved of eight of the following nine trawlers being brought forward temporarily to replace destroyers on non-intervention duties provided additional ratings required can be found from

Mediterranean resources: *Willow, Hawthorn, Lilac, Magnolia, Holly, Laurel, Jasper, Amethyst* and *Agate*.

Request report by signal whether additional personnel required can be provided.

1831/25.

157. *Pound to Admiralty*

[ADM 116/3521] 27 May 1937

SECRET.

[Telegram] Addressed Admiralty, repeated V.A. Gibraltar, R.A.(D), R.A.C. 3rd C.S., C.in-C. Plymouth, C.-in-C. Portsmouth, C.O. Coast of Scotland, V.A.C. Reserve Fleet, S.O. Reserve Fleet Rosyth, S.O. Reserve Fleet Devonport.

253. Your 1831 25th May. The offer of eight trawlers is much appreciated but as it is not possible to withdraw ratings from aircraft carriers, cruisers and destroyers and as H.M.S. *Warspite* and H.M.S. *Malaya* will not have completed working up for some time it only leaves H.M.S. *Barham*, H.M.S. *Hood*, H.M.S. *Repulse* to provide officers and crews and each ship could only provide one crew. The withdrawal of a trawler crew from each of these three ships will only release two destroyers and would cause so much dislocation and loss of efficiency that it is not recommended.

In any case these crews could not be provided until 181 ratings who are steaming minesweepers to Malta have been released unless departure of minesweepers were delayed which is undesirable.

There appears no alternative to continuing to use destroyers for non intervention duty unless some fully manned ships such as those mentioned in my 1433 22nd May are lent temporarily for the purpose.

1723/27.

158. *Lieutenant Commander R.C. Gordon to Rear Admiral Gibraltar*

[ADM 116/3519] H.M.S. *Hereward*, at Gibraltar
7 June 1937

CONFIDENTIAL

I have the honour to submit the following letter of proceedings for H.M. Ship under my command covering the period 30th May, 1937 to 7th June, 1937.

Sunday, 30th May, 1937.
1400. Sailed from Gibraltar and after rounding Europa point, set course 079° for Almeria area.

1700. Passed German A.S. *Deutschland* proceeding on a westerly course towards Gibraltar at high speed. It was not known at that time that anything was amiss with her, and nothing unusual was observed.[1] *Hereward* passed her port side.

Monday, 31st May, 1937.
0500. Arrived in patrol area off Almeria.

0525. German A.S. *Admiral Scheer*, accompanied by destroyers *Albatross* (AT), *Seeadler* (SE), *Luchs* (LW) and *Leopard* (LP), passed close to northward of *Hereward* on a westerly course. The destroyers were stationed one on either bow, and one on either quarter, about 7½ cables from *Admiral Scheer*, and appeared to be sweeping with P.V.s. Their speed was about 15 Knots. A P.V. was later found adrift and recovered by *Hereward*. It was also noticed that ships were cleared for action.

0615. It was observed that the German ships had altered course to the north eastward from a position to the westward of Almeria Bay, and had increased speed to about 25 knots.

0629. *Admiral Scheer* opened fire from a position about 8 miles 197° from Almeria on a course of about 090°. *Hereward* was then about 5 miles to the south east on a westerly course. She fired 27 salvoes from her main armament in three gun salvoes between 0629 and 0646, at which time she altered course about 160° to starboard, and fired 5 more heavy salvoes between 0652 and 0658, when she ceased fire and retired to the south west.

The secondary armament of *Admiral Scheer* and the destroyers' armaments were also observed to be in action during the period 0629 to 0658.

At about 0640 some splashes were observed falling among the German ships, one or two of which appeared fairly large, but it is not thought that any hits were registered and firing from the shore ceased almost as soon as the Germans ceased.

The effect of the German fire has been described elsewhere. Nothing except a dense pall of smoke over Almeria could be seen from *Hereward*. Conditions for bombardment were excellent, the sea being calm and visibility extreme, and the operation seemed well planned and well executed. It is thought that the two rear destroyers attempted to make a

---

[1] On the afternoon of 29 May the armoured ship *Deutschland* had been bombed by Spanish Government aircraft and shelled by destroyers when anchored off Ibiza Island. The ship was hit by two bombs, twenty-three seamen were killed and another eight subsequently died of wounds.

C.S.A.[1] smoke screen to cover the retirement but if this was so it was a complete failure.

Hereward continued her patrol and sighted aircraft at 1100 and 1545, flying from seaward towards Almeria.

* * *

159. *Commander S. Arliss[2] to Rear Admiral Gibraltar*

[ADM 116/3519] H.M.S. *Hero*, at Gibraltar
12 June 1937

CONFIDENTIAL

I have the honour to forward the following letter of proceedings for H.M. Ship under my command for the period June 3rd–11th whilst on Non-Intervention patrol off Cadiz.

* * *

5. At 1215 on 8th June German Submarine *U.26* closed and asked me to stop as he wished to speak to me. He asked me if I would allow him to exercise in some area off Cadiz where he would be safe doing diving practice during daylight hours. I sent an Officer onboard with a chart and gave him an area clear of my patrol. Everything was done in a very friendly manner and the German Captain said that he hoped that in a short time Germany would be resuming N.I. patrol duties,[3] and when the whaler returned to the ship they brought a present of half a dozen bottles of German Lager. Observations made by Lieutenant Hussey whilst onboard the submarine are the subject of a separate letter which is attached.[4]

* * *

---

[1] Smoke floats.
[2] Cdr [later VA] Stephen H.T. Arliss (1895–1954). Commanded destroyers *Sterling*, 1926–29; *Waterhen*, 1929–31; *Thruster*, 1932–33; *Wild Swan*, 1934–36; *Hero*, 1936–37; Naval Attaché, West Coast of South America, 1938–40; commanded destroyer *Napier* and Capt (D) 7th DF, 1940–42; Cdre (D), Eastern Fleet, 1943–44; commanded cruiser *Berwick*, 1944–46; Flag Officer commanding British Naval Forces, Germany, 1947–49.
[3] Following the attack on the *Deutschland* the Germans withdrew from the non-intervention patrols, demanding a guarantee similar attacks would not be repeated.
[4] Not reproduced.

160. *Pound to Admiralty*

[ADM 116/3521] 13 June 1937

SECRET.

[Telegram] Addressed Admiralty.

IMPORTANT.

683. Since my arrival at Gibraltar I have investigated conditions under which destroyers are carrying out non-intervention duties.

2nd Destroyer Flotilla has been employed on this duty for 37 days.

Excluding H.M.S. *Hereward* and H.M.S. *Hunter* the remaining five ships employed on patrol have during this period each been 24 days at sea and average total number of miles steamed by each ship repeat by each ship is 4,800.

The normal distance steamed by a destroyer in a year is 12,000 miles.

If only 2 Flotillas are available each Flotilla will (a) do 30 days on Spanish coast of which 20 days will be spent underway on patrol and during which each destroyer will steam about four thousand miles; (b) be away from Spanish coast for 30 days out of which 7 days will be spent proceeding to Malta and back. The remaining 23 days will be required for rest, making good defects and docking and no time whatever will be available for weapon training.

I am of the opinion that the above state of affairs is one which should not be accepted as apart from loss of weapon training the wear and tear on both machinery and personnel will be excessive. It is not possible to reduce number of ships on patrol or to reduce materially the amount of steaming they do on patrol.

I would suggest therefore that 1st Minesweeping Flotilla which should if possible be brought up to nine ships should work in with First and Second Flotillas so that each flotilla would have one month on Spanish coast and two away from it.

If above is approved in principle it would be convenient if 1st Minesweeping Flotilla could arrive at Gibraltar on 10th July repetition 10th July for 30 days duty on coast of Spain on completion of which they could return to U.K. for nearly two months before their next turn of duty.

2138/13.

161. *Admiralty to Pound*

[ADM 116/3521] 14 June 1937

SECRET.

[Telegram] addressed C.-in-C. Mediterranean, repeated C.-in-C. Home Fleet, Captain F.P. & M., S.O. 1st M.S.F., Rear Admiral (D), Rear Admiral Gibraltar, C.-in-C. Portsmouth, C.-in-C. Plymouth, C.-in-C. Nore, Captain-in-charge Portland.

Reference your 2138/13 to Admiralty only.

Their Lordships fully concur in the importance of relieving for a period the destroyers carrying out the duties of Non-Intervention patrol, not only for the purpose of providing opportunity for Flotilla training but also for giving a necessary rest to personnel and material. Owing to the nature of the service at present involved in the protection of shipping on the North Coast of Spain, this duty must continue to be carried out by destroyers.

Their Lordships have consequently decided that for a period of 30 days the 1st M.S.F. shall carry out Non-Intervention patrol duties in the southern area.

The 1st M.S.F. are therefore to leave Portland for Gibraltar on Monday 12th July and return to the United Kingdom to give summer leave sailing from Gibraltar on Monday 16th August.

As the 1st M.S.F. consists of only 7 vessels these could be augmented if you so desire by *Elgin*, *Lydd* and *Saltash* as there are no other vessels available at home.

2115/14.

162. *Pound to Admiralty*

[ADM 116/3522] 24 June 1937

SECRET.

[Telegram] Addressed Admiralty, repeated C.-in-C. Home Fleet.

374. Reference Admiralty letter M.05465/36 22nd December 1936 para. 12.[1] In view of German Cruiser *Leipzig* incident possibility that a foul attack may be made on one of H.M. Ships cannot be altogether ruled out.[2]

---

[1] Concerning hostile action against British ships by Spanish Government or Insurgent warships outside of Spanish territorial waters.

[2] The German light cruiser *Leipzig* reported that on 15 June three torpedoes had been fired at the ship when off Oran and that on 18 June either a torpedo had glanced off the side of the ship or the ship had come in contact with part of a submarine. There were no Republican

I therefore intend to issue following instructions:

(A). If periscope of a submerged submarine is seen immediate avoiding action is to be taken on the assumption that an attack may follow but no counter attack is to be made unless attacked.

(B). Ships are to consider they have been attacked if track of a torpedo is seen approaching ship.

(C). If a capital ship or cruiser is attacked as in (B) avoiding action is to be taken and two depth charges if carried are to be dropped as near submarine as avoiding action takes them.

(D). If a destroyer or minesweeper is attacked as in (B) the submarine is to be hunted and destroyed if possible.

(E) If a capital ship or cruiser accompanied by destroyer is attacked, capital ship or cruiser is to take avoiding action while destroyer hunts and if possible destroys submarine.

(F). The above rules to apply irrespective of territorial waters.

1803.

163. *Pound to Vice Admiral Gibraltar*

[ADM 116/3521] 25 June 1937

SECRET.

[Telegram] Addressed V.A. Gibraltar, repeated Admiralty, R.A.(D).

384. As it is considered it has now been established that the mine which damaged H.M.S. *Hunter* was a moored mine and as personal observations point to probability that most reports by merchant ships of floating mines are due to mistaking turtles or other floating objects for mines the orders contained in paragraph (a) of my 1155 15th May are cancelled.

The prohibition as regards anchoring is withdrawn in respect of vessels patrolling to westward of Gibraltar or off Malaga.

1953/25.

---

submarines at sea on either day and Konteradmiral Hermann von Fischel, the senior German naval officer in Spanish waters, subsequently confided in private to British officers at Gibraltar that the alleged attack might have been caused by porpoises. See Willard C. Frank, 'Misperception and Incidents at Sea: The *Deutschland* and *Leipzig* Crises, 1937', *Naval War College Review*, vol. 43, no. 2 (Spring, 1990), pp. 40–42.

164. *Captain C. Tait to Pound*

[ADM 116/3520]
H.M.S. *Shropshire*
First Cruiser Squadron, at Chatham
26 July 1937

CONFIDENTIAL.

No. 1580/191B.

I have the honour to forward this my letter of proceedings of H.M.S. *Shropshire* from 15th June to 21st July, 1937. During this time I was Senior Naval Officer on the North East coast of Spain from 15th June to 12th July.

I do not propose to amplify my previous reports in this letter but to invite attention to certain aspects of the Spanish Civil War which touch British interests and may in due course affect the Royal Navy.

2.– Eleven months ago, at the beginning of the Civil War, the merchant ships using the ports of Alicante, Valencia and Barcelona flew the flags of those nations accustomed to trade there. As time went on however, the mercantile flags of those countries not possessing strong Navies became fewer and fewer, until now it can be said that except for an occasional Frenchman or Greek oiler the vessels entering those ports all fly the British flag, and disreputable looking tramps some of them are. I have been informed by masters of the regular shipping lines that these tramps are on time charter to the Spanish Government, and it is a condition of charter (perhaps only implied) that they fly the British flag in order to get the protection of the British Navy.

3.– There is a belief in the mercantile marine that Great Britain is the only nation prepared to supply full protection to ships plying to Spanish Government ports. British ship masters on the regular legitimate run expressed to me their opinion that if the British Navy withdrew from the coast of Government Spain, ships with British crews would refuse to go there either. If this opinion is well founded the results for Government Spain might be serious, and this leads to the thought that the presence of British warships, which is known to be uninterrupted, on this coast is in itself an indirect but very real help to the Spanish Government side. I have no reason to suppose the tramps are doing anything unlawful but it may come about that before this civil war is over the British Navy will be called upon to intervene actively on behalf of such changelings as the *Jenny* where the only British subject on board is the wireless operator, or on behalf of such ships as the *Noemijulia* which carries no British subject at all and whose disreputable appearance alone is an outrage to any self-respecting mercantile marine. Such a possibility has been the subject of much comment in the officers' messes and on the lower deck of the *Shropshire*.

4.– The evacuation of refugees to Marseilles by British warships and the *Maine* continues steadily and is now a recognised institution on the coast, in fact it has almost assumed the proportions of a right. I have emphasised in my telegrams that I do not believe the many thousands of 'Whites' now in Madrid to whom the Government are prepared to give passports can be got to the coast by the existing organisations in a reasonable time. At the moment these wretched people do not appear to be in immediate danger of their lives but their condition is miserable enough. With the railway cut and the main road under frequent bombardment it is a difficult matter to get food for a million people into the city. The army gets what it wants, so do the unions, and the class which supplies the political opponents of the proletariat has to take what is left. Some of the young children refugees show unmistakable signs of malnutrition, but the great danger will come when, if ever, the Nationalists succeed in isolating Madrid from the coast. It is then that people who know Spain and the Spaniards tell me the unions will take charge in Madrid and a wholesale massacre of Whites may take place before the city falls. It is for this reason that Dr. Junod,[1] the Swiss Director of the International Red Cross, is endeavouring to obtain fifty lorries for the transport of refugees to the coast while there is still time. The problem as to how to get them away to France, or some port in White Spain still remains to be solved, and it is urged that plans should be made as soon as possible. Dr. Junod is enquiring as to the feasibility of the International Red Cross chartering a ship, and I have pointed out to him that there are passenger ships laid up in Valencia and Barcelona, because the Government cannot run them in face of the Nationalist sea superiority, any of which would suit his purpose admirably.

The only satisfactory solution will be when the Red Cross takes over the selection of refugees, obtains their passports, transports them by land and finally ships them away.

5.– I have written this last paragraph because the present haphazard evacuation organisation has already created vested interests. Evacuation now goes by favour and takes place under the auspices of some Embassy or ministry. Favours can be bought and much money is being made out

---

[1] Dr Marcel Junod (1904–61). Representative of the International Committee of the Red Cross (ICRC) in Abyssinia, 1935–36; Spain, 1936–39; ICRC delegate in inspection of German and subsequently Japanese prisoner of war camps in accordance with the Geneva Convention, 1939–45; organised relief efforts in Hiroshima, 1945; representative of United Nations Children's Relief Organization (UNICEF) in China, 1948–49; Vice President, ICRC, 1959–61. His account of his experiences is: *Warrior without Weapons*, English trans. (London, 1951).

of the miseries of the Whites. Even the most elect profit by it. A lady refugee told me that her sister, the wife of a rich man in Madrid who had been shot in the early days of the war, paid 60,000 pesetas for the safe passage to a British destroyer of herself and her baby daughter.

The Panamanian Minister himself occasionally accompanies select parties down from Madrid. Before embarking, every refugee is subject to a most rigorous search as no money or valuables are allowed to be taken out of Spain. Refugees can take the clothes they stand up in and a change, and that is about all. The Panamanian Minister accompanies his party to the destroyer to say good-bye. A most polite attention. But when the embarkation boat is well away from the pier he produces packages from his pockets and hands them round; the money and jewellery of his refugees. He himself has diplomatic immunity from search, but he is breaking the law and running considerable risks. Of course it may be that he is actuated by motives of compassion, but the general impression is that the percentage he abstracts from each packet as his commission is high.

General Miaja's[1] nephew, secretary to his uncle in Madrid, has built up a nice little business in selling permits to leave the country. Ostensibly the money goes to the 'cause' but unless some Embassy of repute arranges the whole affair a heavy percentage is demanded by this promising young man. And so the ramp goes on.

6.– When the wireless set is installed in the Consulate General at Barcelona it will probably be unnecessary to station a British man-of-war constantly at that port. Therefore, two instead of three ships should be sufficient to give all the protection that is needed for British nationals in the Government ports of Barcelona and Valencia, and these two should also be able by frequent visits to keep in personal touch with the British Consul at Palma and through him with the Nationalist Naval authorities.

Except as regards prestige it appears to matter little what size or sort our guard ships are. The essential qualification is that they must be able to maintain wireless touch with Valencia and Barcelona all over the area. The evacuation of nationals, if the gangs which control the water front are in a nasty mood and refuse to let them pass, would be a task beyond the capacity of the largest ship in the Mediterranean Fleet. On the other hand when the gangs are in a complaisant mood there would appear to be no urgent need to evacuate foreigners at all.

---

[1] General José Miaja (1878–1958). Spanish military officer who remained loyal to the Republic and chief of Junta de Defensa [Defence Council] de Madrid, 1936–37; commanded Republican forces in central southern zone, 1937–39; joined so-called Casado Coup and was President of the Casado National Defence Council in a desperate and unsuccessful effort to reach a negotiated settlement with the Nationalists, March, 1939; escaped and spent the remainder of his life in exile.

7.– To the monotony of life on board these guard ships has been added the unpleasantness of 'darken ship' at Valencia and Palma. The nights have now become very hot and the discomfort for all hands in consequence has become very real. More than this, the men's recreations in the evening are seriously interfered with. There can be no upper deck dancing, tombola or other games, and the cinema has to be housed in by canvas screens in a stifling atmosphere. For this reason I would recommend that the practice of keeping a ship on the coast for a month at a time be reconsidered and that three weeks would be the maximum.

8.– Of the opposing fleets in this area there is little to note. Spanish Government destroyers convoy oilers up and down the coast, mostly inside territorial waters and the big Insurgent cruisers bombard the Government towns at fairly frequent intervals.

To the outward appearance, discipline on board the *Canarias* and *Baleares* is well maintained. Coming into their harbour at Palma, their crews are always fallen in on the upper deck by divisions, properly dressed and in clean white suits. The ships look seamanlike, no ropes' ends are hanging over the side, guns are properly trained and a general air of efficiency prevails.

On the other hand, to the seaman's eye, nothing can be more offensive than the appearance of the Government destroyers, the battleship *Jaime I* and the submarines. On board these there is no pretence of order or cleanliness, their crews are dressed anyhow and are indescribably dirty.

9.– At Palma the Italians seem to keep permanently three or four destroyers. These ships appear to be clean and well kept but their crews were not as smartly turned out as were those of the *Canarias* and *Baleares*. To my surprise I learnt from the Consul that the Spanish Naval Officers look on these Italians with a certain amount of contempt. But the Italians certainly take themselves seriously. They were constantly at drill, one morning they were even scrubbing decks in gas masks which they kept on for several hours.

The Italian Officers were distinctly cordial towards us, much more noticeably so than they have been before. The Italian Admiral asked me to lunch on the day we arrived and the Italian ward rooms sent calling parties to the *Shropshire* quite early in the day.

10.– At Barcelona the French keep permanently a small ship of the chasseur type, mainly as a wireless link between their Consul General and the outside world. A French destroyer visits Valencia every week and stays for a few days. Added to this the small French merchant ship *Iméréthie II* runs to Valencia roughly once a fortnight and to Barcelona once a week evacuating French nationals to Marseilles. Latterly this ship has, in approved cases, been taking certain Spanish nationals as well.

Outwardly these large French destroyers appear clean and well kept, and on ceremonial occasions and when they came over to the *Shropshire* for games the crew were always well turned out. The officers tried to be friendly and were most polite and punctillious in their calls. I always found the French exceedingly well informed of the state of affairs, and it was most refreshing to listen to the pithy way in which, with a little encouragement, they would sum up the international situation. They are one and all convinced that, if the Spanish Government can tide over this summer without a really heavy military defeat, the campaign of 1938 will finish Franco.[1] They have all the arguments on their side unlike the Italians, who contented themselves with asserting positively, but with I thought a trace of apprehension, that victory for the Nationalists was an imminent event.

11.– An incident which occurred to the destroyer *Verdun* lying in the roads at Valencia nearly gave rise to an international crisis. I was on board her one morning celebrating the birth of a fourth daughter to her Commanding Officer Capitaine de frégate Yves Aubert, a most friendly little man, when a lieutenant burst into the cabin with the news that a submarine had just fired a torpedo at the *Verdun*. We dashed on deck and in the waist joined an excited group round a seaman, who, gesticulating wildly, was pointing to the supposed track and shouting that he, he himself, had seen a torpedo. But I could see no trace nor anything resembling one. Appealed to, a few more seamen after some hesitation at first, but later becoming vehement, also said they had seen the track, and after a search, though no one would confess to seeing a submarine, a man was found who had actually with his own eyes observed a torpedo pass under the forecastle, and from the direction from which it came it must have passed close to the *Shropshire*. I knew that nowhere in that direction was the depth more than seven fathoms and I openly said that I did not think a submerged submarine could be operating in such shallow water. This opinion was received in silence, disappointed silence, but soon the witnesses got going again and became more positive than ever. I returned to the *Shropshire* but no one there had noticed any torpedo or track. By a piece of good luck the *Shropshire*'s aircraft had been sent up that morning

---

[1]Generalissimo Francisco Franco y Bahamonde (1892–1975). Gained an impressive reputation as successful young officer in Spanish Morocco; played leading role during Rif Rebellion in successful amphibious landing at Alhucemas Bay, 1925; commanded suppression of miners' revolt in Asturias, 1934; although outwardly cautious and non-political, sent by a suspicious newly elected Popular Front Government to command the distant and unimportant Canary Islands, February 1936; joined military revolt in July and commanded the Spanish Army in Morocco; proclaimed Generalissimo of Nationalist forces, October 1936; following Nationalist victory in 1939 became head of Spanish state until his death, generally referred to as 'Caudillo' (leader).

to search for a submerged mine that a merchant ship had reported as having seen in the anchorage the evening before. It so happened that at the moment the torpedo was supposed to have passed under the *Verdun* the aircraft was circling over that ship at a height of about 800 feet preparative to descending. Neither the pilot nor observer had seen either submarine, torpedo or track in the clear water. This finally decided me that the thing was impossible and I sent over to Yves Aubert to say so. Back came the little Commandant to inform me that several more men now said they had seen it and he thought he really ought to make a report at once. Politely but definitely I had to tell him that if he did so I should also have to make an immediate report and record my opinion. That shook him and he went away saying that he would think it over. For the next 24 hours the *Shropshire* kept W/T watch on the *Verdun* but no long signals were sent out and so I could conclude he thought better of it. At any rate when he came to say good-bye next day he never mentioned torpedoes and we resumed our conversation about his family history. The *Verdun* left for Toulon that night and I did not see her Captain again.

I should like to pay tribute to the French naval officers of the Flotilla off the coast of Spain. They were always ready to be most friendly and with a little encouragement never failed to be bright and amusing and to take part in anything that was going on. I have not yet met a morose French Naval officer. Their seamen too, although they had little in common with ours, and at first were a little diffident, quickly responded to any advances and were ready to do their part in helping to pass away monotonous time for us all.

* * *

13.– The achievements of Lieutenant Commander A. Hillgarth, Royal Navy (retired), British Consul at Palma are too well known to need recapitulation from me. It is probable that his contacts with the Naval and Military chiefs of the Nationalist Forces in the Balearics are such as would be quite beyond the reach of an ordinary career consul. This is entirely a personal matter beyond the control of the Foreign Office, Admiralty or any outside individual and results in a full and frank disclosure of naval plans and a good deal of information regarding army ones. In fact in the personality of this one man, not only the Navy but Great Britain has a unique asset in the right place whose value cannot be assessed too highly. This unpaid consul seeks nothing for himself. I have not discussed the matter with him but I have the impression that he would appreciate more than anything else promotion to commander on the retired list. Such a step, if the regulations permit or could be stretched sufficiently, would cost nothing and would show at any rate the high appreciation in which the Navy holds his ungrudging services, and I therefore submit this proposal for favourable consideration.

14.– The source of all the official information concerning Nationalist naval forces and movements at Palma is the Port Admiral, Rear Admiral Bastarreche. A portly and most friendly little man, he commanded the *Canarias* on the outbreak of the war but was transferred to Palma on promotion. Frank and open, he does not seem to object to discreet questions and seems to answer them to the best of his ability. He professes an admiration for the British Navy and a liking for British naval officers, as he assured me do all the officers of the Nationalist Navy. It is for this reason and for this reason alone, so he naively informed me, that his side tolerated the activities of the British Navy off Bilbao and Santander. If the French tried the same game it would not be put up with for a moment. He understood the British position, though he did not like it, but all his side knew that the British Navy would not take advantage of the weakness of the Nationalists and go beyond the strict rules. He wished he knew what these rules were, and as he said this I wished that I could hand him a copy, it would make things so much more simple if each side were able to consult the regulations the other party were using.

It came out in the course of conversation that Admiral Basterreche had an elder brother and three nieces in Madrid who are in hiding and concerning whose fate he is extremely anxious. He gave me their address, and on return to Valencia a letter was written to the British Consul at Madrid requesting him to seek out these people and if possible evacuate them to Valencia under the auspices of the British Embassy. If this could be done it would be a courteous gesture and one that would be much appreciated by the admiral if these relatives of his were sent across to Palma in a British destroyer, even if a special trip were necessary.

15.– On 12th July the *Penelope* arrived at Valencia to relieve the *Shropshire*, and after nearly a year on and off that coast we sailed for home to pay off.

* * *

It was a great relief to all hands to know they were sailing for home and to turn their thoughts as well as the ship's head towards a place where Spaniards cease from troubling and gangsters are at rest.

16.– The *Shropshire* had arranged to take White refugees from Valencia to Algeciras up to the number of 200, but owing to General Miaja and general muddle only ten arrived.

* * *

The *Shropshire* arrived at Sheerness on 19th July and went up to Chatham on 21st July, commencing at once to give paying off leave after having been in commission for two years and eight months.

[Minute by Pound]

Forwarded for the information of Their Lordships concurring generally with the remarks of Captain Tait.

2. With reference to para. 6, two ships only are now being maintained in the Eastern Area, vide my telegram 2158 of 17th June. The irksomeness of the periods of duty at Valencia, Barcelona and Palma is fully appreciated and endeavour is made to arrange reliefs so as to alleviate this as much as possible, consistent with meeting the other requirements of the station. With the number of ships available and the length of passage involved from and to Malta it is not at present practicable to reduce these periods to three weeks as proposed by *Shropshire*.

3. With reference to para. 12, I hope something can be done to draw the attention of the appropriate authorities to the really excellent work performed by the Reverend Hubert Cox.[1]

4. With reference to para. 13, Lieutenant Commander Hillgarth has been of very great assistance to H.M. Ships visiting Palma and his name has frequently been favourably mentioned to Their Lordships in reports, etc. He was recently awarded the Order of the British Empire. It would appear to be more appropriately the concern of the Foreign Office to decide whether the nature of his services are considered such as to warrant still further recognition; per se they do not appear to warrant promotion in naval rank.[2]

5. With reference to para. 14, any reasonable degree of assistance in evacuating bona fide refugees or exchanging hostages has been and will continue to be afforded. The question whether the sending of a destroyer on a special trip to Palma would be justifiable, in connection with the Bastarreche family, will depend on the circumstances at the time.

6. I am strongly of opinion that the question of the wholesale transfer of foreign vessels to the British register under circumstances as have been obtaining on the coast of Spain for the past year merits the most careful consideration.

---

[1] In paragraph 12 (not reproduced), Tait praised the work on behalf of refugees transported in British warships performed by the Revd Hubert Cox, Church of England Chaplain at Marseilles.

[2] Captain J.H. Godfrey, who had known Hillgarth while commanding the *Repulse*, subsequently became Director of Naval Intelligence and was instrumental in having Hillgarth appointed as naval attaché in Madrid with promotion to commander and, in 1940, captain.

It would be a travesty of justice if H.M. Ships were called on to defend a so called British ship owned by foreigners, manned by foreigners and operating under charter to one of the belligerents.

[Signed] Dudley Pound,
at Malta [7 August 1937].

165. *Pound to Mediterranean Station*

[ADM 116/3522] 18 August 1937

SECRET.

[Telegram] Addressed Mediterranean Station, repeated Admiralty.

IMPORTANT.

666. H.M. Government has issued warning through press that if a British merchant ship is attacked by submarines without warning, H.M. Ships are authorised to counter attack. Following instructions which apply to whole of the Mediterranean outside territorial waters, are to be brought into force forthwith.

2. <u>British merchant ships torpedoed in daylight.</u>

British destroyer or destroyers which are or come on scene of attack should first see that crew of ship torpedoed are safe and then hunt submarine. If submarine remains submerged it may be taken as a sign of guilt, and if contact is first made within five miles of ship which has been torpedoed, she is to be hunted and sunk.

3. If submarine is detected submerging at a greater distance than five miles from ship torpedoed, she is not to be attacked but should be hunted in the hope that she may be forced to come to the surface and disclose her identity.

4. If a submarine is sighted on the surface in the vicinity of the ship torpedoed she is to be closed and action to be taken as follows.

(a). <u>If she remains on the surface</u> her identity is to be established but she is not to be attacked.

(b). <u>If she submerges after her identity has been established</u> – she is not to be attacked.

(c). <u>If she submerges before she is identified</u>, and position in which she submerges is within five miles of the ship torpedoed, she is to be hunted and sunk.

(d). <u>If she submerges before she is identified</u>, and position in which she submerges is <u>more</u> than five miles from the ship torpedoed, she is <u>not</u> to be attacked, but is to be hunted in the hope that she may be forced to come to surface and disclose her identity.

5. British Merchant ships torpedoed at night.

It is unlikely that attack will have been delivered by submerged submarine. British destroyer or destroyers which are or come on the scene are first to see that crew of torpedoed ship are safe, and are to search for and endeavour to establish identity of vessel that has fired torpedo.

If vessel which is found in vicinity is submarine, the action to be taken is as laid down in (a), (b), (c) and (d) of paragraph 4 above.

6. While these instructions apply primarily to destroyers, their general principles apply to all classes of ships. Capital ships and cruisers, however, should mainly confine their efforts to rescue work, and should not invite attack on themselves by using depth charges or taking other hostile action against submarines, unless destruction of submarine appears certain.

1708/18.

166. *Somerville (at Valencia) to Pound*

[ADM 116/3522]   24 August 1937

SECRET.

[Telegram] Addressed C.-in-C. Mediterranean, repeated Admiralty.

IMPORTANT.

435. H.M.S. *Hotspur*'s 1904 and 1930 of 23rd August[1] which has been passed to C.-in-C. Mediterranean and Admiralty.

I propose instructing Consul Palma as follows: (begins)

Inform Admiral Bastarreche[2] that whilst I recognise and fully appreciate his personal efforts to prevent bombing of British Merchant Ships on the High Seas I must express my deep concern and regret that these efforts have not proved effective. My information shows that British ship *Noemijulia* was attacked outside 3 mile limit. International law stipulates merchant vessels must not be attacked without effective identification and without first ensuring safety of crew. Unless Nationalists abide by these rules safety of British Merchant ships cannot be assured. Warn Admiral Bastarreche that unless immediate and effective action is taken to prevent a recurrence of these incidents I shall be compelled to take offensive action against Nationalist aircraft operating on the High Sea trade routes followed by British Merchant Ships. (ends).

0112/24.

---

[1] Reporting an attack by Nationalist aircraft on the British steamer SS *Noemijulia*.
[2] RA Francisco Bastarreche, commanding Spanish Nationalist naval forces at Palma, Majorca.

167. *Pound to Somerville*

[ADM 116/3522]   24 August 1937

SECRET.

[Telegram] Addressed R.A.(D), repeated Admiralty.

IMMEDIATE.

695. Your 0012. Do not take action proposed. Your proposals require full consideration and would need Admiralty approval.

1003/24.

168. *Admiralty to Pound*

[ADM 116/3522]   26 August 1937

SECRET.

[Telegram] Addressed C.-in-C. Mediterranean, repeated R.A.(D), from Admiralty.

Your 1003/24. Communication to be made by R.A.(D) to Bastarreche should consist of first two sentences of communication proposed in R.A.(D)'s 0012/24, down to '3 mile limit', and should continue as follows. Begins.

Since belligerent rights have not been accorded to either side no attack on a British vessel can be justified in any circumstances, nor can she lawfully be stopped at sea.

Even if, however, belligerent rights have been accorded, recent attacks would still be illegal.

According to international law no merchant ship whether enemy or neutral may be attacked unless she fails to comply with a lawful request to stop in order that she may be visited and her identity established, or unless she thereafter resists lawful capture. In any case it is only permissible to use such force as is absolutely necessary to ensure compliance.

Aircraft are not exempt from these rules, and must refrain from action against merchant ships unless they can comply with them. Ends.

Similar statement of international law position is being made to Insurgent Government through Ambassador Hendaye.

1535/26.

169. *Pound to Somerville*

[ADM 116/3522] 26 August 1937

SECRET.
[Telegram] Addressed R.A.(D), repeated Admiralty.

711. Your 0012/24th August and my 1003 24th August. I have examined practical application of your proposals and I have come to the conclusion it would necessitate either (a) declaration of war on Spanish Nationalists, or (b) convoying ships, for which a large number of cruisers or above with good high angle armament and control would be required, and which would be quite beyond our resources.
2318/26.

170. *Commander M. Richmond[1] to Somerville*

[ADM 116/3519] H.M.S. *Hostile*, at Barcelona
27 August, 1937

CONFIDENTIAL

I have the honour to submit the attached Report of Proceedings from 22nd August to 27th August, 1937, in accordance with King's Regulations and Admiralty Instructions, Article 1132.

* * *

3. At 1029 on Monday, 23rd August, I received a signal from Rear Admiral (D) ordering *Hostile* to listen out on 500 K/cs for distress signals from the British Merchant Vessel *Noemijulia*. I gave orders to come to 15 minutes notice for steam. At 1045 I received a signal from the *Noemijulia* requesting the assistance of a warship. I gave orders to raise steam with all despatch, and originated a signal to *Noemijulia* 'Am proceeding to your assistance'. This signal was never passed.

Shortly after this I received Rear Admiral (D)'s signal timed 1036 ordering me to raise steam and investigate with all despatch. Accordingly I weighed and proceeded at 1104 working up to 26 knots with the two boilers which were connected up on leaving harbour, and increasing to 30 knots at 1130 when the third boiler was connected.

---

[1] Cdr [later VA Sir] Maxwell Richmond (1900–1986). Commanded destroyers *Hostile*, 1936–38; *Basilisk*, 1939–40 (sunk at Dunkirk); *Bulldog* and Senior Officer Escort Group, 1942; COS to Cdre, Londonderry, 1942–43; commanded destroyer *Milne* and Capt(D), 3rd DF, 1944–46; Asst Chf of Supplies, Admy, 1946–48; NLO, Wellington, NZ, 1948–50; SNO, Northern Ireland, 1951; Dep Chf of Naval Personnel (Training), Admy, 1952–55; Flag Officer (Air) and 2nd-in-Command, Med Fleet, 1955–56.

4. From 1104 to 1258 every endeavour was made to get in touch by wireless but without avail. Various ships were heard trying to get in touch with *Noemijulia* and two messages in French were intercepted. These were as follows:–

(a) 'Barirogli Marseilles nous retournons pour vendres sommes bombard Capitan'. This was received at 1112 with no address or originator.

(b) 'NR 83 Sousmarin *Diamante* appareille de Port Vendres 1045/23/8'. This was received at 1221 with no address or originator.

The first message appeared to be from the master to his agent in Marseilles saying he was returning to Port Vendres having been bombarded, but the French was puzzling. The second message clearly stated that the submarine *Diamante* had sailed from Port Vendres at 1045.

5. At 1258 the French submarine *Diamante* made by operating signal to *Hostile* 'Any news of *Noemijulia*?' to which I replied 'Last heard of *Noemijulia* at 1036 GMT'. It transpired later that the wireless operator in *Noemijulia* closed down from 1132 to 1300 without informing anyone. Shortly after this the fuze in the aerial ammeter circuit blew out and although a fault in the transmission was noticed at once the cause was not discovered and repaired until 1350. During this time *Noemijulia* was heard asking for all ships if they had anything for her. Later on after further attempts to raise *Noemijulia* an unknown ship was heard trying to pass our message but with no answer. S.S. *City of Marseilles* was heard working GSJD (*Noemijulia*) and asked if he had received two signals from GXTV (*Hostile*) but there was no reply. It was subsequently learnt that GSJD shifted to 705 metres to answer this signal and say that he had nothing from *Hostile*.

6. At 1400 I arrived in position of 42° 07' North 03° 32' East (the position given in the S.O.S.). The visibility was very good, about 15 miles, and there was nothing in sight. In view of the French submarine's signal 'Any news of *Noemijulia*' at 1258 and the fact that the *Diamante* had sailed from Port Vendres at 1045 I decided to neglect the garbled message given in para. 4(a). I concluded that S.S. *Noemijulia* must have put back to Marseilles and I set course accordingly, reducing speed to 26 knots.

7. From this time onward I repeatedly tried to get in touch with *Noemijulia* and made the signal 'What is your present position' in code and plain language several times without result. This signal was also passed on by GCPY (*City of Marseilles*) and Marseilles radio but without any apparent result.

8. At 1638 I received the message timed 1630 from Rear Admiral (D) 'Does *Noemijulia* require further assistance?'. I was about to reply to this when the wireless office reported a message coming through from

Noemijulia. This message was being passed through Marseilles Radio for *Daily Mail*, London, and was the first intelligible message received from *Noemijulia*.

At 1650 I altered course for Port Vendres and made signal timed 1700 to Rear Admiral (D) repeated to Commander-in-Chief. I arrived at Port Vendres at 2000, and anchored outside. I sent in the Officer of the Guard in the motor boat and permission was obtained to board the *Noemijulia*. The Captain of the *Diamante* was met and stated that he fell in with the *Noemijulia* at about 1330 and came back to harbour with her, where they arrived at 1700.

9. As a result of the boarding officer's report I made my signals timed 2148 and 2300. These referred particularly to the bombing incident. Other points of interest were noted.

(i) The reason for asking for assistance from a warship was the desire for an escort and not that the ship was in any immediate need of assistance.

(ii) Captain Glinsky was a Russian and the remainder of the crew were foreigners including the Wireless operator.

(iii) A subsequent examination of S.S. *Noemijulia*'s wireless log showed that during a great deal of the time when ships were trying to locate her with the intention of proceeding to her assistance she was engaged in receiving and replying to messages from the *News Chronicle* and *Daily Mail* of London. At no time after asking for assistance of a warship did he consider informing us of his position or intentions.

(iv) Two observers were carried, one an Englishman. The latter explained that from the time of the first sighting of the aircraft the whole ship's company were in a state of panic.

(v) After the bombs were dropped the crew demanded that the ship should put back to the nearest French port; in this they were backed up by the Chief Engineer. The Captain, who had joined two days previously and whose last job at sea was seven years ago, did not need much persuasion.

(vi) The crew walked ashore in a body and refused to take the ship to any Spanish Port and Captain Glinsky accordingly had wired his owners requesting to sell the cargo at Port Vendres.[1]

10. While these remarks may not be wholly correct in detail it is clear that ships of this type are not really British Vessels except in name, and that their activities bring no credit to the British Flag.

The cargoes carried (Phosphate in this case) are probably used for war purposes although this cannot easily be proved. The conduct of the British

---

[1] The objections of the crew must have been overcome or a new crew quickly recruited for HMS *Hotspur* reported the arrival of the *Noemijulia* at Barcelona on 28 August. HMS *Hotspur*, Report of Proceedings, 2 Sept 1937, ADM 116/3892.

Navy in this matter is regarded by many as connivance at the safe transport of contraband goods by foreigners and as a result our prestige and that of the British Merchant Navy suffers to a marked degree.

11. It is understood that the provisions of the Merchant Shipping Act which prohibit the employment of aliens as Master, Chief Officer, or Chief Engineer in a United Kingdom registered ship do not wholly apply to vessels which trade habitually abroad. The question of compelling owners of British Vessels to employ British crews was considered unwise, some years ago, owing to the likelihood of owners transferring their ships to another flag, in which case they would not be available in time of war.

Now that the use of the British Flag is all important to owners it might be considered a suitable opportunity to ensure that all British Ships carried at least a proportion of British Subjects in their crews. This would make the protection of British Shipping not only more justified but also more practicable.

* * *

13. On Friday, 27th August, 1937, H.M.S. *Hotspur* arrived Barcelona and I turned over the duties of the port to Commander J.G. Bickford, R.N. At 0800 H.M.S. *Hostile* sailed for Palma.

171. *Pound to Admiralty*

[ADM 116/3522] 28 August 1937

SECRET.
[Telegram] Addressed Admiralty, repeated R.A.(D).

IMMEDIATE.
715. Reference 20001/27 from R.A.(D). I am not clear as to what the policy of H.M. Government as given in Admiralty Message 1535/26 is.

In paragraph 2 it is laid down that a British Ship by which it is presumed is meant any ship flying the British flag cannot be lawfully stopped at sea whereas paragraphs four and five give the impression that it is permissible to order a ship to stop in order that she may be visited to establish her identity.

If we intend to allow the procedure given in paragraph four to be carried out then presumably it will not be an offence if a vessel which has been stopped by the Nationalists on suspicion of wearing false colours proves to be of British registry. This procedure, however, will be tantamount to granting belligerent rights.

If on the contrary we do not intend to permit ships flying the British flag to be stopped at sea the situation will become increasingly difficult for the Nationalists as more and more foreign ships either transfer to British registry or use British colours as a ruse de guerre.

I have already suggested that the transfer of foreign ships to British registry for the purpose of trading on the coast of Spain should be stopped and if it were likely to be any good I suggest that our ships on the coast of Spain should board every ship they met flying British colours to determine her true nationality.

This would be unlikely to catch ships flying false colours as our ships are well known and a foreign ship would not display British colours when approached by one of our ships.

I am unable to suggest any method of establishing the true identity of ships except by examining their papers.

0303/28.

172. *Admiralty to Pound*

[ADM 116/3522]   28 August 1937

<u>SECRET</u>.

[Telegram] Addressed C.-in-C. Mediterranean, repeated R.A.(D) from Admiralty.

Your 0303/28. Policy of H.M. Government is, of course, not to grant belligerent rights at present.

Paragraphs 4 and 5 of my 1535/26 were intended to be statement of law which would be in force if belligerent rights had been accorded, the argument being that, as belligerent rights have not been accorded, Spaniards cannot now claim even such rights to interfere with British shipping, and that in any case, bombing such as has taken place is completely ruled out.

Admiralty fully appreciate your difficulties and steps have already been taken to do what is possible within the limits of existing law to deter transfer of foreign ships to British Registry for the purpose of trading on coast of Spain. Instructions have been issued by Board of Trade to Consuls in Europe and Mediterranean Area to refer applications for provisional certificates of Registry to London.

Admiralty concur in paragraphs 6 and 7 of your telegram and questions involved are at present under examination. In the meantime R.A.(D) should defer reply to Bastarreche's question reported in his 2001/27.

1630/28.

173. *Pound to Somerville*

[ADM 116/3522]   29 August 1937

SECRET.
[Telegram] Addressed R.A.(D), repeated Admiralty.

719. Your 1411. Without prejudice as to question of whether it is desirable or not to agree to proposal of Admiral Bastarreche that we should be prepared to investigate bona fides of any vessel the Nationalists consider suspicious I should like your views as to whether you consider such a policy could be effectively carried out with forces now allocated to coast of Spain.[1] Judging from our experience in Southern Area I believe a very large number of ships would be reported to us as being suspicious and that these suspicious ships would not only be in the vicinity of Coast of Spain but would also include ships reported to Palma by their patrols off North Coast of Africa, off Cape Bon and even possibly ships in Eastern Mediterranean.

0053/29.

174. *Somerville to Admiralty*

[ADM 116/3534]   Rear Admiral (D), H.M.S. *Galatea*, at Palma
4 September 1937

SECRET.
No. 498/B.6.

The attached report of the circumstances attending an attack by an unknown submarine on *Havock* in position 38° 46′ N 0° 30′ E on the night of the 31st August, 1937, is forwarded for the information of Their Lordships.

* * *

(Note:– The narrative as far as *Havock* is concerned has been compiled from relevant signals and a short conversation with her commanding officer).

---

[1] The proposals were: (1) All British ships passing through the Mediterranean should have their name and port of registry painted on the sides in letters six feet high; (2) Nationalist authorities would warn British authorities of course and bearing of any vessel approaching the coast of Spain and believed to be illegally flying the British flag and British naval authorities should then investigate the matter; and (3) British ships should be warned not to approach within 10 miles of the Spanish coast except when they were actually abreast of the port to which they were proceeding.

Havock sailed from Valencia at 1800 31st August, 1937 for Gibraltar at 15 knots. At approximately 2200 she altered course to the Southward to round Cap San Antonio. When the ship had only just steadied on her course the sentry on the lifebuoy aft observed the track of a torpedo approaching from port to starboard ... The torpedo passed a few feet astern of *Havock* and appeared to be set to about 10 feet depth.

2. The sentry informed the bridge at once and Lieutenant Commander R.E. Courage, the Commanding Officer who had just steadied the ship on her new course immediately altered course 180° to port so as to reach the track of the torpedo, increased speed, switched on the searchlight and ordered the A/S set to commence sweeping. When approximately on the track of the torpedo a shock was felt beneath the after superstructure somewhat similar to an underwater explosion. *Havock* made another alteration of 180° to port which brought the ship on to a Southerly course, and, about two minutes later A/S contact was obtained and an attack commenced. About this time the submarine was sighted on the surface at approximately 400 yards distant fine on the port bow, her inclination being between 020 and 050 to the right. The submarine only broke surface momentarily as though her trim had been upset. Her position and course conformed with the position where *Havock* felt the blow.

3. Since the A/S contact obtained was on a slightly different bearing to that on which the submarine was sighted and was not confirmed, *Havock* decided that it would be preferable to attack by A/S rather than by estimation.

4. After passing the position of the submarine, A/S contact was lost. *Havock* circled the position in an endeavour to regain contact but failing to do so shaped course to the North East and searched continuously until she reached the approximate 'furthest out' position at a distance of about 4 miles when she turned and commenced a search to the South and West.

5. At about 0130 A/S contact was regained in position 210° 3 miles from where the torpedo was fired. The Commanding Officer of *Havock* described this contact as being 'as near 100% a submarine echo as it is possible to say'. A full depth charge attack was carried out but nothing was observed after the attack to indicate that the submarine had been sunk or damaged. *Havock* reported that patch of oil was sighted near where the submarine dived and suggested that this might have some relation to the concussion felt in the *Havock* during her first encounter with the submarine but in a subsequent conversation the Commanding Officer inclined to the view that this was not the case and the position of the patch of oil was such that it might have been caused when the submarine was charging on the surface.

6. *Galatea* had sailed from Palma for Valencia at 1930 and was proceeding at 12 knots. The first information of this attack was received

by me about 2230, *Havock* reporting the position where the attack took place and that she was hunting for the submarine. At this time *Galatea* was in position Lat. 39° 25′ N Long. 01° 50′ E. Speed was increased to 25 knots and course shaped to close *Havock*. I ordered *Havock* to continue the hunt and attack the submarine and at 2249 ordered *Active* at Palma to raise steam for 27 knots with all despatch and join me.

7. At 2254 I ordered *Hotspur* at Barcelona and *Hereward* on passage from Gibraltar to Valencia to join me with all despatch. At 2307 I reported to Commander-in-Chief, Mediterranean, Repeated Admiralty, Ships and Authorities Western Mediterranean, that *Havock* had been attacked and that I was hunting the submarine and had ordered *Hotspur*, *Hereward* and *Active* to the scene.

8. At 2315 I received news that *Havock* had lost contact and I ordered Captain (D), Second Destroyer Flotilla and *Penelope* to join me in the position where the attack had occurred.

9. At 2324 I informed the Commander-in-Chief, Mediterranean, Repeated Admiralty and Rear Admiral, Gibraltar, that I intended to carry out an intensive search in order to endeavour to bring to the surface any submarine detected in this area.

10. At 0133 *Havock* reported that she had regained contact and was attacking and that at 0152 had fired a pattern of depth charges and was examining the area.

Havock's searchlight was visible from *Galatea* fine on the starboard bow below the horizon.

11. At 0201 *Havock* was ordered to report the position of the last attack and from this position it appeared that the submarine had moved about 3 miles to the South.

12. In the meantime reports had been received from the ships of the Second Destroyer Flotilla and *Penelope* indicating that my order to rendezvous was being obeyed with the greatest promptitude. *Hostile* at Marseilles was unable to comply as she had not refuelled.

13. By 0215 on 1st September, 1937, *Galatea* had closed *Havock* and was informed by the latter that nothing could be seen in the depth charge area and that she was endeavouring to regain contact. I ordered *Havock* to continue the search in the present area and informed her that I would proceed to the South West and commence an aerial reconnaissance at daylight. *Havock*'s position appeared to be approximately 6 miles to the Eastward of that signalled.

14. I had decided by this time that the course of the search should be to the South for the following reasons:–

(i) The positions given by *Havock* indicated that the submarine was proceeding to the South.

(ii) As the attack had been on a ship steering a Southerly course it appeared possible it was intended primarily to promote an incident since there are few important trade movements to the Southward from Valencia. This suggested that the attacker was a Government submarine.

(iii) The majority of my reinforcements were arriving from the South and it would be possible to start a fairly effective search from the Southward about 1030 next day.

15. To put this decision into effect the following dispositions were made:–

(i) *Hereward, Hotspur* and *Havock* were ordered to rendezvous in position 102 Cap San Antonio 21 and carry out an A/S sweep on a course 230 degrees at 15 knots.

(ii) *Active* not being fitted with A/S was ordered to position Lat. 38° 25' N Long. 01° 19' E by 0600, 1st September, 1937, and to patrol six miles to the North and South of this position at a speed of 12 knots.

(iii) *Galatea* proceeded to position Lat. 38° 24½' N, Long. 00° 07' E in order to fly off her seaplane at daylight and carry out an aerial reconnaissance.

16. The object of this disposition of the forces available at daybreak was:–

(i) To commence a detailed A/S search with 3 destroyers to the Southward from the position in which contact had been lost, and for this search to continue as far as the submarine's 'furthest on' circle[1] if submerged, by which time the remaining destroyers arriving from the South would have joined the search.

(ii) To carry out a surface and aerial search along the submarine's 'furthest on' circle to the South and Eastward assuming the submarine had, after a short dive from the immediate vicinity, attempted to escape on the surface.

17. At 0545 *Galatea* was in position but the seaplane was not ready for service and was not flown off until 0708, when *Galatea* was in position Lat. 38° 15' N, Long 00° 01' E, owing to replacement of engine on the previous day and inability to complete preparation of aircraft for service during dark hours.

18. At 0554 *Hotspur* reported she was in company with *Havock* and intended to commence a search at 0600. At 0630 *Hereward* joined the search.

---

[1] The possible locus of the positions which a submarine could have reached after a given time from a known position. This was prone to error because of the various assumptions that had to be made, notably speed of the submarine, direction of escape and the assumed position of the submarine when the torpedo attack was made. I am indebted to Dr Malcolm Llewellyn-Jones of the Naval Historical Branch for this explanation.

19. At 0515 *Hasty* reported that she was stopped in position 28 miles Cap de Gata and that her port engine was broken down. I ordered her subsequently to return to Gibraltar.

20. At 0900 on 1st September, *Galatea* was in position Lat. 37° 59′ N, Long. 00° 30′ E, and her movements in conjunction with *Active*'s patrol and the air search had covered the submarine's estimated 'furthest on' circle. *Galatea* and *Active* then proceeded to the position of the attack in order to search partially the area lying inside the furthest on circle and to investigate patches of oil which had been reported by the aircraft. These patches were examined, a boat being lowered, but no evidence could be found to suggest damage to or leakage from a submarine.

21. At 1140 on 1st September, 1937, *Galatea*'s aircraft was again flown off to locate and lead *Galatea* to any other patches of oil or debris and to carry out a short search to the Northward to a depth of forty miles. Nothing further was sighted.

22. Meanwhile at 1030 *Hardy* and *Hyperion* joined *Hotspur*, *Hereward* and *Havock*. At 1045 these five destroyers carried out a search to the Northward on a course parallel to and inshore of the area covered on the search South.

23. At 1157 on 2nd September, 1937 I ordered *Penelope* to reduce to economical speed and later I detached her with instructions to intercept the British tanker *Burlington*.

24. At 1227 a signal was received from Captain (D), Second Destroyer Flotilla informing me that all ships searching with him had established contact and that he considered it certain that he was in touch with a submarine. His position 044° Cape Palos 32 miles, suggested this might well be the submarine which fired at *Havock* had she shaped course to the Southward submerged.

25. At 1238 I ordered Captain (D), Second Destroyer Flotilla to fire a depth charge with a view to surfacing, but not sinking, the submarine.

26. At 1235 Admiralty message timed 1212 of 1st September, 1937 was received and decyphered by 1245 and at 1257 a signal was sent to Captain (D), Second Destroyer Flotilla ordering him to take no offensive action but to continue to maintain contact.[1]

27. Captain (D), Second Destroyer Flotilla, had, however, already fired a depth charge at an estimated range of two hundred yards from the submarine. Subsequently Captain (D), Second Destroyer Flotilla lost contact, and failing to regain it, continued his search to the Northward.

---

[1] 'In order to avoid risk of attacking an innocent submarine and in view of lapse of time since last contact and possibility of other submarines being in area further offensive action should not repetition not now be taken.'

Captain (D), Second Destroyer Flotilla later came to the conclusion that this contact could not have been a submarine.

28. The Second Destroyer Flotilla less *Hostile* and *Hasty* rendezvoused with me at 1430 and I proceeded on board *Havock* with Captain (D), Second Destroyer Flotilla in order to obtain further details.

29. Having detached *Havock* and *Active* to Gibraltar, I then commenced an A/S and air search to the Northward with the intention of covering as much ground as possible during the return passage to Valencia having regard to the fuel remaining. Admiralty message 1858 of 1st September was, however, received at 1930 and I discontinued further searching operations at daylight on 2nd September, and anchored at Valencia at 0645 on that day.

REMARKS.

30. The text and handbooks and the instruction given at the A/S School all lay great stress on placing full confidence in the Asdics and carrying out attacks based on information furnished by the Asdic set rather than estimation by eye.

31. In spite of this I consider that the Commanding Officer, *Havock*, showed lack of judgement in not seizing the opportunity furnished by sighting a submarine only 400 yards away and carrying out an immediate full depth charge attack which I feel convinced would have sunk the submarine or brought her to the surface.[1]

32. I consider that my estimation as to the probable course of the submarine after the attack was based on insufficient evidence and that it would have been better if *Galatea* and her aircraft had operated to the North and not to the South in order to cover, so far as was possible, with the forces at my disposal, all avenues of escape.

33. The promptitude with which my orders to rendezvous were complied with by *Penelope* and the Second Destroyer Flotilla is most praiseworthy and merits commendation.

---

[1] The Court of Inquiry held at Gibraltar on 4 September did not agree. They found: 'Although at first sight it might be thought that an immediate depth charge attack delivered on sighting the submarine breaking surface would have offered a fair chance of success, we consider the Captain of H.M.S. *Havock* fully justified in waiting to deliver a deliberate attack, making use of his A/S installation.

The conditions after the final attack bear out this opinion, and the ultimate success achieved in locating and attacking the submarine fully support the Captain's action.' Court of Inquiry to Rear Admiral, Gibraltar, 4 Sept 1937, ADM 116/3534.

175. *Pound to Somerville*

[ADM 116/3534] 1 September 1937

SECRET.

[Signal] Addressed R.A. (D), repeated Admiralty, V.A.C., B.C.S.

IMPORTANT.

733. Assuming evidence is conclusive that H.M.S. *Havock* was attacked by a submarine two situations may arise.

Case A. You gain contact [with] a submarine today in which case she will be followed day and night until she is forced to the surface and identified.

Case B. Should contact not be gained today. In this case all we can do is to create as much fuss as possible, firstly by continuing to operate all your force both by day and night in a considerable area round position of attack for so long as fuel of your destroyers permits. Secondly, subject to Admiralty concurrence, to endeavour to determine by elimination nationality of submarine.

For this purpose both Valencia and Palma should be visited and a protest made in such a manner as will result in both Government and Nationalists providing official evidence that whereabouts of their submarines was such that they could not have been responsible for the attack.

Our ships must not repetition not operate without lights in either Case A or Case B.

1258/1.

176. *Admiralty to Pound*

[ADM 116/3917] 1 September 1937

SECRET.

[Signal] Addressed C.-in-C. Mediterranean, repeated R.A. (D), V.A.C., B.C.S. From Admiralty.

IMMEDIATE.

Continuous contact having proved impracticable consider that the incident should be regarded as closed from the operational point of view.

Proposed action at Valencia and Palma should not repetition not be taken for the present.

1858/1.

177. *Somerville to Pound*

[ADM 116/3917] 1 September 1937

SECRET.

[Signal] Addressed C.-in-C. Mediterranean, repeated Admiralty, V.A.C., B.C.S.

IMPORTANT.

497. Your 1258 and Admiralty 1858. Apart from considerations of oil fuel and effect of increased activity in these areas I consider that present local asdic conditions coupled with lack of practice as result of prolonged turns of duty in Spain and probable delay in searching force reaching scene of attack make it improbable that such searches will prove effective.

Destroyers steaming at night along trade routes where merchant ships are likely to be attacked and thus offer themselves as a target, will offer best chance of successful attack on submarines if risk to destroyers can be accepted.

2305/1.

178. *Somerville to Second Destroyer Flotilla*

[ADM 116/3534] 3 September 1937

SECRET.

[Signal] Addressed 2nd Destroyer Flotilla, repeated C.-in-C. Mediterranean, F.O. 2nd-in-Command, Mediterranean, R.A. Gibraltar, Admiralty.

505. Constant asdic watch is to be maintained by ships while on passage, any contacts obtained are to be held and reported immediately.

0915./3.

179. *Somerville to Pound*

[ADM 116/3522]   3 September 1937

SECRET.

[Signal] Addressed C.-in-C. Mediterranean, repeated Admiralty, V.A.C., B.C.S., R.A. Gibraltar.

IMPORTANT.

506. Admiralty 1656/2.[1] Propose 11th Division take over duties of Southern Area and 2nd Destroyer Flotilla concentrate Eastern Area. Any destroyer operating in Eastern Area should be painted Mediterranean Fleet colour on account of similar Home Fleet and Spanish Government colours.

Forthwith request an oiler of greater capacity and speed than *Viscol* be based Eastern Area, probably Palma where I do not anticipate Nationalist Authorities will raise any objections.

                                               1008/3.

180. *Pound to Somerville*

[ADM 116/3522]   3 September 1937

SECRET.

[Signal] Addressed R.A. (D), repeated Admiralty, V.A.C., B.C.S., R.A. Gibraltar.

IMPORTANT.

747. Your 1008 3rd September. Approved, but when I know what policy of H.M. Government is which has necessitated reinforcements being sent out your proposals may require modification. Will inform you later regarding oiler.

                                               1802/3.

---

[1] The Government decided that for political reasons the destroyer strength in the Western Basin of the Mediterranean would be reinforced by the 11th Division of the 6th Destroyer Flotilla, Home Fleet. The four destroyers would temporarily come under the orders of the C-in-C Mediterranean.

181. *Pound to Admiralty*

[ADM 116/3522] 5 September 1937

SECRET.

[Telegram] Addressed Admiralty, repeated R.A.(D), V.A.C.,B.C.S.

IMPORTANT.

757. In three parts.

PART I.

Important as approval of H.M. Government has been in authorising counter attack of submarines as set out in my 1708 18th August[1] it is obvious months may elapse before a case occurs in which one of H.M. Ships is sufficiently close to scene of attack on a British Merchant Ship to limit counter action to be taken under existing five mile rule.

2. Generally speaking interference with British ships has been directed against ships the genuineness of whose British registry is in doubt. Hence if steps which have now been taken to prevent transfer of doubtful ships to British registry are reinforced by proposals contained in my 1833 3rd September the incentive to attack bona fide British shipping will be much reduced but it would not stop present illegal and inhumane use of submarine to which the genuine British ships may at any time fall a victim.

3. H.M. Government have made it clear that they will not at present contemplate anything in nature of grant of belligerent rights to enable contending parties to determine nationality of ships (See telegram 1630 28th August from Admiralty).

4. There appears to remain only three alternatives:

A. To remove danger to our ships by obtaining concurrence of all parties concerned that submarine warfare shall cease.

B. To take direct action against all submarines operating in area in which unlawful attacks are taking place.

C. To put our ships into convoys.

5. As regards A. This would be the most satisfactory method but there appears no possibility of achieving it.

6. As regards B in the last few days R.A.(D), being convinced that instructions contained in my 1708 18th August do not repetition not go far enough, has proposed that Spanish Government and Nationalists should be treated as hostile and [submarine] attacked which:

(a) is detected submerged.

(b) being on surface, dives on approach of one of H.M. Ships.

---

[1] Doc. No. 165.

(c) being darkened, approaches one of H.M. Ships.

I am of opinion if we go in for direct action we should do so on a large scale using aircraft in addition to destroyers.

7. As seas are free, any nation has the right to exercise their submarines off Spanish Coast, either diving or darkened, and an essential preliminary to such a policy would be to obtain the consent of France, Italy, Germany, Russia and Holland and possibly other nations that they would agree that their submarines would not exercise in those waters. Any form of direct action would have the serious objection that Spanish Government submarines, which as far as I am aware have not made attacks on shipping without warning, would be subject to attack while defending their own coasts. The advantages and disadvantages of direct action are given in part 2. It is unlikely we should be permitted to take direct action alone and one other nation co-operating it would presumably be necessary to detail countries for various areas which would obviously have to be different to those allocated for non-intervention duties. The protection of British Ships on greater part of coast of Spain would pass out of our hands.

8. As regards (C) the advantages and disadvantages of convoy system are given in part 3 from which it will be seen that though undesirable for many reasons it can be put into force without reference to any other country as only those submarines would be counter attacked which had proved their guilt by attacking our ships.

9. Putting our ships into convoy is a poor substitute for clearing seas of piratical submarines by direct action but I have been most reluctantly forced to the conclusion convoy is the only practicable method for protecting our ships. If convoy system is put into force we should make it clear that no repetition no protection will be afforded to ships which do not take advantage of the convoy system.

0018/5.

542. My 0018. <u>PART 2.</u>

Attack method (direct action).

Advantages:

1. Assurance of withdrawal of submarines by other powers would, if honestly kept, largely reduce number of submarine attacks on shipping and therefore the number to be hunted.

2. If only a few submarines are operating, they will probably be in the vicinity of focal area. This would reduce total area to be hunted.

3. Attacks on merchant ships would probably be reduced as submarines would largely be forced to concentrate on self defence.

Disadvantages:

1. The necessity of obtaining assurance from France, Italy, Germany, Holland and Russia that their submarines would not exercise within a specified distance from the coast of Spain.

2. The uncertainty whether certain foreign powers would energetically implement their undertakings.

3. Spanish Government submarines would be exposed to attack.

4. All British men of war would be exposed to submarine attack in the vicinity of coast of Spain and perhaps further afield.

5. Unrestricted attacks on submarines would amount to direct intervention as we should be doing for Spanish Government what they cannot do for themselves.

6. The employment of aircraft off Spanish coast might lead to reprisals and further intervening.

7. Prolonged watch keeping by asdic personnel rapidly reduces their efficiency.

0103/5.

759. My 0018. <u>PART 3</u>.

The convoy system.

Advantages:

1. H.M. Ships would be on spot if merchant ship were attacked by submarine.

2. There would be no doubt of guilt of any submarine that attacked shipping.

3. Innocent submarines of Spanish Government Fleet would not be attacked.

4. There would be no need to attempt to get assurance from other nations that their submarines had been withdrawn from vicinity of Spanish Coast.

5. The undoubted advantage gained by Spanish Government from British convoy might cause Nationalists and their backers to annul unrestricted submarine warfare.

6. If unrestricted warfare was carried out in waters other than vicinity of Spanish Coast it would not be so effective.

<u>Disadvantages</u>:

1. Impossible to prevent foreign and undesirable 'British' ships taking advantage of protection provided.

2. More arms and munitions would reach Spanish Government ports, thereby probably prolonging the war.

3. Indirectly, it would be contrary to our policy of preventing supply of arms and munitions to either side.

4. There would be no guarantee that shipping would not be torpedoed inside territorial waters, though this might be prevented by Spanish Government Fleet.

0218/5.

182. *Somerville to Pound*

[ADM 116/3522] 5 September 1937

SECRET.

[Signal] Addressed C.-in-C. Mediterranean, repeated Admiralty.

IMPORTANT.

Your 0018. Excluding MacAndrews[1] and lines normally trading Spanish Coast it now appears that tankers and other ships proceeding to Spanish Government ports adopt diverse routes escaping observation and thereby arousing suspicion of Insurgents.

Consider convoy of British ship[s] will certainly give maximum degree of security against submarine and air attack.

Exclusion of foreign ships from convoy would be difficult but perhaps not impossible.

Suggest frequency of convoys need not be too liberal in view of dubious character of many ships now engaged in Spanish trade and highly remunerative nature of their business.

Massed arrival of ships at Government Ports may provoke Insurgent air raids but this is a legitimate hazard and would offset indirect assistance afforded Spanish Government.

Possibility that Nationalists may object to our using Palma cannot be ignored.

1133/5.

183. *Admiralty to Pound*

[ADM 116/3522] 6 September 1937

SECRET.

[Telegram] Addressed C.-in-C. Mediterranean, repeated R.A. (D),V.A.C., B.C.S., R.A. Gibraltar.

Your 1802/3 and 0018/5. Arrangements on the lines of your alternatives A and B are under consideration in London, and it is probable that they

---

[1]MacAndrews & Co., Ltd, London. They were the owners of the SS *Carpio*.

will be put forward at the conference which is to be held at Geneva at the end of this week.

In the meantime you should furnish an estimate of forces required for alternative C.

1606/6.

184. *Admiralty to Pound*

[ADM 116/3522] 7 September 1937

SECRET.

[Telegram] Addressed C.-in-C. Mediterranean, repeated C.-in-C. Home Fleet from Admiralty.

IMMEDIATE.

Part One.

It seems probable that as a result of recent Soviet Note to Italy, proposed Mediterranean Conference will not take place owing to Italy's refusal to attend.[1]

Intention was that at this conference attempt should be made to get agreement that no Power would have submarines at sea except in certain areas reserved for exercise purposes and that any submarines met outside these areas could be attacked at sight by warships of Powers concerned. This scheme is no longer possible and as an alternative H.M. Government are now considering what action they can take in conjunction with any Powers willing to co-operate.

Proposal is that right to attack submarines should be extended to any submarine which attacks merchant ships of any nationality and that this right should extend to any submarine found in vicinity in which merchant ship has recently been attacked, term vicinity to remain undefined for purpose of public announcement.

It is important that if a task of this nature is undertaken, it shall be carried out effectively and with our peace strength of destroyers the number that can be made available are very limited. Assuming that all destroyers of Mediterranean and Home Fleets were employed after allowing for refits, etc. an average of about 30 destroyers only is available between now and next January. Working in two watches the number of destroyers employed at one time would be about 15. Part two follows.

2130/7.

---

[1] The Soviet note accused Italy of recent sinkings of Russian ships in the Mediterranean. The accusations were, in fact, true.

<u>Part 2.</u>

It is visualised that some 5 or 6 hunting units, would be formed and that these units would be stationed where required in areas to be agreed between the Powers taking part.

A possible allocation would be for the French Navy to operate generally in the Western Basin north of the 40th parallel, the British Navy south of the parallel, with the possibility of ships having to be detached to the Malta Channel or Aegean.

Other powers are unlikely to do anything effectual though they might be asked to operate in Aegean. Italy and Germany would be asked to co-operate but unlikely to do so.

If Spanish Ports ceased to be available for us French would be asked for use of Algiers.

Operations to be undertaken in the first instance for one month after which their continuance to be reviewed.

You will realise that should this scheme come into force your responsibilities will be greatly increased and should the operations continue it would be necessary for your destroyer strength to be reinforced. Normal fleet training would be further interfered with but to counterbalance this, A/S if these took place would in themselves be most valuable training.

Above is only in outline at present but your very early remarks are requested generally and in particular as to whether you consider 5 or 6 hunting units would be likely to have a moral effect sufficient to prevent the continuance of unrestricted submarine warfare.

2215/7.

185. *Somerville to Pound*

[ADM 116/3522] 8 September 1937

<u>SECRET.</u>

[Signal] Addressed C.-in-C. Mediterranean, repeated Admiralty, V.A.C. B.C.S. From R.A.(D) at Palma.

539. If embargo on use of submarines by contending parties is to be advocated following considerations arise which may influence Italian attitude.

(1) Chief obstacle to Insurgents exercising control by surface ships in Eastern area is Government destroyer force based at Cartagena. Government cruisers and aircraft not considered serious menace.

(2) Diversion of submarine offensive to this objective appears possible particularly in view of effects recent incident have had internationally and

of fact that only one Government destroyer has been fitted with anti-submarine appparatus.

(3) From conversation with Italian admiral in Italian cruiser *Quarto* believe Italians are pressing Insurgents to adopt measures to intercept vessels in vicinity of port of destination as opposed to high seas.

(4) Reliance may be placed in future on air and sea bombardments of Government ports by day to deter merchant ships from trading.

1053/8.

186. *Pound to Admiralty*

[ADM 116/3522] 8 September 1937

SECRET.

[Telegram] Addressed Admiralty, repeated C.-in-C. Home Fleet.

IMMEDIATE.

779. Your 2130 7th September.

1. I am of the opinion that scheme outlined in paragraph 3 which will be called Scheme 'Z'[1] has grave disadvantages in as much as, firstly, it does not permit of every submarine sighted being attacked whatever it is doing. Secondly, it does not possess advantages of the convoy system that every attack must be carried out in the presence of the counter attacking force.

2. Scheme 'Z' must be considered under two headings: (a) Its effectiveness in destroying submarines; (b) Its moral effect on submarines.

3. As regards (a) to achieve success by day it would be essential to have continuous air patrols between General Headquarters and parallel 40° north, otherwise we shall have no idea except possibly by direction finding bearings of where submarines are operating, while enemy can locate our hunting units and warn their submarines.

One Squadron of flying boats at General Hqs. and one at Bougie are considered essential. Algiers and Oran Zones unsuitable. Flying boats to be fully armed and to carry bombs to attack any submarine in the vicinity of ship which has been attacked. H.M.S. *Glorious* or another carrier to be used until flying boats have arrived and then to supplement them as found necessary.

For searching the whole Flotilla is best, while for covering large areas, which ensures early arrival at scene of attack, destroyers stationed singly are best. Best compromise is considered unit of three destroyers.

---

[1] The right to attack any submarine which attacks merchant ships or found in the vicinity of a merchant ship that had been recently attacked.

4. By night problem is more difficult as: (a) Air reconnaissance is not available; (b) with increasing long nights submarines can leave Nationalist Ports in Majorca and Iviza or Spanish Government Ports and arrive at their selected area of operations unknown to us. Having removed themselves from the scene of attack they can return to their base by day without danger of attack from us.

Part 2 follows.

2023/8.

780. My 2023. Part 2.

5. By working in two watches I assume it is Their Lordships' intention that all available destroyers of the Mediterranean Fleet and one Division from Home Fleet, other than those Mediterranean destroyers required for operations in other areas or refitting, should be concentrated in Western Basin and that half this force should be at sea at a time. As this is practically war operation and entirely different to the somewhat boring duties they have been employed on in the past I see no objection to this. It is assumed that non Intervention duties would not be required.

6. Considering distance involved I consider it impracticable to detach destroyers from Western Basin to deal with incidents in Aegean and Malta Channel areas and I would propose two destroyers of 3rd Destroyer Flotilla should be stationed at Mudros and two at Malta. Malta Force should have assistance of Number 202 Flying Boat Squadron, whilst H.M.S. *Repulse* could be stationed at Mudros to provide air reconnaissance.

7. It is assumed French would deal with North African trade routes and Cape Bon area.

8. Assuming four ships of 3rd Destroyer Flotilla were employed as proposed in paragraph six, number of destroyers available in Western Basin at present time would be First Destroyer Flotilla nine, Second Destroyer Flotilla eight, Third Destroyer Flotilla three, Division of Home Fleet four, which would allow four units of three ships each being on patrol at a time. Each unit would then be seven days on patrol and would be at sea for nine out of every fourteen days. Owing to dockings and refits of Mediterranean destroyers, it would be necessary for Home Fleet to provide whole flotilla of destroyers in order to maintain 12 ships on patrol. With additional destroyers more units could be maintained at sea and chance of reaching the scene of attack quickly would be correspondingly increased.

9. As regards moral effect. Factors in order of importance which will produce moral effect are I believe:

(a) Destruction of submarine. From foregoing remarks it will be realized that scheme 'Z' unlikely to achieve this. We may get one by luck.

(b) Depth charging even at a distance is unpleasant. Whenever search for submarine which has made a kill is being carried out, depth charges should be dropped at intervals previous to contact being made.

(c) Hunting them. Every submarine which is sighted by our air patrol, or is detected by our asdics should be hunted for as long as contact can be maintained.

(d) Observing their movements from air.

10. Moral effect of above is difficult to assess without knowing nationality of crew. Submarine Captains are probably volunteers and good men. Discipline of crew probably doubtful and likely to be seriously affected by depth charging.

11. From above Their Lordships will realize that I am not in favour of scheme 'Z', as apart from moral effect which is of uncertain value, scheme is too dependent on luck. Should scheme 'Z' be decided on, however, it is my intention to proceed to Western Basin to conduct operations as in addition to visit of the First Lord my only important commitment is discussion with General Weir in Egypt which must be fitted in later.

12. Convoy requirements follow immediately.

2213/8.

187. *Pound to Admiralty*

[ADM 116/3522] 8 September 1937

<u>SECRET</u>.

[Telegram] Addressed Admiralty, repeated Rear Admiral (D), V.A.C. Battle Cruiser Squadron.

<u>IMMEDIATE</u>.

781. Your 1606 September 6th. Following convoy proposals are for purpose of protecting British ships only. It is not proposed to institute convoy east of Malta unless it is found necessary.

(a) Through traffic between Gibraltar and Cape Bon to Malta unescorted south of line joining Alboran Island to point 000 degrees 30 miles from Cape Cazine and thence to position 000 degrees 20 miles from Galita Island.

(b) Through traffic between Gibraltar and either Marseilles, Toulon or Italian Ports to pass unescorted east of Balearic Islands.

(c) Ships bound from Italy to Spanish ports to join convoy at Marseilles.

(d) (b) from Eastern Mediterranean to Spanish ports to join coast convoy at Gibraltar or Barcelona convoy at Malta.

(e) Convoy G.B. This convoy would sail every four days between Gibraltar and Barcelona being escorted by four destroyers which would permit of vessels dropping off from convoy and joining it, under protection of one destroyer, as far as three mile limit or as far as permissible, on account of mines.

A total of twelve destroyers (one Mediterranean Flotilla and one division from modified Home Fleet Flotilla) would be required for this, each destroyer doing seven days at sea and five days in harbour.

(f) Convoy BM This convoy would sail every four days between Barcelona and Marseilles escorted by two destroyers. Total destroyers required two.

(g) Convoy VB This convoy would sail every twelve days between Valetta and Barcelona escorted by two destroyers. Total destroyers required two. If this convoy were run every eight days a total of four destroyers would be required. Destroyers required for convoys B.M. and V. Barcelona would be provided from a Mediterranean Flotilla.

2218/8.

188. *Admiralty to Pound*

[ADM 116/3522] 9 September 1937

SECRET.
[Telegram] Addressed C.-in-C. Med., repeated C.-in-C. Home Fleet., from Admiralty.

Your 2023/8 and 2213/8. The Conference is taking place in the absence of Germany and Italy and the First Sea Lord with Naval Staff are proceeding to Geneva via Paris this afternoon 9/9/.

2. While Admiralty fully agree that to carry out a convoy system is the only efficient manner of dealing with the problem of unrestricted submarine attack they have had to reject the idea, not only for the reasons given under 'Disadvantages' in your 0218/5, but for the additional reasons:–

(a) That in peacetime a convoy system cannot be made compulsory even for our own merchant ships;

(b) It would not cover merchant ships of other countries unless all other countries agreed to adopt a convoy system, which is highly improbable;

(c) It necessitates convoy assembly ports on the Spanish coast, which is considered politically impracticable.

3. It is clear therefore that in the circumstances the system that we should naturally adopt in war cannot be adopted now. It is for that reason that what you have named 'Scheme Z' was agreed on between the

Admiralty and the Foreign Office. It is recognized that Scheme Z cannot be completely efficient, but once the fact is announced that our forces are limited to our peace strengths that would be publicly understood.

4. With regard to your objection that it does not permit every submarine sighted to be attacked, apart from the fact that such action is not possible without the agreement of all nations concerned, it is considered that to extend the attack to every submarine operating would greatly increase the responsibilities of the Fleet, and any failure to locate and destroy submarines would reduce the prestige of the Fleet. It is considered that the question of the prestige of the Fleet in submarine hunting is going to receive world attention and it is therefore most important that we limit our responsibilities to those for which we have a reasonable chance of expecting our ships to deal with. If we give you a definite area – not to be responsible for but in which you will have the right to hunt submarines carrying out unrestricted submarine attack – and if you can score some success in that area which ships of other nations operating in their areas are unable to effect, the superior efficiency of the British navy in dealing with submarine attack will have been asserted. This may have far-reaching consequences as a deterrent to war. In any public message issued proclaiming the conditions agreed on at Geneva it will be definitely stated that the convoy system, which is the natural system to use, has been rejected for the reasons given above.

5. With regard to your desire to have aerial assistance in the work that will devolve on you: this is fully appreciated. It is complicated, however, by the fact that the French are anxious to have regulations to deal with air attack on shipping. The British view is that the submarine problem shall be dealt with first and separately, and any consideration of the air problem, which would be more difficult to get agreement on, should be considered subsequently. The only effective way of dealing with the air problem would be for mutual agreement not to use aircraft in the war areas, and as this may come into effect it is not desirable to start the anti-submarine campaign by initiating new air measures in the western basin. Nevertheless your point will be borne in mind.

Part II. Practical operation of Scheme Z.

6. The Admiralty agree that a unit of three would be the best compromise and should be adopted. It is the intention that all the Mediterranean and Home Fleet Flotillas shall be considered available for the work, and that generally half of them should be operating at any one moment. It is suggested that the operating period should be one month, as Home Fleet Destroyers will have to be relieved from and return to Home Waters. It is desired, however, that details of this nature

should be worked out conjointly between yourself and the C.-in-C. Home Fleet.

7. It will be further proposed at Geneva that there should be a definite time limit to the operational measure agreed upon, possibly of from one month to 6 weeks; and the continuance of the measures would depend upon the effectiveness in stopping piracy that had been achieved. In view of this it is considered that an increased effort should be made at the commencement. The Admiralty would like the combined proposals of yourself and the C.-in-C. Home Fleet as to this.

8. As regards the bases of operation, it is hoped that the facilities of Spanish ports to our destroyers will continue, and further that they will, if necessary, have the use of French ports. In the conversations to be had with the French before the Conference the following general proposal will be made:

Area 1. The area South of the 40th parallel and west of the 5th Meridian.

Area 2. The Malta Channel East of the 10th Meridian.

The French area to be the whole of the area between the 5th and 10th Meridians and in addition the area west of this north of the 40th parallel.

It will be proposed that the Greek, Turkish and Russian Navies shall be initially responsible for the Aegean and the vicinity of the Greek coast, and it will not be proposed that the British Fleet shall operate in that area except if it becomes necessary in the national interest.

9. It is desired to emphasize that the whole principle being sought at Geneva is to deal with the piratical attacks primarily by submarines and as such the problem is essentially a destroyer problem assisted by aircraft if it were practicable. The absence of large surface ships in the operational areas will obviously be undesirable, firstly because we must always anticipate possibility of S/M retaliation and the sending of such ships would detract from operating destroyer strength. For this reason your presence in operational areas in your flagship would appear undesirable as it would inevitably lead to French heavy ships being also present.

1530/9.

189. *Pound to Backhouse*

[ADM 116/3522] 10 September 1937

SECRET.

[Telegram] Addressed C.-in-C. Home Fleet, repeated Admiralty, British Naval Delegation, Geneva.

IMMEDIATE.

791. Admiralty message No.1530 9th September stresses 2 points:

(a) A demonstration of the efficiency of the British Navy in dealing with submarine attack may have far reaching consequences as a deterrent to war.

(b) An increased effort should be made at commencement by which is presumably meant the initial period which will be agreed on at Geneva which I shall assume to be six weeks.

Taking (a) and (b) together it appears desirable that the greatest possible force should be employed during first six weeks particularly as air co-operation will apparently not be possible.

From paragraph 5 of my 2213 8th September it will be seen that it is my intention that the whole Mediterranean destroyer force should be employed continuously on coast of Spain or in Malta Channel which means half will be at sea at a time.

From paragraph 6 of Admiralty message 1530 9th September I gather that only half of the destroyer force provided by the Home Fleet will be on the coast of Spain and consequently that only one quarter will be at sea at any given time.

The destroyer strength which can be provided by the Mediterranean fleet is given in paragraph 8 of my 2213 8th September but when date of commencement of initial period and minimum duration are known I would see whether possible to defer refits and dockings of the Mediterranean destroyers in order to provide the greatest number of hunting units during initial period.

The reply to Admiralty message 1530 9th September appears dependent on destroyer force which Home Fleet can make available for operating in area one.

1143/10.

190. *Pound to Chatfield*

[ADM 116/3522] 10 September 1937

SECRET.

[Telegram] Addressed British Naval Delegation Geneva.

IMMEDIATE.

Personal for First Sea Lord.

I am afraid Paragraph 11 of my 2213 8th September[1] gave a wrong impression. R.A.(D) will conduct operations at sea. H.M.S. *Barham* would remain at Gibraltar and should it be necessary for me to proceed to sea I should hoist my flag in H.M.S. *Penelope*.

As ultimate responsibility for putting into force any scheme decided on will be mine I consider it essential that I should proceed to Gibraltar forthwith to inaugurate the scheme, request early approval accordingly. Suggest reason for my change of programme should be inauguration of scheme as this would allow of my leaving western basin at any time should this be necessary.

1313/10.

191. *Chatfield to Pound*

[ADM 116/3522] 11 September 1937

SECRET.

[Telegram] Addressed C.-in-C. Mediterranean, repeated Admiralty, C.-in-C. Home Fleet.

Situation is developing here [Nyon Conference] along following lines. The following areas will be established in Mediterranean.

Area 'A'. From Gibraltar as far east as a line joining north end of Majorca and Cape San Antonio. This area will be allocated to Great Britain.

Area 'B'. To eastward of area 'A' as far as – on north – a line joining Franco–Italian territory to north point of Corsica, on east – the west coast of Corsica and Sardinia and Sicily including Cape Bon Channel and Pantelleria. This area to be allocated to France, who is anxious to be responsible for Marseilles–Algiers traffic.

Area 'C'. The Tyrrhenean [*sic*] Sea. To be allocated to Italy.

Area 'D'. The Malta Channel, the eastern end being left undefined. To be allocated to Great Britain.

---

[1]Doc. No. 186.

With regard to Eastern Mediterranean and in particular the Aegean and coast of Greece, strong pressure has been put on British delegation by Riverain powers to take over this area to avoid Russian assistance, which they dislike.

It is proposed therefore that in this area each participating power will carry out agreed measures in its own territorial waters, but on high seas up to entrance to Dardanelles the British and French Fleets will carry out the measures in areas where there is reason to apprehend damage to shipping. In this connection participating powers will give these Fleets all possible assistance, including use of bases and permission to take action in territorial waters.

Part 2.

A draft declaration will probably be agreed to to-day and will be communicated to you.

Part 3.

The French intend to use aircraft to assist in their patrol as fully as possible. You will therefore be able to do the same. I am taking immediate steps through Admiralty to induce Air Ministry to agree to provide two flying boat squadrons forthwith. French are anxious to give us any assistance they can with their air reconnaissance and this can be arranged direct between you and French C.-in-C. when final decisions have been promulgated.

French also offer full use of their ports, including Algiers if desired, and I still assume Spanish ports will be at your disposal as at present.

French will accept defence of their territorial waters in area 'A' by British Navy, but this would be a staff arrangement only, as a declaration to that effect would affect French public opinion.

Part 4.

With regard to your personal message 1313/10, I fully understand and approve of your intention to conduct operations, and you have full authority to do so in any manner you desire, and to proceed to Gibraltar or elsewhere as necessary. It is recognized by both France and ourselves that certain cruiser forces will be necessary in support of destroyers. As it is not intended that scheme would come into force before 20th September it will be preferable that you should not move to Gibraltar at present, but as soon as international policy has been settled and promulgated complete freedom of movement for yourselves and your Fleet will of course be necessary.

1700/11.

192. *Chatfield to Pound*

[ADM 116/3522]   11 September 1937

SECRET.

[Telegram] Addressed C.-in-C. Mediterranean, repeated Admiralty, C.-in-C. Home Fleet.

Following are action clauses of draft Declaration referred to in my 1700 September 11th, which was agreed to to-day. Begins

1. The participating Powers will instruct their Naval Forces to take action indicated in paragraph 2 and 3 below with a view to protection of all merchant ships not belonging to either of conflicting Spanish parties.

2. Any submarine which attacks such a ship in a manner contrary to rules of International Law referred to in the Treaty for limitation and reduction of Naval Armaments signed in London on 22nd April 1930 and confirmed in protocol signed in London on 6th November 1936, shall be counter-attacked and, if possible, destroyed.

3. The instructions mentioned in preceding paragraph shall extend to any submarine encountered in vicinity of a position where a ship not belonging to either of conflicting Spanish parties has recently been attacked or sunk in violation of rules referred to in the preceding paragraph, in circumstances which give valid grounds for belief submarine was guilty of attack

4. In order to facilitate putting into force of above arrangements in a practical manner, the participating Powers have agreed upon following arrangements:

(i). In Western Mediterranean and in Malta Channel, with exception of Tyrrhenian Sea which may form subject of special arrangements, the British and French Fleets will operate both on high seas and in territorial waters of participating Powers, in accordance with division of areas agreed upon between the two Governments.

(ii). In Eastern Mediterranean, (a) each participating Power will operate in its own territorial waters; (b) on high seas, with the exception of Adriatic Sea, the British and French Fleets will operate up to entrance to Dardanelles in those areas where there are reasons to apprehend danger to shipping, in accordance with division of areas agreed upon between the two Governments. The other participating Governments possessing a seaboard on the Mediterranean undertake within limits of their resources to furnish to those fleets any assistance that may be asked for; in particular they will permit them to take action in their territorial waters and to use such of their ports as they shall indicate.

(iii). It is further understood limit of zones referred to in sub paragraphs (i) and (ii) above and their allocation shall be subject to revision by participating Powers in order to take account of any change in situation.

5. The Participating Powers agree, in order to simplify operation of above measures, they will for their part restrict the use of their submarines in the Mediterranean in following manner;

(a). except as stated in (b) and (c) below, no repetition no submarines will be sent to sea within Mediterranean.

(b). Submarines may proceed on passage after notification to other participating Powers, provided they proceed on the surface and are accompanied by a surface ship.

(c). Each participating Power reserves for purposes of exercise certain areas in which its submarines are exempt from restrictions mentioned in (a) or (b).

The participating Powers further undertake not to repetition not to allow presence in their respective territorial waters of any foreign submarines except in case of urgent distress; or where conditions prescribed in sub paragraph (b) above are fulfilled.

6. The participating Powers also agree, in order to simplify problem involved in carrying out measures above described, they will severally advise their merchant shipping to follow certain main routes in Mediterranean to be agreed upon between them.

7. Nothing in present agreement restricts right of any participating Power to send its surface vessels to any part of Mediterranean. Ends.

2140/11.

193. *Chatfield to Pound*

[ADM 116/3522]   12 September 1937

SECRET.

[Telegram] Addressed C.-in-C. Mediterranean, repeated Admiralty, C.-in-C. Home Fleet.

IMMEDIATE.

Reference paragraph 5 (c) of declaration contained in my 2140 11th September the areas which it is proposed to reserve for ourselves are as follows:–

Area 1. A sector to southward to Malta of 40 miles radius between bearing of 135° and 270° from Delimara Point.

Area 2. A sector to eastward of Cyprus of 40 miles radius between bearing of 045° and 135° from Famagusta.

Your remarks as to suitability of these areas are requested.

0900/12.

194. *Pound to Chatfield*

[ADM 116/3522] 12 September 1937

SECRET.

[Telegram] Addressed British Naval Delegation, Geneva, repeated Admiralty, C.-in-C. Home Fleet.

IMMEDIATE.

811. Reference telegrams 2045 11th September and 2148 11th September from First Sea Lord. Intended initial disposition.

1. North Aegean. At Mudros R.A. 1st B.S. in H.M.S. *Malaya* with four float planes lent from H.M.S. *Glorious*. Two destroyers of 3rd Destroyer Flotilla.

2. South Aegean. At Suda Bay or Navarin. H.M.S. *Repulse* with four float planes. Two destroyers of 3rd Destroyer Flotilla.

3. No repetition no flying boats in Aegean.

4. Any extra destroyers the French can provide to work with the northern and southern Aegean patrol line would be welcome.

5. Malta Channel and Eastern Medn. At Malta V.A.C.B.C.S. in H.M.S. *Hood* in charge. Three destroyers of 3rd Destroyer Flotilla and H.M.S. *Douglas*. 1st M.T.B. Flotilla. Flying Boat Squadron No. 202 at Malta. H.M.S. *Glorious* unless required in Area 'A' until arrival of flying boats. A flying boat squadron at Alexandria is desirable now and may be essential later. Subsequently three destroyers of 2nd Destroyer Flotilla will proceed to Malta after relief by 1st Destroyer Flotilla in Area 'A'.

6. Area 'A'. C.-in-C. Medn. in H.M.S. *Barham* in general charge. R.A.(D) in H.M.S. *Galatea* and Commodore (D) in H.M.S. *Cairo* to take charge of patrol at sea. 1st Destroyer Flotilla. Six ships of 2nd Destroyer Flotilla. All available Home Fleet Destroyers. H.M.S. *Woolwich* at Oran. Destroyers working in Area 'A' to be based on Gibraltar and Oran. Two flying boat squadrons in addition to that referred to in paragraph 5.

7. The two flying boat squadrons in Area 'A' to be both based on Oran if possible. Otherwise one at Gibraltar and one at Algiers or Bougie.

8. Submarines exercising in area for working up destroyers in antisubmarine in following areas (a) west of Mudros, (b) vicinity of Malta, (c) Tetuan Bay.

9. Balance at Barcelona. H.M.S. *Delhi* for keeping in touch with British interests and escorting Hospital Ship *Maine*.

10. At Gibraltar. H.M.S. *Penelope* for supporting patrol and general duties.

11. Submarines for keeping A/S of destroyers up to date to be stationed at Gibraltar, Malta and Mudros.

12. Will Italians be responsible for Dodecanese area?
1418/12.

195. *Chatfield to Pound*

[ADM 116/3522]   12 September 1937

SECRET.

[Telegram] Addressed C.-in-C. Mediterranean, repeated Admiralty, C.-in-C. Home Fleet.

My 1700 11th September, part 3. Anglo French co-operation has been under discussion with French Delegation, and a short Anglo French agreement on the subject is being drafted, text which will be communicated to you when completed.

It will be necessary for many details to be settled between yourself and French Admiral Esteva[1] who will be in command of French area. Admiral Esteva is proceeding to Oran, which will probably be his Headquarters, arriving Thursday, 16th September. Request you will proceed to Oran, arriving if practicable Thursday 16th September so that sufficient time may be available for consultations before Monday 20th September when measures will probably come into force.

French Delegation suggests arrangements concerned with your arrival should be made direct with Commandant de la Marine Algiers.
1430/12.

196. *Chatfield to Pound*

[ADM 116/3522]   13 September 1937

SECRET.

[Telegram] Addressed C.-in-C. Mediterranean, repeated Admiralty, C.-in-C. Home Fleet.

IMPORTANT.

3. My 1430/12. Following is text of Anglo French Agreement. Date of entry into force will be communicated later.

---

[1] Amiral Jean Pierre Esteva (1880–1951). Commanded French naval forces in Far East, 1935; Inspector-General Maritime Forces, 1937; Amiral-Sud (C-in-C naval forces at Toulon and then Bizerte), 1937–40; Resident General in Tunisia, 1940–43; expelled by Germans, May 1943; condemned by High Court for failure to resist German occupation of Tunisia following Allied landings in North Africa and sentenced to life imprisonment, 1945; freed for reasons of health, Aug 1950.

1. In order to facilitate the putting into force of the Nyon arrangement the French and British Naval Staff have agreed as follows:–

2. Zones of operation in the Mediterranean

A. Western Mediterranean with the exception of Tyrrhenian Sea is divided into 2 zones;

Zone A. From the Straits of Gibraltar Eastward as far as the following limits on the North a straight line joining Cape San Antonio and Dragonera Point in Majorca as far as Cape Pera. On the East a straight line joining Cape Pera, Majorca and Algiers.

Zone B. From Zone A eastward as far as the following limits, on the North a straight line joining Mentone and Cape Corse, Corsica. On the east the west coast of Corsica and Sardinia as far as Cape Spartivento and thence the following limits; a straight line joining Cape Spartivento and Maritime Islands longitude of 12° 5′ east, as far as latitude 36° 40′ north and thence along a parallel of 36° 40′ north to the African coast.

B. Malta Channel. Zone D. Consists of the channel between Sicily and the African coast to the eastward of Zone B.

The Aegean Sea and Coast of Greece bordering on the Ionian Sea are divided into two zones – Zone E the Aegean Sea to the northward of latitude 38° 10′ north. Zone F. the remainder of the Aegean Sea and Greek coast bordering the Ionian Sea.

3. Allocation of Zones. The French Navy will operate in Zones (E) and (F) and the British Navy in Zones (A), (D) and (E).

4. The above allocation of zones does not exclude action by French or British Naval and Air Forces in a zone allocated to the other party if the circumstances are such as to render such a course of action advantageous.

5. Neutral Assistance. During the time the Nyon arrangement is in force French and British Forces employed in carrying out the Nyon arrangement may visit each other's ports and fly over and land on each other's territory without previous notice and without diplomatic authorisation. The French and British local authorities concerned will facilitate by all means in their power the operations of the warships or aircraft in question.

6. Collaboration of the Fleets. The practical measures concerning the collaboration of the two fleets will be arranged in consultation between the British and French Admirals respectively, charged by their governments with the responsibility for the conduct of the operations under the Nyon arrangement. for this purpose these two officers will place themselves in direct touch. Among the measures concerning which it will be necessary for arrangements to be made are the following:–

1. The drawing up of a simple reporting code for use between the two Navies.

2. Arrangements in connection with wireless transmission.

3. A simplification of formalities connected with visits and salutes.

7. With object of facilitating the collaboration of the two Fleets each party will place a liaison officer at the disposal of the Admiral of the other Fleet charged with the responsibility for the conduct of operations.

8. <u>Entry into force of the present arrangement</u>. The arrangement will enter into force on ..... and will remain in force so long as operations are carried out by the British and French Fleets in accordance with the Nyon arrangement.

1300/13.

197. *Admiralty to Pound*

[ADM 116/3522] 13 September 1937

SECRET.

[Telegram] Addressed C.-in-C. Mediterranean, repeated British Naval Delegation, Geneva.

<u>IMMEDIATE</u>.

Your 1523/12.

2. Provision of flying boat squadron now under consideration by Air Ministry.

3. Question of provision of suitable depot ship for two squadrons in Western Basin presents difficulty.

4. To commission *Hermes* for this purpose would involve considerable interference with other manning commitments.

5. To convert suitable merchant ship would involve delays.

6. It has been suggested that *Cyclops* might be employed for this purpose, temporary arrangements being made for accommodation of Mediterranean Fleet submarines.

7. Depot ship would be required to accommodate some 450 R.A.F. personnel. Arrangements would be made to take up petrol carrying ship in addition.

8. Your early remarks on this proposal are requested.

1730/13.

198. *Chatfield to Pound*

[ADM 116/3522] 13 September 1937

SECRET.

[Telegram] Addressed C.-in-C. Mediterranean, repeated Admiralty, C.-in-C. Home Fleet.

Your 1418/12 and 2353/12. you will see from my 1300/13 that French are now taking over southern half of Aegean Sea. It is not anticipated that operations will be necessary from Alexandria at any rate in initial stages, and in consequence it is not proposed to ask Air Ministry for flying boat squadron at Alexandria at present. French delegation suggests that Arzeu is preferable to Oran for operations of flying boat squadrons, but you will no doubt discuss this with Admiral Esteva.

If Italians cooperate it is probably that Zone F will either be taken over by them or divided so as to allow them to operate in the vicinity of Dodecanese.

The 3 areas you propose for anti-submarine exercises and also one at Alexandria have been included in the draft agreement.

With reference to the area near Gibraltar you will appreciate that Spanish submarines are not debarred from the Mediterranean and it is undesirable at present to discuss the matter with them formally. It may be possible later to arrange with Admiral [at] Palma informally that his submarines will not be used in this area.

2045/13.

199. *Chatfield to Pound*

[ADM 116/3522] 13 September 1937

SECRET.

[Telegram] Addressed C.-in-C. Mediterranean, repeated Admiralty, C.-in-C. Home Fleet.

You will no doubt be wondering how far continuous patrol will be expected of your units when patrol scheme starts, as it is obvious that small number of units available will be insufficient for a complete patrol even if all are at sea simultaneously. It will probably be necessary for the first two or three weeks that you should operate in force within the limits of refuelling.

If a reasonable period elapsed without any incident taking place it could be considered that piracy has been deterred.

If this proves to be the case the patrol could be reduced and gradually stopped, pending the recurrence of incidents. If this were not the case the patrol would anyhow be reduced to that number which experience had shown could be maintained continuously.

I should be glad of your remarks.

2110/13.

200. *Pound to Chatfield*

[ADM 116/3522] 14 September 1937

SECRET.

[Telegram] Addressed British Naval Delegation, Geneva, repeated Admiralty, C.-in-C. Home Fleet.

IMMEDIATE.

835. If I may I should like to defer reply to your 2045 13th September until I have discussed the matter with Admiral Esteva, R.A.(D) and Commodore (D). I am taking V.A.C.B.C.S. to Oran in H.M.S. *Barham* as he will be in charge in Area D and will have to co-operate with the French in Area B. He will return to Malta in a destroyer. Reduction of British area in proposals approved is much appreciated.

I can now detail four destroyers for Area E but it has occurred to me that from a political point of view it might be desirable to ask both the Turks and the Greeks to lend two destroyers each to R.A. 1st B.S. in Area E so long as Russians need not be asked for similar contribution.

1118/14.

201. *Pound to Admiralty*

[ADM 116/3522] 14 September 1937

SECRET.

[Telegram] Addressed Admiralty, repeated British Naval Delegation, Geneva.

IMMEDIATE.

836. Your 1730 13th September.

By removing all naval personnel except repair staff and navigating party H.M.S. *Cyclops* can accommodate R.A.F. personnel as follows:– Officers, 48; senior ranks, 130; junior ranks, 272.

Removal of submarine spare gear to make room for R.A.F. stores will be necessary.

H.M.S. *Cyclops* ready to receive R.A.F. personnel will be ready to leave Malta Thursday, 23rd September.

For petrol carrier see my 0041 14th September …

1128/14.

202. *Director of Naval Intelligence*[1] *to Chatfield*

[ADM 116/3525] 14 September 1937

[SECRET]

[Telegram] Addressed British Naval Delegation at Nyon.

Personal for C.N.S.

Most secret information suggests that Italian submarines of Aegean patrol have been ordered to take no repetition no offensive action against units of any flag whatsoever. C.-in-C. Mediterranean and R.A. (D) have been informed.

1229/14.

203. *Pound to Chatfield*

[ADM 116/3522] 14 September 1937

SECRET.

[Telegram] Addressed British Naval Delegation, Geneva, repeated Admiralty.

IMMEDIATE.

846. Probably Trebuki Bay, Skyros Island will be most convenient base for H.M.S. *Repulse* to control southern part of Area E. I do not know whether Turkish Government are offering or can be induced to allow use of their harbours but if so it seems a good opportunity to break down their prejudice to the use of their harbours. Suggest permission be obtained for the force controlling southern part of Area E to use the Bay of Eritra on the west side of Kara Burnu Peninsula when found necessary.

1743/14.

---

[1] RA [later VA Sir] James Andrew Gardiner Troup (1883–1975). Commanded cruiser *Cairo*, 1926–28; Capt, Navigation School, Portsmouth, 1928–30; Flag Capt and CSO to VA, 1st BS, Med Fleet, 1930–32; Dir of Tactical School, Portsmouth, 1933–35; DNI, Admy, 1935–38; retired, 1938; Flag Officer-in-charge, Glasgow and District Shipyard Controller, 1940–46.

204. *Chatfield to Pound*

[ADM 116/3522]   15 September 1937

SECRET.
[Telegram] Addressed C.-in-C. Mediterranean, repeated Admiralty.

Your 1743 14th September. I concur it will be advisable to ask for use of a Turkish harbour.

According to Nyon arrangements, signatories and riverain powers agree to place their harbours at our disposal, but as an act of courtesy they will be asked through diplomatic channels for use of particular harbours we require.

Request you will confirm following facilities which should be asked for: from Greece – Lemnos Island and Skyros Island; from Turkey – Bay Eritra.

When confirmed it is requested Admiralty will take necessary action through diplomatic channels.
1105/15.

205. *Chatfield to Pound*

[ADM 116/3522]   15 September 1937

SECRET.
[Telegram] Addressed C.-in-C. Mediterranean, repeated Admiralty, C.-in-C. Home Fleet.

8. Your 1118 14th September. Original proposal made by French and ourselves at commencement of conference envisaged employment of Turkish and Greek destroyers in Aegean outside territorial waters. Arrangement for this area embodied in final arrangements was, however, made at request of smaller powers, particularly Greece and Turkey, and it is not desired therefore to ask them for assistance at any rate at present.
1120/15.

206. *Pound to Admiralty*

[ADM 116/3525] 17 September 1937

SECRET

[Signal] Addressed Admiralty.

IMPORTANT.

890. (A). Have just left Oran after very successful meeting with Admiral Esteva and visit to Arzeu.

(B). French had no proposals and willing to adopt all ours which may be summarised as follows.

1. H.M.S. *Woolwich*, oiler *Plumleaf* and destroyers based on Oran.

2. H.M.S. *Cyclops* and *Petrella* and nine flying boats based on Arzeu.

3. H.M.S. *Woolwich* to be link ship between C.-in-C. and also between all S.N.O.'s of areas for secret messages.

4. International code to be used for non-confidential messages.

5. A simple reporting table produced by us has been adopted.

(C). French dispositions apparently will be:–

Area B.

1. Area north of line joining Cape Sebastian and Straits of Bonifacio. Float planes and three destroyers.

2. Zone about 120 miles deep north of African coast between Algiers and Cape de Fer. Ten seaplanes and two destroyers.

3. Bizerte Cape Bon area. Local aircraft and four destroyers.

4. Area between one and two and coast of Spain and Sardinia. Nine Leaders, no regular air patrol but long range aircraft held in reserve in south of France to deal with incidents at sea.

Area F. Extends from parallel 038 degrees 10 minutes north to a line which runs from Corfu close round Cephalonia, Zante, Crete and Scarpanto to Seven Cape. No aircraft, one destroyer and one patrol vessel detached from Beirut command.

(D). French will deal with S.O.S. calls in Area C but will not hunt submarines in that area.

(E). Particulars of Arzeu follow in separate telegram.

1933/17.

207. *Pound to Admiralty*

[ADM 116/3525] 17 September 1937

SECRET.

[Signal] Addressed Admiralty, repeated V.A.C.B.C.S., V.A. Malta, A.O.C. Mediterranean.

IMMEDIATE.

891. V.A.B.C.S. pass to A.O.C. Mediterranean.

A. Very urgent to commence operating flying boats from Arzeu so as to release H.M.S. *Glorious* for the Eastern Mediterranean.

B. Arzeu inspected by myself and Fleet Aviation Officer and found very suitable.

C. 12 moorings already in place and hotels available as temporary accommodation.

D. Lubricating oil can be sent from Gibraltar to destroyers if required before H.M.S. *Cyclops* arrives about 26th September. Local oil unsuitable.

E. Aviation spirit is the only snag. Petrol Carrier *Petrella* cannot arrive at Arzeu before 30th September. Large available stocks at Arzeu belonging to the French Army supplied by the Shell Company and used by the French flying boats. This aviation spirit is 77 octane, but it is uncertain whether it is of correct quality. Specific gravity is point 735.

F. If Air Ministry ascertain from the Shell company that this spirit is suitable there appears no reason why Flying boat Squadron No. 210 should not be sent to Arzeu forthwith.

1943/17.

208. *Pound to Admiralty*

[ADM 116/3525] 17 September 1937

SECRET.

[Signal] Addressed Admiralty, repeated V.A. Battle Cruiser Squadron, R.A. First Battle Squadron.

IMPORTANT.

892. I do not think that French take area F very seriously.

Route No. 8 to west of Cape Matapan[1] appears to be likely place for submarine attack on munitions ships and oilers from Black Sea. It may

---

[1] Route No. 8 of the *Routes Advised To Be Used By British Merchant Ships In The Mediterranean–Nyon Arrangement* (Admiralty Notice to Mariners, No. 1696, 20 Sept 1937) concerned the Aegean and Black Sea to the Western Mediterranean and passed through a point 35° 47′ North, 16° 40′ East and the Cervi and Doro channels.

be necessary to send H.M.S. *Repulse* to Navarin after she has delivered aircraft to H.M.S. *Malaya*.

Request: (a) early information <u>of presence of submarines</u> west of Cape Matapan; (b) Greek permission for use of Navarin and Suda Bay if found necessary.

2128/17.

209. *Pound to Admiralty*

[ADM 116/3525] 17 September 1937

<u>SECRET</u>.

[Signal] Addressed Admiralty.

894. At today's conference Admiral Esteva and I agreed to ask our respective Governments to endeavour to arrange that British and French merchant ships on the high seas in the Mediterranean should, on sighting a submarine, hoist a black ball at the masthead and keep it hoisted until 4 hours have elapsed since the submarine was last seen.

2. Object to give early indication to man of war of the presence of a submarine in the vicinity and put other merchant ships on their guard.

3. If this proposal is approved it is requested that the necessary steps may be taken to implement it forthwith.[1]

2351/17.

210. *Admiralty to Pound*

[ADM 116/3525] 18 September 1937

<u>SECRET</u>.

[Signal] Addressed C.-in-C. Mediterranean, repeated V.A. Battle Cruiser Squadron, R.A. First Battle Squadron, from Admiralty.

<u>IMPORTANT</u>.

Your 2128/17. Admiralty agree that Route No. 8 to the west of Cape Matapan appears to be a very likely place for submarine attack. As insufficient ships are available to patrol the whole of the Mediterranean, it was necessary to state in the Nyon arrangement that in the Eastern Mediterranean the British and French Fleets would operate in those areas where there is reason, based on present experience, to apprehend danger

---

[1]The Admiralty approved and ordered the signal to be transmitted immediately on commercial wave lengths as a Notice to Mariners. Admiralty to Pound, 1448/21.

to shipping in accordance with the division of the area agreed upon between the two Governments. The French zone 'F' includes the Matapan Channel area to an undefined distance to westward. Similarly, our area 'D' continues for an undefined distance to the eastward from the Malta Channel. If attacks take place in the Matapan area or in any other unpatrolled area, it will become necessary to redispose forces and arrange for a modification of the zones with the French. It is also just possible that if Italy came in she would assume responsibility for the area from the east coast of Italy to Greece, which lies to the northward of zone 'D'.

As Navarin and Suda Bay are in a French area, it is thought preferable to wait until the occasion arises before asking the Greeks for facilities. This could always be done while our forces were being redisposed.
1535/18.

211. *Pound to Admiralty*

[ADM 116/3525] 19 September 1937

SECRET.
[Signal] Addressed Admiralty.

IMPORTANT.
909. Reference 2010 13th September from 1st Sea Lord. During first 48 hours that [Nyon] patrol is in force R.A.(D) in H.M.S. *Galatea* and 18 destroyers working in six units of three each will be on patrol in area A. During this time 8th Division Destroyers and three ships of 12th Division Destroyers will be painted muddy grey. After first 18 hours intend working destroyers in area A in two watches as I consider that this is maximum which should be attempted considering alertness required and strain on submarine detectors and wireless ratings.

When most of the reinforcements have arrived from Home Fleet I shall probably send five more destroyers to area D which will leave 27 to work area A in two watches.

H.M.S. *Vanoc* not counted in report as she is required to escort submarines to and from exercise areas, etc.
0028/19.

212. Pound to Admiralty

[ADM 116/3525]   20 September 1937

SECRET.

[Telegram] Addressed Admiralty.

IMPORTANT.

920. I informed Spanish Admiral at Palma that a large number of British aircraft will be operating south of Balearic Islands and west of three degrees east but they could not fly within 10 miles of Spanish Territory. In reply he has requested that our aircraft shall not fly north of parallel 038° 21' North to avoid causing false air alarms in his island and that he will inform me if any of his submarines are sent south of this latitude. I shall not make any reply but in practice shall keep aircraft south of 038° 28' North and see how situation develops.

0042/20.

213. Admiralty to Pound

[ADM 116/3525]   20 September 1937

SECRET.

[Telegram] Addressed C.-in-C. Mediterranean, repeated V.A.C.B.C.S., H.M.S. *Glorious*, From D.N.I.

IMPORTANT.

British Naval Attaché Berlin was recipient today of following communication which German Admiralty described as friendly communication between German and British Navies. Begins:–

Press reports state that aircraft are to be used in conjunction with surface patrols established under Nyon arrangement. German Admiralty wish to observe that German warships in Mediterranean have instructions to open fire on approaching aircraft and accordingly desires once again to warn the aircraft taking part in patrols under Nyon arrangement against flying over German warships. Ends.[1]

2100/20.

---

[1] Pound accordingly made a general signal that as warships of 'certain foreign powers' may open fire on aircraft flying over them, aircraft operating in conjunction with the Nyon arrangements should not normally approach foreign warships within a distance of 4,000 yards. If they had to approach, the approach should be made in such a manner as to indicate clearly no attack was intended by bomb or torpedo. 1647/23. The Germans issued another warning in December after a British aircraft allegedly approached German torpedo boats in the vicinity of Sardinia. German Admiralty instructions were that any aircraft approaching German naval forces was to be fired upon and 'every submerged submarine is to be regarded as hostile'. DNI to C-in-C Mediterranean and VACBCS, 13 Dec 1937, ADM 116/3529.

214. *Pound to Chatfield*

[CHT/4/10] Commander-in-Chief, Mediterranean Station
n.d. [20 September 1937]

[Holograph]

I do hope you managed to get your leave over before you had to go to Geneva. Here we are in the first day of Nyon after a lull of a week as far as s/m piracy is concerned. I must say I do not know what to make of things. There can be little doubt I should imagine that the Italians are feeling that they have got themselves into a humiliating position because after all their talk about *Mare Nostrum* they now find they are left out in the cold. I have been wondering what their reaction would be and was quite prepared for increased piracy and we may get it yet. On the other hand there are indications that some of the s/m activities have been called off and the information which I wired to you last night – that the Admiral at Palma would let me know if he was going to send a s/m to operate south of 38° 28′ may mean that the Nationalists are going to rely more on their surface craft to intercept trade.

I rather imagine, however, that the majority of the submarines are controlled by the Italians and what they will do is uncertain.

From what I know of what the French are doing I should think there will be a tendency for submarines to operate in Area B rather than A. I gather this effort has come at a bad moment for the French as their Mediterranean ships are changing their crews.

I am very grateful to the Admiralty for bringing pressure to bear on the Air Ministry to get their flying boats out quickly. I am just off to Arzeu, which incidentally is a much better place than it looks on the chart – to discuss air patrols with O.C. 210 Squadron and *Glorious* and see the former installed. If the Italians are controlling most of the submarines it is possible their activities may be more in Eastern Medn than in Western Basin and as soon as I feel able to do so I shall transfer the remaining 5 ships of 2nd Flotilla to Area D – this will also give them a change of scene as they have already been 6 weeks on this coast. The *Repulse* and *Malaya* each with 4 seaplanes form two very useful mobile units for this kind of game.

As you know I was not in favour of this present scheme but unless the submarines are a great deal more determined than it looks as if they were going to be I am very glad Scheme Z has been put into force instead of convoy because one can ease up on this scheme if there is no s/m activity whence with convoy it is very difficult to drop it once it has been started.

Warspite gives one many shocks and this insubordination will I am afraid have undone a great deal of the good work of the Fleet's during the last 18 months in putting the Navy back into its proper place in the esteem of the country.[1] I wish it had been possible to give as the reason for relieving three [?] officers that they had not acted with sufficient firmness rather than that they had mishandled the situation as the latter can be interpreted in two ways. Her continual breakdowns are annoying at the moment as I have not room for all the staff in *Barham* and *Resource* has to follow me round with the remainder. If *Warspite* is not out here in time to do her working up before *Barham* has to go home about 1st week in December the situation will become more than annoying as I do not think there should be less than two efficient battleships on the station, with the Mediterranean situation as it is. Also, if the s/m piracy is not killed by the end of October a possible relief for *Glorious* may arise.

As you will have seen I have urged that First Lord should stick to his visit to Eastern Medn in which case I think it might be the best arrangement if I met him at Alex. so that I could go straight on with the meetings with the other two services about the plans for the Defence of Egypt. The meeting of the heads of the 3 services is the first step of the chain which finally brings the plan before the C.I.D. and it is so important that delay should not be caused that I ought not to be the cause of delay in the first step. As soon as it is seen how the s/m campaign goes I will put proposals before you officially. If I do not get the Egyptian decision over in October I might have to defer them until December as I have got the French C.-in-C. and Greek Fleet for the greater part of November.

I hope very much that it may be possible to approve my proposal that 'Tribals' should be incorporated in 3rd C.S.[2] as I am sure this will be the most efficient arrangement ...

There have been several firm rumours that Roger Backhouse is to be relieved in January. If this is so I shall have to make out arrangements for the Combined Exercises and should have to commence to do so almost at once because as you know it is customary for a Staff Officer from Home Fleet to come out to Malta in November.

---

[1] The mechanical difficulties encountered by the *Warspite* resulted in prolonged trials and delays in sailing for the Mediterranean. An order to return early from weekend leave resulted in a small demonstration on the mess deck that was quickly ended but the news was leaked to the press and resulted in undesirable publicity. The incident is covered in S.W. Roskill, *H.M.S. Warspite: The Story of a Famous Battleship*, paperback edn (London, 1974), pp. 166–9.

[2] Pound's proposal concerning the Tribal-class destroyers just entering service was not approved.

I am sure we ought to have a flying boat squadron with depot ship attached to the Fleet permanently – they are really a necessary part of a Fleet. This opinion is not the result of the present emergency. I will put up something officially about it.

I hope you will find things are not so strenuous this autumn. I got a few quail and duck at Kalamanta but Samothraki has deteriorated badly and we only got 6½ brace after 6 hrs. walking – whereas 10 years ago 5 of us got 17 brace before 9 a.m. I had only one afternoon at Mudros before I had to dash off. We had a very successful regatta between *Barham* and *Malaya* the day before we left Mudros – extraordinary good feeling in *Malaya* who were handsomely beaten by *Barham*, cheered the latter spontaneously as we left.

All good luck and kindest regards to Lady Chatfield.

215. *Pound to Mediterranean Station*

[ADM 116/3524] 21 September 1937

[Telegram] Addressed Med. Station, repeated Admiralty.

<u>IMPORTANT</u>.

933. Following received from Admiralty 21st September and is to be embodied in I.N.A. as article 26, headed 'Attacks by surface ships and aircraft.' Begins:

By an arrangement signed at Geneva on the 17th September the following provision supplementary to the Nyon arrangement are to enter immediately into force.

(1) The present agreement is supplementary to the Nyon arrangement and shall be regarded as an integral part thereof.

(2) The present arrangement applies to any attack by a surface vessel or aircraft upon any merchant vessel in the Med. not belonging to either of the conflicting Spanish parties when such attack is accompanied by a violation of the Humanitarian principles embodied in the rules of International Law with regard to warfare at sea which are referred to in Part Four of the Treaty of London of April 22nd 1930, and confirmed in the Protocol signed in London on November 6th 1936.

(3) Any surface war vessel engaged in the protection of merchant ship in conformity with the Nyon arrangement which witnesses an attack of the kind referred to in the preceding paragraph shall:

(A) If the attack is committed by an aircraft open fire on the aircraft.

(B) If the attack is committed by a surface vessel intervene to resist it within the limits of its powers, summoning assistance if such is available and necessary.

In their territorial waters each of the participating powers concerned will give instructions as to the action to be taken by its own war vessels to comply with the spirit of the present arrangement. (1845/20) Ends.
1414/21.

216. *Captain G. Muirhead-Gould[1] to Pound (Excerpt)*

[ADM 116/3681] H.M.S. *Devonshire*, at Alexandria
22 September 1937

CONFIDENTIAL

No. 245/0191

LETTER OF PROCEEDINGS.

* * *

11. I had already arranged to carry out sub-calibre practices during the afternoon [Wednesday, 15th September] and *Protector* brought out a Pattern VI target. On completion of this practice *Protector* proceeded into harbour with *Devonshire* about 4 miles on her port quarter steaming at slow speed. *Protector* signalled that 'one Officer and some ratings had sighted a torpedo passing astern of target'. I ordered all ships to take A/S precautions, and to manoeuvre at speed and zigzag until ready to enter harbour. On arrival in harbour I sent my Torpedo Officer to *Protector* to investigate the report. Lieutenant-Commander Peterson and four or five ratings independently had sighted what they thought was the feather of a submarine periscope, or the splash of a torpedo running on the surface. It crossed astern of *Protector* and vanished. It was in sight for several seconds, but no one saw it through glasses, and it was not seen from the bridge. One seaman thought he saw 'something like a brown shape' under the feather. The weather at the time was fine and clear, but there was a Northerly wind, force 4, and a lot of white horses were visible. There was a four foot swell running from the Northward.

In these circumstances I have come to the conclusion that it is most unlikely that what was seen was either a submarine or a torpedo. Porpoises are plentiful and sharks have been reported. *Protector* had just finished dinner and it is established that a quantity of mess scraps had just been thrown overboard, which may have attracted a shark.

---

[1]Capt [later RA] Gerard C. Muirhead-Gould (1889–1945). Commanded sloop *Bluebell*, 1926; served in Naval Intelligence Dept, 1931–33; naval attaché in Berlin, 1933–36; commanded cruiser *Devonshire*, 1936–39; Tactical School, Portsmouth, 1939; lent to Royal Australian Navy and Cdre-in-command, Sydney, 1940–44; Flag Officer-in-charge, Western Germany (accepted surrender of Heligoland), 1945.

Captain D.J.R. Simpson, Royal Navy, of H.M.S. *Protector* concurs in my opinion.

12. I have instructed H.M.S. *Coventry* to draw depth charges from R.F.A. *Churraca*. My own depth charges were already in place on the chute, and I have given orders that all ships are to place anti-submarine lookouts and to keep anti-submarine guns' crews closed up when at sea. Also that ships are not to remain stopped outside when this can be avoided. When *Devonshire* is recovering Queen Bees *Coventry* will act as an anti-submarine screen. I shall arrange for marking aircraft to carry out a reconnaissance before and after practices.

13. Ships' companies have been told not to talk about the incident ashore, but I understand that Reuters have discovered something about it, and I have asked their representative to come and see me.

* * *

217. *Admiralty to Pound*

[ADM 116/3524] 1 October 1937

SECRET.

[Telegram] Addressed C.-in-C. Mediterranean, repeated C.-in-C. Home Fleet.

My 1546/23/9.[1] Agreement has been reached between naval representatives for inclusion of Italy in Nyon arrangement patrols. Text of agreement follows in my 2329. No action is to be taken in connection with this agreement at present and you will be informed if and when it is to come into force. You will observe that the zones now allocated to Britain are greater than those which you are now undertaking. The main reason for this is that the whole Mediterranean is now divided up into zones whereas previously large parts were left unallocated. You will notice however that the extra zones which will now come under you are those in which it is improbable that it will be necessary to undertake any serious patrol work.

With regard to zone (4) you will observe that Matapan and Kithera channels have been excluded but your remarks are requested as to whether you would like arrangements made with Greece for the use of Navarin Bay or a port in Crete. The fact that north coast of Crete is in French zone would present no difficulties.

With regard to zone (6) Italian zone was extended north to point N on shipping route at their express request. From that point boundary was

---

[1] Informing Pound of the impending talks to bring Italy into the Nyon arrangements.

taken to Khios Island for the purpose of leaving Eritra Bay in your zone and available for you.

It will be desirable that when agreement comes into force a meeting should be arranged with French and Italian admirals and it is hoped that it may be possible for you to do this during your stay at Malta.
2045/1.

218. *Admiralty to Pound*

[ADM 116/3524] 1 October 1937

SECRET.

[Telegram] Addressed C.-in-C. Mediterranean, repeated C.-in-C. Home Fleet.

Following are operative clauses of Agreement between French, Italian and British naval representatives:–

I. The main routes to be recommended to merchant shipping in the Mediterranean will remain as defined in Annex 2 to the Nyon Arrangement with the following amendment:

Route No. 7.– Marseilles–Messina–Port Said. Pass through the following points: J. 25 miles south of Marseilles; the Straits of Bonifacio; the Straits of Messina; [point] E 35° 47' North, 16° 40' East.

II.– Zones of Operation.

Those portions of the Mediterranean Sea which under the Nyon Arrangement were entrusted to the supervision of France and Great Britain, and the Tyrrhenian Sea, which was reserved for special arrangement, will be sub-divided as follows:–

Zone I. From the Straits of Gibraltar as far as the line C. San Antonio–Dragonera Point then along the territorial waters of the S.W. coast of Majorca to Cape Salinas and thence to Algiers.

Zone II. From Zone I eastward as far as the following limits:–

From the Franco–Italian frontier to the following points:– 10 miles N. of Giraglia Light; 10 miles west of Pianosa Light and then south along this meridian as far as the Parallel 41° 20' N; westward along this parallel to 5° 45' E; thence by lines joining the points:– 20 miles east of Aire Island Light; 15 miles 229° from Toro Island Light; latitude 37° 55' N, longitude 11° 30' E; then south on this meridian as far as the parallel of Port Pantellaria; then through a point 60 miles 090° from Sfax; then to the Tripoli–Tunis border.

Zone III. To the East of Zone II, to the limits:–

The parallel of Saseno Island; the territorial waters of Albania and Greece, as far as the parallel of the S. point of Cephalonia Island; then to the point 36° 25′ N, 10° 20′ E; westward along parallel 36° 25′ to 13° 40′ E; south along meridian 13° 40′ E as far as 33° 30′ N; eastward along the parallel 33° 30′ N as far as 20° 00′ E; then to Kavo Krio (Crete); along the territorial waters of the south, east and north coasts of Crete as far as longitude 25° 06′ E; thence along this meridian to latitude of Skyros Island Light and thence to the N. point of Khios Island; to the territorial waters of Turkey; following these territorial waters to the southward of Cape Khelidonia; then to Port Said; from Port Said to a point five miles north of Damietta Light and from there to a point 33° 55′ N, 27° 50′ E, and thence the boundary between Egypt and Libya in the Gulf of Sollum.

Zone IV. Bounded on the east by a line Kavo Krio (Crete) to Sapienza Island Light, thence along the territorial waters of Greece as far as the parallel of the S. point of Cephalonia Island and bounded on the north, west and south by Zone III.

Zone V. Bounded on the west by Zone IV and the territorial waters of Greece, on the south by the territorial waters of Crete, on the east by Zone III and the north by the parallel of C. Doro.

Zone VI. The Aegean outside of territorial waters north of Zones III and V.

Zone VII. Bounded on the south by the Egyptian territorial waters between the Egyptian–Libyan boundary and Port Said, and on the north by Zone III.

Zone VIII. The Mediterranean, outside territorial waters, east of Zone III.

III.– Allocation of Zones.

The French Navy will operate in Zones II, V, VIII.

The Italian Navy will operate in Zone III.

The British Navy will operate in Zones I, IV, VI, VII.

IV. The above allocation of Zones does not exclude action by French, Italian or British Forces in a zone allocated to another party if the circumstances are such as to render such a course of action advantageous.

V. Collaboration between the Fleets.

Practical measures for the collaboration of the three Fleets will be arranged in consultation between the French, British and Italian Admirals responsible.

Among the measures concerning which it will be necessary for arrangements to be made are the following:–

(i) The drawing up of a simple reporting code for use between the three Navies.

(ii) Arrangements in connection with wireless transmission.

(iii) The simplification of formalities connected with visits and salutes.

VI. <u>Submarine Exercise Areas</u>.

The Italian Navy accepts the restrictions on the movements of submarines agreed to in paragraph V of the Nyon Arrangement. They reserve for their exercises the areas defined below, which complete the list of areas reserved by the Powers parties to the Nyon Arrangement. The list is given in Annex I of that arrangement.

VII. <u>Areas reserved for Italian Submarine Exercises</u>.[1]

* * *

VIII. The present Agreement is confidential. It will come into force when approved by the three Governments and will continue in force as long as the measures of protection agreed upon in the Nyon Arrangement themselves remain in force.

2329/1.

219. *Chatfield to Pound*

[ADM 116/3529] 5 October 1937

SECRET.

[Telegram] Addressed C.-in-C. Mediterranean, from First Sea Lord.

PERSONAL.

I am somewhat concerned at the *Basilisk* incident of yesterday, which has been given the same prominence in the press as the previous *Havock* incident. In both cases a submarine was proved to be in the vicinity of a destroyer and had fired a torpedo at her and subsequently escaped.

You will recall my telegram 1530 of 9th September, Part I[2] in which I emphasised the importance of our prestige both at home and abroad in this matter. There has always been a school here inclined to cast doubts on the Admiralty expectations as regards submarine detection – a school which has expressed its intention of watching to see the results of the piracy patrol. It was for that reason that I felt that we did not want to have any repetition of the *Havock* incident, and that the best way to avoid it was to group our destroyers so that if a submarine was attacked she would have no chance of escape.

I of course fully understand that by grouping your destroyers the chances of obtaining contact with a submerged submarine are much reduced, but in my opinion it is less important to obtain contact with a

---

[1]The locations are given for areas at Spezia, Maddalena, Naples, North Sicily, Taranto, Brindisi, the Upper Adriatic, Tobruk and two areas in the Aegean around the Dodecanese.
[2]Doc. No. 188.

submerged submarine that is doing no damage than to ensure that if a submarine is attacked, owing to some hostile act against a warship or piratical act against a merchant ship, she is destroyed. I should therefore be glad to have your reactions to this occurrence, particularly:

(i) Are we wise to use asdics against a submerged submarine for a prolonged period in order to force her to the surface, as at present laid down by the Admiralty to you in their instructions?

(ii) Is such an act liable to be so alarming to the submarine that she may fire a torpedo in self-defence, thereby laying this country open to the charge that she has provoked an incident?

(iii) Is it desirable to concentrate destroyers in groups of three, accepting the necessary limitations of discovery of submerged submarines that such a step entails?

1350/5.

220. *Pound to Chatfield*

[ADM 116/3529] 6 October 1937

SECRET.
[Telegram] Addressed Admiralty, repeated V.A.C.B.C.S.

IMPORTANT.
PERSONAL.

62. Personal for 1st Sea Lord and V.A.C.B.C.S.

(a) I share your concern but should like to defer answering your questions until V.A.C.B.C.S. has made report ordered in my 2208 4th October which will be repeated to Admiralty and should be received tomorrow Thursday.

(b) Though I have great faith in Commander Dangerfield I am not convinced that H.M.S. *Basilisk* was attacked by a submarine. H.M.S. *Boreas*, the other ship of the unit, never confirmed confirmation of a contact. Two other ships have reported torpedoes being fired at them but on investigation this was disproved.

(c) As regards maintenance of our prestige which is always in my mind there is direct method of destroying submarines and indirect method making our zone so unhealthy for submarines that they will operate in other zones with consequent deduction by the world. Though I am now in agreement with decision to adopt scheme Z I have always been doubtful whether it would result in a kill owing to unlikelihood of any of our patrolling units being able to arrive at scene of an act of piracy in time to get on terms with submarine. Hunting submarines until they come to the surface, dropping depth charges when approaching scene of an attack and covering our zone with patrol units and aircraft have all been part of indirect method.

(d) The attack on H.M.S. *Havock* took place when on passage before Nyon patrol commenced and if it is proved H.M.S. *Basilisk*'s submarine was non-existent, then there is nothing proved that indirect method has failed.

(e) I purposely reduced patrolling units to two ships rather than having a smaller number of three ship units for following reasons: (i) to keep as many units as possible in zone A in pursuance of indirect method; (ii) to allow systematic anti-submarine training with our two submarines at Gibraltar; (iii) not to overwork destroyers while situation was calm.

(f) The reduction from three ships to two ships in each unit does not reduce killing power of that unit against any submarine except those which are very expert in anti-asdic tactics. In fact against average submarines two ships are better than three as any possibility of third ship causing interference is removed.

(g) The advantages of three ship units are: (i) cover a wider front when searching; (ii) if one destroyer is torpedoed there are still two to hunt.

(h) The efficiency of asdics in Summer in Mediterranean is always reduced. Water conditions were bad up to three weeks ago but I had hoped they would have improved as a result of bad weather experienced lately.

(j) I realise at the moment as far as our critics are concerned we are in a cleft stick. If a submarine was depth charged by H.M.S. *Basilisk* it ought to have been destroyed and if there was no submarine we cannot say so without acknowledging unreliability of asdics.

(k) I am sure V.A.C.B.C.S.'s investigation will if possible discover cause of failure if there was failure.

1748/6.

221. *Vice Admiral A.B. Cunningham to Pound*

[ADM 116/3529] 7 October 1937

SECRET.

[Telegram] Addressed C.-in-C. Mediterranean, repeated Admiralty, Commodore (D).

IMPORTANT.

339. Your 0943 6th October [1748/6]. I directed Captain (D) 1, assisted by two Commanders to investigate the H.M.S. *Basilisk* incident of Monday 4th October calling any witnesses necessary.

2. Having read reports from H.M.S. *Basilisk* and H.M.S. *Boreas* and report of Captain (D) 1, I find:–
   (a) No torpedo attack was made on H.M.S. *Basilisk*.
   (b) No submarine was in fact present at the time of the H.M.S. *Basilisk* reports.
3. H.M.S. *Basilisk* acted under a mistaken impression due to lack of experience of her personnel in anti-submarine work particularly under Mediterranean conditions. His actions in mistaken circumstances were proper and well carried out.
4. Following facts support finding:–
   (i) Alleged torpedo approached down sun from Green ten degrees when about 200 yards away altered course to port passing down starboard side at a distance of 100 yards or less than altered course to starboard passing astern at broad angle.
   (ii) Alleged torpedo was stated to have been seen by four officers and four men and approached at high speed from about ten degrees on starboard bow as described in (i) but neither First Lieutenant wearing bridge head phones nor two anti-submarine cabinet watch keepers heard any hydrophone effect.
   (iii) If hydrophone effect of a submarine had been heard the hydrophone effect of a torpedo passing close to the ship at high speed would have been heard extremely loudly and unmistakably.

1838/7.

340. My 1838. Part 2.
   (iv) No periscope was seen.
   (v) At the time hydrophone effect was heard no change of doppler was noticed.
   (vi) From 0833 to 1050 submarine appeared to remain within 500 or 600 yards of her position when first detected.
   (vii) Judging from anti-submarine bridge log (admittedly never a very accurate document) the relative speed of approach during first attack at some 16 knots was only four and a half knots, though the submarine was reported by the anti-submarine cabinet to be closing.
   (viii) Though the first attack at 0940 was apparently as well carried out as can ever be expected, no tangible results were obtained. Further, if submarine was undamaged contact should have been regained before 1049.
   (ix) H.M.S. *Boreas* with an experienced Petty Officer anti-submarine detector instructor was unable to confirm any contact.
   (x) On occasions when contact was temporarily lost the position of echo when re-gained was not consistent with submarine's movements just prior to losing contact.

5. Your 2208 4th October, experience of last two weeks and of this incident emphasizes need for more training in all flotillas in anti-submarine work.

With this in view patrols should not be increased since this would be at the expense of anti-submarine training.

6. Commodore (D) concurs.

1839/7.

222. *Pound to Chatfield*

[ADM 116/3529] 8 October 1937

SECRET.

[Telegram] Addressed Admiralty (for First Sea Lord), C.-in-C. Home Fleet, V.A.C. Battle Cruiser Squadron.

IMMEDIATE. PERSONAL.

Reference 1350/5th October from Admiralty. My 1748 6th October and V.A.C. Battle Cruiser Squadron's 1839 of 7th October.

My answers to the questions contained in Admiralty's 1350 5th October are as follows.

(A). Question 1 and 2. H.M.S. *Havock* was attacked without warning and H.M.S. *Basilisk*'s submarine has been proved to be non-existent. Hence situation envisaged has not yet arisen nor do I think submarine which has not recently carried out piratical attack would act in the manner contemplated.

There is no doubt that the eleven hours hunt carried out by the Second Destroyer Flotilla on 4th September has produced very creditable effect at Palma and probably elsewhere.

I am still in favour of hunting submarines to make them disclose their identity as it is important part of indirect methods referred to at C. in my 1748 6th October. Only exception would be when Admiral [at] Palma informs me where his submarines are operating – when Rear Admiral (D) recently saw Admiral [at] Palma I directed him to inform latter that if he informed me where his submarines were going to operate as he did volunteer to do, I would not hunt these submarines as long as they behaved themselves, but that I should keep a patrol not far off and should concentrate heavy attack on any offending submarines.

(B). Question 3. For reasons given at E. of my 1748 of 6th October I am inclined to adhere to two ship units but if Their Lordships prefer smaller number of three ship units I see no objection to its adoption. I am entirely in agreement with V.A.C. Battle Cruiser Squadron that we should keep number of destroyers on patrol sufficiently low to allow systematic anti-submarine training.

(C). As regards our prestige. It is my intention to give orders which, whilst not discouraging depth charge attack in cases such as H.M.S. *Basilisk* incident, will prevent alarming reports reaching the Press before thorough investigation of incident has taken place.

As it might be deduced from above orders that H.M.S. *Basilisk* was not attacked I shall not issue it until I know whether Their Lordships intend to make any further announcement regarding H.M.S. *Basilisk* incident. So far I am in favour of further announcement both from point of view of Fleet and also because it would be in contrast to German attitude over *Leipzig* incident.

(D). As regards efficiency of asdics. In the case of H.M.S. *Basilisk* anti-submarine personnel were undoubtedly short of practice as ship had been refitting and the new anti-submarine control officer only joined 19th September. C.-in-C. Home Fleet did all he could to remedy this by sending them to Portland but only day practice on a submarine was possible before Nyon patrols had to commence.

In the case of H.M.S. *Havock* ship was on passage and acting under I.S.C.W. before Nyon Conference and unfortunately anti-submarine installation was not manned. Good results cannot be immediately expected from men called in the middle of the night.

1438/8.

223. *Admiralty to Pound*

[ADM 116/3529] 9 October 1937

SECRET.

[Telegram] Addressed C.-in-C. Mediterranean.

Reference your 1438/8th October. Concur in your proposed action at (A).

In regard to (B) Their Lordships agree that anti-submarine training is a pressing requirement and prefer to leave composition and number of units entirely to your discretion.

Any action you take to implement (C) will be very welcome.

1455/9.

224. *Pound to Cunningham*

[ADM 116/3524] 11 October 1937

[Telegram] Addressed V.A.C.B.C.S., repeated Admiralty, C.-in-C. Home Fleet, R.A.(D), Commodore (D).

94. It is desirable that flotilla training should be resumed as soon as possible and that wear and tear on destroyer machinery should be reduced. I know C.-in-C. Home Fleet concurs in this view. If further experience in near future shows that submarine piracy on high seas has been dropped, Admiralty might consider following procedure could be adopted whilst keeping large destroyer force ready to resume intensive patrol if necessary.
   A. If there are three flotillas available:–
      (a) One flotilla to be duty flotilla, remaining two flotillas to carry out flotilla training.
      (b) Duty flotilla to have one division working from Oran and other from Palma. These two divisions to take it in turn to be at sea in Zone A.
   B. If there are two and half flotillas available:– The same procedure as at A but during every third period to have only one division working from Oran with only two destroyers at sea at a time.
   C. I have reason to believe Nationalists would have no objection to our destroyers using Palma as a base.

1803/11.

225. *Pound to Admiralty*

[ADM 116/3530] 14 October 1937

SECRET.

[Telegram] addressed Admiralty, repeated C.-in-C. Home Fleet, V.A.C.B.C.S., R.A.(D), R.A.C. 1st Battle Squadron, R.A. 3rd Cruiser Squadron, R.A. 2nd Cruiser Squadron, Commodore (D), V.A. Malta, R.A. Gibraltar.

IMMEDIATE.
   118. Your 2105/12.[1] Should Salamanca not acquiesce in views of H.M. Government it would appear necessary to reinforce western Mediterranean

---

[1] The Admiralty insisted that a warship must not employ force unnecessarily when dealing with merchant ships and that the same rule applies to action by aircraft. On 8 October the British steamer SS *Cervantes* proceeding to Tarragona had been intercepted by a Nationalist seaplane when 12 miles from the coast and ordered by lamp signal to stop and proceed to Palma. *Cervantes* complied, but when the seaplane disappeared, thereby relinquishing

and it might be thought desirable to commence this reinforcement even before a reply is received in order to show that we mean business.

If above view is concurred in by Their Lordships I would undoubtedly send H.M.S. *Repulse* to Gibraltar forthwith as I consider H.M.S. *Malaya* with her four aircraft and four destroyers can look after Zone E. The question whether two ships of 1st Cruiser Squadron should also proceed to western basin will I suggest depend on whether we have to protect British ships from this new menace in both Zones A and B or whether the French will do it in Zone B.

If either battle cruiser or 8-inch cruiser patrols coast of Spain I consider each ship should have a screen of four destroyers. Unless anti-submarine practices are suspended or destroyers are over worked this will mean either 4 or 8 of the 8 destroyers now at sea for anti-submarine patrol will have to be diverted to this new duty.

The larger ships doing this special patrol would be based on Gibraltar but the two ships which are now in eastern area to maintain contact with Valencia, Barcelona and Palma would continue to use Palma.

It is possible, however, that Nationalists under the new conditions which may arise will not permit us to do so. This would remove one point of contact we have with Nationalists and we should no longer get news of area in which their submarines were operating.

1553/14.

226. *Godfrey to Cunningham*

[ADM 116/3681] H.M.S. *Repulse*, at Mudros
14 October 1937

SECRET.

No. 191/7111.

With reference to King's Regulations and Admiralty Instructions, Article 1132, I have the honour to forward the following report of proceedings for H.M. Ship under my command for the period 19th September, 1937 to 8th October, 1937.

2.– Having embarked officers and a draft for H.M.S. *Repulse* and H.M.S. *Malaya* from *Somersetshire*, H.M. Ship under my command left

---

'continuous control', *Cervantes* resumed course for Tarragona. Approximately 40 minutes later, another seaplane appeared and though the steamer altered course for Palma bombed the ship. Over three hours later and with *Cervantes* on course to Palma another seaplane dropped a torpedo which passed a few feet astern. By this time the British destroyer *Firedrake* had arrived on the scene and witnessed the attack and subsequently escorted the steamer into the port of Valencia.

Grand Harbour, Malta at 0800 on Sunday, 19th September, and shaped course and speed so as to reach Mudros at 0900 on Tuesday, 20th September.

3.– Whilst on passage anti-submarine and anti-aircraft lookouts and armament were closed up and the opportunity was taken on the Sunday night of exercising the organisation for the second degree of readiness outlined in B.C.S. 27/19 of 10th September, 1937.

4.– The sea conditions on Monday were calm enough for aircraft to alight on the water, and a patrol was maintained during the day along route number 8 laid down in the Nyon agreement.

5.– <u>Tuesday, 20th September</u>, *Repulse* arrived and anchored at Mudros at 0900. Ships present were *Malaya* flying the flag of Rear Admiral B.H. Binney, C.B., D.S.C., Rear Admiral First Battle Squadron, H.M.S. *Inglefield* (Captain D.3), H.M.S. *Despatch* (flying the flag of Rear Admiral L.V. Wells, D.S.O., Rear Admiral Commanding, Third Cruiser Squadron) and H.M.S. *Severn*.

6.– After receiving my orders from Rear Admiral Binney and having transferred officers, ratings and stores to *Malaya*, *Repulse* sailed for Port Trebuki, Skyros at 1315, which was reached at 1830.

7.– In his operation orders, Rear Admiral Binney divided Zone E (Aegean, north of latitude 38° 10′ N) into two areas north and south respectively of latitude 39° 5′ N. It was for the patrols by aircraft and destroyers in the southern area that *Repulse* was made responsible. Of the four destroyers of the 5th Division operating in Zone E one was detached to the southern half of the area, to patrol the trade route by day and then by night at my discretion either to anchor with steam at short notice or else to be employed as patrol outside the entrances to Port Trebuki. The destroyer doing this duty from Tuesday until Thursday, 23rd September, was H.M.S. *Imogen*.

8.– At Mudros, in order to safeguard the security of ships in harbour, a power boat is kept continuously on patrol at the entrance to the harbour by day; by night a destroyer patrols outside the harbour. *Malaya* is anchored in a position behind SANGRADA POINT, where she is secure from attack by a submarine firing a torpedo from outside the harbour.

The operation orders directed the Senior Naval Officer at SKYROS to take such steps as are practicable to guard against torpedo attack in harbour.

9.– The following precautions were therefore taken, work on them being completed by Saturday, 25th September –

(a) Indicator nets were laid in Piato Channel and Sarakino Strait, with picket boats fitted with depth charges and usually at anchor watching them day and night.

(b) A dummy obstruction consisting of carley floats moored to a jackstay was established in Marmora channel.

(c) Shore look-out and signal stations, manned respectively by Royal Marines and Seamen were established on Plati and Sarakino Islands. *Repulse* anchored in the best position to render attacks from seaward difficult.

10.– <u>Wednesday, 22nd September</u>. The routine was begun, which continued on subsequent days, of carrying out three air patrols of the merchant shipping routes daily. The aircraft took off at 0530, 1000 and 1530, and each patrol lasted three hours.

11.– *Imogen* secured alongside at 1730 and completed with oil fuel. It had been intended to keep her alongside all night but on receipt of a signal from Rear Admiral, First Battle Squadron, ordering the flying off of an aircraft to search the vicinity of PSARA for a possible submarine whose presence there was thought to be indicated by D/F wireless, *Imogen* cast off at 2300 and proceeded to sea. An aircraft was hoisted out shortly afterwards to comply with Rear Admiral First Battle Squadron's signal but returned at 0120 without having sighted anything.

* * *

14.– <u>Saturday, 25th September</u>. During the forenoon a signal was received from the Naval Attaché at Athens, Captain H.A. Packer,[1] saying he was arriving that afternoon at Port Kumi, Euboea, a distance of 25 miles from Port Trebuki and asking that a boat night be sent for him. Weather conditions being unsuitable for such a boat journey, *Ilex* was ordered to close *Repulse* and I then went over in the destroyer to meet Captain Packer, who returned with me to Skyros in *Ilex* and spent the night onboard *Repulse*.

15.– The object of Captain Packer's visit was to explain the attitude of the Greek Government regarding the landing of armed parties of men and the laying of nets and obstructions in Greek harbours used by H.M. Ships in execution of the Nyon agreement. It appears that General Metaxas[2] stated before the signing of the agreement that the granting of such facilities was undesirable as it might give rise to similar requests from other Powers. I accordingly told the Naval Attaché that such indicator

---

[1] Capt [later Adm Sir] Herbert Annesley Packer (1894–1962). Naval Attaché Athens, Ankara and Belgrade, 1937–39; commanded cruisers *Calcutta*, 1939–40; *Manchester*, 1940–41; CO, HMS *Excellent*, Gunnery School, Portsmouth, 1941–43; commanded battleship *Warspite*, 1943; Cdre (Admin) to C-in-C Med, 1943–44; COS to C-in-C Med, 1944–45; RAC 2nd CS, 1946–48; a Lord Commissioner of the Admy and Chf of Supplies and Transport, 1948–50; C-in-C South Atlantic Station, 1950–52.

[2] General Ioannis Metaxas (1871–1941). PM, Apr 1936; suspended Greek Parliament and after August ruled as dictator until his death.

nets, etc., as had been laid in Port Trebuki could be removed at very short notice and in fact would be weighed during the next 48 hours. The nets had been drawn from the dockyard with a view to carrying out a net laying exercise while in the Aegean which was now complete. I have since heard from Captain Packer that this was the explanation he gave to Admiral Sakallarion, the Chief of the Naval Staff who received it with every appearance of satisfaction.

* * *

17.– <u>Monday, 27th September</u>. Orders to weigh the nets were received from Rear Admiral First Battle Squadron and as *Repulse* was to visit ERITRA BAY (Turkish waters) to investigate the possibility of operating aircraft from there; it was also decided to withdraw the shore signal station. The day was spent carrying out these tasks and at 1815 the ship sailed for Eritra. Valuable experience was gained both in laying and weighing the indicator nets and no serious difficulties were encountered, the work being carried out expeditiously.

Icarus left harbour at 0530 on patrol and returned at 1830.

18.– <u>Tuesday, 28th September</u>. *Repulse* arrived and anchored off Eritra at 0700.

I landed at 0900 to get in touch with the shore authorities, and in accordance with instructions received from Commander-in-Chief to endeavour to obtain permission for H.M. Ships to use Turkish waters in that neighbourhood as a base from which to carry out Nyon patrols.

* * *

31.– <u>Monday, 4th October</u>. *Repulse* arrived and anchored off Chesme at 0845.[1] The Consul General from Smyrna Mr. Greig came on board immediately, accompanied by Lieutenant Hilmi Okcugit, Turkish Navy, who had arrived at Chesme the previous Wednesday, just after *Repulse*'s earlier visit, having been sent as liaison officer by the Commander in Chief at Smyrna.

32.– The object underlying this second visit of *Repulse* to Turkish waters is indicated in the following telegrams from the Commander-in-Chief:–

To:– R.A.1 (R) Admiralty, *Repulse*, H.M. Representative at Istanbul.

'Your 1125/29th September. It is most important that we should make the most of this opportunity of breaking down the reluctance of the Turks

---

[1]Capt Godfrey found little at Eritra, reconnoitred the area and called at Chesme but did not receive an answer from Turkish authorities in Smyrna on the question of landing recreational parties and returned to Skyros 29 Sept–4 Oct.

to allow our ships to use their harbours and grant the usual facilities. It is necessary however to move slowly and progressively. The present situation as I understand it is as follows.

(a) Turkish Government concurred in use of Eritra Bay provided that aircraft carried by ships shall not carry out flights inside the bay.

(b) H.M.S. *Repulse* visited Eritra which was found in ruins but was received very courteously by local gendarmerie who informed H.M.S. *Repulse* that the local centre of authority was Chesme.

(c) H.M.S. *Repulse* then visited Chesme where the authorities were very friendly and readily gave permission for ships and aircraft to use without restriction any anchorage in Eritra Bay and on the coast as far south as Port Mersin with the exception of Chesme itself and the Gulf to the North-Westward of line joining Mavro Vuni Point and Keras Point. No objection was made to ships visiting Chesme for short periods to buy fresh provisions. Question of landing men for recreation was referred to Smyrna and you have not yet reported the result.

(d) Port Egrilar appears more suitable than Eritra Bay and is included in part of the area which local authorities at Chesme said might be used.

(e) It is desirable that *Repulse* should again visit a Turkish port in the near future and if possible should arrange to meet representative of Consul General at Chesme when the following points could be discussed with him:–

(i) Whether courtesy visit by Commanding Officer of *Repulse* made on authorities at Smyrna would be appreciated.

(ii) Whether Smyrna authorities would agree to his paying a visit either in *Repulse* or a destroyer.

(iii) Whether permission could be obtained for a destroyer occasionally to obtain fresh provisions.

(f) The ultimate object to be achieved is that our ships should be allowed to visit Smyrna occasionally for recreation and relaxation of ships' companies in similar way that we now use Athens. This will require tact and patience and we must work at it very gradually and be prepared to discontinue advances towards our object at any moment should it become apparent that it would do more harm than good.'

To:– *Repulse* (R) Admiralty, R.A.1, Ambassador Istanbul.
'My 1913/30. you should be guided by following:–

(a) Idea of an official visit by *Repulse* or a destroyer to Smyrna should not be pursued.

(b) Question of an occasional private visit of destroyer is being dealt with by ambassador and should not be raised locally.

(c) Call on Vali by Commanding Officer proceeding by car from Chesme has apparently been agreed to by Vali [but] confirmation should be obtained from Consul General before visit is paid.

(d) Officers can land for exercise on beach provided that arrangements were made beforehand with local authorities.

(e) No request for landing men for recreation should be made locally but question might be raised with Vali if you visit him and if a favourable opportunity occurs.

(f) Permission to land shooting parties should not be asked for at present.

(g) We have gained a great deal so do not attempt to rush things.'
[1533/3]

33.– The Consul General had arranged for me to proceed to Smyrna by road (wearing plain clothes) the same afternoon, and then to pay official calls on the Vali and on the Commander-in-Chief the following afternoon. I accordingly left the ship at 1400 and reached Smyrna 2½ hours later. On landing at Chesme, I was met by a deputation of farmers who presented a bouquet and cheered us on departure. *Repulse* meanwhile left Chesme Roads, which is an uncomfortable anchorage during northerly winds and at 1600 anchored off Port Egrilar.

* * *

36.– After the [official] calls were over, a meeting was held at the Consulate General at which the Deputy Vali, General Officer Commanding, Consul General and I were present. In accordance with instructions received from the Commander-in-Chief, I broached the question of whether permission for the use of certain Turkish harbours as bases for Nyon patrol work might be extended to include Ali Agha which, from inspection of the chart, looked as if it might be suitable for the operation of floatplanes with almost all directions of wind, an advantage not possessed by Eritra Bay or Port Egrilar. The British point of view was thoroughly appreciated by both the Deputy Vali and the General Officer Commanding, the latter of whom agreed to take the matter up with Ankara by telegram. Consent was given to the landing of officers and men for recreation at Chesme and at Ilija Bay, where there is a good bathing beach, 3 miles East of Chesme.

37.– I returned to Chesme by car after lunch, and was taken round to Port Egrilar in *Imogen*.

38.– *Imogen* sailed on patrol at 0530, and at 0800 *Repulse* weighed and proceeded as requisite to rendezvous at 1400 for a full 15″ full calibre throw off firing by *Malaya* and then proceeded to Kalamitza Bay [Skyros] where the ship came to anchor at 1650. At anchor already were H.M.S. *Inglefield* and H.M.S. *Severn*, and later arrived H.M.S. *Malaya* wearing the flag of Rear Admiral T.H. Binney, C.B., D.S.O.

39.– Thursday, 7th October. H.M.S. *Enchantress* with the First Lord of the Admiralty onboard arrived and anchored at 0700. During the course of the forenoon, the First Lord visited all ships present. Whilst on board *Repulse*, Mr. Duff Cooper[1] took the salute at a march past of the ships' company and then witnessed a demonstrating of the catapulting of four aircraft.

40.– The First Lord in *Enchantress* left for Famagusta at 1500, the ships' companies of all ships present being fallen in at Divisions as *Enchantress* left harbour. The Rear Admiral, First Battle Squadron in *Malaya* sailed for Athens at 2045.

* * *

227. *Pound to Admiralty*

[FO 371/21169] 16 October 1937

SECRET.

[Telegram] Addressed Admiralty.

IMMEDIATE.

127. PERSONAL for First Sea Lord and for C.I.G.S. and C.A.S.[2]

2. Local Commanders held their first meeting this Saturday forenoon and were in agreement that as international situation appeared to have deteriorated considerably during last few days it was desirable that a report should be made now on steps which should be taken immediately to strengthen armament position vis-a-vis Italy should H.M. Government be in agreement with our appreciation of the situation.

3. We are aware that the policy of H.M. Government has been that German menace has been considered greater than that from Italy but it is for consideration whether Italian attitude generally, the massing of troops in Libya and information contained in telegrams No. 552 and 555 from British Minister to Mr. Eden do not at least temporarily call for special measures being undertaken for the defence of Egypt.

---

[1] Alfred Duff Cooper (1890–1954). Created 1st Viscount Norwich, 1952. MP (U) Oldham, 1924–29; MP (C) St George's div of Westminster, 1931–45; Financial Sec, WO, 1928–29; 1931–34; Financial Sec to the Treasury, 1934–35; Sec of State for War, 1935–37; 1st Lord of the Admiralty, 1937–38; Minister of Information, 1940–41; Chllr of Duchy of Lancaster, 1941–43; Representative of British Government with French Cttee of National Liberation, 1943–44; Amb to France, 1944–47.

[2] The parts of the telegram dealing with military and air details are not reproduced.

4. The outstanding deficiency is the total absence of anti-aircraft guns and searchlights in either Egypt or Palestine, and we would urge most strongly that a complete anti-aircraft brigade should be sent out now.

5. Naval situation. The present distribution of ships is generally favourable for opening moves as set forth in Interim appreciation J.P.233.[1] If H.M.S. *Repulse* were sent to Gibraltar the two battle cruisers, one Mediterranean and one Home Fleet Flotilla would be available for escorting H.M.S. *Furious* with fighter aircraft for Egypt. The remaining forces in Aegean could be moved to Egypt for protection of Canal and Alexandria. Net defences in British ship *Thistlegarth* and H.M.S. *Protector* and ammunition in A.S.I.S. *Churruca* are now in Eastern Mediterranean.

6. The question of H.M.S. *Glorious* requires immediate consideration. The present arrangement is that H.M.S. *Glorious* should leave Malta for United Kingdom on 25th October with three Torpedo Spotter Reconnaissance and all her fighter aircraft, the remainder being disembarked at Malta. If the situation is considered serious it is recommended that H.M.S. *Glorious* should proceed to Alexandria forthwith which would allow two alternatives according to development of the situation:

(a) If delay in arrival of H.M.S. *Glorious* in United Kingdom can be accepted then H.M.S. *Glorious* should retain her aircraft on board.

(b) If H.M.S. *Glorious* must adhere to her present programme then all her aircraft are to be disembarked in Egypt. If the latter course were adopted the air situation might be further strengthened by sending out H.M.S. *Courageous* to replace H.M.S. *Glorious*.

* * *

228. *Admiralty to Pound*

[FO 371/21169] 18 October 1937

SECRET.

[Telegram] Addressed C.-in-C. Mediterranean (and for G.O.C. and A.O.C.). personal FROM First Sea Lord.

IMMEDIATE.

Your 1603/16 and 0003/17. Chiefs of Staff do not consider military situation has become more serious during the last two months. Movement

---

[1] Mediterranean and Middle East Appreciation prepared by the Joint Planning Sub-Committee. The primary naval responsibility in the event of an emergency would be escorting convoys with personnel, stores, equipment and aircraft to reinforce Egypt. Extensive correspondence and memoranda in ADM 1/9533.

of troops to Libya has been long foreseen and is not considered to have brought about new situation. Political situation remains uncertain pending result of international discussions now taking place. No immediate action such as proposed in your telegram therefore considered necessary due to existing military situation. You will receive early instructions informing you whether anticipated developments in near future justify reinforcements being sent you. Meanwhile no action on your proposals in paragraphs 8, 9 and 10 of your 1603/16 should be taken.[1]

1830/18.

229. *Admiralty to Pound*

[ADM 116/3530] 20 October 1937

SECRET.

[Telegram] Addressed C.-in-C. Mediterranean, repeated C.-in-C. Home Fleet, R.A.(D), Commodore (D), V.A.C.B.C.S.

Your 1803/11 to V.A.C.B.C.S. Admiralty concur that it is very desirable to relax the strain on the flotillas and provided there is no recrudescence of submarine piracy it is left to your discretion to modify your orders and plans as you think fit.

2024/20.

230. *Pound to Cunningham*

[ADM 116/3525] 21 October 1937

SECRET.

[Telegram] Addressed V.A.C.B.C.S., repeated Admiralty, C.-in-C. Home Fleet, R.A.(D), Commodore (D).

IMPORTANT.

153. Reference Admiralty's 2024 of 20th October. Request you will bring into force as soon as convenient a system of patrol outlined in my 1803 11th October allowing normally for one Division to be on patrol at a time. If there are five or four destroyers in duty division they are to be organised as two striking forces, otherwise as one striking force.

Duty flotilla to work from Oran. Arrangements to be based on 8th Division Destroyers arriving in United Kingdom 2nd December when any

---

[1] These concerned movements of British and Egyptian troops forward to Mersa Matruh, establishment of coast defences at Alexandria, construction of petrol bomb dumps at Mersa Matruh and possible transfer of an RAF squadron from Iraq to Egypt.

modifications to patrol necessary on account of reduced numbers can be considered.

A cruiser from Palma as suggested by R.A.(D)'s 1611 16th October will not be available.

1138/21.

231. *Admiralty to Pound*

[ADM 116/3533] 22 October 1937

SECRET.

[Telegram] Addressed C.-in-C. Mediterranean, repeated C.-in-C. Home Fleet, V.A.C.B.C.S., R.A.(D), R.A.C. 1st C.S., R.A. 1st B.S., R.A.C. 3rd C.S., R.A.C. 2nd C.S., Commodore (D), V.A. Malta and R.A. Gibraltar.

Your 1553/14.[1] It is not repetition not considered necessary to reinforce Western Basin at present. Reply from Salamanca to protest has not yet been received.

2015/22.

232. *Admiralty to Pound*

[ADM 116/3834] 23 October 1937

SECRET.

[Telegram] Addressed C.-in-C. Mediterranean. From Admiralty (First Sea Lord).

Your 1603/16. Following is extract from telegram from Chiefs of Staff to G.O.C., A.O.C. and yourself which has been forwarded from War Office to G.O.C.

Begins. With reference to paragraphs 5 and 6 of C.-in-C. Mediterranean's 1603/16 the Chiefs of Staff do not consider that any change in the Fleet movements, or as regards *Glorious*, need take place and that the programme of that ship should proceed as arranged. Ends.

Request you will act accordingly.

1205/23.

---

[1]Doc. No. 225.

233. *Rear Admiral Binney to Pound*

[ADM 116/3525] 23 October 1937

SECRET.

[Telegram] Addressed C.-in-C. Mediterranean, repeated Admiralty, Captain (D)3.

543. Following incident has occurred. At 1211 today, Saturday, H.M.S. *Ilex* made contact with submerged object in position 039 degrees 34 minutes north 025 degrees 44 minutes east. H.M. Ships *Inglefield*, *Imogen* and two aircraft immediately closed position of contact and destroyers carried out a hunt. An oil patch was observed in the vicinity and has persisted and observer in one of the aircraft reported he sighted submerged object resembling submarine. Object has not moved for 8 hours and it appears most probable that it is a wreck. Captain (D)3[1] dropped 3 depth charges set to 500 feet in the vicinity at 1705 and still object did not move. Captain (D)3 is satisfied that object is not submarine, but I am arranging contact to be maintained during the night.

2119/23.

[Holograph] I cannot conceive why she dropped DC's. Suggest we enquire.

C. [Chatfield]
25/10.

234. *Pound to Binney*

[ADM 116/3525] 25 October 1937

SECRET.

[Telegram] Addressed R.A. 1st Battle Squadron, repeated Admiralty, Captain (D)3.

IMPORTANT.

174. Your 2119 23rd October and your 1607 24th October. Unless there was some reason for doing so of which I am not aware, the dropping of

---

[1] Capt [later VA] Arthur George Talbot (1892–1960). Commanded destroyer *Inglefield* and Capt (D), 3rd DF, 1937–39; Dir of Anti-Submarine Warfare Div, Admy, 1939–40; commanded aircraft carriers *Furious*, 1940–41; *Illustrious*, 1941–42; *Formidable*, 1942–43; Naval Cdr, Force S (Eastern Assault Force at Normandy invasion), 1943–44; RAC Force X, Pacific, 1944–45; Head, British Naval Mission to Greece, 1946–48.

depth charges was quite unjustified. The instructions contained in I.P.A. paragraph 13 should have been followed.[1]

1038/25.

235. *Binney to Pound*

[ADM 116/3525]   25 October 1937

SECRET.

[Telegram] Addressed C.-in-C. Mediterranean, repeated Captain (D)3, Admiralty.

IMPORTANT.

547. Your 1038 25th October. There was not, in my opinion, sufficient reason for dropping depth charges and I informed Captain (D)3 of this at the time. Captain (D)3 reported subsequently that he had dropped the depth charges after being convinced that there was no submarine present and in an effort to disperse echo effect after nine ounce charges had failed to do so. Full report will be forwarded by letter.

1411.

236. *Admiralty to Pound*

[ADM 116/3525]   25 October 1937

SECRET.

[Telegram] Addressed C.-in-C. Mediterranean, repeated R.A. 1st Battle Squadron, Captain (D)3.

Your 1038 is fully concurred in. If a Spanish submarine had been resting in that position on the bottom she would have been acting entirely legitimately and an international incident of first class importance might have resulted.

1215/25.

---

[1] *Memorandum I.P.A., 13. Action to be taken by SURFACE SHIPS outside Spanish Territorial Waters.* In the case of a submarine sighted on the surface or submerged, which is not taking and as far as is known, has not taken action against shipping, 'the submarine is to be shadowed and reported until destroyers arrive on the scene but is not to be attacked'.

237. Pound to Admiralty

[ADM 116/3525] 30 October 1937

SECRET.

[Telegram] Addressed Admiralty.

IMMEDIATE.

216. I have just had a very satisfactory and pleasant meeting with French and Italian Admirals.

2. Everything which could be settled locally was easily disposed of without argument or difficulty.

3. Italian and French agreed to accept code and signalling procedure which we had prepared, copies of which will be forwarded by post. Italians also agreed to adopt black ball signals for merchant ships sighting submarines.

4. There is only one snag. Article 4 of Nyon Agreement gave French and British ships authority to operate inside territorial waters of Signatory powers. This authority is not referred to in Admiralty message 2330 October 1st giving me details of Paris Agreement. Italian Admiral raised this point very definitely and asked whether same facilities were accorded to Italian as to French and British. He pointed out that it would be impossible to obtain permission on each occasion that it was required if a submarine was being hunted and if they did not have the right it would hamper them in case of routes 12 and 13 which pass close to Crete. Both Admiral Esteva and I agreed this was a question for decision by our governments. I am sure that its refusal would cause Italians serious umbrage and if for any reason this facility is withheld from them, I consider it should also be surrendered by French and ourselves. We do not really require it to be laid down as up to the present we have not made use of it except in the case of Turks where it has caused protest. If our destroyers were hunting a submarine to destroy it I would see that they continued to do so irrespective of territorial waters. Admiral Esteva concurred in this view.

5. As regards routes both Italian and French agree that on route 12 point L was situated 30 miles 249 degrees from north point of Zante Island and that point M was situated 033 degrees 30 minutes north, 025 degrees 00 minutes east, consequently I agreed to accept this.

6. French Admiral desired to extend French submarine exercise area one to read 'a zone bounded by coast and following line: Cape Guardia to a position two miles north of Cape Zebia to a position two miles north

of Plane Island to Island of Zecbra to Ras el Amer'. This was agreed by Bernotti[1] and myself.

7. Admiral Esteva referred to recent attacks by aircraft on shipping on high seas and expressed opinion that its continuance might compel us to adopt a system of escort. I pointed out some of the disadvantages which this would entail but added that if such attacks continued we might be forced to consider it.

8. Question of date of agreement coming into operation was raised by Italian Admiral who suggested 10th November. It was agreed that subject to consent of all necessary Powers having been obtained by 10th November, it should become operative on that date. If however Admiral Bernotti found he could not complete his plans by then, notice to this effect would be given to British and French Governments by Rome.[2]
2033.

238. *Admiralty to Pound*

[ADM 116/3525] 3 November 1937

SECRET.
[Telegram] Addressed C.-in-C. Mediterranean, repeated V.A.C.B.C.S., C.S.1.

IMMEDIATE.
My 1125. Admiralty are in general agreement with the views expressed in first and last paragraphs of A.C.Q.'s [Cunningham] 1400/2.[3] With regard to the second paragraph,[4] however, Admiralty view is that although violation of the laws of war has taken place, scope of Nyon Arrangement

[1] Ammiraglio di Squadra [later Ammiraglio di Squadra designato d'Armata] Romeo Bernotti (1877–1974). Founder and director of Institute for Maritime War for Senior Officers, 1922–35; commanded 2nd Squadra, 1936–37; commanded Italian patrols established under Paris agreement supplementing Nyon arrangement, 1937–38; President of Consiglio Superiore di Marina, 1938; President of Comitato Ammiraglio, 1938; retired and named Senator of the Realm, 1939–42; President of Lega Navale, 1939–42. Bernotti was a well-known writer on naval warfare whose works were also translated into other languages.

[2] Bernotti arrived at Bizerte in his flagship, the light cruiser *Emanuele Filiberto Duca d'Aosta*, the same ship which flew his flag when she bombarded the Republican port of Valencia in February 1937, an act supposedly committed by insurgent warships.

[3] Concerning the illegality of the sinking by Nationalist aircraft on 30 October of the British-flagged steamer *Jean Weems* whether or not she was in territorial waters, unless as was not the case she had refused to stop or resisted control. The *Jean Weems* was on passage from Marseilles to Barcelona with a cargo of wheat and condensed milk and carried two Non-Intervention Control Officers.

[4] The above principle might be applied to all non-Spanish ships and might, possibly, be incorporated in the IPA.

is limited to attacks contrary to the dictates of humanity which, in practice, means attacks contrary to the rule enunciated in sub-paragraph (2) of Part IV of London Naval Treaty. Object of wording of preamble and paragraph II of Nyon Arrangement and of paragraph II of Supplementary Agreement was to secure this result and thus avoid impossible extension of our commitments, whilst at the same time avoiding the inference that the rules actually enunciated in London Naval Treaty are the only rules applicable to action at sea against merchant ships. As you will be aware rules enunciated in Treaty deal only very partially with this matter and in particular make no reference to important rule that excessive force must not be employed.

2. It follows that Nyon Arrangements should only be applied where an attack has taken place on a non-Spanish ship in circumstances which involve a violation of London Treaty rules. In the present case, although there may be some doubt whether sufficient time was given before bombing commenced, the crew did in fact get away in safety, and we do not therefore propose to contend that a violation of London rules occurred.

3. Communication to be made to Admiral [at] Palma should be on the following lines. As the Insurgent Naval authorities are well aware, since belligerent rights have not been accorded to either side no attack on a British vessel can be justified in any circumstances, nor can she lawfully be stopped ast sea. A.C.Q. should then refer to the communication which was made to Admiral Bastarreche by R.A.(D) as a result of Admiralty telegram 1535/26/8, in which it was explained that, even if belligerent rights had been granted, no merchant ship, whether enemy or neutral, may be attacked unless she fails to comply with a lawful request to stop in order that she may be visited and her identity established or unless she thereafter resists lawful capture.

4. In the case of the *Jean Weems* no suggestion arose that the ship resisted in any way. Even, therefore, on the basis that belligerent rights had been granted, a plain illegality was committed in that no attempt was made to visit or divert the ship in order to establish her identity or her liability to capture as a prize, but instead the ship was deliberately sunk by bombs in the open sea.

5. Admiral Moreno[1] must be aware that under International Law the sinking, even of an enemy merchant ship which has been captured as a prize, is illegal unless compelled by military necessity, and that the conditions under which a neutral prize may be sunk (even if at all) are even stricter. In this case no question of military necessity can have arisen

---

[1] Vice Admiral Francisco Moreno had replaced Bastarreche as Nationalist naval commander at Palma. The two were brothers-in-law.

and no attempt whatsoever was made to ascertain the nationality of the ship or the voyage on which she was engaged. In fact she was a British ship carrying a lawful cargo upon the High Seas.

6. For your own information, Salamanca authorities have admitted that the aircraft which attacked *Jean Weems* was Spanish, i.e. not one of the Italian machines operating from Majorca and they have stated that if it should transpire that the pilot had exceeded his instructions the Nationalist Government would be prepared to make reparation. After the information as to the instructions which has been furnished by Admiral Moreno (See A.C.Q.1105/31/10 and *Aberdeen*'s 1916/1) there seems little doubt that the pilot did in fact exceed his instructions.

1930/3.

239. *Cunningham to Pound*

[ADM 116/9948] 4 November 1937

SECRET.

[Telegram] Addressed C.-in-C. Mediterranean, repeated Admiralty, R.A.C. 1st Cruiser Squadron.

IMPORTANT.

392, Admiralty message 1930 November 3rd. I interviewed Admiral Moreno today Thursday in company with H.B.M. Consul. Admiral handed me his reply to protest re *Marvia*.[1] He admits pilot, who states he dropped bombs to make ship display her colours, was completely in the wrong. He regrets what happened and has reiterated his orders that British ships are not to be interfered with outside territorial waters. I then handed the Admiral protest re British ship *Jean Weems*. Facts reported on both sides agree except in regard to position of attack. Insurgent pilot claims *Jean Weems* was two and a half miles from shore. Visibility assessed by aircraft at two miles. I drew attention to fact that boats took between six to seven hours to reach shore. Admiral appreciated force of argument in protest and will refer the case to Salamanca. Meanwhile until issues are settled he will give orders that British ships are in no circumstances to be attacked.

1431.

---

[1] On 21 October the British steamer *Marvia* bound from Barcelona to Varna in ballast was bombed and strafed by a Nationalist seaplane in the position 41° 00′ N., 3° 00′ E.

240. *Admiralty to Pound*

[ADM 116/3533] 10 November 1937

SECRET.

[Telegram] Addressed C.-in-C. Mediterranean, repeated C.-in-C. Home Fleet.

Provided that submarine piracy remains quiescent it is considered that a considerable reduction in patrol force can be accepted. It is hoped therefore that you will shortly be able to dispense with one or both Home Fleet Flotillas, so that the ships' companies can be given Christmas Leave in United Kingdom. In any case it is very desirable for the Fourth Destroyer Flotilla to return to United Kingdom about the end of November.

1635/10.

241. *Pound to Somerville*

[ADM 116/3533] 11 November 1937

SECRET.

[Telegram] Addressed E.A.(D), repeated Admiralty, C.-in-C. Home Fleet, Commodore (D).

IMPORTANT.

274. Reference 1635 10th November from Admiralty. Assuming that I agree to the release of both divisions of 4th Destroyer Flotilla on 29th November and release of 5th Destroyer Flotilla not later than 10th December, request your views on the following points: (a) When only one Mediterranean Flotilla of either seven or eight ships is available for zone one what patrol arrangements would you propose in order that ships might not be overworked?; (b) Would patrol arrangements you propose under (a) permit of Mediterranean Flotilla in western basin doing any practice?

Answer to this may affect dates on which you would propose to release 4th and 5th Flotillas as Home Fleet destroyers might do all the patrolling until they leave in order to leave Mediterranean Flotilla free for exercises. Necessity for Home Fleet Flotilla maintaining their efficiency must however also be allowed for.

(c) Should submarine activity increase and require more than one flotilla in zone one it would not be possible to send additional Mediterranean destroyers to western basin and I should therefore ask that a Home Fleet flotilla should at all times be available to come out at short notice.

(d) After considering (a) and (b) you will be in a position to state for how long you consider one Mediterranean flotilla can continue to look after zone one supposing submarine activity does not increase.

1358/11.

242. *Somerville to Pound*

[ADM 116/3533] 11 November 1937

SECRET.

[Telegram] Addressed C.-in-C. Mediterranean, repeated Admiralty, C.-in-C. Home Fleet, Commodore (D).

634. your 1358.

(i) Present arrangement for relief allows for H.M.S. *Kempenfelt* and 8 Division destroyers to sail for United Kingdom in accordance with programme issued by Commodore (D) and approved by V.A.C.B.C.S. Propose this should stand.

(ii) Answer to your (a). One division of three or four ships at Oran. One unit of two or preferably one ship on patrol. Remaining division at Gibraltar. Home Fleet Flotilla have had little opportunity of exercising together and I should depreciate any restriction of facilities by giving them additional patrol. [Question] (d). Consider period in zone 1 should not exceed six weeks. Presume absence of all incidents in Eastern Mediterranean may permit of destroyers being withdrawn in which case substantial progress in training could be achieved if commitments are reduced to one flotilla in Spanish waters.

(iii) So far as patrols are concerned and subject to above, no objection seen to releasing 5th Destroyer Flotilla and 7 Division Destroyers on 29th November.

2323/11.

SPANISH CIVIL WAR AND NYON AGREEMENTS, 1936–1937   355

243.   *Pound to H.M.S.* Repulse

[ADM 116/3530]   12 November 1937

SECRET.
[Telegram] Addressed H.M.S. *Repulse*, repeated Admiralty, R.A. 1st B.S.

IMPORTANT.
27. Under existing conditions I do not consider it necessary to have any destroyers on patrol in Zone 6 but one destroyer should have steam at short notice. Air patrols should also be reduced to what can comfortably be carried out by machines available.
1228.

244.   *Godfrey to Cunningham*

[ADM 116/3681]   H.M.S. *Repulse*, at Malta
29 November 1937

SECRET.
No. 191/7439.
I have the honour to forward the following report of proceedings for H.M. Ships under my command …

* * *

4-. <u>Monday, 1st November, 1937</u>.
Repulse left Malta at 0920 and shaped course and speed as requisite to arrive at Kalamitza Bay, Skyros at 0600 Wednesday, 3rd November.

5-. <u>Wednesday, 3rd November</u>.
H.M.S. *Malaya*, flying the flag of Rear Admiral T.H. Binney, C.B., D.S.O., anchored in Kalamitza Bay at 0640, half an hour after *Repulse* had arrived. Rear Admiral Binney turned over to me the duties of Senior Officer, Area 'E', and then at 0815 sailed for Malta in *Malaya*.

* * *

7-. <u>Thursday, 4th November</u>.
Repulse arrived at Mudros at 1330, ships present being *Clyde*, *Ilex* and *Icarus* having left for Port Egrilar at 0500. *Isis* arrived at 1400.

8-. Owing to the decreased likelihood of piratical activity in Area 'E', it was decided to modify temporarily the system of patrols, as follows.

Each day the patrol of the shipping route will be carried out either by aircraft or else by destroyers. This will sufficiently fulfil the purpose of 'showing the flag' while conserving personnel and materiel for times of greater emergency.

On the days that the patrols are being done by aircraft, only one will be carried out, but it will usually be done by three or four aircraft in company with the idea of creating a greater impression on the merchant ships that sight them.

* * *

16-. <u>Wednesday, 17th November.</u>
At about 0245 Admiralty's signal timed 2116 of 16th November was received. This signal addressed to the Commander-in-Chief, Mediterranean, and repeated to C.S. One, R.A.(D), V.A.M. and *Repulse* was as follows:–

'Information has been received from a secret source that British Ship *Euphorbia* and Steamship *African Mariner* have passed Bosphorus bound for Spain with large quantities of war material in addition to other cargo. *Euphorbia* passed Bosphorus on 11th November, Steamship *African Mariner* on 15th November. It is desirable that these vessels should be intercepted and sent to Malta for investigation under Carriage of Munitions to Spain Act. Vessels should be escorted to Malta if possible to prevent illicit cargo being discharged en route.'

17-. The 2nd Destroyer Division (*Greyhound*, *Grenade* and *Glowworm*) who were at Mudros at half an hour's, two hours' and six hours' notice respectively, were ordered to raise steam with all despatch and proceed to sea to search for S.S. *African Mariner*. Two aircraft were flown off from *Repulse* at daylight and the ship raised steam for full speed and proceeded at 0800, as soon as libertymen had been embarked.[1]

18-. Aircraft 089 located *African Mariner* in the Zea Channel at 0745. *Grenade* and *Glowworm* were thereupon ordered to return to Mudros, and *Repulse* closed the merchant ship and maintained contact until *Greyhound* arrived on the scene. The latter, having put an armed party of one Lieutenant and ten men onboard, was ordered to escort S.S. *African Mariner* to Malta.[2]

19-. On conclusion of this service *Repulse* shaped a northerly course, and on termination of a night encounter exercise with *Grenade* and *Glowworm*, anchored in Mudros Bay at 2230.

* * *

[1] *Repulse* had been anchored in Salamis Strait off Athens since the 13th.
[2] The ship, bound from Odessa and Novorossisk to Barcelona, was the object of an intensive search for ten days at Malta with much of the cargo of salt fish and sulphate of ammonia discharged and wheat in the deep holds prodded by rods. Nothing illicit was found and she was allowed to proceed to Barcelona. Heaton, *Spanish Civil War Blockade Runners*, pp. 74–5.

25-. [20th November] As a result of the *African Mariner* episode and the possibility of again having either to intercept a merchant ship at short notice or to be able to state whether or not a certain ship had passed southbound through the zone, it was decided to modify the patrols.

For the immediate future *Repulse* would be based at Skyros. One aircraft would take off at daylight and start a patrol at the southern end of the route and so sight all merchant ships that had left the Dardanelles during the night and who were steaming at not more than 10 knots. On completion of the patrol the aircraft would alight at Mudros and remain there for the day secured astern of the destroyer until it was time to take off for an evening patrol starting at the northern end of the route and returning to *Repulse* at Skyros before dark.

Consequent upon this decision *Repulse* left Mudros at 1730 for Port Trebuki.

* * *

31-. Thursday, 25th November.
Rear Admiral Binney in *Malaya* arrived at Kalamitza Bay at 0700 and assumed command of Zone VI. *Repulse* sailed for Malta at 0815.

32-. Friday, 26th November.
After passing Cape Matapan at midnight, the wind gradually freshened from ahead until by 1600 it was blowing a strong gale. Heavy seas were breaking over the catapult deck from the starboard side, and in an endeavour to save the three aircraft stowed there speed was progressively reduced and course altered until by 1700 the ship was hove to on a north-westerly course, revolutions for 6 knots.

33-. In spite of these precautions however, the aircraft stowed on the starboard side of the catapult deck received considerable damage during the night and it is feared can never be made serviceable again.

Apart from this one aircraft the damage sustained in the gale was negligible.

34-. It would seem that with five aircraft onboard the risk of writing one of them off in the event of a gale arising must be accepted as there is no possible alternative stowage for the machines where they could be placed out of reach of the heavy water that comes over the catapult deck under these conditions.

35-. Saturday, 27th November.
By noon the weather had sufficiently moderated for course to be shaped for Malta once more, speed 13 knots, and anchorage was found in Marsaxlokk Bay at 2200.

* * *

245. *Pound to Chatfield*

[CHT/4/10] Commander-in-Chief, Mediterranean Station
17 November [1937]

[Holograph]

I think my last letter was sent with the Heads of Services comments on the appreciation.[1] Since then I have had my meeting with Esteva and Bernotti at Bizerta. The whole thing was over in an hour and the Italians as one was sure they would be were very pleasant and fell in with all our ideas. The 'Paris' zones are more satisfactory than the original Nyon ones as our areas are more clearly defined. It looks as if the Italians were more concerned in getting areas which would enable them to report shipping than to hunt submarines which they know well will probably not be there.

The French visit went off very well and both officers and men were very popular – much more so than I expected them to be.[2] Probably the officers in *Algérie* were rather a picked lot, anyhow they were very different to the crowd we had in 1926 when the French Fleet visited Malta in Roger Keyes' time.

I was very sorry the Greeks had to put off their visit but it would have been a mistake to have only one ship – they could not have lasted the course. There are rumours that the real reason was that they could not afford the cost of the oil.[3]

The reduced patrols which you have approved will allow of the Home Fleet destroyers getting back for their Xmas leave and as Roger Backhouse is sending a Flotilla and Com. D [Commodore (D)] out early in January we should get on well with our *Centurion* firings and be able to give ships adequate time for working up before they have to fire. I am afraid I shall have to enter a protest against the proposed programme for *Centurion* because, unless I give up or seriously curtail the Spring Cruise on the South coast of France, it will mean that we shall have to get over all the *Centurion* firings, both day and night, in February, which I am sure you will agree would unduly rush things.

How much of this Spring cruise we shall be able to bring off must remain in doubt so long as the Spanish business continues but officers and

[1] Not found, possibly referring to the defence of Egypt.
[2] Vice Admiral Abrial, commanding 1re Division de Croiseurs with his flag in the first-class cruiser *Algérie*, France's newest and best, visited Malta 9–11 November.
[3] The Greek Government informed the British Minister in Athens that owing to the requirements of the anti-submarine patrol there would be practically no vessels available except the armoured cruiser *Averoff* for the projected visit of the Greek Fleet to Malta in mid-November. Waterlow to Foreign Office, 19 Oct 1937, copy in ADM 116/3536.

men enjoy it so much that it would be heartless to do them out of any of it. You will remember that the Medn Fleet did not get *Centurion* at all last autumn.

Turkish Ports. I think we have got all we shall get – though they are still 'considering' the question of the use of a port from which we can operate aircraft. We are making use of the facilities which have been granted us so as to get them used to our presence there.

Leave for R.A.F. at Arzeau. A.O.C. (Maltby) is over there now and I have told him that if they would like to send a limited number of men home in the H.F. Flotillas I will try and arrange this as I do not want them, now that they are under our orders, to think that we only look after our own people. All the R.A.F. at Arzeau are on 'Home Service'. Everything is working very smoothly there.

Basilisk affair. I could not quite understand Winston's[1] cryptic reference to it at the Navy Club diner.[2] No doubt there was some very good reason for not giving a more detailed explanation but had it been possible I think it would have prevented people thinking we had something to hide, which various references in the papers suggest. The Italian papers took the line that we knew it was a 'Red' s/m and therefore pretended there was no s/m there.

Egypt. The situation is gradually improving due to the various preparations which have been approved but I feel that I ought to be there if any of the larger questions are under discussion otherwise the Ambassador must turn to A.L.O. Port Said who is not in a position to advise him over larger questions.

D.2's depth charges in the Aegean. This was a shifty [?] business. Talbot is bountiful [?] of ardor and initiative but as I have just told him he has always before had someone over him to 'vet' his proposals – now he has got to do his own vetting and not go off the deep end. He will soon settle down all right and will I am sure do very well. It is much easier to curb excessive initiative than to put it into anybody.

---

[1] Winston S. Churchill (1874–1965). Home Sec, 1910–11; 1st Lord of the Admy, 1911–15; Chllr of the Duchy of Lancaster, 1915; Min of Munitions, 1917; Sec of State for War and Air, 1919–21; for Air and the Colonies, Feb–Apr 1921; and for the Colonies until Oct 1922; Chllr of the Exchequer, 1924–29; 1st Lord of the Admy, 1939–40; PM, 1st Lord of the Treasury and Min of Defence, 1940–45; PM and 1st Lord of the Treasury, 1951–55.

[2] The reference is not clear but it may be the speech given by Winston Churchill at the Navy League Nelson Day Dinner, Grosvenor House, London on 19 October. Churchill stated: 'There are opposite the British naval power on which our lives and fortunes depend two ugly notes of interrogation – the submarine and air bomb. We are led to believe, and I accept it, that the Admiralty have great confidence in their ability to cope with the submarine menace. Their technical methods are far superior to anything available in the Great War.' Robert Rhodes James (ed), *Winston S. Churchill. His Complete Speeches*, vol. VI: *1935–1942* (New York and London, 1974), p. 5897.

Servaes[1] of *Resource*. I am sending a separate chit about him in case it may be of use when you see him when he gets home.

The Governor[2] is much better and now hobbles about on sticks.

You asked whether there was anything more you could do which would help to maintain the fighting efficiency of ships during these troublesome times. I do not think there is. We are getting in a fair amount of exercises of all kinds and if it was not for the continual changes in control personnel there would not be anything to prevent a high state of efficiency being maintained.

We have just had a taxi strike (3 days) because taxi meters are being made compulsory. It has done no harm to anyone except the taxi owners. They naturally hate not being able to go on robbing everyone as they have been used to doing.

Geoffrey Blake. I still hear very good news of him and hope to see him out here again in the New Year.

I hope you are able to get away for a shoot occasionally. I live in hopes of getting as spell at Corfu either before or after Xmas.

246. *Chatfield to Pound*

[CHT /4/10] 23 November 1937

[Carbon]

I have been meaning to write to you for some time in reply to your two letters of the 13th and 25th October,[3] but I have been so busy that I have really not been able to find time to do so. I had to spend the best part of last week in full dress when the King of the Belgians visited London. I find also that the work of the Chiefs of Staff Committee grows monthly, and I spend more time out of the Admiralty than in it. While this is a compliment to the Chiefs of Staff, that nothing can be done these days without asking our opinion, it makes life extremely laborious.

The Egyptian situation is of course a source of an immense amount of papers and study. I find it very difficult to move my two colleagues on the subject of the 'Defence of Egypt', and we have been hampered by the absolute priority the Government have given to the anti-German problem,

---

[1]Capt [later VA] Reginald Maxwell Servaes (1893–1978). Commanded repair ship *Resource*, 1937; Dep Dir of Local Defence, Admy, 1938–39; Dir of Local Defence, 1939–40; commanded cruiser *London*, 1941–42; Dir of Trade Div, Admy, 1943; ACNS (Foreign), 1943–45; RAC 2nd CS, British Pacific Fleet, 1945–46; Flag Officer commanding Reserve Fleet, 1947–48. Pound gave a glowing report on Servaes.

[2]Gen Sir Charles Bonham-Carter.

[3]Not found.

especially as regards the Air Force and anti-aircraft guns for the A.D.G.B. scheme. All these things are behind hand, and as anything sent out to Egypt means a weakening of our position at home, there has been general obstruction in the C.I.D. to the measures which the three Commanders-in-Chief and the High Commissioners have so frequently represented as essential. However, the last telegrams from Lampson have at last had some effect, not so much from the military point of view since nobody feels here that there is really a definite Italian threat at the moment, but rather from a political point of view, that our prestige in Egypt is going down because we are doing nothing to defend her when we have promised to do so. I think therefore, that there is little doubt that this week some definite steps will be taken to improve the position.

Of course everything we do will lessen the responsibility of getting out reinforcements in an emergency, i.e. it will considerably affect the question of the fast and slow convoys. I quite agree with you that we do not want to have to take a fast convoy through the Mediterranean, still less a slow one. The line I have always taken is that the fast convoy is a possible military operation, but since it would have to take place at the beginning of the war before we were able to check the Italian submarines or test the efficiency of the Italian aircraft over the sea, it must be a risk, and when you have two risks to face you must take the lesser one.

The matter, however, is only important during the next year or two, and the difficulty will be largely solved if we are able to induce those concerned to send out material and personnel before trouble starts. As soon as our A.D.G.B. scheme and the Royal Air Force are ready, say by the end of next year, then we shall be able to get places like Malta, Gibraltar, Alexandria and Aden properly equipped, but until then, and while the German fear exists and we are unready to meet it, the C.I.D. will give the Mediterranean second place. However, as I have said above, I very much hope that something will be done this week to improve our position in Egypt, on which the Chiefs of Staffs have just made a further report and recommendations. I am not going into further details in this letter.

One of the Naval points under consideration was the question of sending out the *Coventry* or *Curlew* now, mainly from the point of view of Red Sea convoys, the defence of Aden, etc. The difficulty is that we have nobody to man them without paying off some other vessels, and I am loath to pay off any cruisers. I am taking steps to see that *Coventry* and *Curlew* are placed in a state of readiness, so that in case of emergency they can be rushed out before mobilisation.

Another question is whether the minesweepers at Malta should be transferred to Alexandria, but we are communicating with you about that separately.

I am a little concerned about the M.T.B. question. Your telegram to me recommended twelve squadrons, and if you add on the squadron to Hong Kong and Singapore, a depot ship here and there, and then carry the same principle into the North Sea, it will mean great expense. While I agree that they are a useful craft for the offensive defence of harbours, they do not really take the place of anything else [*phrase crossed out*: 'i.e. it means building up a new arm of the Navy'] and I am still unable to see how we are going to finance the arms we have already got, or rather intend to have. [*Added in margin*: 'But see P.S.'] I notice in your telegram you say that they would be of great value for the attack of harbours, but in a subsequent paragraph you state that, as all harbours would be defended against them, they could not be relied on for this purpose and it would be better to trust to aircraft. There is a strong school in the Admiralty in favour of developing these M.T.B.s, and of course it is not easy to decide what to do. When you realise the enormous expense of our present light forces: our 144 modern destroyers of ever increasing size, plus a large number in reserve being re-armed with A.A. guns for convoy work; the immense cost of the modern cruiser and the very large number that we need so long as we have to face the chance of war both in the East and in the West; and the expense of Fleet Air Arm now added to our Estimates; I view with some alarm the enthusiasm for these M.T.B.s. However, I am investigating the question of cost to see where it will place us.

Looking through your letter again of the 13th October I see that you say you have got the impression that the Government are averse to consolidating our position for fear of hurting the Italians' feelings, but I do not think that is really the case. There are two points of view about the Italians which have to be considered. The first is that they are thoroughly untrustworthy and probably little better fighters than they used to be, are indolent and bombastic, and the best thing would be to teach them a lesson and answer threat by threat; and the second is that the Mediterranean position is only one of three anxieties (Germany and Japan are the other two), and as we are still hopelessly weak to meet the responsibilities of all three services, and so long as we cannot come to terms with either of our two chief opponents, it will be better in the long run to get an agreement with Italy because we have no basic cause of enmity with that country, as we have with the other two. I think this latter factor, together with the unreadiness I have referred to, has really dominated the thoughts of the C.I.D.

There is only one more point and that is about Cyprus, which you very often refer to in your letters and telegrams. Cyprus is out of the question for the immediate needs of the moment. It is a very long range problem of ten years and twenty millions, so Cyprus as a base can give us no satisfaction at the present time. It would indeed imply that we have no hope of coming to an agreement with Italy during those ten years. On the contrary we have the immediate problem, and that is why I press for the dock at Alexandria. To ask the Government to sanction both Alexandria and Cyprus is not practicable, but Cyprus will undoubtedly be reverted to if Alexandria fails.

I also have on hand schemes for the development of Aden into a harbour which could be used by big ships. Another point is that, although Cyprus might be of value for a Mediterranean war when we are not engaged with Japan, it would be an extra responsibility if we were. It is quite possible that if we were engaged in war with Japan, and Germany was threatening, that we might have to largely abandon the Mediterranean, but we should always have to keep submarines and flotillas in the vicinity of Egypt for its defence. Under those circumstances Cyprus might fall as well as Malta. However, nothing is settled pending Lampson's conversation with the Egyptians.

P.S. Re M.T.B.s. Henderson[1] is getting on hopefully with his trials of them for Asdic work and Minesweeping. If they can replace Coastal Patrol Sloops and Minesweepers there will be a different light on the question. I intend to build one more M.T.B. squadron next year.

247. *Somerville to Pound*

[ADM 116/3892] H.M.S. *Galatea*, at sea
24 November 1937

CONFIDENTIAL

No. 992/26/2/5.

I have the honour to submit the following report regarding the visit of *Galatea* to Cadiz.

2. *Galatea* sailed from Gibraltar at 0500 on 23rd November and arrived at Cadiz 1000 on the same day anchoring in position 030° 2.05 miles from Cadiz Cathedral.

* * *

---

[1] VA [later Adm] Sir Reginald Guy Hannam Henderson (1881–1939). Member of the Anti-Submarine Div, Admy War Staff and considered one of the originators of the convoy system, 1917; Flag Capt and CSO to C-in-C China, 1919–21; commanded carrier *Furious*, 1926–28; Naval Mission to Roumania, 1929; RAC Aircraft Carriers, 1931–33; 3rd Sea Lord and Controller of the Navy, 1934–39.

6. At 1040 the British Vice-Consul, Mr. R.A. Black, accompanied by an Officer representing the Captain of the Port, came onboard and I landed in company with them and my Flag Lieutenant. After calling on the Captain of the Port, I proceeded by motor car to the Naval Headquarters at San Ferdnando, some ten miles distant from the port of Cadiz.

7. On arrival I was received by a guard and by Admiral Bastarreche in person. The Admiral expressed great pleasure at seeing me again and was most cordial throughout the interview, which lasted over an hour.

I conveyed the message of appreciation and thanks contained in your 1338/30th October, 1937. In replying the Admiral said he would like to express his appreciation of the consideration and good feeling displayed by the British Senior Officers visiting Palma.

8. Referring to the proposals contained in my 1759/1st November, 1937, the Admiral said that night exercises and night firing by H.M. Ships in the area suggested, viz. south of latitude 35° 30′ N. and west of longitude 06° 40′ W. would be unlikely to interfere with any operations by Nationalist ships. He would be grateful, however, if a few days' notice could be given of the intention of carrying out such practices in order to remove any danger of encountering darkened Nationalist war vessels.

9. Other matters referred to during the interview are as follows:–

* * *

<u>Carriage of munitions to Spain</u>. The Admiral considers that a number of ships are now trading from the Baltic and North European ports to Spain with munitions and enquired how, in the case of British ships, information concerning their movements could be conveyed to the British Naval Authorities. I replied that intelligence of this nature could be transmitted via Algeciras to the S.O.(I), Gibraltar, and that this line of communication with Cadiz has been in use for some time. In connection with these reports I referred to the case of the *Euphorbia* and pointed out that such large expenditures of fuel could not be justified unless suspicions concerning the cargo carried were based on positive and accurate information.[1] I asked that in future it should be stated clearly whether a suspicion only existed or whether definite information was available that ships had munitions onboard. The Admiral appeared to appreciate that we could not be expected to embark on many more of these expensive wild

---

[1] The *Euphorbia* had not been intercepted on leaving the Dardanelles (see Doc. No. 244) and became the object of an extensive search and dragnet off the Balearics and eastern coast of Spain involving at least ten British destroyers and three cruisers as well as flying boats. The ship was finally intercepted on 18 November by the *Galatea* and brought into Gibraltar. Search described in H.M.S. *London*, Report of Proceedings, 27 Nov 1937, ADM 116/3891.

goose chases.[1] He is establishing a daylight air reconnaissance over the western approach to the Straits in order to locate and report ships which are believed to be carrying munitions.

* * *

Mining operations off the Eastern Spanish Coast. The Admiral had received no news up to date of any ships striking the mines laid at Valencia and elsewhere and had not heard of the sinking of the small Spanish freighter *Caleelcofauba* outside Castellon Harbour, reported in the Wireless Press dated 23rd November.

Safety Anchorages. The Admiral again raised the question of the use which is being made by Merchant Ships at Valencia and Barcelona of the safety anchorages and asked what action could be taken in the matter. I suggested that a reminder, generally promulgated to the effect that these safety anchorages were intended only for foreign war vessels and that no assurance could be given concerning the safety of Merchant vessels anchored in these areas, would probably have the desired effect. (It is, of course, obvious that if foreign men-of-war are present, adjacent Merchant ships will be safe, but this point was not raised by the Admiral).

Attacks on ships without lights. The Admiral asked if it would be in order to attack without warning any ships who were not burning navigation lights on the grounds that it was not lawful for ships to navigate in this manner. I told the Admiral that this would be quite unjustifiable and that the only penalty such vessels could incur would be as the result of collisions caused by lack of displaying the authorised navigation lights.

Exchange of Red and White families. The Admiral asked me to assist if possible in promoting the exchange of certain Red families from Ferrol and Cadiz whose husbands are at Carthagena, with White families of Naval Officers who had been murdered at Carthagena. I promised to do what I could in the matter.

* * *

---

[1] The *Euphorbia* was actually bound for Gdynia, Poland rather than a Spanish port and her cargo was manganese ore and cotton. Heaton, *Spanish Civil War Blockade Runners*, p.75.

248. *Somerville to Admiralty*

[ADM 116/3533] 29 November 1937

SECRET.

[Copy of Telegram] Addressed Admiralty, repeated C.-in-C. Mediterranean, V.A. Gibraltar.

IMMEDIATE.

664. Following is result of interview with Admiral [at] Palma.

2. Admiral [Moreno] expressed great concern at form in which warning had been promulgated.[1] Has asked Salamanca to suspend promulgation and is taking no action himself to enforce the warning.

3. Nationalists maintain that safety zones are being abused by merchant ships and their object is to check this abuse.

4. Nationalists claim that so far as merchant ships are concerned safety zones are part of adjacent ports. Admiral will be glad to receive views of the British Government on this point but does not bind himself to accept these views.

5. Intention of Admiral is to persuade Salamanca to issue amended warnings in sense that merchant ships using safety zones cannot be regarded as immune from attack though form of attack will not be specified. Suggestion is that vessels in such proximity to ports will be in danger zone created by military operations in the vicinity.

6. Admiral agreed that terms of blockade and threats to attack had been used injudiciously in Cadiz broadcast. He was fully alive to legal and international aspect and in this respect differed markedly from his predecessor.

1242/29.

249. *Admiralty to Pound*

[ADM 116/3525] 3 December 1937

SECRET.

[Telegram] Addressed C.-in-C. Mediterranean, repeated R.A.(D), V.A. Gibraltar, H.M.S. *Penelope*.

IMPORTANT.

R.A.(D)'s 1242/29 (not to *Penelope*). As safety anchorages are still used by H.M. Ships it seems to be desirable that R.A.(D) should seek

---

[1] The Nationalist warning broadcast from Cadiz apparently used the term 'blockade' and threatened a general attack on merchant ships in Republican ports, including those anchored in the so-called safety zones reserved for foreign warships.

confirmation, (unless he is satisfied on these points as a result of his interview) that there is still no intention of mining in safety anchorages and that suggestion in last sentence of paragraph 5 of R.A.(D)'s 1242/29 does not imply that a risk to warships exists.

It is proposed to defer any warning to merchant shipping in respect of these anchorages until text of amended warnings, referred to in same paragraph of R.A.(D)'s telegram, is received.

Although Admiral Moreno's interview with R.A.(D) does not suggest that blockade similar to that carried out off Bilbao was in mind when warning was broadcast from Cadiz, it is necessary to bear possibility of such action by Insurgents in mind. Later telegrams give text of correspondence with Hendaye[1] which bears on this question.[2] In these circumstances Admiralty are considering desirability of re-establishing arrangement by which advance movements of British merchant ships bound for Spanish Government ports would be telegraphed to you and they would be glad of your comments on this suggestion.

As regards question raised in R.A.(D)'s 1138/29 (not to R.A. Gibraltar or *Penelope*) paragraph 3 of Admiralty telegram 1930/3/11 (not to R.A.(D), R.A. Gibraltar or H.M.S. *Penelope*)[3] contains text of communication which was to be made to Admiral [at] Palma. Instructions as regards degree of protection to be afforded by H.M. Ships to British merchant ships inside Spanish territorial waters are in paragraph 20 of Admiralty Memorandum M.05465/36,[4] as amplified in respect to air attack by paragraph 5 of my 1825/22nd October (not to R.A.(D), R.A. Gibraltar or *Penelope*)[5] but it was of course never envisaged that it would be practicable to protect ships actually lying in a Spanish port. Question of where to draw the line in case of ships not actually lying in port but lying at anchor off the port is a practical one, which can best be decided by Naval Authorities present at the time, on the basis that protection should be given whenever practicable against any attacks coming within the

---

[1] Sir Henry Chilton, the British ambassador to Spain, had left Madrid and established the embassy just over the frontier in Hendaye, France.
[2] As the Spanish Nationalist Note Verbale of 24 November appeared intended as a definite notification of a blockade of the whole of the Spanish coast now in Government hands, the British Government informed the Nationalists that belligerent rights had not been granted to either side in the conflict and the British Government did not admit their right to declare any blockade. Furthermore, the Spanish Nationalists were warned against any interference with British shipping trading with Spanish ports and such shipping would continue to be protected as before. Foreign Office to Sir H. Chilton, Hendaye, Telegram No 409, 3 Dec 1937, copy in ADM 1/9948.
[3] Doc. No. 238.
[4] Doc. No. 126.
[5] If the attack is witnessed, aircraft should be treated on exactly the same basis as a Spanish warship.

scope of the instructions referred to above which are deliberately aimed at a British ship.

Natural answer to Admiral Moreno's question in paragraph 4 of R.A.(D)'s 1242/29 would seem to be that any anchorage which is regularly used by ships visiting a port should be regarded as part of the port in the wider sense of the term. At the same time it may well be the intention of Spanish Insurgents to attack merchant ships lying in safety anchorages and Admiral Moreno's question may have been prompted by this possibility. Question should therefore be ignored and a suitable opportunity taken to make it clear to Admiral Moreno that His Majesty's Government would regard as unjustified any attack deliberately directed at merchant ships whether on the high seas, in territorial waters, or in a Spanish port and that they would take a serious view of the matter if it turned out that any ships thus attacked were British.

1430/3.

250. *Pound to Admiralty*

[ADM 116/3525] 4 December 1937

SECRET.

[Telegram] Addressed Admiralty, repeated V.A.(D), R.A. Gibraltar, R.A.C. 3rd Cruiser Squadron, H.M.S. *Penelope*.

403. Reference Admiralty message 1430 3rd December. While non-intervention patrols were in force a knowledge of movements of British merchant ships was of some value but this was largely offset by the additional congestion in wireless traffic with consequent delay.

2. I am not clear what is the object of present proposal re-introducing signalling of movements of British shipping unless it is with the idea that individual ships should be afforded protection. Without largely increasing forces in the Eastern Area it would be quite impracticable to afford such protection.

3. Up to the present time we have always refused to escort ships and if we once started it we should I am sure find it most embarrassing.

4. The fact that Admiral [at] Palma is also keen on our ships being escorted is I think an additional reason for not doing so as it might lead to attacks on ships which were not under escort.

5. For the reasons given below I do not consider instructions contained in I.S.C.W. Article 3 paragraph 20 are sufficiently explicit to justify putting onus on S.N.O. present of what action should be taken in the event of British merchant ships at anchor being attacked by aircraft.

6. When an aircraft is seen approaching it would be quite impossible to determine whether she intended to attack the town or ships in the safety anchorage. Neither would it be possible to determine whether aircraft was a Government or Insurgent aircraft. It appears clear therefore that action can only be taken after bombs have been released.

7. Another difficulty arises due to nationality of merchant ships in anchorages. If they are all British then retaliatory action is obviously justified if it is considered that the attack was a definite one. Should however British and foreign ships be anchored in close proximity it would not appear to be justified to take retaliatory action unless a British ship was actually observed to be hit by a bomb. Even this would appear to be unduly stretching the terms of deliberate attack as bombs might well have been directed at one of the foreign ships.

1958/4.

251. *Admiralty to Pound*

[ADM 116/3530] 6 December 1937

SECRET

[Telegram] Addressed C.-in-C. Mediterranean.

Your 1708/5. Question of general reduction to a minimum of sea and air patrols under Nyon Agreement is under consideration with Foreign Office. In view of the state of maintenance of G.R.1 Wing, it is intended as a first measure to withdraw this Wing, one Squadron being held in readiness to return to Mediterranean at short notice, equipment and spares being left at Gibraltar.

It is appreciated that the effectiveness of the patrol will be much reduced, but it is hoped in a few days to arrange for a general reduction.

1434/6.

252. *Admiralty to Pound*

[ADM 116/3529] 9 December 1937

SECRET.

[Telegram] Addressed C.-in-C. Mediterranean, repeated R.A. Gibraltar, C.S.3, R.A.(D).

Your 1958/4. Proposal to supply information of movements was not intended to suggest that arrangements for escorting British ships should be made. Their Lordships' view was that further measures of naval protection for British ships trading with Spanish Government ports might be necessitated by recent threats of Spanish Insurgents, and it seemed to them that if interference with British ships on the high seas or attacks inside territorial waters contrary to paragraph 20(c) of I.S.C.W. Section A were to be apprehended, protection would be facilitated if advance information of movements were available. Recent telegrams from R.A.(D) seem, however, to indicate that Admiral Moreno intends to act correctly although possibility must not be overlooked that he may be overruled by General Staff.

It is, however, not proposed to supply the information in question unless you ask for it.

As regards protection of British merchant ships at anchor off Spanish ports, here again verbal statement reported in para. (iv) of R.A.(D)'s 1930/6 seems to indicate that problem may not arise.[1]

If it does Admiralty view is that it is important, particularly from the point of view of our own interests when belligerent, to maintain principle that neither belligerent nor neutral ships may be attacked whether from the air or otherwise unless they refuse to submit to a lawful order. It is of course impossible with certainty to identify merchant ships, except by visit. Unrestricted attacks on merchant ships must, therefore, be a matter of concern to us and to all seafaring States, and if they are made in such a way that British ships are endangered we have a just cause for action.

Instructions should therefore be that whenever deliberate attack is made on merchant ships lying at anchor off a Spanish port when British merchant ships are actually present in the anchorage, H.M. Ships should intervene if it is practicable to do so whenever there is a genuine risk that British ships may be damaged. It is of course realised that H.M. Ships could not intervene until the aircraft had demonstrated their intention either by dropping bombs or by opening fire with machine guns on merchant ships.

There is, of course, no intention of affording protection against incidental damage to British ships due to their presence in areas where rival Spanish forces may be taking place.

---

[1] '(iv) Operations against Republican ports will never be directed specifically against British Merchant Ships lying in or in immediate vicinity of those ports. Right of entry into those ports also not denied but vessels entering or lying in or near those ports render themselves liable to damage from mines, bombs, shells or military operations directed against military objectives in such ports. Nationalists will not hold themselves responsible for any damage which may result.' RA(D) to Admiralty, 6 Dec 1937, copy in FO 371/21365.

As R.A.(D)'s 1639/4 envisages further communication from Salamanca, no warning is being given to merchant ships as regards safety anchorages at present.[1]

1450/9.

253. *Pound to Rear Admiral Binney*

[ADM 116/3533] 14 December 1937

SECRET.

[Telegram] Addressed R.A. 1st Battle Squadron, repeated V.A. Malta, A.S. Malta, V.A.C.B.C.S., Admiralty, H.M.S. *Barham*, H.M.S. *Malaya*, Captain (D)1, R.A.(D), H.M.S. *Greyhound*, H.M.S. *Glowworm*, H.M.S. *Griffin*, H.M.S. *Grenade*.

IMMEDIATE.

445. As the only likely task for force in zone six at the present time appears to be interception of suspicious vessels and as use of aircraft in the winter months in the northern Aegean is always uncertain, I have decided to reduce forces in zone six to two destroyers.

H.M.S. *Malaya* and either H.M. Ships, *Greyhound*, *Glowworm* or *Griffin* should return to Malta in time for *Malaya* to be docked for cleaning and coating bottom A.M. 17th December.

H.M.S. *Barham* will be docked during week commencing 20th December.

1020.

254. *Pound to Chatfield*

[CHT/4/10] Commander-in-Chief, Mediterranean Station
22 December [1937]

[Holograph]

Thanks so much for your letter of 23rd Novr just received. Knowing how busy you are I hope you will never feel that I expect an answer to my letters but I do appreciate your letters more than I can say.

---

[1] An expanded version of Nationalist naval policy referred to in RA(D) 1930/6 transmitted to the Admiralty on 24 December included the statement: 'In this connection [proximity of merchant vessels to military objectives] the safety anchorages Valencia and Barcelona are intended solely for use of foreign warships. The safety of merchant vessels using these anchorages can in no way be assured.' C-in-C Mediterranean to Admiralty, 1643/24, copy in FO 371/21365.

I can quite imagine how much extra work the meetings of the Chiefs of Staff Committee gives you but I am sure you do not regret it as it is such a healthy sign that the opinion of the Fighting Services is considered necessary.

I am very grateful to you for all the help you have given as regards Egypt – it is a great thing to have got the defences onshore there and if we can only get the dock, the Great Pass dredged and Minesweepers, then the situation will be much more healthy. I hope that the reorganisation of the Army Council will result in the soldiers improving their position. Their attitude is so extraordinary that one can only imagine that they suffer from an inferiority complex.

I can imagine how terribly difficult it must be to steer the correct course with the German menace in the West, the Japs going off the deep end in the East and an unreliable Italy in the middle. Only today the Press people here whom I have in fairly good control brought me a telegram which showed that Italy is watching our reaction to the Eastern situation very closely (as one naturally imagined they would be) and it pointed out (for what it is worth) that any weakening of the Fleet in the Mediterranean would be seized upon by Italy to try and stiffen Germany up. This treble situation has caught us at a bad moment with so many capital ships hors de combat.[1] It will be a great relief when the aircraft carriers are once more efficient.

It is good news that as soon as the situation at home has been righted that Malta, Gibraltar, Alexandria and Aden will be taken in hand. One only hopes that we shall be given time to do this.

Your letter has made me realise that the German menace must be much more acute than one realises here and that the C.I.D. consequently gives priority to everything anti-German. It seems to me, however, that this priority should only be taken into account where fighting equipment is concerned and that it should not apply to finance. The situation would appear to be so serious that we should get on with everything that is necessary whatever it costs. For instance, excavating shelters for s/m's at Malta costs money but does not retard any anti-German measures.

As regards the M.T.B. question, I think you have missed my telegram 2313 of 8th Novr. which I have by me as I write. In your letter you say that in one place I say that they would be of great use for the attack of harbours. I cannot however find anything in my telegram to this effect. In Par. 2 (v) I say that they would be of value for offensive [?] operations off enemy coasts. By this I meant against the enemy craft which would

---

[1] Notably the battleships *Queen Elizabeth*, *Valiant* and *Warspite* and battlecruiser *Renown*.

be found off enemy coasts – patrol craft – ships entering or leaving harbours.

I am afraid it was the wording of my telegram, which might have been clearer, which gave you an erroneous impression. Like any other weapon one would always be glad to have them but I would not for one moment suggest building them at the expense of the standard fighting types such as cruisers, destroyers or submarines. I am of the opinion that whenever you have them you require numbers and I would much rather have a large number at Malta than split the same number between say Gibraltar, Malta and Alex. The great thing about M.T.B.'s is or should be their speed – it is not high enough at present but once you go in for minesweeping and a/s work very high speed is unnecessary, and it seems to me somewhat illogical to adapt an essentially high speed craft to such things as minesweeping or a/s work. Also, in their present state of development they are very fair weather craft. In any sea it would I consider be quite impossible to have hands on deck to work a sweep and I cannot imagine that when they are 'bumping', as they do in any sea, that an asdic operator will be able to function. I should as far as my present experience goes be very sorry to be dependent on M.T.B.'s in winter to either sweep a passage for the Fleet or clear the approach of submarines.

If it is a question of money, then a much more vital requirement is a depot ship for flying boats. The Fleet ought to be certain of having <u>really extended</u> reconnaissances round it wherever it goes and only flying boats can do this. We shall never be certain of this flying boat reconnaissance until we have a mobile unit ready to move with the Fleet. At the moment I have a Committee presided over by A.C.Q. going into this question and whether the depot ship should not be utilised for both flying boats and M.T.B.'s. The advantage of this would be that the Fleet would take with them their long distance reconnaissance and their harbour defence. I would surrender a cruiser to be certain of knowing what is going on in the ocean within 500 miles of my base but I would not give up a cruiser for an M.T.B. force.

All ships have now completed docking and we are ready to move to the Far East at short notice. My short cruise to the West coast of Greece saved the situation as far as Press comments on the alteration in docking dates as I was able to say that it would have been out of the question the Fleet Flagship leaving Malta if an early move to the Far East was contemplated. As you can imagine Malta was seething with rumours. Though we are all very fit the 10 days away from the social activities of Malta did us all a lot of good. The *Barham* celebrated her return by winning the Capital Ships Marathon. The *Barham* was always a very good ship under Wodehouse's[1] command but Horan[2] has put a wonderful spirit into them. He is also an extraordinarily fine ship handler and it is a pleasure to go alongside the Mole at Gib. or come to our berth in Malta. We had some very sporting days in Albania but our bags were disappointing as there was no snow in the hills and weather was very warm, also there had been unprecedented floods all over Albania which affected the duck and snipe.

As there will be no winter cruise for reasons given in my telegram I have proposed going to Egypt, Cyprus and Haifa as I very much want to visit these places before next Autumn and I am sure there will be a good many outstanding points to settle in Egypt.

I had a letter from Geoffrey Blake containing the very worst of news – that there was no chance of his coming back as A.C.Q. and that in view of the uncertainty of when he would be fit for service he was proposing to retire. I do hope it will not come to this as he will be such a tremendous loss to the service. As I wrote to you before I have no hesitation in saying that even with the limited experience he obtained out here he would be entirely capable of taking on C.-in-C. of one of the Main Fleets later on. I rather fancy his wife's health is entering very largely into his feelings.

I have withdrawn the large ships from the Aegean because there was really nothing for them to do there and in the Winter in the Northern Aegean their aircraft could rarely function.

I am afraid it is too much to hope that under existing conditions there will be any lessening of your labours and responsibilities but it may help you to know what supreme faith the Fleet has in the Admiralty under your direction.

With all good wishes for Xmas and the New Year to you and your family.

---

[1]Capt [later VA] Norman Atherton Wodehouse (1887–1941). Commanded cruisers *Ceres*, 1928–29; *Calypso*, 1930; CO, RN College, Dartmouth, 1931–34; Flag Capt, HMS *Barham*, 1935–37; Head of British Military Mission to Portugal, 1938–39; RA-in-charge and Adm Superintendent, Gibraltar, 1939; retired, 1940; Cdre of Convoys, 1941; lost with SS *Robert L. Holt* sunk by submarine, 4 July 1941.

[2]Capt [later RA] Henry Edward Horan (1890–1961). First Naval Member of NZ Naval Board and CNS, 1938–40; commanded cruiser *Leander*, 1940; RA Landing Craft and Bases, Combined Operations HQ, 1941–43; RAC Combined Operations Bases (Clyde), 1943–46.

255. *Chatfield to Pound*

[CHT/4/10] 30 December 1937

[Carbon]

MOST SECRET & PERSONAL.

I have been meaning to write to you but we have been so busy and Christmas intervening that I have not had time. The whole situation as regards the Fleet going East is at present very uncertain. Naturally I am adverse to sending it if it can be avoided but I am making all preparations as far as I can. Neither am I forgetting the difficult questions of maintenance, ammunition, etc. Obviously the Fleet that you will have to take out is not very satisfactory, but if it did go out I think we should be certain to have the American Fleet as well and that will make a great difference.[1] Of course in that case everything would be arranged beforehand as regards strategical movements and you will be fully informed. I have had some pressure put on me to send out 'reinforcements' but I have stated that nothing less than a Fleet equal to the Japanese would be sufficient. Actually, of course, as you will have seen in my telegram to you the Fleet that we could send out would in many respects not be equal to the Japanese but the *Hood* would make up in some way for the inferiority in 8″ Cruisers and anyhow the Americans have plenty of those.

All talk, however, of any action by the United States is taboo and highly secret, but we wouldn't mention it to anybody else. Anyhow one can never be sure what they will do so we cannot rely on them absolutely. I wish I could send the *Repulse* as well but as it is there will only be three ships left at home and with the '*Deutschlands*' to prey on our trade routes I have had strong pressure put on me to keep both the *Hood* and *Repulse* at home, but this I have resisted, as the Japanese have some fast battleships which you will not have. A difficult question will arise about the *Warspite* and if you have hoisted your flag in her before anything happens then no doubt

---

[1] On 12 December Japanese aircraft bombed and sank the American gunboat USS *Panay* in Chinese waters. Chatfield was probably influenced by the impending arrival in London of the US Navy's Director of the War Plans Division, Captain Royal E. Ingersoll, for talks with his counterpart at the Admiralty, Captain T.S.V. Phillips. Director of Plans. The talks were highly secret in both the United States and Great Britain but President Roosevelt had ordered Ingersoll not to commit the US to any joint action. See Stephen Roskill, *Naval Policy between the Wars*, vol. II: *The Period of Reluctant Rearmament, 1930–1939* (London, 1976), pp. 366–8; Ian Cowman, *Dominion or Decline: Anglo–American Naval Relations in the Pacific, 1937–1941* (Oxford, 1996), pp.132–7; and Michael Simpson (ed.), *Anglo–American Naval Relations, 1919–1939* (Farnham, 2010), Docs Nos 200–201, pp. 260–78.

she will have to go out with you and the Second-in-Command would have to hoist his flag in such ship as you decided. It will, of course, always be possible for you to shift your flag to the *Nelson*; that is a matter which I will leave to your discretion.

Naturally if the *Warspite* went out I should have to keep another ship back. A good deal might happen and changes might be made also as the result of staff conversations with the United States. It is quite possible of course that now the Japanese have had such a warning of the danger they are running to, they will be more careful and their military officers will be less inclined to run amuck and endeavour to confine the war to the east and centre of China. If they decide to attack Canton, danger of incidents will increase. At any rate I intend to send the Home Fleet on their Spring Cruise and the sooner you can get them together for exercises the better it would seem to be. Backhouse can always come home to organise the Home Fleet. Between you and me he is designated as my successor which will be public about the middle of January. I do not expect to give up my post before August or September and I am afraid if the international situation with them is still dangerous, they may refuse to let me go even then. As you well know I am not staying on by my own wish but of course I am a servant of the State and must do what is wanted. The advantage of having Backhouse at home is that if we went to war, Forbes could relieve him at any time and he could come to the Admiralty as Deputy First Sea Lord and thus be ready to step into my shoes at any time. This is all for your very personal and confidential information.

The situation at Singapore is not at all satisfactory as the 15″ gun defences are by no means finished. One battery will be ready by next midsummer but the other will not for sometime later. The facilities of the Singapore Harbour Board for repairs are however considerable and the Controller thinks you will be able to get on all right as far as ordinary maintenance is concerned. We are taking further steps to hasten out machinery, etc. to the dockyard and taking every possible step we can but of course the question of action damages will be most difficult. No doubt, however, if both we and the United States send powerful Fleets to the East, war would not arise and the object would be an overwhelming display of combined forces.

By combined forces I do not wish to give you the impression that I mean combined action tactically; that I think would be most undesirable as it would raise all the problems of command and methods which would be most undesirable to try and effect suddenly. Meanwhile I am doing everything I can to hasten on the completion of the *Renown* and *Valiant*, and I think they should both be available in about a year's time. The laying up of those ships and the *Queen Elizabeth* was of course done with our eyes open and with the hope that our foreign policy would keep us out of risk during that time. Unfortunately that has not happened and means we intend to increase rather than decrease our enemies.

PART IV

THE APPROACH OF WAR, 1938–1939

The Spanish Civil War dragged on throughout 1938 and for much of the year it seemed as if neither side had the strength to bring it to a successful conclusion. This meant that the Mediterranean Fleet was obliged to continue the onerous duty of detaching cruisers and destroyers to the eastern coast of Spain to protect British shipping. The rotation of destroyer flotillas was particularly disliked because it interfered with training and flotilla work with the remainder of the fleet. At the beginning of the year it seemed as if there would be some relief. The Nyon agreements were apparently working, attacks by unknown 'pirate' submarines on shipping in the Mediterranean apparently ceased. This was not surprising since the perpetrator of those attacks, the Italians, had become part of the agreement. The British in early January were able to secure French and Italian agreement to a reduction in the patrols with the proviso that a small force was ready in each area to resume them if necessary [256–8]. The period of relief was short, for by mid January there were reports of attacks on British and other ships by submarines working outside of territorial waters. The Spanish Nationalists had only a pair of submarines obtained from Italy, but the Admiralty suspected the Italians might have been behind the attacks as they knew the Nyon patrols had been reduced [259]. Acting on Admiralty orders, Vice Admiral Cunningham, at the moment senior naval officer in the western basin of the Mediterranean, delivered a stiff protest to Admiral Moreno, the Nationalist naval commander at Palma. Cunningham sarcastically summed up Moreno's eventual reply after investigation: 'The ships which were not sunk were attacked by Insurgent submarines whose Commanding Officers disobeyed their orders, whereas those which were sunk were not attacked by them' [260].[1] The formal protest by the British Government to both sides in the war contained a threat to attack any submerged submarine encountered in the British zone [261, 262]. The protests and threats had their effect, Admiral Moreno agreed to restrict activities of the Nationalist submarines and

---

[1] Writing after the war, Cunningham was harsh in his judgement of Moreno (whose name he misspelled): 'However, he was a very weak man with little real control over the Italian and few Spanish aircraft supposed to be acting under his orders.' Viscount Cunningham of Hyndhope, *A Sailor's Odyssey* (London, 1951), p. 186.

Lieutenant Commander Hillgarth, the industrious consul in Palma was confident similar incidents would cease except, perhaps, for British merchant ships risking damage from air attack in Republican ports [264]. Attacks by aircraft on merchant ships, particularly Italian-manned aircraft according to Moreno, were likely to remain a problem. It was a candid admission that the Nationalists did not have complete control over their allies [270].

The threat to sink submerged submarines in the British zone soon faced a major test. On the night of 5–6 March the Nationalist cruiser *Baleares* was torpedoed and sunk by Republican destroyers. The British destroyers *Kempenfelt* and *Boreas*, who had been close to the scene, rescued large numbers of survivors, a humanitarian action which earned them the gratitude of the Nationalists [270] but caused them to be bombed by Republican aircraft while transferring survivors to the Nationalist cruiser *Canarias*. Five seamen in *Boreas* were injured, one of whom soon died of his injuries.[1] The Nationalists, understandably, now sent one of their submarines to patrol off the Republican port of Cartagena, accepting the risk of British action [265]. What would happen if the submarine was detected by British patrols? The Foreign Office advised the Admiralty the submarine need not be hunted provided its activities were confined to 'legitimate objects', meaning no attacks on merchant shipping [266]. How could this be reconciled without modifying the policy regarding submerged submarines in the British zone? Pound suggested that the senior British naval officer at Palma find some 'pretext' to see Admiral Moreno and then drop 'a strong hint' any submarine would be ill advised to cross certain positions [267].

The British policy on attacking submerged submarines was modified once again as a result of the European crisis caused by the German occupation of Austria which began on 12 March. A German submarine had been reported passing through the Gibraltar Strait and in order to avoid an incident at this tense time Pound ordered that submerged submarines encountered in the British zone west of longitude 5° West were not to be attacked [268, 269].

In April there was a proposal in the Admiralty to withdraw the two remaining cruisers (*Arethusa* and *Penelope*) of the 3rd cruiser squadron. The older 'D' class cruisers that had belonged to the squadron had already been withdrawn. The justification was a desire to reduce the percentage of personnel on foreign service. Pound not surprisingly argued against it and in doing so gave an interesting comparison of his fleet with the Italian

---

[1] An abridged version is in Michael Simpson (ed.), *The Somerville Papers* (Aldershot, 1995), Docs Nos 14 and 15, pp. 25–7. The full text of the Aide Memoire is in ADM 1/9949.

fleet. He demonstrated that the British were outnumbered 7 to 4 in 8" gun cruisers and 12 to 3 in 6" gun cruisers and, among other points, explained that the new Tribal-class destroyers joining his fleet were much weaker in AA armament compared to the *Arethusa* [271]. Pound won the argument. Chatfield decided no changes would be made until the end of the year when the question would be reviewed but the international situation might be clearer.

In the spring of 1938 the continued absence of submarine attacks on Mediterranean shipping led Pound to propose discontinuance of Nyon patrols in Zone One after 27 June and retention of only a small force in the western basin [273]. He was anxious to do this because the Admiralty was disinclined to approve his request for temporary relief of the Mediterranean Fleet destroyers by ships detached from the Home Fleet. The Admiralty agreed, but specified the action should not receive any publicity [274]. However, just like the preceding January, barely three days later the Admiralty reconsidered the decision because of a recurrence of the bombing of British ships at sea and the apparent intention of the Nationalists to stop all traffic to Republican ports. These actions might lead to a recurrence of incidents on the high seas [275]. Pound adjusted to the situation as best he could. He pointed out that the Nyon dispositions had been adopted to deal with the submarine menace but were uneconomical because they required a whole flotilla at sea and were not well placed for dealing with incidents at sea involving surface ships or aircraft. He now proposed new dispositions that could be maintained by two cruisers and four destroyers based on Palma and two destroyers at Gibraltar. This would eliminate the necessity of keeping a whole flotilla in the western basin of the Mediterranean and would not require assistance from the Home Fleet [276–8]. The Admiralty approved the new dispositions as adequate and 'the best that can be devised in the circumstances' [279].

By mid-1938 the Spanish Civil War was far from the major problem likely to be faced by the Mediterranean Fleet. The aggressive policy of Nazi Germany, typified by the recent occupation of Austria, the uncertain outcome of attempts by the British Government to reach an accommodation with Italy,[1] and the openly aggressive policy of Japan in the Far East all meant that, unlike the situation a few years before, a major war in the near future was a serious prospect. In the event of trouble with Japan the Mediterranean Fleet under Pound's command was to proceed to the Far

---

[1] On the Anglo-Italian 'Easter Accords' of April 1938 see Reynolds M. Salerno, *Vital Crossroads: Mediterranean Origins of the Second World War, 1935–1940* (Ithaca, 2002), pp. 49–51; and John Gooch, *Mussolini and His Generals: The Armed Forces and Fascist Foreign Policy, 1922–1940* (Cambridge, 2007), pp. 385–7.

East. Pound hoped it would never be necessary, for the absence of British naval forces in European waters would be welcomed by the Germans and Italians. On the other hand, to merely reinforce the Far East without sending the main fleet 'would be simply playing into the hands of the Japanese' [263].

In case of a war with Germany, a capital ship force from the Mediterranean Fleet under the Commander-in-Chief would move to Gibraltar and in the early stages of the war be responsible for protection of the trade routes in the Atlantic. Pound hoped that this would only be temporary and later, when major threats to Atlantic trade had been removed, the Gibraltar force could move east to protect the Suez Canal and discourage intervention by other Mediterranean powers [280]. The potential operations in the Atlantic raised a number of new questions including the extent of the Atlantic command and relations with the French [282, 285, 289]. The planning became more urgent by August when it became apparent the next European crisis would involve Czechoslovakia and Pound expressed his concern the Mediterranean Fleet's cruise programmes in the Aegean would be interrupted once again [281]. On the last day of the month the Admiralty advised Pound that some time in September the Germans would be fully mobilised for war and it was necessary to consider what precautions to take in case of a sudden emergency. These included sending the battlecruiser *Repulse* home earlier than anticipated and the battlecruiser *Hood* and 3rd Destroyer Flotilla to the western Mediterranean with the cover story that the displacement was due to the Spanish situation [283]. The planned and much anticipated regatta at Navarino Bay would not be cancelled, again to avoid excessive comment [286, 288]. Pound was fully alert to the Czechoslovak situation, in fact his first telegram concerning precautionary moves actually crossed the Admiralty message on the same subject [284–6]. In staff conversations with the French during 1938 it had been agreed that, in the event of an Anglo–French war against Germany, the British would be responsible for the North Sea and control of its exits and the French for general control in the Mediterranean.[1] The Mediterranean Fleet would not generally operate in the western Mediterranean. Battleships of the Mediterranean Fleet with appropriate light forces would be based on Gibraltar to control the entrance to the Mediterranean and assist in protection of Atlantic trade [287].

---

[1] A useful account of the prewar Anglo-French staff conversations is in N.H. Gibbs, *Grand Strategy*, Vol. I: *Rearmament Policy* (London, 1976), chaps 16–17. Important French documents are in M.A. Reussner (ed.), *Les conversations franco-britanniques d'État-Major (1935–1939)* (Paris, 1969). See also the essays in Martin S. Alexander and William J. Philpott (eds), *Anglo-French Defence Relations between the Wars* (Basingstoke, 2002).

The Czechoslovak crisis deepened throughout the month of September with a German deadline of 1 October to invade if Hitler's demands were not met. The Cabinet ordered naval mobilisation on the 28th. For a time war seemed inevitable, averted only at the last minute by the controversial agreement at Munich.[1] Acting on Admiralty instructions, Pound ordered a concentration of the main portion of the Mediterranean Fleet at Alexandria by 23 September. However, when Pound requested authorisation for the British ambassador in Egypt to obtain the permission of the Egyptian Government to immediately lay underwater defences at Alexandria and initiate an examination service, the Admiralty declined. There was still insufficient reason to go beyond a precautionary concentration and the approach to the Egyptian Government would need Cabinet sanction [290, 291]. Moreover, the Admiralty regarded Pound's proposals as 'definitely of a warlike nature' and if carried out would clearly 'invest the movement of the Fleet with the character of a preparation for a war with Italy'.[2]

The cruiser *Devonshire* and destroyers *Afridi* and *Cossack* were on a visit to Istanbul when Pound's order to concentrate at Alexandria was received on the 21st. Captain Muirhead-Gould, who had already been restricting lengthy shore excursions the preceding week in Greek waters because of the international situation, described how he prepared for war and the precautions he took when passing through Italian waters in the Dodecanese en route to Alexandria [292]. At the other end of the Mediterranean, the *Hood* and destroyers of the Gibraltar force were ordered out into the Atlantic to meet and escort the large Cunard liner *Aquitania* which had been chartered to bring reservists out to the Mediterranean. The most immediate danger appeared to be the armoured ship *Deutschland* and two German submarines in Spanish waters. Rear Admiral Layton, commanding the Battle Cruiser Squadron, described precautionary actions at Gibraltar and *Hood*'s movements in his report of proceedings [293]. Before the crisis eased, the First Sea Lord Roger Backhouse had apparently made up his mind to send one of the reserve destroyer flotillas to Gibraltar so as to release the Mediterranean Fleet destroyers at Gibraltar for service with Pound in the eastern Mediterranean. He recognised how 'terribly short' of destroyers the Mediterranean Fleet was [294]. This was quite true: of the 40 destroyers Pound estimated he needed for operations in the central and eastern Mediterranean he had only 23 [295].

---

[1]The situation in London and Admiralty actions are described in Roskill, *Naval Policy between the Wars*, Vol. II, pp. 438–43.
[2]Minute by E.A. Seal for Head of M [Military Branch], 21 Sept 1938, ADM 116/4161.

After the crisis Pound submitted a long report on the situation in the Mediterranean on 1 October, the day on which German forces had been scheduled to enter Czechoslovakia and for war to break out had the Munich conference not taken place [295]. He had to consider the worst case, that Italy would join Germany the moment war began. With modern communications and aircraft, the Mediterranean was now 'a very small place'. The advantageous geographical position of Italy and her strong air force meant that after war broke out movements in the Mediterranean could not take place without strong escorts and, for example, the movement of even a single auxiliary between Malta and Alexandria would 'become a major operation'. This was indeed an accurate prediction of what would eventually occur during the Second World War. Moreover, had war broken out on 1 October the *Aquitania* with so many of the men necessary to bring the fleet up to war strength would have been compelled to take the long route around the Cape of Good Hope rather than risk a passage through the Mediterranean. This demonstrated a paradox: on paper the Mediterranean was the closest of the Navy's foreign stations, but when forced to use the Cape route it was, with the exception of China, the most distant. As the British would not have the luxury of completing their preparations in safety once war had broken out in the Mediterranean, Pound concluded they needed a more definite strategical policy [296]. This would essentially take the form of an advance into Libya as soon as possible by British forces in Egypt from the east and French forces, possibly assisted by British and Dominion forces, from Tunisia in the west. This rapid offensive action was necessary because, even with the reinforcements Pound had deemed necessary, they would still be operating in the face of superior Italian forces and with the lack of repair facilities at Alexandria they could not stand for long the steady attrition that would accompany the major objective of cutting Italian communications. With the limited resources at their disposal, they could not afford to waste them in dispersed operations not part of a general plan and the mere attrition of Italian forces as an object in itself was not sufficient.

Pound soon learned from the Admiralty that his recommendations were unrealistic. The limited air and land forces in Egypt 'were likely to be inadequate' for such an offensive operation and the most the British forces could do would be to hold up an Italian attack on Egypt [307]. The French were also likely to stand on the defensive in Tunisia at first.[1] There was, however, a disturbing aftermath to the crisis. This concerned gossip about pessimism among certain officers caused by the fleet having to leave

---

[1] On planning against Italy see Christopher M. Bell, *The Royal Navy, Seapower and Strategy between the Wars* (Stanford, 2000), pp. 119–23.

Malta each time there was a crisis, a feeling reinforced by Malta's admitted deficiencies in anti-aircraft defences [297, 298]. Backhouse reported it to Pound who naturally reacted strongly. He made 'exhaustive inquiries' and concluded the alleged state of mind was 'entirely without foundation' [299, 303].

By the autumn of 1938 the Spanish Civil War was slowly moving towards its conclusion. The Nationalist forces had actually reached the Mediterranean in April 1937, severing the remaining Republican territory in two. The following June they captured the sizeable port of Castellon raising premature hopes that the war might end. However, a stalemate followed. In 1938 the last major offensive by Republican forces took place along the Ebro River but, despite initial gains, ended in failure after three months of fighting. Unfortunately, incidents involving British merchant ships on the high seas continued to occur, notably attacks by Nationalist aircraft on ships that were well outside of territorial waters. The Foreign Secretary, Lord Halifax, feared a strong reaction in the House of Commons should the attacks receive much publicity, especially since British protests appeared futile. He had no wish to become involved in a war with Franco but the attacks were 'quite intolerable'. Halifax therefore sounded the First Lord on 13 December about possible counteraction such as increasing the number of ships on patrol [300]. The various options and possible degrees of reprisal were examined at the Admiralty. Cunningham, now Deputy Chief of Naval Staff, characteristically emphasised that no threat should be made without the full intention of carrying it out in case of necessity [301]. The First Lord's reply to Halifax shows a clear lack of enthusiasm for any action likely to be effective which could result in additional heavy commitments or even an expansion of the Spanish war through renewed intervention by other powers [302]. Given the threatening international situation and the anticipation 1939 was likely to be a year of crisis, this reluctance to add to the Navy's many potential challenges is understandable.

There was another reason to avoid complications. It was becoming evident that the Nationalists would soon bring the war to a victorious conclusion. Barcelona fell to Franco's forces at the end of January. The cruiser *Devonshire* and the destroyers on patrol along the north-eastern coast of Spain evacuated the British minister, embassy and consulate staff, archives, British nationals and political refugees whose lives were thought to be in danger. The evacuations took place with difficulty from the small ports of Caldetas and Arenys north of Barcelona [304]. Shortly afterwards, Captain Muirhead-Gould, acting on a request from the Nationalists through Consul Hillgarth and with Foreign Office approval, carried a representative of the Nationalists to the island of Minorca which had remained loyal to the Republic throughout the war. The *Devonshire* served as a neutral meeting ground in negotiations to arrange the surrender of the island thereby avoiding the bloodshed that was certain to follow an impending Nationalist invasion. The negotiations were difficult, particularly over the question of allowing political refugees to leave in the British warship [305, 306].

The possibility of incidents at sea continued to the end of the war. There is an interesting report from the cruiser *Sussex* on patrol in the waters off Valencia during March in which an apparently 'British' ship turned out to be in the hands of a Nationalist prize crew. The latter were, however, allowed to proceed when it was determined the ship had been stopped within territorial waters and therefore a legitimate prize, although their continued flying of the red ensign was improper. Shortly afterwards, the *Sussex* frustrated a Nationalist attempt to capture another British ship despite the Nationalist claim she had first been ordered to stop while still in territorial waters. The British captain disagreed [309]. During the same month, Rear Admiral Tovey in the cruiser *Galatea* also frustrated a Nationalist attempt to capture a British merchantman, this time determining that the Spanish had not 'established effective control' while the ship was still in territorial waters [310]. These reports are another example of the Mediterranean Fleet's activities during the war in which naval officers might be called upon to make fine judgements of international law and the 'cat and mouse' games that were played with Nationalist warships.

The Nationalists continued their now unstoppable advance and in the south Nationalist forces entered Valencia on 30 March and occupied Almeria and Cartagena the following day. In the closing days of the war, British warships evacuated from Gandia a mix of British subjects, political refugees, Italian prisoners-of-war in the hands of the Republicans, and assorted members of international commissions who had been active in Republican Spain [310]. The Spanish coast ceased to be a major preoccupation for the Mediterranean Fleet.

By the end of March 1939 any hopes that the Munich agreement of the preceding September had ended German ambitions and would result in a real period of peace ended when the Germans began the occupation of what remained of Czechoslovakia on 15 March and annexed the Baltic port of Memel, which had been in dispute with Lithuania, on the 23rd. The British and French countered with the announcement of a guarantee to Poland on 31 March, followed by a British and French guarantee to Romania and Greece on 13 April. There was also an Anglo–Turkish joint declaration of mutual aid in April and a similar Franco–Turkish declaration in June. In the light of the deteriorating situation, Pound obtained

Admiralty permission to frame the traditional summer cruise programme so that the fleet should be immediately ready for any emergency without having to make a large number of movements at the last minute that might exacerbate a tense international situation [311]. A long cruise to Black Sea ports or a normal summer cruise to Adriatic and Italian ports would be cancelled. The fleet would now remain in the eastern Mediterranean operating from Port Said, Alexandria or Haifa and, if the international situation permitted, undertake short cruises to Greek or Cypriot ports.

The next crisis was not long in coming. On 7 April the Italians began the occupation of Albania. There were at the moment British ships in Italian ports on visits that had been arranged before the Italian action. The immediate problem was what action the Mediterranean Fleet should take [312–15]. Pending further instructions, Pound ordered the ships to act as if nothing had happened but to take certain precautions such as recalling all officers and men from extended leave [316]. The Vice Admiral at Malta was ordered to take certain unostentatious steps, largely involving getting ships ready for sea [317]. The following day the Admiralty reassured Pound that there was no reason to believe the Italian occupation of Albania was part of a larger plan and there was no need for British forces in the Mediterranean to take special precautions. Nevertheless, due to public opinion in Britain the visits to Italian ports should be ended [318]. Pound did not remain reassured for long. Only a day after receiving the encouraging message from the Admiralty, he received word on 9 April from the British minister in Athens that the Greek government had good reason to believe the Italians were about to occupy Corfu. In advance of any instructions from London he ordered all available ships at Malta to be ready for war at 4 hours' notice for steam [319]. Despite Italian assurances that no aggression against Greece was planned, the situation remained unclear. The primary impression gained from the messages of 9 April is that the Admiralty and Pound were determined the fleet should not be caught napping at Malta, a short distance from the Italian air bases in Sicily [320–25]. By 10 April the Admiralty was confident enough that Italian reassurances were genuine and authorised the fleet to reassemble at Malta for normal docking, refits and training. However, as a precautionary measure a proportion of the anti-aircraft guns were to be manned [326, 327]. While the situation had obviously eased, Pound was ordered to keep the fleet as far as possible in a state of readiness to move to the eastern Mediterranean [328]. The concentration of ships at Alexandria that had been visiting other Mediterranean ports was moved to a later date by the Admiralty on 22 April on the assumption tension would not become more serious [340, 342]. The revised date was on or about 15 May.

The tense situation raised another question, that of naval families at Malta. The Admiralty, however, did not think it was time to take immediate steps to remove families of officers and men from Malta, as Pound had recommended. They should, instead, be encouraged to leave voluntarily and others stopped from coming [329, 335]. He also advised when reviewing the state of shore defences at bases in the eastern Mediterranean that the Navy, Army and Air Force receive similar instructions regarding the state of tension, something that was apparently not the case at the moment [334]. Pound was not satisfied with the state of anti-aircraft defences at Alexandria, largely because there were insufficient British as opposed to Egyptian personnel available to man the guns and searchlights [349, 351]. He also wanted rapid firing 40mm anti-aircraft Bofors guns and the erection of lattice masts to improve the range of radar in providing advance warning of air attacks [350, 352]. The assumption that docking facilities at Malta might not be available in the event of war also resulted in plans to transfer the Portsmouth floating dock to Alexandria until a new graving dock could be completed there [330]. The transfer began in late June with the Mediterranean Fleet prepared to provide a close escort from Gibraltar through the Mediterranean should the international situation deteriorate [354, 355]. The dock arrived in Alexandria in August, less than a month before the war began.[1]

The continued presence of German submarines in Spanish waters and the announcement that the German navy would conduct month-long manoeuvres in Spanish waters was another source of anxiety for Pound at the time of the Albanian crisis in April. He wondered if this might herald another 'stunt' on the part of the Axis powers and wanted the fleet concentrated at Alexandria by 27 April so that no ship would be caught unprotected by a surprise attack in an unprotected anchorage [332, 337]. The Admiralty did not attach any special significance to the German manoeuvres and Pound's anxieties may seem exaggerated [336]. However, after a series of surprise moves by the Germans and Italians – Austria, Czechoslovakia, Memel, and Albania – it was understandable how there was a strong sentiment of 'what next?'

In this atmosphere minor events could seem to hold significance. On 26 April a submarine exercise off Malta became very real when the destroyer *Afridi* gained contact with an unknown submarine, almost certainly Italian. Continuous contact was maintained for over two and a half hours before Pound called off the hunt not wishing to give the

---

[1]Naval Staff, *The Royal Navy and the Mediterranean* Vol. I: *September1939–October 1940* (London, 2002), pp. xiii–xiv. Further detail on the Alexandria base is in I.S.O. Playfair, *The Mediterranean and Middle East* vol. I: *The Early Successes against Italy (to May 1941)*, facsimile edn (Uckfield, 2004), pp. 74–7.

submarine 'further practice in evasion' [343]. In May there was a different example of suspicious activity when a commercial airliner, with the large German registration letters characteristic of the late 1930s, on a few occasions flew low over the harbour at Gibraltar [353].

The British in 1939 would not be fighting a Mediterranean war alone. The French would be an ally and staff talks with the French took place. On 21 February, in a private letter to the First Sea Lord, Pound had suggested some outstanding points of joint interest to the British and French, especially regarding lines of demarcation for the operations of their respective fleets [308]. When the naval staff talks finally began the British and French agreed that the line of demarcation would run in a line from Port Empedocle in Sicily to a point 50 miles east of Tripoli in Libya [331]. On 21 April Pound received word that conversations were authorised between the various British and French military and naval authorities in the Mediterranean area [339]. The British intended to propose that the general line of action in the Mediterranean and Middle East would be to render the Italian position in Libya and eventually Ethiopia untenable. The naval aspect would be to cut communications between Italy and Libya while retaining control of the Red Sea and entrances to the Mediterranean. There was, however, a complication regarding cooperation with the French. The direction of French naval operations was much more centralised than was the case with the British and the French commander-in-chief had far less scope for initiating operations. The French naval staff believed that coordination of operations in the eastern and western basins of the Mediterranean, if desired, should be arranged through the naval staffs in Paris and London [345]. Consequently, Pound was not enthusiastic about meeting his French counterpart as it was not likely to be useful. He did not think it was possible to synchronise operations and did not think collaboration was necessary as long as the dividing line between their respective areas was what he had suggested [341].

Pound continued to emphasise the importance of an early offensive in the Mediterranean, but before he could complete his plans there were certain issues involving government policy and international law that required clarification. These included the territorial waters of Spanish Morocco and whether or not according to past treaties the Sultan of Morocco rather than Spain retained sovereignty [338]. Greece would presumably remain neutral but could, in certain areas, Greek territorial waters be treated 'somewhat lightly', perhaps even to the extent of sinking a merchant ship to block the Corinth Canal? Could the threat of mining be magnified by declaring 'dangerous areas' at the beginning of war? Other questions, such as whether or not merchant ships in certain areas could be sunk without warning by submarines when in escorted Italian convoys, seem almost quaint in the light of what would occur during the coming war.

The strategy that Pound planned to pursue ran along the following lines. The relative inferiority in manpower and aviation compared to the Italian forces would require the British to remain on the defensive in Egypt until heavily reinforced. Consequently, the fleet was the only British force in the Mediterranean able to immediately undertake the offensive against the Italians [346–8]. This would take the form of severing Italian sea communications and attacking Italian naval forces caught at sea, supplemented by bombardments of Libyan and possibly Sicilian ports and perhaps raids by the Fleet Air Arm on Italian ships in harbour. The Italian surface forces might 'to some extent be discounted' but the large Italian submarine fleet 'cannot be disregarded' and the Italian air force was 'obviously a very serious obstacle' although the effect of air attacks on ships at sea was 'largely an unknown factor'. Within a year the answer to that question would be clear. Pound recognised the vulnerability of Malta and accepted the French were not likely to have sufficient air strength to neutralise the Italian air force. He therefore recommended that before a war broke out a force of British fighters and bombers be moved to Algeria and Tunisia (or at least ground arrangements made) where they would be in a position to bomb vital targets in Sicily and Italy and perhaps act as a deterrent on Italian policy. Operations following the opening moves in the war would, however depend on the success in neutralising Italian submarine patrols and the 'experience of air menace' [348]. Pound did mention the possibility of carrier attacks on Augusta and Taranto, but he seemed to hope shore-based aircraft would be available for this. Pound apparently expected high casualties for British aircraft but thought he ought to get some use out of his single carrier, HMS *Glorious*, before the ship was put out of action by air or submarine attacks in the relatively confined waters of the Mediterranean.[1] The preceding year, at the time of the 1938 Czechoslovak crisis, he asked Captain Arthur Lyster of the *Glorious* to prepare plans for an attack on Taranto.[2] On one point the British were wrong in this pre-war planning. There was an assumption Turkey would probably be an ally [344]. Turkey in fact remained neutral until the closing days of the war.[3]

---

[1] Pound to Cunningham, 18 Aug 1939, in Michael Simpson (ed.), *The Cunningham Papers*, vol. I: *The Mediterranean Fleet, 1939–1942* (Aldershot, 1999), Doc. No. 5, p. 31.

[2] There was an earlier plan in 1935. At the time of the Abyssinian crisis the Mediterranean C-in-C Admiral W.W. Fisher asked that one be prepared. For further detail, see: David Wragg, *Swordfish: The Story of the Taranto Raid* (London, 2003), pp. 24–5; John Winton, *Carrier* Glorious*: The Life and Death of an Aircraft Carrier* (London, 1986), pp. 25, 75, 83–4; and B.B. Schofield, *The Attack on Taranto* (London, 1973).

[3] On relations with Turkey see Playfair, *The Mediterranean and Middle East*, vol. I, pp. 25–6, 50–53; and N.H. Gibbs, *Grand Strategy*, vol. I, pp. 713–14.

Admiral Sir Roger Backhouse, the First Sea Lord, was forced to relinquish his post in June as a result of a brain tumour. He died the following month. Pound, who had expected to be C-in-C Portsmouth when his time in the Mediterranean was over, returned to the Admiralty as First Sea Lord. Andrew Cunningham replaced him in command of the Mediterranean station.[1] He discovered that completion of his plans for offensive action on the outbreak of war were dependent on answers to his predecessor's questions concerning Spanish Morocco, submarine attacks on ships in convoy and the use of Greek territorial waters [356]. The Admiralty decision was a very strict interpretation of the law. Spanish Moroccan territorial waters were to be treated as if they were Spanish and the only penalty a ship incurred by travelling in a convoy would be chance damage during attacks on legitimate targets, that is ships that could be positively identified as belonging to enemy forces [359]. As for Greek territorial waters, the disadvantages of treating them lightly from the outbreak of war might well outweigh the advantages. The Admiralty's policy was to conduct war in the opening stages in strict conformity with international law 'in order that we can have a clear case for reprisals when and if international law is violated by the enemy'. There would be similar restrictions on bombardments of shore targets that might involve what in the twenty-first century would be termed 'collateral damage'. Cunningham pointed out that there were few places in Italy that could be bombarded without risk to civilian life and the disregard of the latter by the totalitarian powers in Spain gave an indication of how they were going to act [360]. The Admiralty policy on bombardments would in theory restrict the British to legitimate targets but again there was the expectation the enemy would probably commence indiscriminate attacks from the air that would release Cunningham from legal restrictions [361].

The proposal to counter Italian air attacks by basing British air forces in Tunisia could not be carried out as long as British fighter strength was below the 800 considered necessary for the air defence of Great Britain and British bomber strength was half that of Germany [358]. The French, in turn, could also not reinforce their air forces in North Africa because of the menace of the German air force. It would require British air commitment, especially in fighters, to operations in north-eastern France to allow the French to reinforce North Africa but this was not possible because of the needs in fighter strength for the air defence of Great Britain.

---

[1] Interesting correspondence between Pound and Cunningham in the weeks before the outbreak of war is reproduced in Simpson, *The Cunningham Papers*, vol. I, Docs Nos 1–5, pp. 22–31.

The British and French Mediterranean commanders met at Malta between 27 and 29 July. The Commander-in-Chief of the French Mediterranean Fleet, Vice Amiral Emmanuel Ollive, arrived in his flagship the battleship *Provence* escorted by four destroyers. The French provided a plan of their proposed operation and Cunningham did the same in somewhat more detail [362]. The two commanders-in-chief agreed that directly synchronised action was not feasible but that collaboration could take place. The line of demarcation between their forces was altered somewhat. In the discussions, the French seemed alarmed at the lack of fast and powerful aircraft at Malta, especially when they heard that none would be available until late in the year. Both sides exchanged information on the disposition of their fleets and agreed that their subordinate commanders should meet.

Immediately after the departure of the French squadron Cunningham in the *Warspite* escorted by four Tribal-class destroyers sailed for Istanbul. Simultaneously, he arranged for his wife and niece to cruise in the sloop *Aberdeen* employed as the Commander-in-Chief's yacht. The officers of the sloop were ordered to take advantage of the cruise to quietly investigate possible anchorages in western Crete and southern Greek ports and islands for British tankers supporting possible future operations against Italy. The cruise included an overland trip to Ankara for Cunningham and his staff, another attempt to win over the Turks.[1] Cunningham was back in Alexandria by the middle of August, having conducted with the fleet an exercise simulating the attack and defence of an Italian convoy on the return voyage.[2] The cruise to Turkish waters was to be the last of these characteristic aspects of life in the Mediterranean Fleet before the war.

The fleet had not been back in Alexandria for more than a few days when the international situation turned even more threatening with German demands on Poland. On 19 August the first of a series of warning orders came from the Admiralty followed by a series of others that culminated, on 3 September after the German invasion of Poland, with the order to start hostilities against Germany.[3] The outbreak of war proved to be an anticlimax for the Mediterranean Fleet because of the decision by Mussolini not to immediately join his German allies. Cunningham, along with the other British Commanders-in-Chief on distant stations, was ordered by the Admiralty to ensure any defensive measures against

---

[1] A copy of the British ambassador's report of the visit is Knatchbull-Hugessen to Halifax, 11 Aug 1939, ADM 116/3877.
[2] Cunningham, *A Sailor's Odyssey*, pp. 212–16.
[3] The August–September precautionary measures in the Mediterranean are listed in Naval Staff, *The Royal Navy and the Mediterranean*, vol. I, pp. 82–3.

Italy would be as non-provocative as possible [363]. The Mediterranean was a backwater for the better part of a year before Mussolini decided to take advantage of the collapse of France to enter the war in June 1940. The Mediterranean Fleet would then experience some of the severest fighting and trials of the Royal Navy during the Second World War.

256. *Admiralty to Pound*

[ADM 116/3530] 1 January 1938

SECRET.

[Telegram] Addressed C.-in-C., Mediterranean, repeated C.-in-C., Home Fleet. From Admiralty.

My 2024/20 October.[1] Question of reducing the patrols under the Nyon Arrangement and Paris Agreement has been under discussion with French Government, who concur in principle that patrols should now be reduced to a minimum provided a small force is held available in each area ready to re-introduce patrols should it become necessary. The French Naval Authorities have not hitherto considered a reduction of their Naval and Air Patrols but might seize a favourable opportunity to reduce temporarily the number of French vessels at sea at one time. The matter has not repetition not been discussed with the Italian Government so previous discussion between British and French Governments should not be referred to. Request you will approach the French Admiral and the Italian Admiral with a view to the reduction of patrols. Request you will report action taken and decisions reached. Sixth D.F. will relieve 3rd D.F. in Western Area on 11th January as already arranged.

1621/1.

257. *Pound to British Ambassador Rome*

[ADM 116/3525] 2 January 1938

SECRET.

[Telegram] Addressed H.B.M.A. Rome, repeated Admiralty.

544. Please pass following telegram to Italian Ministry of Marine for transmission to Admiral Bernotti. Begins:
To Admiral Bernotti from British Naval C.-in-C.

1. Under the terms of the Nyon and Paris Agreement, France, Italy and Great Britain are under certain circumstances responsible in their respective zones for the safety of merchant shipping of all participating powers. The maintenance of adequate forces is not therefore the concern only of the power responsible for that zone.

2. When I had the pleasure of meeting you and Admiral Esteva at Bizerta in October, I think I am right in saying that it was the general opinion that it was desirable to have considerable forces immediately

---
[1] Doc. No. 229.

available for the security of each zone and that regular patrols should be carried out in each zone.

3. Experience during the last month has I think shown that forces and patrol arrangements we then severally considered desirable in our respective zones has proved to be on an unnecessarily large scale under present conditions.

4. I would suggest therefore that whilst the present conditions last, continuous patrols by a considerable number of vessels is no longer necessary and that situation would be adequately met by:

    (a) as a general rule, the maintenance in each zone of a small force which is immediately available.

    (b) occasional as opposed to continuous patrols.

    (c) arrangements for the reinforcing of any zone at short notice.

5. The force to be maintained in each zone and the patrol arrangements in that zone are obviously at the discretion of the power responsible for that zone, but purely as an illustration of what I have in mind, it would be my intention to deal with zones for which Great Britain is responsible as follows:

    <u>Zone One</u>. A force of nine destroyers immediately available. Patrols by two destroyers. No aircraft immediately available but reinforcements held in readiness.

    <u>Zone Four</u>. A force of four destroyers immediately available. Patrols by one or two destroyers at intervals. Occasional air patrols.

    <u>Zone Six</u>. Two destroyers immediately available. Occasional patrol by one destroyer.

    <u>Zone Seven</u>. It would not be my intention to maintain any forces (immediately available) in this zone but reinforcements are immediately available.

6. I have ascertained from my Government that they see no objection to proposals contained in paragraph (4) and I should be grateful if Your Excellency would inform me whether you agree to their adoption.

7. I have sent an identical telegram to Admiral Esteva.

8. In view of the contents of this telegram I have considered it desirable to send it via British Ambassador Rome and Italian Ministry of Marine instead of direct either in plain language or international code.

                                       2008/2.

258. *Admiral Bernotti to Pound*

[ADM 116/3530] 5 January 1938

[Copy of Telegram]

1. In compliance with instructions received from my Government, I have to communicate to Your Excellency that the Royal Italian Navy is ready to adhere to the proposals detailed by Your Excellency regarding the forces to be employed on patrol service in the zones assigned to it and the method of employment of such forces.

It is agreed that:–

(a) The Italian Navy will maintain in each sector of Zone 3 a small nucleus of ships immediately available.

(b) Occasional patrols along the recommended routes will be carried out.

(c) Aerial patrols will be dispensed with.

(d) Naval and air forces will be stationed at convenient bases to reinforce patrols in case of necessity.

2. The nucleus assigned to the various sectors shall be adequate for the length of the route to be watched.

259. *Admiralty to Pound*

[ADM 1/9948] 25 January 1938

SECRET.

[Telegram] Addressed C.-in-C. Mediterranean, repeated V.A.C.,B.C.S., C.-in-C. Home Fleet. From Admiralty.

IMMEDIATE.

Recent reports of torpedo attacks on shipping in the Mediterranean, culminating in V.A.C.,B.C.S.'s 2151/24 (not to C.-in-C. Home Fleet)[1] some of which have undoubtedly taken place a certain distance outside territorial waters, leave no room for doubt that Nationalist submarines are now carrying out attacks contrary to London Naval Treaty, with which the Nyon arrangement was designed to deal. Although there is no evidence that Italian submarines are participating, it is possible that this departure has taken place at the instigation of Italian advisers, observing that the Italian Government is well aware of the fact that we have recently reduced the Nyon patrols.

---

[1] Not reproduced.

2. It is important that Admiral Moreno shall be made to realise that H.M. Government are very concerned at this recrudescence of such attacks. V.A.C.,B.C.S.[1] should therefore take an early opportunity of interviewing Admiral Moreno on his return from sea, and draw his attention to the repeated reports which leave no room for doubt that Nationalist submarines are acting in a manner contrary to the London Naval Treaty.

3. He should add that if this tendency continues, H.M. Government will be obliged to take steps to re-establish the Nyon patrols at full strength, in which event a Nationalist submarine would very probably be engaged and sunk by the patrolling warships – a contingency which we all desire to avoid.

4. He should leave no doubt in Admiral Moreno's mind that this communication has been made on the instructions of H.M. Government, and should urge Admiral Moreno to ensure that immediate instructions are issued to the submarines in question to desist from attacks contrary to the London Naval Treaty.

5. If Admiral Moreno represents that attacks have not taken place on the high seas, the case of *Clonlara* can be cited to the contrary. It could also be emphasised to Admiral Moreno that torpedoing of ships inside territorial waters is no less reprehensible than on high seas, and would be likely to lead to a strong revulsion of feeling amongst maritime powers having interests in Mediterranean. He is therefore mistaken if he imagines that Nyon powers would be prepared to sit down passively whilst such attacks were carried out.

6. The above is of course distinct from interference with British shipping on the high seas in a manner not contrary to the London Naval Treaty, and therefore not covered by the Nyon Arrangement. V.A.C.,B.C.S. should leave no doubt in Admiral Moreno's mind that H.M. Government is equally anxious that all such interference shall cease. See my 1320/22 (not to C.-in-C. H.F.).[2]

2110/25.

---

[1] VA A.B. Cunningham.
[2] The British Government would regard the capture of any British ship as illegal even inside territorial waters and would demand release of the captured ship whether or not she was infringing the Carriage of Munitions to Spain Act. The persons responsible would be prosecuted under British law if a breach of the Act could be proved.

260. Vice Admiral A.B. Cunningham to Pound

[ADM 116/3679] Hood, at sea
4 February 1938

SECRET

B.C.S. 59/040 S.

I have the honour to report that *Hood*, flying my flag, left Malta on Wednesday, 5th January, 1938, and after completing a 15-inch full calibre firing, proceeded to Palma arriving at 0800 on Friday 7th January, and I took over the duties of Senior Officer, Western Basin, from the Vice Admiral Commanding, First Cruiser Squadron.[1]

* * *

3. *Devonshire* arrived at Palma on Tuesday, 11th January, and *Hood* sailed the same day for Barcelona.

4. Whilst at Barcelona it was reported that the Dutch S.S. *Hannah* had been torpedoed and sunk in a position approximately 6 miles south east of Cape San Antonio on Tuesday, 11th January.

Enquiries were made in Barcelona, but little more information could be ascertained concerning this incident except that some of the crew stated that they saw a torpedo track and that the ship sank after 7 hours.

* * *

5. *Hood* left Barcelona on Thursday, 13th January and arrived Valencia on Friday, 14th January, 1938, subsequently arriving at Palma on Saturday, 15th January, 1938. On arrival I despatched *Devonshire* to Marseilles with mails.

* * *

11. On 23rd January, the Insurgent Fleet carried out a bombardment of Valencia and the coastline to the southward over a distance of 15 miles, a number of shells falling close to the Consular residence at Perello.

A protest was lodged by H.B.M. Consul at Palma and the naval authorities have undertaken not to bombard in the vicinity of these residences in future.

---

[1] VA [later Adm Sir] Charles Edward Kennedy-Purvis (1884–1946). Dir of Signal Dept, Admy, 1927–29; commanded aircraft carrier *Glorious*, 1931–32; Asst CNS, 1934–36; commanded 1st CS, 1936–38; Pres RN College Greenwich, 1938–40; C-in-C America and West Indies Station, 1940–42; Dep 1st Sea Lord, 1942–46.

12. Information was received from Valencia on 20th January, 1938, that the British S.S. *Clonlara* was attacked by an unidentified submarine 10 miles north of Castellon de la Plana at 1330 on 19th January, 1938, as reported in my signal 1448 of 20th January, 1938.

As the *Clonlara* sailed for Liverpool before H.B.M. Consul could ascertain further details, I directed Rear Admiral-in-Charge, Gibraltar, to obtain full particulars if the ship touched at Gibraltar. His report is contained in his telegram 2302 of 23rd January, 1938.

13. On 20th January, 1938, it was reported that the British S.S. *Thorpness* was bombed during an air raid on Tarragona when 2 men were killed, 7 wounded and 5 missing.

* * *

16. On Monday, 24th January, a further report was received from H.B.M. Consul at Valencia that the British S.S. *Lake Geneva* was attacked by an unknown submarine 3½ miles south east of Valencia. A torpedo was fired, passing under the ship, and at the time of firing the submarine partly rose to the surface some 300 yards away and submerged quickly.

17. All reports of attacks on ships off Cape San Antonio and Valencia were passed to Admiral Esteva at Oran.

18. On 25th January, 1938, the Insurgent Fleet returned to Palma, and I at once arranged an interview with Vice Admiral Moreno and informed him of the grave concern with which H.M. Government viewed these renewed attacks on merchant shipping.

He expressed his distress that there had been any incidents with British ships and assured me that his orders are most definite that British ships are not to be interfered with.

I conveyed to him the substance of Admiralty telegram 2038 of 17th January, 1938.[1]

Vice Admiral Moreno was bitterly discontented with the attitude of the French who, he stated, caused him to release ships captured inside the three mile limit and have themselves escorted these ships into Government ports.

* * *

---

[1] Even if belligerent rights had been recognised, mere compliance with the London Treaty rules would not justify sinking of a neutral prize even inside territorial waters. Furthermore, the torpedoing of merchant ships even inside territorial waters and in circumstances complying with London Treaty rules was likely to lead to a strong revulsion of feeling against the Nationalists among maritime powers having interests in the Mediterranean. That revulsion would be even stronger if reports ultimately showed the attack was carried out in such a manner as to endanger the safety of the ship's company.

19. Vice Admiral Moreno visited me again on Wednesday, 26th January, 1938, and I conveyed to him the purport of Admiralty telegram 2100 of 25th January, 1938.[1]

He informed me that he could give no more information concerning the incidents relating to the S.S. *Clonlara* and *Lake Geneva* until the submarines returned from patrol but that in the meantime he had given definite orders by W/T not to fire at any ships except Government warships.

A copy of an aide memoire which was read to Vice Admiral Moreno on the subject contained in my telegram 1930 of 19th January, 1938, is enclosed with the turn-over notes handed to Rear Admiral Commanding, Second Cruiser Squadron, which are enclosed herewith.

He appeared much gratified by this statement.[2]

* * *

28. On Monday, 31st January, H.M. Consul General at Barcelona reported S.S. *Endymion* was sunk by torpedo at 0700 on Monday, 31st January, 16 miles south of Cape Tinoso.

29. I immediately arranged, as approved in your message 2348/31st January, a sweep by the 11th Destroyer Division from a position just outside the 3-mile limit in latitude 37° 20' N, following the coastline to a position terminating at latitude of Cape Nao with the object of making a demonstration and, if possible, locating and identifying the submarine.

On completion of this sweep, I directed patrols by sub-divisions to be established approximately 10 miles off the coast, one between Cape Nao and Cape Cervera, and one between Cape Cervera and latitude 37° 20' N.

Meanwhile, one Division was ordered to proceed from Gibraltar to Oran.

Neither this sweep nor the patrol reported any contacts or sighting of submarines.

30. I interviewed Vice Admiral Moreno on Tuesday, 1st February, and again on Wednesday, 2nd February, 1938. A resumé of these interviews is given in my messages 1225/1, 1230/2 and 2149/2.

The gist of these reports is that the ships which were not sunk were attacked by Insurgent submarines whose Commanding Officers disobeyed their orders, whereas those which were sunk were not attacked by them.

---

[1]Doc. No. 259.

[2]The British Government would not tolerate interference with British shipping on the high seas. They would regard as illegal the capture of British ships in territorial waters but in compliance with their non-intervention policy did not intend to intervene by force if the capture took place in territorial waters provided excessive force was not used by the Spanish authorities in effecting the capture. Provided the capture took place in territorial waters without undue force used and the ship was kept under control, there was no objection to her being brought to a Nationalist port.

The incident of the S.S. *Hannah* ... however, disproves this.

31. On Thursday, 3rd February, the Rear Admiral Commanding, Second Cruiser Squadron,[1] in *Southampton*, arrived and, after discussing the situation fully with him, I did not consider it necessary to proceed to Gibraltar to acquaint the Commander-in-Chief, Home Fleet, further of the situation.

I transferred my duties at Palma to the Rear Admiral Commanding, Second Cruiser Squadron, and proceeded to Malta at 1130 on Thursday, 3rd February, 1938.

261. *Admiralty to Admirals Pound, Backhouse and Calvert*

[ADM 116/3530] 5 February 1938

[Telegram] Addressed C.-in-C. Mediterranean, C.-in-C. Home Fleet, R.A. 2nd Cruiser Squadron. From Admiralty.

IMMEDIATE.
Following is text of communication to Barcelona and Salamanca referred to in my 1930. Begins:
You should as soon as possible make the following communication to the Spanish Government (Insurgent Authorities).
Begins: H.M.G. have noted with indignation the recent recrudescence of attacks on merchant shipping by submarines in the Western Mediterranean.
(For Salamanca only). In addition there has occurred on February 4th the wholly unjustified bombing and sinking of the British ship *Alcira*, as to which H.M.G. will be addressing a further communication to General Franco.[2]
(For both). Such attacks made against merchant ships which were carrying on lawful trade on the high seas are in flagrant violation of the rules of international law referred to in Part 4 of the Treaty of London of April 22nd, 1930, with regard to the sinking of merchant ships. They could therefore in no circumstances be justified.
The Spanish Government is (the Insurgent Authorities are) aware that the naval forces of certain Powers with special interests in the Mediterranean

---

[1] RA Thomas Frederick Parker Calvert (1883–1938). Commanded cruiser *Frobisher*, 1926–28; Dir, RN Staff College, Greenwich, 1930–32; commanded battle cruiser *Renown*, 1932–33, COS, Home Fleet, 1933–35; commanded 2nd CS, 1936–38.
[2] The British steamer *Alcira* (1923, 1,400 tons) was bombed and sunk approximately 20 miles south-east of Barcelona by two low-flying aircraft with Nationalist markings in spite of the fact she was flying the red ensign and had red, white and blue stripes painted on the top and sides of the bridge.

were authorised under the Arrangement concluded at Nyon, to take certain measures for the better protection of merchant ships, not belonging to either of the Spanish parties in conflict, from interference on the high seas in a manner contrary to the rules above referred to. Recent experience has, however, shown that these measures as at present applied are not sufficient to protect merchant shipping against illegal attack. In these circumstances, H.M.G. are forced to the conclusion that, in order to secure adequate protection for British shipping and in the interests of the safety of the shipping of other nations, it is necessary to proceed to further measures in addition to those provided for under the Nyon Agreements. They therefore desire to inform the Spanish Government (Insurgent Authorities) that if from now onwards a submarine is detected submerged in the zone in the Western Mediterranean in which the British Fleet operates in accordance with the division of the area agreed upon between the French and Italian Governments and H.M.G., it will be considered as contemplating an attack on merchant shipping. H.M.G. will not tolerate submarines being submerged in this zone and orders have accordingly been given to H.M. warships that if a submarine is found so submerged henceforth it shall be attacked. The zone above referred to is constituted as follows: from the Straits of Gibraltar as far as the line joining Cape San Antonio–Dragonera Point, then along the territorial waters of the southwest coast of Majorca to Cape Salinas and thence to Algiers.

A similar communication has been made to the Insurgent Authorities (Spanish Government). Ends.

1950/5.

262. *Rear Admiral Calvert to Admiralty*

[ADM 116/3532] 6 February 1938

SECRET.

[Telegram] Addressed Admiralty, repeated C.-in-C. Home Fleet, C.-in-C. Mediterranean.

IMMEDIATE.

922. Your 1930/5.[1] I personally handed communication to Moreno. He stated note entirely clear. Also that he was expecting something of the kind. While outwardly maintaining normal demeanour – he was nevertheless shaken. No indication of protest. He remarked that note did

---

[1] Directing Admiral Calvert to call on Moreno in order to hand him a copy of the Foreign Office Note (Doc. No. 261).

not include action against his submarines submerged in territorial waters to which I replied the note defines zone clearly.
1014/6.

263. *Pound to Chatfield*

[CHT/4/10]   Commander-in-Chief, Mediterranean Station
7 February [1938]

[Holograph]

Very many thanks for your letter of 30th Decr which helped very much to put one into the picture. I am naturally delighted at having been selected to command the Fleet in the Far East should we send a Fleet out but leaving aside the personal side one hopes it will not be necessary as we can ill afford to practically denude European waters with the situation as it is. The German and Italian forces would be delighted to see the Fleet the other side of the Canal. To send out reinforcements only and not the Main Fleet would be simply playing into the hands of the Japanese. According to the War Plan the *Hood* would not go out after July but I rather gather from your letter that it is your intention that she should go out in any case to make up for the lack of 8″ cruisers. I very much hope this is the case.

<u>Combined exercises</u>. I quite appreciate the difficulties in cancelling them, but I am still of the opinion that it would be the right thing to do if the opportunity occurred. The re-introduction of the Nyon patrols on a more intensive scale has strengthened my opinion in this as the H.F. destroyers will be unable to do much training with only one Flotilla out of 3 doing exercises and one Flotilla at least will be required on patrol and one will be resting. One way and another the Medn cruisers and destroyers get a good deal of practice in their various functions both by day and night as we have carried out a regular weekly programme of exercises throughout the winter. It is not that part of their training which is adrift. Nor do I wish to be an alarmist as regards the weapon training as the general efficiency is I think reasonably good and will I hope by the end of the *Centurion* firings be considerably better. I only hope we shall have time to complete all these before either the yellow or red white and green balloons go up. The unsatisfactory thing of course is that having carried out an intensive G programme and I hope reached a high standard of efficiency we almost immediately have to disgorge both officers and men from key positions in the control team and down goes our efficiency with a bump. If the situation becomes really critical I

think the Admiralty should accept the fact that changes in the control team must cease.

Warspite. General opinion is that her appearance is much improved as her silhouette is so much smaller. At fighting ranges the *Warspite* looks about half the size of the *Barham* and *Malaya*.

It is a great relief that the cause of the trouble with the couplings has been discovered. Everyone seems very pleased with her fighting equipment – possibly with the exception of the searchlights which the beams [?] prevent the Director Layers seeing the target. However no doubt we shall find some way of overcoming this – and anyway we have not had sufficient experience yet.

I witnessed a march past of *Warspite*'s ships' company yesterday and as I could not just step into a Fleet Flagship with a record like *Warspite* as if nothing had happened I told them very plainly what we all felt out here – that whilst we were all struggling – and making great headway by our work on the coast of Spain – to put the Navy back where it was before the mutiny of 1931 they had let the side down and retarded the progress we had made.[1] That closes the incident and they start fair. Naturally I did not go into any details of the incident but dealt with it on the lines of the effect it had had. All the signs, in the short time that they have been out here, are that they will do well.

Barham leaves on Thursday and will take this letter. I shall be very sorry to lose her. At one time or another she has been my Flagship over [?] a year – longer than *Queen Elizabeth* was. She has played up to the part magnificently. Wodehouse with his thoroughness and methodical ways gave the ship a very sound foundation, but he had not the power of inspiring great enthusiasm. Then Horan came with his Irish temperament and taking full advantage of what Wodehouse had done has put them absolutely at the top of their form and there is nothing they could not do at the present time.

Norman[2] the Commander has done very well throughout and will make a v.g. Captain and I do hope it will be possible to promote him next time as he not only thoroughly deserves it but is getting on in age.

R.B. [Backhouse] as your successor. I am very glad this has been given out as it will put a stop to rumours by all kinds of people who did not know enough to know that it was a foregone conclusion. The Service will

---

[1] A reference to the disturbances in *Warspite* at Portsmouth (see Doc. No. 214). Pound's speech may well have succeeded in motivating the ship's company to work hard to show his impression of them was wrong. See Roskill, *H.M.S. Warspite*, p. 188. Roskill subsequently emphasised its harmful effects in his *Naval Policy between the Wars*, vol. II, p. 465.

[2] Cdr [later VA Sir] Horace Geoffrey Norman (1906–92). Commanded battleship *Queen Elizabeth*, 1943–45.

have the greatest confidence in him but at the same time the Service, recognising as <u>everyone</u> does all that you have achieved and the difficult times you have seen us through, would never think of accusing you of holding on to office should the Govt as is only rational wish you to remain on until the situation is clearer.

Western Mediterranean. It is hard luck R.B. coming in for this new 'spasm' just as he takes over the Western Basin. However, as far as the s/m business is concerned, the return of the 3 s/ms to Italy is a good sign. It would be interesting to know who was in command of these s/ms. It is the same thing as in the Far East – undiscipline amongst subordinates. It remains to be seen whether the German pilots will behave themselves in future. I am very grateful to R.B. for taking over the Spanish duties again as it gives us a chance of working up my Flotillas.

* * *

Flag Officer Appointments. I was very interested in the list of future appointments which Naval Secretary sent me for my personal information. I shall be very sorry to lose Cunningham but no doubt it is necessary for him to get Admiralty experience. Layton should do very well in his place. Binney I am sure will do admirably at I.D.C. Kennedy-Purvis will not I imagine be over pleased at A.C.R.

It is very good news that you are coming out for part of the Combined Exercises and I hope nothing will crop up to prevent.

264. *Rear Admiral Calvert to Admiralty*

[ADM 116/4084] 12 February 1938

SECRET.

[Telegram] Addressed Admiralty, repeated C.-in-C. Home Fleet, C.-in-C. Mediterranean, R.A. Gibraltar.

944. Following for Foreign Office from Consul Palma:
Telegram No. 17. Admiral Moreno has informed me as follows:
(1) Bombing of cities and ports is suspended until further orders. Bombing of isolated military objectives will continue as necessary.

(2) All attacks on cities had military objectives such as forts, ammunition factories or fuel depots and ministries. These are frequently surrounded by civilian houses. The term open town presumably means without such objectives.

(3) The only two submarines still available named *Sanjurjo* and *Mola* will remain in Soller until further orders. Salamanca wish submarines

stationed off Cartagena for legitimate object of attacking Spanish Government warships but Moreno has replied this can only be done if British acquiescence is obtained through diplomatic channels.

(4) Captain of *Sanjurjo* has been summoned to Salamanca for cross examination about his movements on the day S.S. *Endymion* was sunk.

(5) Report called for by Salamanca on reprisals of Tarragona when British ship *Thorpness* was hit states categorically that ship herself was not object of attack.

(6) Action by German seaplane pilots is now considerably restricted.

My impression is that there should be no new incidents in the immediate future of a similar nature to those that have occurred lately except possibly in the case of merchant ships risking damage in a Republican port. Hillgarth. Ends.

2255/12.

265. *Captain F.E.P. Hutton[1] to Pound*

[ADM 116/3530] 10 March 1938

SECRET.

[Telegram] Addressed C.-in-C. Mediterranean, repeated R.A.C. 3rd Cruiser Squadron, R.A. Gibraltar, Admiralty.

IMPORTANT.
PERSONAL.

Most secret. Following information was given in strict confidence to British Consul by Admiral Moreno. In view of sinking of Spanish cruiser *Baleares* a submarine left Majorca PM on 8th March to patrol off Carthagena with the sole object of observing and reporting movements of enemy warships. Admiral Moreno fully appreciates risk involved.

1345/10.

---

[1] Capt [later RA] Fitzroy Evelyn Patrick Hutton (1894–1975). Commanded cruiser *Penelope*, Med Fleet, 1936–39; commanded aircraft carrier *Hermes*, 1939–40; COS to C-in-C China, 1940–41; commanded battleship *Warspite*, 1942–43; Cdre Algiers, 1943–44; Cdre and Flag Officer, Belgium, 1944–45; Flag Officer, Western Germany, 1945–46.

266. *Admiralty to Pound*

[ADM 116/3530] 11 March 1938

SECRET.

[Telegram] Addressed C.-in-C. Mediterranean, repeated C.-in-C. Home Fleet, R.A.C. 3rd Cruiser Squadron, H.M.S. *Penelope*, R.A. Gibraltar. From D.C.N.S.

<u>PERSONAL</u>.
<u>IMMEDIATE</u>.

Your 2123/10 And *Penelope* 1345/10 not to C.-in-C. Home Fleet. Have discussed these telegrams with F.O. [Foreign Office]. They agree that provided the submarine confines her activities to the legitimate objects stated by Admiral Moreno she should not be hunted. This was on the understanding that our patrol vessels would be immediately available to deal with any attack on merchant shipping.

1245/11.

267. *Pound to Rear Admiral L.V. Wells*

[ADM 116/3530] 12 March 1938

SECRET.

[Telegram] Addressed R.A.C. 3rd Cruiser Squadron, repeated Admiralty, C.-in-C. Home Fleet, R.A. Gibraltar.

850. We do not wish to sink Nationalist submarines on passage to patrol off Carthagena provided they behave in a legitimate manner but at the same time there is no intention to modify our attitude that every submarine submerged will be attacked.

2. The only way in which a possible incident can be avoided is for the submarine to take a route to Carthagena which does not cross our patrol line.

3. You should therefore arrange to see Admiral Moreno on some other pretext on Tuesday Morning 14th March and in course of conversation drop him a strong hint that any submarine would be ill advised to cross lines joining points on Nyon routes A and F; and F and a position 038 degrees 04 minutes North, 001 degrees 42 minutes East.[1]

1213/12.

---

[1] As Admiral Moreno was at sea, Wells conveyed the gist of the message to the Captain of the Port at Palma on the morning of the 14th. Wells, Letter of Proceedings, 28 Mar 1938, ADM 116/3891.

268. *Admiralty to Pound*

[ADM 116/3530]　　　　　　　　　　　　　　　　　　19 March 1938

SECRET.

[Telegram] Addressed C.-in-C. Mediterranean, repeated C.-in-C. Home Fleet. From Admiralty.

Vanoc's 1734/18 and 1808/18.[1] Having regard to present political situation,[2] it is important to avoid an incident with this German submarine should she submerge in Zone 1.

　　　　　　　　　　　　1507/19.

269. *Pound to Rear Admiral Gibraltar*

[ADM 116/3530]　　　　　　　　　　　　　　　　　　20 March 1938

SECRET.

[Telegram] Addressed R.A. Gibraltar, repeated Admiralty, C.-in-C. Home Fleet, R.A.(D), Commodore (D), R.A.C. 3rd C.S., 2nd Destroyer Flotilla, Capt.(D)1, Capt.(D)3, Capt.(D)4, Capt.(D)5.

877. I.P.A. 12A and Home Fleet temporary amendment I.P.A. article 13 (not issued to 2nd Destroyer Flotilla).

Submarines encountered submerged in Zone One to west of longitude 5 degrees West are not repetition not to be attacked.

　　　　　　　　　　　　1513/20.

270. *Somerville to Pound*

[ADM 1/9949]　　　　　　　　　　　　　　　　H.M.S. *Galatea*, at sea
　　　　　　　　　　　　　　　　　　　　　　　　　27 March 1938

SECRET
No. 359/26/2/6.

I have the honour to submit the following report of the proceedings of *Galatea*.

2. *Galatea* sailed from Gibraltar for Palma at 0700 on the 25th March, 1938, and arrived without incident at 0900 on 26th March, 1938.

---

[1] Reporting the passage of the German submarine *U.35* through the Gibraltar Strait west to east and the boat's subsequent entry at Ceuta.
[2] Probably a reference to the *Anschluss*, the German occupation of Austria which began on 12 March.

3. Whilst on passage instructions were received that an Aide Memoire of the verbal communication respecting the rescue of survivors from *Baleares* should be handed to Admiral Moreno. This was prepared accordingly and a copy is attached (enclosure No.1).[1]

4. At 0935 Admiral Moriondo Di Morenco[2] of the *Quarto* paid me a visit. He was verbose and inquisitive. I informed him that I had come to say goodbye to my many friends at Palma. Referring to the sinking of the *Baleares* he said this should never have happened and that the Spaniards were quite hopeless. I refused to be drawn into a discussion of their inefficiencies in view of a report I had received from Lieutenant Commander Hillgarth, the British Consul at Palma, to the effect that Admiral Moriondo had been active in circulating stories at Palma to the effect that, in me, the Nationalists had their worst enemy.

5. At 1015 I landed in company with the British Consul, Surgeon Commander W.A. Hopkins, and my Flag Lieutenant, and proceeded to the British Consulate where I presented gifts of silver to the Commandant and Staff of the Military Hospital and to Dr. Porcel, the Port Medical Officer, in recognition of the services rendered by them to the officers and men of H.M. Ships visiting Palma. (Admiralty letter M.64/38 dated 11th January, 1938). The presentations were sincerely appreciated and attached is copy of a formal reply handed me by the Commandant.[3] Dr. Porcel, who speaks English expressed his gratitude coupled with admiration of the British Navy. A Spanish Naval doctor who was present described his escape from Cartagena in H.M.S. *Arrow* and expressed his warmest admiration of the British Navy and his gratitude to them for saving his life.

6. After completing the presentation I called on Admiral Moreno and read to him an account of the rescue work carried out by *Kempenfelt* and *Boreas*. This account undoubtedly enlightened the Admiral on many points which had hitherto been obscure. He was emphatic in his expression of gratitude at the manner in which British officers and men risked their ships and their lives in their endeavours to save the officers and men of *Baleares*. I am convinced that this accurate narrative has served a most useful purpose. The British Consul informed me that the Italian papers reporting the event stated that *Canarias*[4] was responsible for rescuing survivors, no mention being made of the British Destroyers.

---

[1]Partially reproduced along with an excerpt from this report in Michael Simpson (ed.), *The Somerville Papers* (Aldershot, 1995), Docs Nos 14 and 15, pp. 25–27.

[2]Contraammiraglio [later Ammiraglio di divisione] Alberto Moriondo Di Morenco commanding the division of Italian warships on the west coast of Spain. Commanded 4th Division at Battle of Punto Stilo, 1940.

[3]Not reproduced.

[4]The Nationalist cruiser *Canarias* was a sister ship of the *Baleares* and now the major surviving ship in the Nationalist Navy.

This so infuriated a party of Spanish Naval Officers that they seized bundles of these papers awaiting sale and burned them in the streets.

* * *

9. Referring to 'incidents' I pointed out that, as the Admiral was probably aware, the effect of these was to stimulate trade by British ships to Spanish Government ports, since incidents invariably led to increased activity by our patrols and consequently increased security to British ships on the High Seas. The Admiral stated he was fully aware of this and both deplored and condemned attacks on British ships on the High Seas. He gave a most emphatic assurance that such incidents would not arise owing to operations of submarines. Only two of these were in service now and engaged solely on patrolling the approaches to Cartagena. Both submarines were under his direct orders and his control of their operations was completely assured. He informed me in confidence that the only possibility of incidents would arise from operations by Italian manned aircraft. Although these were theoretically under his orders the situation was extremely difficult and he did not pretend he was able to exercise adequate control. For example the bombing of Barcelona had been carried out whilst he was at sea, without his knowledge, and in direct defiance of his orders. He had asked to have all Italian airmen replaced by Spaniards but it was doubtful whether this could be accomplished.

10. Referring to interference with ships in the Straits of Gibraltar, I suggested to the Admiral that immediate steps should be taken to stop this quite useless and unnecessary irritation. He fully agreed and said the matter would be represented to Salamanca on the occasion of a visit he was contemplating shortly. He was anxious however, that no publicity should be given to this matter since otherwise ships might be encouraged to carry munitions to Government ports.

* * *

12. As a result of this interview I gained the impression that no further submarine incidents are likely to occur. Italian manned aircraft continue to be the main difficulty but I have reason to believe that, as a result of the outcry caused by the bombing of Barcelona, Italians will feel the need for greater circumspection. On the other hand they obviously dislike the good feeling which exists between the Spanish and British Naval Officers at Palma. To provoke incidents with British ships may well be a part of Italian policy in the hope that the subsequent protest and action of our patrols will tend to antagonise the Nationalists. I judged that Admiral Moreno is under no illusions on this score, and it seems probable that, during his visit to Salamanca, General Franco will be fully informed of the situation.

13. After the interview had terminated I returned to *Galatea* and entertained to luncheon the Military Governor, Admiral Moreno, the Captain of *Canarias*, the Captain of the Port, and three other Spanish officers including Captain the Marquis de Nules, together with the British Consul and Mrs. Hillgarth. The general atmosphere on this occasion was most cordial and continued reference was made to the long period during which *Galatea* had been so frequent a visitor to Palma. Sincere regret was expressed at our departure together with invitations to return to Majorca when peace has been established.

14. *Galatea* proceeded for Malta at 1700 arriving at 1030 on the 28th March, 1938 without incident.

15. As this visit marks the termination of my association with Palma I wish to remark on the valuable services rendered throughout the Civil War by Lieutenant Commander Hillgarth, the British Consul. In spite of the occasional strained relations which have existed between the British and Nationalist Governments and the many occasions on which British policy has, in Nationalist eyes, been directed against them and in favour of the Spanish Government, Lieutenant Commander Hillgarth has always retained the respect and confidence of the Nationalist authorities. This he has accomplished by unusual tact and discretion coupled with great firmness. He thoroughly understands the Spanish character and mentality, is trusted by them, and looked upon as a man of his word.

* * *

271. *Pound to Chatfield.*

[ADM 1/9534]
H.M.S. *Warspite*, at Villefranche
4 April 1938

MOST SECRET

Herewith are some comments on the proposal to withdraw the remaining ships of The Third Cruiser Squadron.[1]

2. <u>The Italian and British Mediterranean Fleets.</u>

	BRITISH	ITALIAN	ITALIAN SHIPS BUILDING
Battleships	3	2	2 (by end of 1938)
Battlecruisers	2*		
Aircraft Carriers	1		
8″ Cruisers	4	7	

[1] The cruisers *Arethusa* and *Penelope*. The proposal came from a desire to reduce the percentage of naval personnel on foreign service as opposed to home service.

6" Cruisers	3	12	
'Tribals' or Scouts	8[n]	18	13
Destroyers	29	40	12
Torpedo Boats		44	16
Submarines	7	76	22

* = Late in 1938.

[n] = For a considerable portion of the next 12 months this will be reduced to 1.

3. The table of comparison between the two fleets shows the weak position of the Mediterranean Fleet in the case of a war with Italy, with Germany unfriendly.

In such circumstances, it would probably be necessary to retain in Home Waters a force to counter the German Fleet, and no immediate reinforcements could be expected in the Mediterranean. At the same time it is unlikely that any immediate assistance could be expected from France, whose fleet would inevitably be fully occupied in the protection of troop movements between North Africa and her southern coast.

4. Although it is improbable that the individual efficiency of the Italian men-of-war will approach that of the British Fleet, it is evident that the naval situation in the Mediterranean would be serious until reinforcements arrived.

The modernised Italian capital ships are reported to be capable of 25½ knots. When these are joined later in the year by the two 31 knot, 15 inch capital ships now building, it may be very difficult to bring them to action in open waters, except with battlecruisers and cruisers.

Under these circumstances, our carrier borne air striking forces will be of the greatest importance, and every endeavour will have to be taken to provide adequate A.A. protection for the *Glorious*.

5. At present, seven cruisers are stationed in the Mediterranean. Of these the *Galatea* is permanently employed with the flotillas.

Of the four 8" cruisers, only three are normally available for 50% of each year. This is due to the short and long self-refits at the six-monthly docking periods, and the policy whereby ships have to give foreign service leave at their home ports before re-commissioning.

6. Were the two remaining ships of the Third Cruiser Squadron withdrawn from the station, the cruiser strength of the fleet would be reduced to four.

From this number, all the cruiser requirements of the station would have to be met. With ships on duty in Palestine or Spanish waters, absent from dockings, self-refits etc., it would seldom, if ever, for the ships of the squadron to work together as a whole. Furthermore, such a shortage of cruisers would result in an unbalanced fleet consisting of four or five capital ships, one aircraft carrier, one flotilla of 'Tribals', three flotillas of destroyers, but only four cruisers, as opposed to the two well balanced 'squadrons' of the Italian Fleet.

With such a force, fleet training would prove difficult and unrealistic.

7. As conditions of weather and light are very different in the Mediterranean from those in home waters, it is considered that light forces should carry out their training in the waters in which they are most likely to have to fight.

8. In the event of war in the Far East, or the fleet being sent to Singapore as a precautionary measure, the two remaining ships of the Third Cruiser Squadron have important duties to perform in connection with control of the Suez Canal, which could not be adequately carried out by smaller ships with a lighter armament and smaller crews.

The Canal Control Force is the first of all units to proceed to their war stations. Were these duties taken over by ships of the First Cruiser Squadron, the advance force in the Indian Ocean would be weakened to an unacceptable degree.

The *Arethusa* and *Penelope* are visiting Port Said this month, in order to gain first hand knowledge of the special duties to be carried out by the Canal Control Force.

9. In the case of war with Italy, *Arethusa* and *Penelope* will in all probability have to be sent to the Red Sea to escort the Indian convoy from Aden to Suez.

In spite of the reinforcements which have recently arrived or should shortly arrive in Egypt and Palestine, reinforcements from India will undoubtedly be required in Egypt, and these will require protection against air attack from Italian East Africa.

10. Neither of these special duties assigned to the Third Cruiser Squadron could be taken over efficiently by vessels of the 'Tribal' class, as the latter are not fitted with the powerful long-range A.A. armament and duplicated control systems of the 'Arethusas'.

It is understood that the fire control system fitted in the 'Tribals' does not allow for their main armament guns to be controlled in A.A. fire.

11. The long-range A.A. batteries of the *Arethusa* and *Penelope* will be of great importance, whether they are employed at the Canal Ports; on escort duties in the Red Sea; in providing A.A. protection for the *Glorious*; or on other fleet duties.

12. Though it is agreed that the percentage of naval personnel at present on foreign service is high, the situation is gradually improving as more ratings become trained and the increased entry is making itself felt.

The withdrawal of the two remaining ships of the Third Cruiser Squadron would result in the transfer of not more than one per cent of personnel from foreign to home service.

13. It is desired to point out that the crews of the vessels of the First 'Tribal' Flotilla have been provided by paying off the two 'D' class cruisers of the Third Cruiser Squadron and the *Resource*. Were the ratings concerned not serving in the 'Tribals', they might well be expected to be serving in the *Resource* and the *Delhi* and *Despatch*, or the *Emerald* and *Enterprise* who were due to relieve them.

14. The situation may be summarised as follows:–

(i) The governing factor should be the correct strategical disposition of the fleet.

(ii) The withdrawal of the *Arethusa* and *Penelope* would have the following results:–

(a) The Mediterranean Fleet would be weakened at a time when important units are joining the fleet of its potential enemy.

It would appear that a reinforcement of, rather than a reduction in the strength of the Mediterranean Fleet is likely to be required in the near future and hence any temporary reduction at the present time would accentuate the reinforcement later on.

(b) The Mediterranean Fleet would become an even more unbalanced force than it is at present, and adequate fleet and squadron training could not be carried out.

(c) In the event of the fleet being ordered to the Far East, the most important units of the Canal Control Force could not be provided at the crucial moment.

(d) In a war with Italy, the Red Sea escort force for the Indian Convoy would be inadequate.

(e) The long-range A.A. fire power of the fleet would be considerably reduced.

(f) Not more than one percent of personnel would be transferred from foreign to home service and the effect of this individually would be negligible.

(iii) In view of the above, and as two cruisers and the fleet repair ship have been sacrificed in order to provide crews for the First 'Tribal' Flotilla, it is considered that no further reduction in the strength of the Mediterranean Fleet should be contemplated, either before or after the arrivals of the 'Tribals' on the station.

[Minute by D.C.N.S.]

C.N.S.

Once more I have carefully examined this vexed question of the Third Cruiser Squadron, in consultation with D.P.S., D. of P. and D.O.D. The only principle in the matter is that the Fleet must be correctly disposed strategically. The other factors are concerned more with our own internal administration.

Taking the strategical principle first, the C.-in-C.'s arguments, based on strategical grounds, have lost some force with the signing of the Italian agreement. But these two cruisers have a very special function in war because they provide the personnel for the Suez Canal defences in the additional stage of a Far Eastern emergency, and the Mediterranean War Orders envisage the despatch of these ships to the Canal area within 48 hours of the receipt of orders to prepare for despatch to the East. The C.-in-C. advances good reasons why these duties should not be undertaken by other forces. Thus, if war in the Far East is still considered a principal preoccupation, it is correct strategically to retain these cruisers in the Mediterranean. Eventually their place will of course be taken by four of the new cruisers.

But there is just one further point. Our one great asset is our mobility and though the absolutely correct strategical position for these cruisers may be the Mediterranean area, on account of their Canal duties, it might be considered that, provided they are in commission and in the Home Fleet they would not be too far away from their war station if our relations with Japan and Italy improve steadily and Japan is exhausted by her present war.

Turning next to the internal administration aspect, though C.-in-C. Mediterranean, very naturally, wishes to keep his Fleet well balanced by retaining two cruiser squadrons, the argument that their withdrawal will unbalance the Fleet applies equally to the Home Fleet. Until and unless both Fleets can enjoy the presence of two cruiser squadrons each, one Fleet must be unbalanced.

On the other hand it is true, as experience has shown, that the C.-in-C. Mediterranean, has far more extraneous calls on his available forces than the C.-in-C. Home Fleet. The last two or three years have been exceptional but even in times of profound peace there are always special calls on the ships of the Mediterranean Fleet.

But all through the discussion on these cruisers the major question has been that of retaining the proportion of home and foreign service. The original intention was to pay off a flotilla in order to man the First 'Tribal' flotilla, but the flotilla has been kept in commission additional and to compensate for this H.M.S. *Delhi* and *Despatch* were withdrawn and H.M.S. *Resource* reduced to reserve. This produced an adequate number of E.R.A.'s to man the 'Tribals', but the total effect when the 'Tribals' are commissioned will be an increase of 600 foreign service billets.

It is this that disturbs the Personnel Departments, and D.P.S. holds the view that the high percentage of foreign service is definitely a cause of discontent. One of the principal points is that though on the total the foreign service percentage is 41%, the percentage is much higher in certain categories, particularly those which are composed of the staid and married class of rating.

The problem fines down to this. In view of the present state of world affairs, can we sacrifice a feature of our immediate readiness for war (i.e. Canal defence measures) in order to preserve a more correct balance between home and foreign service? In view also of the world situation, does the mobility of our ships sufficiently offset the sacrifice of not retaining them in the immediate vicinity of their first war task?

Taking all the factors into consideration, I feel that until the situation in the Far East is clearer these ships should remain in the Mediterranean, but when we can see more clearly what our commitments in the Far East are likely to be we should then give the greatest weight to the personnel aspect.

W.M.J. [James]
27th April, 1938.

[Minute by Chatfield]

Until (a) the war in Spain has come to a conclusion, (b) experience has been gained of the improvement of our relations with Italy, and (c) the Far Eastern situation is clearer, I do not consider the present composition of the Mediterranean Fleet should be reduced.

Four *Dido* Class are to join the Mediterranean Fleet in 1940, and the very slight relief that would ensue in the interval by bringing home *Arethusa* and *Penelope* would not appear to be really justified. Further, *Arethusa* and *Penelope* are commissioned with foreign service crews, and might in any case finish their foreign commissions.

I do not intend, therefore, to make any change in the composition of the Mediterranean Fleet at the present time. The matter should be considered again at the end of this year when the Annual Review of foreign service percentage is held and when the International situation is bound to be clearer.

It is regrettable that, although 41% foreign service is in my opinion quite a reasonable one, in certain classes of rating the percentage is considerably higher. However, even the latter position must be growing better every day and although there may be some unhappiness among these particular classes I understand there is no active discontent or cause for it which would necessitate undesirable strategic steps being taken.

I have sent a reply to the Commander-in-Chief giving the above decision,[1] and these papers should now be made official.

C. [Chatfield]
29th April, 1938.

[1] Chatfield to Pound, 29 Apr 1938, ADM 1/9534.

272. Pound to Cunningham

[ADM 116/3530] 14 April 1938

SECRET.

[Telegram] Addressed V.A.C. B.C.S., repeated Admiralty, C.-in-C. Home Fleet, R.A. Gibraltar, R.A.(D).

IMPORTANT.

979. (a) Whilst at Gibraltar I discussed the question of provision of destroyers for Nyon patrol with C.-in-C. Home Fleet and pointed out that if the present strength of patrol was to be maintained without the assistance of a Home Fleet Flotilla until 27th June it would necessitate the Second Flotilla being continuously on the coast of Spain for a period of three and a half months as the First Flotilla would not be available from 27th May until they had completed working up in September after recommissioning. In my opinion this threw an undue strain on the Second Flotilla.

(b) C.-in-C. Home Fleet informed me that it would be possible for one Home Fleet Flotilla to arrive at Gibraltar on 16th May but that this would prevent this Flotilla carrying out their *Centurion* firings and missing the visit of His Majesty The King to the Home Fleet. I am reluctant therefore to ask for the assistance of Home Fleet until 27th June.

(c) As the present Nationalist policy excludes submarine attacks on British ships and as the only two Nationalist submarines are now employed solely on an observation patrol off Cartagena and have given no trouble I am of the opinion that the number of destroyers on patrol can now be reduced to four working in two pairs. This will enable destroyer strength in western basin to be reduced to one and a half flotillas and will not necessitate help from Home Fleet Flotilla until 27th June.

(d) As soon as the patrol can conveniently be readjusted one division of Second Destroyer Flotilla is to be sailed for Malta. This division will spend a fortnight at Malta and then relieve the other division of Second Destroyer Flotilla which will return to Malta for approximately 14 days.

(e) Should there be a recrudescence of submarine activity or the general situation demand the presence of destroyers at Spanish ports this must be met by destroyers then in the western basin by reducing their rest period until reinforcements can be sent should this be considered necessary.

1428/14.

273. *Pound to Admiralty*

[ADM 116/3530] 17 June 1938

SECRET.

[Telegram] Addressed Admiralty.

IMPORTANT.

276. As pointed out in my 1543 10th May, my 1513 19th May and my 1803 19th May, the Mediterranean Destroyer arrangements have been based on only maintaining one division and H.M.S. *Vanoc* in Western basin after 27th June, the other half of the flotilla being provided by Home Fleet. As I understand reinforcements from Home Fleet will not now be available request permission to discontinue active Nyon patrol in Zone One from 27th June onwards and to only maintain one division and H.M.S. *Vanoc* in Western Basin.

1958/17.

274. *Admiralty to Pound*

[ADM 116/3530] 21 June 1938

SECRET.

[Telegram] Addressed C.-in-C. Mediterranean, repeated C.-in-C. Home Fleet. From Admiralty.

Your 1958/17 (not to C.-in-C. Home Fleet). Approved to suspend active Nyon patrol in Zone One from 27th June. It is very desirable this should not receive publicity.

1852/21.

275. *Admiralty to Pound*

[ADM 116/3530] 24 June 1938

SECRET.

[Telegram] Addressed C.-in-C. Mediterranean, repeated C.-in-C. Home Fleet. From Admiralty [D.C.N.S.].

IMMEDIATE.

In view of the serious situation that has arisen consequent on further bombing of British ships which has occurred immediately after protest to Nationalist authorities H.M. Government have had to reconsider the suspension of the Nyon patrol which was approved recently.

It is of course appreciated that the Nyon patrol can have no effect against bombing inside the three mile limit, but the signs of increased determination on the part of the Nationalists to stop all cargoes reaching Government ports may result in a recurrence of incidents on the high seas.

The matter is now before ministers but meanwhile the patrol must be maintained at present strength and it appears most likely that final decision will be against any reduction at the present time.[1] You will be informed at once of ministerial decision. You may find it possible to maintain the patrol with a division and Leader plus H.M. Ships *Vanoc* and *Afridi*, to which H.M.S. *Cossack* can be added by July 12th.

Request your remarks on future dispositions as though it may be possible to strengthen your force by Home Fleet destroyers, Their Lordships are most anxious to avoid the serious disturbance to the Home Fleet programme and exercises which would result.

1857/24.

276. *Pound to Admiralty*

[ADM 116/3530] 25 June 1938

SECRET.

[Telegram] Addressed Admiralty, repeated C.-in-C. Home Fleet, R.A.C. 3rd Cruiser Squadron, R.A. Gibraltar, R.A.(D).

IMMEDIATE.

312. Your 1857 24th June …

(a) Gist of Admiralty message 1857 24th June is that Nyon patrol cannot be discontinued on account of serious situation that has arisen due to further bombing of British ships and possible recurrence of incidents on the high seas due to increased determination on part of Nationalists to stop all cargoes reaching Government ports.

(b) Present Nyon patrol on line joining position A.F. in I.P.A. Appendix 2 were instituted to deal with submarine menace. It is not only uneconomical as it entails maintenance of a whole flotilla in the Western Basin as pointed out in my 1513 19th May but, as experience has shown, it is not well placed to deal with incidents at sea.

---

[1] The Foreign Office had been prepared to agree to the suspension of the Nyon patrol but at the last minute the Board of Trade raised objections. Duff Cooper, the First Lord pointed out to Admiral James that feeling in the House of Commons was running high over the bombing of ships in harbour and if by chance there was a submarine attack followed by the admission that the Nyon patrol had been withdrawn, the Government 'would be in an impossible position'. Chatfield agreed that there was no doubt their national policy must be implemented even if the combined exercises had to be cancelled. Minutes by James, 24 June and Chatfield, 25 June 1938, ADM 116/3530.

(c) To deal with incidents at sea following dispositions would appear to meet requirements: (i) one ship in area Cape Salinas, Cape Palos, Cape San Antonio, Dragonera Island either at sea or at Alicante at immediate notice for steam; (ii) one ship in area Dragonere Island, Cape San Antonio, Calbetas, Dragonera Island, either at sea or if at Gandia or Calbetas at immediate notice for steam; (iii) one ship at Palma at immediate notice for steam.

(d) It is considered that dispositions given in (c) could be maintained by two cruisers and four destroyers based on Palma. This would leave H.M.S. *Vanoc* and one destroyer in Gibraltar area as H.M.S. *Vanoc*'s refit will be deferred. R.A.C. 3rd Cruiser Squadron is to signal comments on paragraphs (c) and (d).

(e) Above arrangements would obviate having to maintain a whole Mediterranean Flotilla in Western Basin which it is particularly desired to avoid for reasons given in my 1513 19th May and my 1803 19th May. Also it would not require assistance from Home Fleet. End of Part 1 …
1633/25.

313. Part 2 of my 1633 25th June.

(f) It would however have the following disadvantages:–

(i) Ships when on the coast of Spain spending practically the whole of their time in the Eastern Area instead of part at Gibraltar. With occasional visits to Marseilles and with better weather at Palma this can be accepted.

(ii) Evacuation of refugees will be more difficult and further commitments in this direction should not be accepted.

(iii) Cruisers in the Eastern Area will have to spend more time at sea and make less frequent visits to Marseilles. In view of the short time they spend on the coast compared to destroyers this is acceptable.

(iv) Destroyers will have to work singly instead of pairs. This they are already doing except on patrol.

(g) Above proposals which it is considered should be given a trial do not entail abandonment of the Nyon patrols but only altering them to suit present conditions. If experience shows that ships in Eastern Area are being too hardly worked they can immediately be reinforced by Mediterranean destroyers but should this be necessary I hope some assistance may subsequently be provided by Home Fleet.

(h) H.M.S. *Afridi* is already detailed to arrive in Eastern Area on 6th July to carry out duties of one of the two cruisers.

(j) In view of the international situation I am averse to using H.M.S. *Cossack* on the coast of Spain until she has completed working up.

(k) Until R.A.C. 3rd Cruiser Squadron comments have been received and Their Lordship's approval for above proposals or some other arrangement has been received present Nyon patrol will be continued.
1723/25.

277. *Rear Admiral 3rd Cruiser Squadron[1] to Pound*

[ADM 116/3530]　　　　　　　　　　　　　　　　　25 June 1938

SECRET.

[Telegram] Addressed C.-in-C. Mediterranean, repeated Admiralty, C.-in-C. Home Fleet, R.A. Gibraltar, R.A.(D).

422. Your 1633/25 para. (c). I have no personal indication that incidents at sea which are rare at present will increase. Nothing short of close escort can protect ships at sea from bombing or surface attacks. If ship is merely diverted to a Nationalist port as in the case of the *African Trader* a warship at normal notice for steam can deal with the case. If some sort of patrol in place of present Nyon patrol is desired suggest one ship working from Cape Palos to Caldetas would be adequate. One ship to run routine mail trip for Embassy Barcelona, etc., which takes seven days, and two ships at Palma at normal notice. Consider this force ample to deal with incidents at sea. Moreover if we have a larger force based on Palma which Nationalists may consider is working against their interests they may deny us altogether use of Palma as a base.

Reference paragraph (d). If dispositions in paragraph (c) of your message are still considered essential it will be necessary to base an oiler on Palma and this may lead to further difficulties with Nationalist Authorities.
2335/25.

278. *Pound to Admiralty*

[ADM 116/3530]　　　　　　　　　　　　　　　　　26 June 1938

SECRET.

[Telegram] Addressed Admiralty, repeated R.A.C. 3rd Cruiser Squadron, R.A. Gibraltar, C.-in-C. Home Fleet, R.A.D.

314. My 1633 25th June and R.A.C. 3rd Cruiser Squadron 2335 25th June.

[1] RA L. Wells.

A. I propose that the disposition stated in paragraphs (c) and (d) of my 1633 25th June should be adopted.

B. Although additional ships will be using Palma it does not necessarily mean that a much larger number will be in harbour at any given period. Moreover if Nationalists raise objections the obvious answer is that forces can be reduced when bombing of British ships ceases. A less convenient alternative would be to base two destroyers on Gibraltar or Oran for patrol work south of Cape San Antonio but in this case this patrol would only be in operation two thirds of the time so as to avoid overworking the destroyers.

C. Adequate arrangements can be made for fuelling without basing an oiler at Palma.

D. As early a decision as possible is requested.

2058/26.

279. *Admiralty to Pound*

[ADM 116/3530] 27 June 1938

SECRET.

[Telegram] Addressed C.-in-C. Mediterranean, repeated C.-in-C. Home Fleet, R.A.C. 3rd C.S., R.A. Gibraltar, R.A.(D). From Admiralty.

Your 1633/25, 1723/25 and 2058/26. Your interpretation of the responsibility of the forces in the Western Mediterranean is correct and the Admiralty consider your proposals adequate and the best that can be devised in the circumstances.

1730/27.

280. *Pound to Admiralty*

[ADM 1/9543] 23 August 1938

SECRET.

[Telegram] Addressed Admiralty.

546. I assume dispositions in Admiralty messages 1230, 1235, 1241 26th May and 2130 2nd June are those on which to base war plans and can be regarded as amending Admiralty letter M.00599/38.[1]

Investigations of defence of trade by Gibraltar Force in early stages of war shows following needs.

---

[1] War Memorandum (Germany).

(A) Clarification of Atlantic Station Command while a capital ship force under C.-in-C. Mediterranean is based at Gibraltar.

(B) Method of liaison with French Mediterranean Command.

(C) Knowledge of intended French dispositions.

(D) Additional destroyers for A/S patrol of Straits of Gibraltar one flotilla being minimum requirement for trade defence. Patrol(s) should be of primary interest to French.

(E) Flying boats at Gibraltar, Dakar and if possible Oporto for patrol of focal area.

Presume inter staff conversations have covered items (B), (C), (D) and (E) (Dakar) and that Flying Boat Squadron No. 202 could be transferred to Gibraltar if none could be spared from home.

At a later stage when major threats to Atlantic trade have been removed it seems Gibraltar Force would be better positioned in Eastern Mediterranean to safeguard Canal and discourage intervention by Mediterranean and Far Eastern powers.

1848/23.

281. *Pound to Chatfield*

[CHT/4/10]

at sea
24 August [1938]

[Holograph]

We are on our way back to Malta after a very pleasant cruise in the Adriatic, except that the weather has been exceptionally hot though personally I have not felt it.

* * *

We are really making good progress with the various War Plans now that we have one person continually at it.

The Italian visit was a great success and I think they would have liked to stay longer in an official capacity. I liked Riccardi[1] the Admiral very much and we got on very well together. I think our opinion of their efficiency as a result of the visit was exactly as it had been before – just second rate.

The visit was a splendid bit of anti-Italian propaganda as far as the Maltese were concerned as the latter thought very little of the physique and bearing of their men and what hit the Maltese hardest was that they

---

[1] Ammiraglio di Squadra [later Ammiraglio di Armata] Arturo Riccardi (1878–1966). Dir Gen of Personnel, Ministry of Marine, 1935–37; commanded I Squadra navale, 1938–39; Sottosegretario di Stato and Capo di Stato Maggiore, Dec 1940–July 1943.

had practically no money to spend – what little they had they spent chiefly on bread and other foods.

Now everything is staged for the Greek visit but it will be a bit trying in September. If we could only collect the Yugoslavs, Albanians and Bulgars our record would be complete!

I am very glad they are coming as we owe them so much for the use of their ports and all the facilities they give us.

We have had a great entente with the King[1] at Corfu ... I suppose after all his time he spent in England his tastes have become very English because he seems to harken after English society and one way and another we found ourselves doing something with them nearly every day. We enjoyed it because he was content with simple things. We finished up yesterday by taking him 'incognito' to Buitrino Lake which he had always wanted to visit but for some reason his Government wouldn't approach the Albanian Govt. about his going there.

Yes, Godfrey has done very well in Palestine. I have a great opinion of him.

I had a very nice letter from Haining,[2] whom I have not yet met, saying how much they appreciate what he has done for them in Palestine. I am afraid it looks like having to keep a ship there for some time but after October they will have to be satisfied with a cruiser as the ship must be inside [Haifa] and I am not prepared to berth a capital ship inside during the winter months.

We had the sailing regatta at Navarino and the weather was very good to us. The standard of sailing was very high. *Hood* rather swept the board due I think to the enthusiasm of Cunningham and Pridham.[3]

* * *

I hope we shall not be dragged away from the Aegean like we were last year but do not feel at all certain with the Czecho-Slovak business on hand.

* * *

[1]George II (1890–1947), King of the Hellenes, 1922–24, 1935–47.
[2]Lt Gen [later Gen Sir] Robert Hadden Haining (1882–1959). Dep Dir of Military Operations and Intelligence, WO, 1933–34, Commandant Imperial Defence College, 1935–36; Dir of Military Operations and Intelligence, WO, 1936–38; Gen Officer commanding British Forces in Palestine and Transjordan, 1938–39; Gen Officer Commanding-in-Chief, Western Command, 1939–40; Vice Chf Imperial General Staff, 1940–41; Intendant Gen, Middle East Forces, 1941–42; retired, 1942.
[3]Capt [later VA Sir] Arthur Francis Pridham (1886–1975). Commanded cruisers *Calliope*, 1928–29, *Curlew*, 1929; Dep DNO, Admy, 1932; commanded HMS *Excellent* (Gunnery School), 1933–35; Flag Capt (HMS *Hood*) and CSO to VAC Battle Cruiser Sqdn, 1936–38; Asst controller, 1938; Flag Officer, Humber area, 1939–40; Vice Pres Ordnance Board, 1940–41; Pres Ordnance Board, 1942–45.

282. *Admiralty to Pound*

[ADM 1/9543] 27 August 1938

<u>MOST SECRET AND PERSONAL</u>.
[Telegram]

Your 1848/23 August.
(A) This matter is under re-consideration. Request your proposals.
(B) Agreement has been reached with French for Liaison Officers to be exchanged. British to Oran and in French flagship; French to Gibraltar and in your flagship. Until special appointments can be made these officers must be provided from station resources.
(C) These are in most secret stage. It is under consideration to send details to you and Commander-in-Chief, Home Fleet.
(D) All light forces which can be spared from Home Waters are shown in M.00663/38 which you should by now have received.
(E) *Albatross* with seaplanes is detailed for Freetown or Dakar. Air Ministry is being informed of your requirement for Flying Boat Squadron No.202.
Staff conversations have not specifically dealt with (D) and (E) but we have accepted responsibility for Gibraltar Straits, leaving French in general control of Mediterranean. Dakar is mentioned as available for our use. Movements referred to in your final paragraph would depend upon development of situation.

283. *Admiralty to Pound*

[ADM 1/9542] 31 August 1938

<u>MOST SECRET AND PERSONAL</u>.
[Telegram] Addressed C.-in-C. Mediterranean Fleet, repeated C.-in-C. Home Fleet. From Admiralty.

IMPORTANT.
There is some uncertainty and consequently some anxiety, as regards possible German intention in connection with the Czechoslovakian situation.
You are aware that a large part of the German Army is being mobilized, and we are now informed that the German Fleet is to carry out large scale manoeuvres in the North Sea sometime in September, and that some of the ships of the German Merchant Navy are to take part. The manoeuvre area extends as far north as a line from the Faroes to Trondhjem. It has not been possible to find out between what dates the naval exercises are to take place.

In view of the fact that, at some date in September, the Germans will be practically mobilized for war, it is necessary to consider what precautions we should take in case of sudden emergency. It is for this reason that H.M.S. *Repulse* is being required to come home sooner. In addition, it has been decided that H.M.S. *Hood* and the 3rd Destroyer Flotilla are to proceed to Gibraltar and Western Mediterranean (not necessarily in company), leaving Malta not later than the 5th September. It may be necessary to send two 8-inch cruisers and H.M.S. *Glorious* to Gibraltar also, but no action in regard to this should be taken at present.

As these precautionary measures are most secret, the ostensible reason for the movement of H.M.S. *Hood* and the 3rd Destroyer Flotilla should be attributed to the requirements of the Spanish situation, and the real reason kept secret.

You will be kept informed of any new developments, but in the meantime no other alteration need be considered in the Mediterranean Fleet programme.

1652/31.

284. *Pound to Admiralty*

[ADM 1/9542] 31 August 1938[1]

SECRET.

[Telegram] Addressed Admiralty.

Personal for First Sea Lord.

582. I am not aware when result of Lord Runciman's[2] investigations may be expected to be made public but a crisis appearing possible at that moment I desire to call attention to the following points in case any modification of the autumn cruise programme is considered desirable.

(a) The greater part of the Fleet including H.M.S. *Hood* will proceed to Eastern Mediterranean on 10th September.

(b) H.M.S. *Devonshire*, H.M.S. *Afridi* and H.M.S. *Cossack* will go to Black Sea.

(c) H.M.S. *Glorious* will proceed direct to Alexandria after Fleet exercises on 11th September to disembark No. 3 A.A. Co-operation Unit having left one fighter squadron at Malta.

---

[1]The message apparently crossed the preceding telegram and was not received at the Admiralty until 0212 on 1 September.

[2]Sir Walter Viscount Runciman, President of the Board of Trade, 1931–37, had been sent by the British government in early August to act as a mediator in the dispute between the Sudetanland Germans, supported by Hitler and Germany, and the Czechoslovak government.

(d) H.M.S. *Malaya* is at Haifa.
(e) I could send H.M.S. *Hood* instead of H.M.S. *Arethusa* to Eastern Area [of the Spanish coast] on 6th September to get her nearer England but I do not recommend this as H.M.S. *Hood* would be exposed to smoke, bombs and submarine attacks in open Spanish ports.
1848/31.

285. *Pound to Admiralty*

[ADM 116/4068] 31 August 1938

SECRET.
[Telegram] Addressed Admiralty.

IMPORTANT.
580. Most Secret and Personal.
Part I.
1. Until convoy is instituted I consider I should be responsible for a definite area using Gibraltar forces. Proposed area is Atlantic Station north of line between Freetown and position 004° 20′ North 040° 00′ West.
2. In early days of convoy before Reserve Fleet cruisers become available I would suggest that I retain 1st Cruiser Squadron under my orders and be responsible for escort of convoy in area mentioned in para. 1 as a more powerful escort than a cruiser may be required. R.A. Gibraltar would be kept fully informed and could deal with any question while I am at sea with W/T silence in force.
3. Before war has been in progress very long it will become apparent whether Gibraltar force will be required for service elsewhere and question of turning 1st Cruiser Squadron over to C.-in-C. Atlantic and making latter responsible for whole of Atlantic Station could be considered.
1853/31.
581.
Part II.
4. Should it appear desirable after I have reached Gibraltar propose that V.A. Malta should take over duties of Senior British Naval Officer Mediterranean at a date to be proposed by me and that he should have a French Liaison Officer on his staff.
5. Consider that Gibraltar may be threatened by enemy air attacks from Spain and should be provided with an A.A. ship until adequate defences are installed.

6. Request that a limited amount of self destructive pom pom ammunition may be supplied for use by ships of Gibraltar Force when in harbour.

7. Request that Air Officer Commanding Mediterranean may be authorised to co-operate at short notice in:

(a) transfer to Gibraltar of Flying boat Squadron No. 202 personnel and stores and necessary F.A.A. stores for Gibraltar Force.

(b) preparation of Gibraltar landing ground for any temporary disembarkation required by H.M.S. *Glorious* aircraft.

8. Intend sending H.M.S. *Protector* to Oran when liaison becomes necessary and supplying her with Naval cyphers generally in addition to S.P.'s now carried.

2203/31.

286. Pound to Admiralty

[ADM 1/9542] 1 September 1938

SECRET AND PERSONAL FOR FIRST SEA LORD
[Telegram] Addressed Admiralty.

IMPORTANT.
PERSONAL.
583. Your 1652 31st August.

(A) 6th Division Destroyers is now in Western Basin and according to present arrangements remains there until 21st September.

(B) Remainder of 3rd Destroyer Flotilla comprising H.M.S. *Inglefield* and 5th Division Destroyers will arrive in Western Basin on 5th September.

(C) H.M.S. *Hood* will leave Malta on 3rd September in company with H.M.S. *Inglefield* and 5th Division Destroyers ostensibly to visit Tangier in place of H.M.S. *Repulse* and in order to have a large ship in Gibraltar area in view of activity of rival Spanish forces in that area. The question whether H.M.S. *Hood* shall actually visit Tangier must depend on whether it is considered safe for her to do so and can be decided later.

(D) My 1848 31st August. H.M.S. *Glorious* will not disembark No. 3 A.A. Co-operation unit unless I am informed situation has cleared up. Also H.M.S. *Glorious* if necessary will be retained at Malta after 10th September on plea of carrying out training of pilots in night landing.

(E) H.M. Ships *London*, *Sussex* and *Shropshire* are immediately available and H.M.S. *Devonshire* on 6th September. According to present arrangements 1st Cruiser Squadron leaves Malta 10th September for pulling regatta

at Navarin about which there is great keenness. Hence cancelling visit to Navarin will cause considerable comment. By existing programme H.M. Ships *London* and *Shropshire* return to Malta 17th September.
0048/1.

287. *Admiralty to Forbes and Pound*

[ADM 1/9543] 1 September 1938

MOST SECRET AND PERSONAL.
[Telegram] Addressed to C.-in-C. Home Fleet, C.-in-C. Mediterranean. From Admiralty.

The following are general dispositions and intentions for French Naval Forces in event of war with Germany.

2. Atlantic and Mediterranean Squadrons form high sea forces and will act together or separately as necessary to cover French lines of communication and territory. Covering of troop transports will take precedence over trade protection in early stages.

3. It has been proposed to French that in event of Anglo-French war against Germany:–

(a) British shall be responsible for control of North Sea and both its exits.

(b) French shall be responsible for general control of Mediterranean.

(c) Battleships of British Mediterranean Fleet, with appropriate light forces based on Gibraltar, will control entrance to Mediterranean and assist in protection of Atlantic trade.

4. As corollary to above, British will generally not operate in Western Mediterranean nor French in North Sea or English Channel. It is hoped that French battlecruisers based on Brest, or if they desire on Gibraltar, will be available to assist in Atlantic trade protection.

5. Local forces with headquarters at places indicated are as follows:–

(a) Brest area. Seaplanes and light craft for coast defence of Bay of Biscay.

(b) Oran area. Older battleships, light cruisers, destroyers, submarines and seaplanes to hold Western Mediterranean and protect troop transports.

(c) Toulon and Bizerte areas. Normal local defence forces.
1845/1.

288. *Admiralty to Pound*

[ADM 1/9542] 6 September 1938

SECRET.
[Telegram] Addressed C.-in-C. Mediterranean. From Admiralty.
IMPORTANT.
MOST SECRET AND PERSONAL.
Your 0048/1st September.
 D. Concur that H.M.S. *Glorious* should remain at Malta whilst present situation continues.
 E. Visit to Navarin should stand.
1230/6.

289. *Admiralty to Commanders-in-Chief*

[ADM 116/4068] 6 September 1938

MOST SECRET.
[Telegram] Addressed C.-in-C. Mediterranean, C.-in-C. Home Fleet, C.-in-C. Plymouth, C.-in-C. Africa, C.-in-C. America and West Indies. From Admiralty.

MOST SECRET AND PERSONAL.
 Reference Admiralty letter M.00663/38 dated August 1938, paragraph 3, while Commander-in-Chief Mediterranean, or other senior Flag Officer afloat is based on Gibraltar, he will be responsible for an area to be known as the North Atlantic Station with the following limits.

Northern boundary – From Cape Finisterre along the parallel of 42 degrees 54 minutes North to long. 40 degrees 00 minutes West.
Western boundary – Meridian of 40 degrees West.
Southern boundary – Parallel of 20 degrees 47 minutes North from Long. 40 degrees 00 minutes West to African coast.
Eastern boundary – Spanish and Portuguese coastline from Cape Finisterre to the meridian of 5 degrees 17 minutes West. Thence across Straits of Gibraltar to Almina Point and following African coast to lat. 20 degrees 47 minutes North.

 2. C.-in-C. Africa will be responsible for an area to be known as South Atlantic Station whose northern limit is parallel of 20 degrees 47 minutes North from longitude 40 degrees 00 minutes West to African coast.

Remaining boundaries as shown in Admiralty Letter M.00599/37 page 63.
3. It still remains desirable for C.-in-C. Africa to administer South Atlantic Station as in Admiralty Letter M.00599/37, page 64.
4. Necessary amendments to Memoranda will be promulgated by letter.

1645/6.

290. *Pound to Admiralty*

[ADM 116/4161] 21 September 1938

SECRET.
[Telegram] Addressed Admiralty.

IMPORTANT.
685. Request following action may be taken.
(A) Authorise Ambassador to obtain permission of Egyptian Government for immediate laying of under water defences and institution of examination service.
(B) Instruct G.O.C. British Troops Egypt that sea front and A.A. defences Alexandria should be fully manned forthwith.
(C) Fortress Commander and staff for Alexandria to be sent out in order that they may be in readiness to take over.

1055/21

291. *Admiralty to Pound and Forbes*

[ADM 116/4161] 21 September 1938

SECRET.
[Telegram] Addressed C.-in-C. Mediterranean, repeated C.-in-C. Home Fleet.

IMPORTANT.
SECRET AND PERSONAL.
Your 1055/21. Not to C.-in-C. Home Fleet.
Movement of Fleet to Alexandria was ordered because of undesirability of ships being spread over a large area at this time of uncertainty when future developments may depend very greatly on result of Prime Minister's visit to Germany tomorrow Thursday. Until result of this second visit is known nothing can be taken for granted. Two recent speeches of Mussolini appear to indicate that, in the event of failure to

obtain a peaceful situation, Italy will support Germany, although this might not be immediate. Possibility would therefore exist, even if remote, of Italian warlike action against unprotected ships. At the present time there is insufficient reason to go further than we have already gone in concentrating the fleet, and Cabinet decision would be necessary before any action could be taken with Egyptian Government. While you may consult our Chargé d'Affaires in any manner you wish, Admiralty cannot authorise anything further without specific Cabinet sanction. You may be sure that we shall keep you as well informed as possible, and take any further steps that are permissible without delay.

1914/21.

292. *Captain Muirhead-Gould to Pound*

[ADM 116/3903]   H.M.S. *Devonshire*, at Alexandria
23 September 1938

No.4397/192.

I have the honour to submit the following report of proceedings of H.M. Ship *Devonshire* under my command from 16th September to 23rd September, 1938, and of H.M. Ships *Afridi* and *Cossack* under my orders from 18th September to 23rd September, 1938.

* * *

3. St. Nikolo (16th–18th September).

* * *

St. Nikolo offers no facilities for the amusement or recreation of the ship's company. There is a bus service to Athens, buses leaving St. Nikolo twice daily and returning the same day, but owing to the international situation I was unable to grant leave to parties wishing to make this excursion Shooting is dealt with in a separate letter.

* * *

8. Stay at Istanbul.
Third Day (Wednesday, 21st September).

I received the Commander-in-Chief's cypher telegram at 0545 and at once ordered the Squadron to raise steam. Unfortunately a shooting party had been arranged and had left the ship at 0500, and was not expected to return until 0900. I took advantage of the delay to send an officer with a note to the Chargé d'Affaires and with mails, requesting him to cancel the 'At Home' on board *Devonshire*, and sent an Officer of the Guard to the *Hamidiye*[1] to say goodbye and to the Vali to cancel a large sightseeing tour which he had arranged for 0900. As soon as these officers and the shooting party were on board the Squadron weighed and proceeded at 20 knots.

9. Although the Chargé d'Affaires had no news of any serious development I considered that the concentration of the Fleet at Alexandria might indicate a grave situation, and I decided, therefore, that the ships under my orders should prepare for war and gave orders that in *Devonshire* shell-rings were to be filled with S.A.P., warheads fitted, and 300 4-inch A.A. shell were to be fuzed, and general preparation for war to be carried out. The Captain (D), First 'Tribal' Destroyer Flotilla, took similar measures in so far as they were appropriate for the 'Tribals'.

10. In view of the fact that the Squadron had to pass through Italian waters and within a short distance of an Italian air base I gave orders for Upper Deck lighting to be reduced to a minimum and burned dimmed navigation lights. The Commander-in-Chief's telegram 1750 was received at 1907 and these instructions were thereupon cancelled.

11. A.A. defence stations, General Action stations, rangefinding and inclination exercises, destroyer torpedo (dummy) attacks, and concentration exercises have been carried out during the passage from Istanbul. The new organisation between decks, based on C.B. 1997(38) Damage Control, has been thoroughly tested, and while several details still require examination in general the organisation is satisfactory.

* * *

16. The squadron rejoined your Flag at 0600, 23rd September, at Alexandria.

---

[1] An old cruiser serving as a training ship in which the Rear Admiral commanding the Turkish Reserve Fleet was flying his flag.

293. *Rear Admiral G. Layton*[1] *to Pound*

[ADM 116/3903]   *Hood*, at Gibraltar
3 October 1938

SECRET.
B.C.S. 686/41/4
In continuation of my letter B.C.S. 651/41/4 of 12th September, 1938, I have the honour to forward the following report of proceedings.

The week ending 18th September passed without incident at Gibraltar, though intelligence continued to emphasise the uneasy nature of the international situation.

On 15th September, the relief of the 3rd Destroyer Flotilla in Spanish waters by the 1st Destroyer Flotilla was postponed.

The German submarine *U.30* joined *U.27* at Cadiz on 11th October [September?], and the German cruiser *Emden* was reported leaving Funchal northward on the 15th.

The Spanish Nationalist warships maintained their patrols in the vicinity of Gibraltar during the week.

2. On Tuesday, 20th September, I proceeded to sea in H.M.S. *Hood* with the destroyers of 6th Division for exercises. The *Hood* carried out sub-calibre firing, turning trials, and torpedo firing at destroyer targets, and the destroyers carried out torpedo attacks on *Hood*. Night exercises which had been arranged were cancelled on representations from the Spanish Nationalist authorities that the movement of darkened ships might possibly give rise to 'incidents' owing to Spanish patrols being at sea in the same area.

In leaving harbour, the *Hood* touched the ledge abreast the detached mole. A separate report has been rendered on this occurrence ...

3. On Wednesday, 21st September, I learnt of the concentration of the main part of the Fleet at Alexandria, and on this day H.M.S. *Sealion* was ordered from Gibraltar to Malta and H.M.S. *Thames* ordered to return to Gibraltar from her African cruise.

4. On Friday, 23rd September, the situation showed considerable deterioration after the Prime Minister's second visit to Germany, and further precautionary measures proved necessary.

---

[1]RA [later Adm Sir] Geoffrey Layton (1884–1964). COS, China Station, 1931–33; commanded battle cruiser *Renown*, 1933–34; Cdre RN Barracks, Portsmouth, 1934–36; DPS, Admy, 1936–38; commanded Battle Cruiser Sqdn, 1938–39; commanded 1st BS, 1939–40; C-in-C China Station, 1940–41; C-in-C Eastern Fleet, 1941–42; C-in-C East Indies, 1942–45; C-in-C Portsmouth, 1945–47.

H.M.S *London*, which arrived at 1100 from Malta on her voyage home, was ordered to remain at Gibraltar.

The remaining ships of the Third Flotilla were ordered by the Admiralty to be withdrawn from the Eastern area and concentrated at Gibraltar.

In the course of the evening, instructions were received from Admiralty (Admiralty message 2033/23) to take every precaution against submarine attack. In pursuance of these, and in order to cover the period before boom defences could be placed as ordered in your message 2227, I ordered the following action to be taken:–

(i) An anti-submarine destroyer patrol to be instituted in Gibraltar Bay during daylight hours;

(ii) Boat patrols to be instituted off each entrance to the harbour by day and night;

(iii) *Hood* and *London* to man half H.A. armament and searchlights;

(iv) Oilers or B.P. targets to be placed so as to protect *Hood* and *London* from torpedo attack.

In taking these precautions I had also in mind the possibility of surprise aircraft attack, possibly from Spanish Morocco.

I also requested the military authorities, through Rear Admiral, Gibraltar, to man the guns and searchlights of the breakwater defences, which was done.

5. On Saturday, 24th September, *Inglefield*, *Imogen* and *Ivanhoe* arrived from the Eastern Area at 1500.

The destroyer anti-submarine patrol commenced at daylight.

In view of the general situation, all ships were ordered to prepare in all respects for war.

6. On Sunday, 25th September, H.M.S. *Imperial* arrived and the 3rd Flotilla was then complete. The Flotilla was ordered to fit Type 15 (Staybrite) Asdic domes with the aid of diving parties from *Hood* and *London*, and this work was completed by a.m. Thursday, 29th September, the 6th Division being complete before sailing on the 28th.

7. On Monday, 26th September, H.M.S. *Thames* arrived at 1500, being escorted into harbour by a destroyer of the 3rd Flotilla.

The 3rd Destroyer Flotilla proceeded to sea for high speed exercises during the day, and in consequence of a verbal report from the Captain (D) on their return to the effect that the speed and fuel consumption of all ships of the Flotilla was likely to be much affected by foul bottoms, I gave instructions for all ships to be docked for bottom cleaning only as quickly as possible. This work was completed by noon 29th September.

The boom defences were placed in position on 26th September, that for the southern entrance a.m. and that for the northern entrance p.m. In consequence of this, the boat patrols were withdrawn at 1325 and the destroyer A/S patrol at sunset.

H.M.S. *Vanoc* was ordered to the Eastern Mediterranean during the day, and left p.m. at 25 knots, taking with her some of the officers required for commissioning minesweepers.

As it was evident that the special arrangements made for the purposes of the Spanish Civil War were no longer applicable, after consultation with Rear Admiral, Gibraltar, and with his complete agreement, I decided to assume the normal duties of Senior Naval Officer of H.M. Ships at the port, informing you that I had done so (message 1941/26).

During the day, information was received from the Admiralty that in view of the presence of the German armoured ship *Deutschland* at Vigo, it would be necessary to employ heavy ships on trade protection and convoy in this area, and that the *Hood* would be employed on this service. Further telegrams emphasised the critical nature of the situation, and pointed out that the presence of the *Deutschland* and the German submarines at Cadiz would necessitate rapid action on the Warning Telegram being received or orders being given to watch them.

8. Tuesday, 27th September.

In consequence of the information received the previous evening, orders were prepared and issued for *Thames* to carry out a diving patrol off Cadiz when ordered, with a view to the destruction of *U.27* and *U.30* as soon as possible after the outbreak of hostilities.

On this day, ships were ordered to paint in accordance with T.T.M. 13.

In the evening, instructions were received from the Admiralty (message 1827/27) for *Hood* and one division of destroyers to be employed in escorting S.S. *Aquitania* to Gibraltar,[1] taking over the convoy from Home Fleet units in the vicinity of Finisterre.

Warning was also received by telegram from the Admiralty that naval mobilisation would be ordered the following day.

9. Wednesday, 28th September.

Naval mobilisation was ordered, and instructions issued that time-expired men should be detained. No action was, however, necessary as far as the Rear Admiral Commanding, Battle Cruiser Squadron, was concerned in consequence of these instructions, as the relief of time-expired men was already suspended.

---

[1] The large Cunard liner *Aquitania* (45,650 tons, 1914) had been chartered to carry Reservists in order to bring the Mediterranean Fleet up to full war-time strength.

Orders were prepared and issued for the operation of escorting the *Aquitania*. The latest intelligence at the time of their issue gave *Deutschland* as still being in the vicinity of Vigo, *U.30* at Cadiz, *U.27* as having left Cadiz that morning, and the German flying boat depot ship *Schwabenland* at Fayal in the Azores.

A rendezvous was arranged for 1700 G.M.T. Friday, 30th September, in a position 270 degrees 60 miles from Cape Finisterre. To reach this position I decided to proceed well outside the normal trade route along the coast, employing the 6th Division as the A/S screen, and keeping the ships in the 4th degree of readiness, with half the H.A. armament closed up in *Hood*, and depth charges at instant readiness in the destroyers. Orders were, however, given that depth charges were not to be dropped except by order of the senior officer. At night, ships were to be darkened, but navigation lights burnt.

At 1345, H.M.S. *Manchester* arrived from England, and was ordered by you to proceed Eastward with all despatch as soon as she had fuelled.

At 1700, I proceeded in *Hood*, the 6th Division carrying out a preliminary A/S sweep of Gibraltar Bay before *Hood* left, and then forming an A/S screen.

10. Thursday, 29th September.

At 1038, information was received that the convoy's sailing would be delayed, and the rendezvous was therefore altered to a position 219 degrees 90 miles from Finisterre at 0800 G.M.T. 1st October. In consequence of this information, speed was reduced to 11 knots and courses and speeds adjusted as requisite during the remainder of the day in order to keep the revised rendezvous. Manoeuvres with destroyers were also exercised.

The convoy, with H.M.S. *Royal Oak* and two destroyers as escort, sailed from Spithead at 1745. H.M.S. *Revenge* and two destroyers as covering force left Plymouth at 2100.

11. Friday, 30th September.

Agreement having been reached at the Munich conference at 0130, instructions were received from Admiralty at 1200 to withdraw the escort from the *Aquitania*, and I accordingly turned back and shaped course for Gibraltar with the destroyers in company.

12. Saturday, 1st October.

At 0800 I exercised manoeuvres with the destroyers of the 6th Division.

In the Straits of Gibraltar, the Spanish Nationalist cruisers *Canaris*, *Almirante Cervera* and *Navarra* were sighted proceeding westward. Four Nationalist destroyers were also reported arriving at Ceuta and left later in the day for the westward. The three Nationalist minelayers who had recently been employed on the patrol in Gibraltar Bay consequent on the presence of the *Jose Luis Diez* had also left Algeciras.[1] In accordance with my suggestion, Rear Admiral, Gibraltar, had asked the Nationalist authorities to withdraw the patrol now that the boom defences of Gibraltar were in place and egress for the *Jose Luis Diez* was practically impossible, and this request had been complied with. Two of these three minelayers subsequently returned to Ceuta, but their westward movement of practically the whole effective strength of General Franco's fleet is interesting, in view of the strong naval position he had achieved in the Mediterranean.

At 1100, *Hood* and 6th Division arrived at Gibraltar.

13. Sunday, 2nd October.

Aquitania arrived during the night, and drafts from her were temporarily accommodated in *Hood* and *London*, pending their return to England in S.S. *Montcalm*. At 1300 *Aquitania* sailed for Malta.

14. The emergency measures necessary during this period, though happily they were not required to stand the test of actual hostilities, have proved a valuable exercise for all the personnel. The work which had to be undertaken in ships at short notice, such as painting, docking and fitting A/S domes in destroyers, was well and expeditiously performed.

---

[1]The Spanish Government destroyer *Jose Luis Diez* had been intercepted by superior Nationalist naval forces while attempting to pass through the Strait of Gibraltar on the night of 26–27August. The ship was damaged in the ensuing engagement and entered Gibraltar harbour in a sinking condition with a number of dead and wounded. The British government as a neutral power were then presented with the dilemma of whether or not to intern the ship or to allow emergency repairs, their extent and possible dockyard assistance, before requiring the destroyer to put to sea. There were considerable diplomatic and political complications including the allegation that the Nationalists had threatened retaliation to families of workmen on the mainland who repaired the ship in Gibraltar. Moreover, sympathies of the workmen in the dockyard itself were divided into pro-Republican and pro-Franco supporters thereby raising the prospect of turmoil in the colony. A summary internment of the destroyer, however, had potential repercussions in Parliament where pro-Republican sentiments were strong among certain groups. The affair dragged on for months while diplomatic representatives of both sides argued their case. In the early hours of 30 December the destroyer, faced with an approaching deadline for internment, attempted to escape but was intercepted by Nationalist warships and after a short engagement grounded herself in Catalan Bay. The entire action took place in British territorial waters and shells from the Nationalist ships caused casualties on shore. The ship was subsequently refloated, towed back into harbour, her crew interned and in January repatriated to Government-held Almeria in the destroyers *Greyhound* and *Glowworm*. Protests were lodged by the British with the Nationalists over the violation of territorial waters and by the Spanish Republicans over the internment of the ship and repatriation of its crew. Lengthy correspondence on the affair is in ADM 116/3947.

294. *Backhouse to Pound*

[ADM 205/3]　　　　　　　　　　　　　　　　　　11 October 1938

[Carbon]

I have to thank you for several letters, which I am afraid I have not previously had time to answer. As you can imagine, we have had considerable pressure here, and during the week of the acute crisis it did not stop for one moment. Being plunged into it all after so short a time did not make it any easier from my point of view, but I think on the whole it all worked out as well as could be hoped.

There are some things of course that I am not at all happy about, and one of them is our present weakness in Destroyers. It is true that the position will have improved a great deal when all the Destroyers now building are completed, but that will not be for some time to come. In the meantime I realise most fully how terribly short of Destroyers you are in the Mediterranean.

I had actually made up my mind to send one of the Reserve Flotillas from home to Gibraltar so as to release your 'I's to rejoin you, but we never quite got to that as the crisis began to ease up.

Another thing that I had not realised so fully before was that there would be so many requirements for escorting transport, and with two German Submarines at, or near Cadiz, I was naturally anxious about the Gibraltar area.

It may interest you to know that we have had fairly authentic reports of three German Submarines being off the coast of South America, which makes one think a bit. This information will be useful to us, however, when we come to reconsider disposition.

I have heard various stories about the probability of Italian intervention on the side of Germany, but, if words are anything, there can be no doubt that Mussolini intended to support Hitler at once. Some say that the King and Italian people would not have stood for this, but that is not a presumption which I think it would be safe to accept.

The Mobilisation went off extremely well and was a wonderful experience for everybody. I wish it had taken place a few days earlier, as that would have enabled us to have completed it rather more thoroughly. We felt we could not go on holding our Reserves once the situation had completely eased and it had been stated that the crisis was past, especially as everyone began to wonder why the Admiralty did not follow the example of the rest of the Country. Also, I am sure that officers and men, called up from Reserve, had lost interest, and the great majority of them wished to get back to civil employment. The men are always suspicious that they may lose their jobs while they are away.

Anyway, the test has provided us with a lot of information and experience which we could not have had otherwise, and we can certainly improve our organisation a great deal in consequence.

In the easing up process I have tried to make it as easy as possible for you and the other Commanders-in-Chief. I hope you have not felt handicapped in this respect. I am daily expecting to hear that the whole time of precaution is over, and that we can withdraw the extra men sent to the Mediterranean. It will take a week or two, however, before we can provide for their transport.

* * *

I think a considerable crusade is going to be started at once to speed up our defence measures, and there has been some talk of it already, although most of the Ministers have now gone away for a short holiday. We are already looking into things that could be hurried on, and other steps that could be taken to increase preparedness.

I feel myself that 1939 will be a critical year, as it is most unlikely that Hitler will not have some great scheme he wants to put through. It seems to me that the most likely one is the demand to get back his Colonies. Judging from what went on in the recent crisis, he is not likely to be easily satisfied unless he gets 99% of his own way – not an easy man to negotiate with.

A great block in our preparations vis à vis Germany is the state of our Air Force by comparison with hers, and I should think it would take at least two years to get into a happier position. The German aircraft production is now prodigious – 700 to 800 machines a month with growing capacity.[1] Unless that power is over-rated, I think we should have a terrific experience here if war broke out. What is more, the A/A gun defences of this country are far behind what they should be. As you know, the A/A defences at Gibraltar, Malta and Aden are woeful. It remains to be seen whether they could be speeded up in the next six months – I shall do my best.

---

[1] Backhouse demonstrates the overestimation of German aircraft production common at the time of the Munich agreement. German monthly aircraft production was only 450–500 at the end of 1937 and, while it rose to 700 in 1939, did not exceed 800 until after the outbreak of the war and 1940. Air Ministry (A.C.A.S.(I)), *The Rise and Fall of the German Air Force (1933–1945)* (Air Ministry, 1948; reprinted, Kew: Public Record Office, 2001), p. 19. There is a large literature on the subject: see, for example, Wilhelm Deist, *The Wehrmacht and German Rearmament* (Toronto and Buffalo, 1981), pp. 66–9; and the tables in Williamson Murray, *Strategy for Defeat: The Luftwaffe, 1933–1945* (London, 2003), pp. 24–6; and Richard Overy, *War and Economy in the Third Reich*, paperback edn (Oxford, 1995), p. 204.

295. *Pound to Admiralty (Excerpts).*

[ADM 116/3900] Office of Commander-in-Chief, Mediterranean
Station
At Malta, 14 November 1938

[Carbon]
SECRET.
Med.01060/0708/8.
SITUATION IN THE MEDITERRANEAN – 1ST OCTOBER, 1938.
Be pleased to lay before the Board the enclosed report on the situation in the Mediterranean had war broken out on 1st October, 1938.

This date has been taken as that on which German forces would have entered Czecho-Slovakia had not the final conference at Munich been called.

Naval Mobilisation had been ordered on 28th September, 1938.

2. As we must be ready to deal with the worst situation which may arise, it is assumed that Italy will have entered the war with, and at the same time as, Germany. All the other conditions in this review of the situation are those which were actually prevailing on the date in question.

3. When considering war in the Mediterranean, it is essential to keep two facts clearly in mind:–

(a) That with present day communications and long range aircraft the Mediterranean has become a very small place;

(b) That Italy will not fail to make full use of her advantageous geographical position and of her strength in long range aircraft.

4. It is, therefore, evident that from the moment war is declared, or the enemy takes hostile action against us, there can be no question of allowing either weak naval forces, transports or fleet auxiliaries to make passages in the Mediterranean without escorts of sufficient strength to deal with any forces, either underwater, surface or air, which the enemy may bring against them.

In effect, the movement of a single auxiliary from, say, Malta to Alexandria, will become a major operation.

5. In this respect, the Mediterranean differs entirely from other stations, where our preparations for war can generally be completed with comparative safety after hostilities have commenced.

Moreover, under present day conditions it seems improbable that the normal procedure of warning telegram, mobilisation, and the sending of war complements and defence personnel, will have been completed before war breaks out.

Thus the Central and Eastern Mediterranean, though seemingly one of the nearest of the foreign stations, becomes when using the Cape route, the most distant of all, except China, where it has long been considered necessary to maintain our personnel at war strength in peace time.

6. In the enclosed report the situation on the Mediterranean Station is dealt with ...

* * *

7. As a result of this analysis I have come to the conclusion that a more definite strategical policy in the Mediterranean is required and this forms the subject of a separate letter which is being forwarded for the consideration of Their Lordships under Med.01061/0708/9 of 14th November, 1938.

PART I – STATE OF READINESS FOR WAR ON 1ST OCTOBER, 1938.

* * *

ALEXANDRIA – CONCENTRATION OF NAVAL FORCES.

I. The Mediterranean naval forces allocated to the Eastern Mediterranean Fleet were at Alexandria on 1st October, with the following exceptions:–
Arethusa and 4 destroyers – in Red Sea.
Penelope   – at Port Said (Suez Canal Defence Plan).
Douglas   – at Haifa.
3 Malta reserve T.S.M.S.[x] – at Malta – ready to sail 2nd October.
Wolsey[x]   – at Malta – ready to sail 5th October.
Manchester   – at Sea – due Alexandria 2nd October.

Maidstone, allocated to Aden in the war dispositions, was at Alexandria, and her retention there (in view of the absence of *Resource* for fleet repair duties and of there being no depot ship for the Flying Boat Squadron) had been concurred in by the Commander-in-Chief, East Indies.

[x] These ships were ready to sail at this early date because the necessary personnel had been sent overland to Marseilles and transported thence to Malta by *Shropshire*, arriving on 29th September.

ALEXANDRIA – NAVAL PERSONNEL FOR THE FLEET.

2. *Aquitania* with war complements for the Eastern Mediterranean Fleet, crews of Alexandria minesweepers and relief personnel for the Suez Canal Defence Plan, had left England on 29th September en route to Gibraltar and was due to arrive there on 2nd October.

Had war broken out on 1st October, it would have been necessary for this vessel to proceed via the Cape and Red Sea, and there is no certainty that the personnel she carried would have reached their destination.

RECOMMENDATION.
(a) That while there continues to be any possibility of hostilities against Italy, the Mediterranean Fleet should be manned on a scale that would enable the fleet, with the addition of the ratings under training in *Resource* (vide Admiralty telegram 1545/2nd October), to be at war strength without relying on transports from home which may have to be routed via the Cape in order to reach the fleet.

(b) That, in order to avoid supply by the fleet at a time when they can least be spared, personnel for the following purposes should be sent out from England so as to reach their destination before hostilities commence:–
(i) Suez Canal.
(ii) Minesweepers in reserve at Alexandria.
(iii) Local Naval defences and Fortress Headquarters at Alexandria.

ALEXANDRIA – LOCAL NAVAL DEFENCES.

BAFFLES.
3. The outer A/T baffle has been laid and preparation of the inner A/T baffle was in hand.

The Boom Defence Officer was at Alexandria, but no other provision of special personnel had been made and the work had been carried out entirely by officers and men of the Fleet, assisted by shore labour.

RECOMMENDATIONS.
That the A/T baffles should be at 7 days' notice and the A/B Boom and gate at 14 days' notice.

* * *

For Personnel – see under ALEXANDRIA – NAVAL PERSONNEL FOR THE FLEET.

INDICATOR NETS.
4. The adjustment of the indicator nets on shore for the depths required was well advanced and the first line of nets, i.e. 1½ miles normally stowed in H.M.S. *Protector*, could have been laid at 7 hours' notice.

Even though these nets are now adjusted for the depths required, it would take *Protector* 14 days to lay the whole of the indicator net defence scheme.

* * *

MINESWEEPING.
7. Two of the four Alexandria reserve T.S. Minesweepers were in commission and ready for sweeping. They were manned by the personnel allocated to all the sweepers in reserve with a balance of Fleet personnel and of Egyptian ratings who had received some training.

The remaining two sweepers were being brought forward, but could not have been commissioned until the arrival of the personnel in *Aquitania*.

The tugs *Brigand* and *Roysterer* had been prepared for bottom sweeping in the Great Pass.

RECOMMENDATIONS.

See under ALEXANDRIA – NAVAL PERSONNEL FOR THE FLEET.

* * *

ALEXANDRIA – FORTRESS DEFENCES.

* * *

A.A. Defences.

11. Available A.A. guns and lights were in position and manned.

Air attack warning system by observation posts was not organised and no vessels, other than those of the fleet, were available to extend the observation line to seaward.

The use of destroyers for this purpose is most undesirable, particularly in view of the small numbers of light craft in the fleet. The intention (now included in the Alexandria Defence Scheme) was to take up on the outbreak of war a number of Italian motor trawlers which normally work from the port, equip them with commercial W/T sets obtainable locally, and man them with Egyptian coastguard crews. The disadvantages of the scheme are that, at the best, it cannot be initiated until after war had broken out.

'Black-out' arrangements in the town were not organised.

* * *

RECOMMENDATIONS.

13. In many respects Alexandria was better equipped for defence on this occasion than at the start of the Abyssinian emergency in 1935.

It was only at a meeting which I held immediately on my arrival that the arrangements were made at my request to put the A.A. guns and lights in position, man the seafront defences and establish the fighter squadron at Amriya. These steps were all taken on the initiative of the General Officer Commanding, British Troops in Egypt, and of the Air Officer Commanding, Middle East, and not as a result of any instructions from War Office or Air Ministry.

It is recommended, in the first place, that whenever the fleet receives orders to concentrate at Alexandria in an emergency, the War Office and Air Ministry should simultaneously give orders for the necessary steps to be taken for the protection of the fleet in that harbour.

* * *

That the uncertainty and delay in taking up enemy fishing vessels for air attack warning duties could be overcome by the provision in peace of at least 8 seagoing launches (4 on patrol continuously) equipped with W/T and manned by Egyptian coastguard or naval ratings.

It is suggested that such vessels would form a more useful part of Egyptian Naval Forces to be provided than Motor Torpedo Boats.

ALEXANDRIA – AIRCRAFT.
Fleet Air Arm Aircraft.
14. All fleet aircraft at Alexandria were serviceable on 1st October though, owing to shortage of trained observers, two of the battleships could each operate only one catapult aircraft at a time.

Six reserve aircraft were at Alexandria and a further sixteen were to be despatched shortly from Malta.

There were no Mark XI torpedoes available in the fleet, other than those embarked in *Glorious*, and there was also a shortage of flame floats, reconnaissance flares, etc.

The stocks brought from Malta in *Maidstone* prevented a most serious shortage of Fleet Air Arm bombs and pyrotechnics.

* * *

Royal Air Force Aircraft.
15. No. 202 Squadron (including reserve aircraft brought in *Maidstone*) was at Alexandria, moored at the Imperial Airways moorings with personnel accommodated in *Maidstone*.

RECOMMENDATION.
The provision of a depot ship for flying boats has been recommended in Med.0336/032 of 15th April, 1938, and this was an occasion on which such a ship would have been invaluable. *Maidstone* normally would have been sent to her war station at Aden and would not have been available either as a carrier or a depot ship for the flying boats.

No. 80(F) Squadron of 21 Gladiators was based at Amriyeh and ready for service in connection with the defence of the Fleet at Alexandria.

* * *

ALEXANDRIA – FLEET FUEL SUPPLIES.
17. Briefly, the oil fuel situation in the Eastern Mediterranean on 1st October was as follows:–
at Port Said (on shore)   15,600 tons.
At Alexandria (in tankers)   20,000 tons.
    Total 35,600 tons.

The fleet was complete to full stowage.

One freighting oiler was in the Red Sea, due Port Said 3rd October, and two in the Indian Ocean, due Port Said 8th October.

Shore tanks at Alexandria were available at short notice, but were not taken up.

It is estimated that the Eastern Mediterranean Fleet when at sea for war operations, would expend 3,400 tons per day approximately.

The operation of the Fleet would, therefore, have been seriously restricted unless a steady supply through the Red Sea and from Port Said to Alexandria could safely have been maintained.

RECOMMENDATIONS.

20. (i) Oil Fuel.

Since the tanks at Alexandria cannot be made available in peace time, it is recommended that, until the new Admiralty tanks approved for Port Said are completed, the five tanks at Haifa harbour should be leased and stocked in peace with fuel for the use of the fleet. They are vulnerable to attack from the sea and a submarine could effect damage by gunfire. It is considered, however, that in the circumstances this risk must be accepted.

Arrangements must be made to camouflage all oil tanks which, with their present aluminium colour are most visible from the air and from seaward.

* * *

DOCKING SITUATION.

25. By 1st October all destroyers more than 3 months out of dock had been given 'a quick scrape and paint'. In addition *Shropshire* was given a quick docking. The majority of ships on 1st October were between 3 and 4 months out of dock.

RECOMMENDATION.

That full use be made, as was done on this occasion, of the docking facilities available at Alexandria for the smaller ships of the fleet.

As regards docking the larger ships, all that can be done is to get the Egyptians to complete the new graving dock at Alexandria as soon as possible.

It is further recommended that a Constructive and Engineer Officer to serve on the Staff of the Rear Admiral, Alexandria, should be sent as soon as possible to co-ordinate repair work ashore, and deal with docking arrangements.

* * *

MALTA.

NAVAL FORCES.
34. The forces allocated to Malta in the war dispositions were present on 1st October with the exception of *Clyde* at Alexandria and *Thames* at Gibraltar.

Shropshire was also at Malta and ready for service.

Wolsey and the 6 Malta reserve minesweepers had just commissioned with crews sent out overland via Marseilles and thence in *Shropshire*.

Bideford and *Cherryleaf* were refitting.

NAVAL PERSONNEL AND LOCAL NAVAL FORCES.
35. Officers and ratings for *Wolsey* and the reserve minesweepers had arrived via MARSEILLES in *Shropshire* on 29th September.

Officers from the Fleet, replaced by R.N.R. Officers then under training, had been sent to Malta from Alexandria and Gibraltar to commission the reserve minesweepers, the Commanding Officer of *Bideford* being appointed to command the M/S Flotilla.

Requirements in personnel for boom defence and other local services were partially met by reducing *Bideford* to dockyard control.

LOCAL NAVAL DEFENCES.
36. The boom defences for the Grand Harbour had been laid.

MILITARY DEFENCE PERSONNEL.
37. No reinforcement of the garrison had taken place. Admiralty letter M.00657/38 shows these reinforcements as leaving England in a 15 knot convoy on the 9th day after mobilisation.

Recommendation.

That a high priority be given to Malta in the defended ports abroad to be manned in peace on the higher colonial establishment.

Furthermore, it is considered that the policy as regards Malta should be that it is defended not only in such a manner that it cannot be captured but also that it should be most strongly defended against air attack, with both the latest guns and with adequate air forces so that the dockyard may be in existence (though possibly damaged) and available as a Fleet base in the later stages of a war.

If air attacks on Malta are on a large scale, arrangements must be made to move the bombing squadrons from Malta to Tunisia at short notice so that operations against the enemy may be continued from that area. Fighter squadrons must be retained to assist in the defence of the Island.

FUEL.
38. One 'Leaf' class (refitting) and two Fleet attendant oilers were at Malta, also the petrol carrier *Petrella*. The latter had been on the coast of

Greece with the M.T.B. Flotilla when the concentration of the Fleet took place and it had been decided to send her to Malta rather than delay Fleet movements or leave her unescorted on her passage to Alexandria.
Naval stocks at Malta were:–

Oil Fuel	46,000 tons.
Diesel Oil	14,000 tons.
D.T.D. 230	27,000 Gallons.
D.T.D. 224	57,000 Gallons.

* * *

GIBRALTAR.

NAVAL FORCES.

41. H.M. Ships *Hood*, *London* and the Third Destroyer Flotilla were based on Gibraltar on 1st October.

Thames and *Plover* were also at Gibraltar on this date.

NAVAL OPERATIONS.

42. *Hood* and 6th Division were at sea for the purpose of escorting *Aquitania* to Gibraltar.

Patrol of the Strait and a skeleton form of Contraband control Service were ready to be put into operation provided a division of destroyers remained available.

LOCAL NAVAL DEFENCES.

43. The booms had been in place in both entrances since 26th September.

LOCAL NAVAL FORCES.

44. The two minesweepers from the Fishery Protection and Minesweeping Flotilla had not yet arrived.

MILITARY AND AIR FORCES.

45. The garrison was still at its peace establishment and no aircraft were based at Gibraltar.

Recommendation.

It is most urgent that Gibraltar should be strongly defended against air attack and should have air forces available. A separate report has been forwarded. (Med.01058/0729/2G of 12th November, 1938).

* * *

GENERAL ARRANGEMENTS.

Allied Naval Plans and Dispositions.

49. Apart from the general division of the Mediterranean Station into areas of French and British naval responsibility I was not aware of any

intended French naval dispositions or plans for hostilities against Germany and Italy.

Such knowledge if available would have been of special value to the Eastern Mediterranean, Gibraltar and Malta commands in their early war operations and it is requested that I may be kept in close touch with the progress of any naval inter-staff conversations which have been or will in the future be held.

Reinforcement of Egypt.

50. I am entirely in agreement with the Commander-in-Chief, East Indies, in his No. 923/E.I. 5186 of 11th October, 1938 in which he recommends investigation and all possible improvement of any practicable methods of transporting military personnel from INDIA to EGYPT via BASRA and overland routes through IRAQ, TRANSJORDAN and PALESTINE.

I foresee numerous calls on the Eastern Mediterranean Fleet in the early stages of war for extra troop convoy escorts through the RED SEA, and any steps which can be taken to reduce such commitments will help to maintain the strength, particularly in badly needed light forces, of the Eastern Mediterranean Fleet.

* * *

PART II – ADEQUACY OF NAVAL FORCES.
MAIN FLEET.

1. The adequacy of a force in any area depends on:–
    (a) What the force is required to do.
    (b) The strength of the enemy forces which may be encountered in the area.

2. As regards 1(a), the tasks of the Eastern Mediterranean Fleet may be summarised as follows:–
    (a) The interception of Italian trade between the Black Sea and Italy both via Cape Matapan and the Corinth Canal.
    (b) The control of the Suez Canal.
    (c) The control of Italian lines of communication between Italy and Libya.
    (d) Assisting the Army and Royal Air Force in the defence of Egypt.
    (e) In case of invasion, assisting the Army and Royal Air Force to defend Malta.
    (f) Reinforcements of Naval forces in the Red Sea to give more adequate protection to Red Sea convoys.
    (g) The protection of supply ships proceeding to Malta.
    (h) The protection of supply ships and auxiliaries between Port Said and Alexandria.

3. <u>Interception of Italian trade between the Black Sea and Italy.</u>

(a) It seems probable that the Italian forces available for the protection of this trade will consist of a small number of submarines and light forces operating from Leros.

(b) To attack this trade, we shall, at the outset, be able to do no more than make occasional sweeps from Cape Matapan to the northeastward into the Southern Aegean. Such sweeps would probably be made by a division of destroyers supported by a cruiser detached from the main fleet, the operation being covered by the main fleet on passage to and from the Central Mediterranean. The interception of trade through the Corinth Canal would be effectively achieved if the blocking of the Corinth Canal could be accepted.

4. <u>The control of the Suez Canal.</u>

(a) Enemy action against the canal and the terminal ports, apart from the main operation of advancing into Egypt from Libya may be expected to consist of blocking the canal, the landing of armed forces at the terminal ports before the outbreak of war and submarine offensive patrols off the terminal ports.

(b) The forces we should require to counter these possible courses of action must at present be detached from the Main Fleet. Their duties however are confined to local defence and will be considered in Section II of this part.

5. <u>Control of Italian lines of communication between Italy and Libya.</u>

1. <u>Italian forces to be taken into account.</u>

(a) (i) Judging by the Italian dispositions during the late emergency, so far as they are known, the following Italian naval forces would have been available for operating in the Central and Eastern Mediterranean:–

2 Battleships (modernised)
4 eight-inch cruisers
8 six-inch cruisers
30 leaders and destroyers
25 torpedo-boats
57 submarines
An unknown number of M.T.B.'s

(ii) It must be assumed that the French Navy, operating in the Western Mediterranean would generally have contained the balance of the Italian Fleet. Though there can be no certainty that the Italians would not concentrate the whole of their fleet for operations in the Central or Eastern Mediterranean at their own selected moment, our requirements cannot be based on the latter possibility and we need only consider the force enumerated above.

2. British forces required.
(b) (i) Large ships
On the outbreak of war and until reinforcements can arrive our requirements in heavy ships, carriers and cruisers to deal with the Italian forces given in paragraph 5(a)(i) above would be:–
3 Battleships
1 Carrier
4 eight-inch cruisers
4 six-inch cruisers

It is considered that the extra battleship and aircraft carrier would balance the four six-inch cruisers we would be short, and also would help to balance to some slight extent the weight of attack that may be delivered by Italian shore based aircraft.

(ii) Destroyers
For screening the forces given in paragraph 5(b)(i) against submarine and air attack and to balance the Italian light forces given in paragraph 5(a)(i) we shall require 36 'Tribals' or destroyers, i.e.–
18 for advanced Asdic Screen, Surface warning zone against air attack or minesweeping ahead of the Fleet.
8 for intermediate A/S screen for the battleships.
5 for A/S screen for the four eight-inch cruisers.
5 for A/S screen for the carrier and her cruiser escort.

This number does not allow for any destroyers refitting, boiler cleaning or detached for minor operations.

(iii) Submarines
In view of the large number of Italian submarines in the Mediterranean we must expect the fleet to sustain damage through their activities; moreover any damage we may sustain will be difficult to make good owing to the lack of repair facilities in the Eastern Mediterranean. It will be necessary therefore at least to ensure that we are in a position to inflict similar if not greater damage to the Italian Fleet. To achieve this, it would be necessary to maintain continuously at least four offensive submarine patrols, one off Tripoli and 3 off Southern Italian ports. In order to allow for refit periods, time taken to and from patrol etc., the total number of submarines required to operate these 4 patrols would be 14. In addition it is considered that four minelaying submarines will be required to carry out the mining policy (see section 3). All the above submarines should be based at Malta. It would also be necessary to base three 'River' class submarines at Alexandria for operating strategically with the fleet. These submarines might also be frequently made available for A/S exercises with destroyers. Hence, the total number of submarines required in the Mediterranean is as follows:–

3 'River' Class
4 'Porpoise' Class
14 small submarines.

(iv) <u>Shore based reconnaissance aircraft</u>

It is probable that many occasions will arise when it will be necessary to conduct minor operations with light forces both in the Central and Eastern Mediterranean. We may be tolerably certain that the departure of any of our forces from Alexandria will become known to the Italians before long, either through agents or through air reconnaissance. Owing to the distance of Alexandria from the areas of many such minor operations it would be a simple matter for the Italians to intercept and destroy our light forces engaged in them unless proper precautions were taken to guard against such action.

Our most economical method of giving security to our forces detached from the main fleet, and to no less extent, the security of our main fleet, against surprise attack by surface forces, is to provide air reconnaissance of large areas in the Central and Eastern Mediterranean. To provide this reconnaissance the following numbers are considered to be a minimum requirement:–

At Malta – 8 flying boats or the equivalent number of G/R land planes.

At Alexandria – 8 flying boats.

It would be quite impracticable, however, to base the above aircraft at Malta unless air superiority over the Italians could be obtained. For this reason as well as for the defence of Malta, the existing situation, in which no bomber or fighter aircraft are provided at this base, is considered most unsatisfactory. In the meantime it is suggested therefore that 16 flying boats should be based at Malta in peacetime, the whole being transferred to Alexandria in an emergency and until the scale of Italian air attack on Malta is under our control to an acceptable degree.

(v) <u>Depot Ships</u>

For the maintenance, repair and accommodation of certain of our forces operating with the main fleet, facilities in the form of depot ships or shore establishments will be required.

For the flying boats that it has been proposed should be based at Alexandria, a depot ship will be necessary and it is strongly urged that the ship recommended in Med.0336/032 of 15th April 1938 should be laid down at an early date and in the meantime a ship should be taken up immediately for this purpose.

H.M.S. *Maidstone* will be required at Alexandria as a depot ship for the three 'River' class submarines and it is proposed that other arrangements should be made for a depot for submarines at Aden.

For the submarines to be based at Malta a depot ship is unsuitable owing to vulnerability to air attack. Moreover it seems unlikely that a depot ship will be available for Malta. It is recommended therefore that arrangements be made to establish a shore submarine depot at Malta; a scheme has been worked out for such a base and forms the subject of a separate report which will be forwarded for the consideration of Their Lordships' approval in the near future; the scheme envisaged would be inexpensive and it is hoped that early approval may be given to put it into effect. In Med.076/01221 dated 24 March 1937 proposals were put forward for the provision of bomb shelters for submarines. This proposal was put in abeyance and in the scheme now proposed this question has not been re-raised. It is considered that quite apart from cost, the time taken to provide bomb shelters is unacceptable, and it is considered that on report of enemy aircraft approaching submarines must obtain the necessary protection against bombing by diving in certain positions inside the harbour.

For the maintenance of the Fleet Air Arm in the Eastern Mediterranean, a large quantity of stores must be transported from Malta to Alexandria before the outbreak of war. At present as Their Lordships are aware, 50% of F.A.A. reserves are stored at Malta and 50% in Egypt; the total reserves are being raised to 150% and eventually 75% will be stored at Malta and 75% in Egypt. With the main part of the Fleet proceeding to the Eastern Mediterranean, it will be necessary to transfer the major portion of F.A.A. reserves from Malta to Alexandria. On the first indication of an emergency arising therefore, arrangements must be made for the necessary transport.

6. Assisting the Army and Royal Air Force in the defence of Egypt.

This task must be undertaken by the operations of our main fleet in controlling the lines of communication between Italy and Libya. Minor operations on the Libyan coast will also be necessary. These duties must all be carried out by the force enumerated in paragraph 5(b) and the effect is unlikely to be sustained.

7. Assisting in the defence of Malta.

(a) The operations of our main fleet in the Central Mediterranean will act as a deterrent provided the local defences of Malta are efficient.

(b) Local defences at Malta. This will be dealt with in Section II.

8. Reinforcement of Naval Forces in the Red Sea to protect Red Sea convoys.

With the present reduced number of cruisers on the East Indies station and with the small number of A/S vessels immediately available for convoy duties in the Red Sea it seems probable that on or before the outbreak of war it will be necessary to detach one six-inch cruiser and four destroyers for convoy duties in the Red Sea. These forces will not be available for operations with the Eastern Mediterranean Fleet and are therefore required in addition to the forces enumerated in paragraph 5 (b) (i) and (ii).

9. <u>Protection of Supply ships proceeding to Malta</u>.

It has been estimated that approximately 5 ships per month will be required to keep Malta in essential commodities. It might be possible for supply to be effected from Susa on the African coast; more probably however it must reach Malta via Port Said. If there is to be any hope of any ships getting through from Port Said, they must be either convoyed or at least covered by the Eastern Mediterranean Fleet. Fast freight ships running between Susa and Malta would appear to give a better chance of success in the early stages of a war; an escort provided by light forces based on Malta, and air reconnaissance from Malta, should provide reasonable security.

10. <u>The protection of supply ships and auxiliaries between Port Said and Alexandria</u>.

(a) Merchant ships on this route should be immune from surface ship attack, being adequately covered by the main fleet. Attack by submarines however is highly probable.

(b) In years to come the Egyptian Navy may be in a position to provide A/S escorts; in the meantime it is considered that 2 patrol vessels should be provided for the protection of shipping on this route.

11. <u>Total requirements – main fleet</u>.

To fulfill all its tasks given above the composition of the main fleet in the Central and Eastern Mediterranean on the outbreak of war would be as follows:–

3 battleships
1 aircraft carrier
4 eight-inch cruisers
5 six-inch cruisers
40 'Tribals' or destroyers
21 submarines (3 'Rivers', 4 minelayers and 14 small)
16 flying boats
1 flying boat depot ship
1 submarine depot ship

12. The composition of the main fleet available for operating in the Eastern and Central Mediterranean and the Red Sea on 1st October 1938 was as follows:–

3 battleships
1 aircraft carrier
3 eight-inch cruisers
4 six-inch cruisers

23 'Tribals' and destroyers (including *Wishart*)
5 submarines
4 flying boats
1 submarine depot ship

Section II – LOCAL DEFENCE.

1. ALEXANDRIA.
(a) The defence of Alexandria is mainly the concern of the Army and the Royal Air Force.

* * *

(b) The Royal Air Force must be in a position to neutralise the Italian Air Forces in Libya. In addition, it is considered that two fighter squadrons should be available for the protection of the fleet base at Alexandria.

(c) In addition to the various naval services mentioned in Part I, it is considered that at least two local defence destroyers are required. Their duties would be to:–
    (i) Prevent the blocking of the Great Pass.
    (ii) Carry out A/S duties in the approaches to the port.
    (iii) Protect minesweepers against attack by detachments of light forces.
    (iv) Prevent minelaying operations off the port.

2. SUEZ CANAL.
(a) At present it is necessary to detach a six-inch cruiser on or before the outbreak of hostilities to assist in the protection of the canal. It is considered that this ship must be made available at the expense of the main fleet. Any measures that can be taken in peace to reduce the time taken to mount the coast defence guns at the terminal ports will help to reduce the time that it will be essential to immobilise a cruiser and employ her as guard-ship in the canal; it is emphasised that it might be necessary to remove the guard-ship from the canal at an early stage in a war, should her presence with the fleet be essential for the conduct of operations against the enemy.

(b) Other duties that will be required at the terminal ports consist of:–
    (i) Enforcing the examination service until the examination batteries are ready.
    (ii) A/S operations in the approaches.
    (iii) The control of shipping outside the three mile limit in connection with the contraband Control Service.

To fulfill these duties it will be necessary to employ two vessels at each terminal port. It is suggested that *Aberdeen* would be a suitable ship for Suez; in the event of war with Italy it is unlikely that she will be able to be fitted with A/S and minesweeping gear at Malta, and it is considered that she should proceed straight to Suez on an emergency arising. For the remaining three vessels required, patrol vessels are considered to be a most suitable type.

3. HAIFA.

As already mentioned in Part I, the presence of one of H.M. Ships was considered essential in order to commence the laying of the boom defence and act as guard-ship.

It is considered that at least one destroyer will always be required at Haifa at the outbreak of hostilities. No naval authority exists at Haifa in peace time, and though the arrangements for the establishment of the Examination Service and Contraband Control Base appear to have been satisfactory, the presence of a naval officer at the outset will ensure that these services commence without delay.

The presence of one of H.M. Ships is likely to have a steadying effect on the local population.

A ship capable of dealing with a hostile submarine in the approaches is most necessary.

4. MALTA.

(a) Remarks on the defence of Malta from the naval and military aspect have already been made in Part I. The importance of powerful anti-aircraft defences for the island cannot be overestimated. Equally important is the provision of large numbers of bombers and fighter aircraft. These aircraft, it is suggested, should be based at Malta during peace time, a large proportion of the bombers being transferred to Tunisia on or before the outbreak of war; from this position they may be well placed for operating against Sicily and Sardinia and a fair degree of dispersion will be effected.

The first requirement at Malta is in fact a sufficient air force and sufficient anti-aircraft defences to act as a powerful deterrent to enemy air action. Until provision for this is made, Malta will, in war, suffer both from the point of view of material and morale to such extent as to become a liability rather than an asset, and little hope can ever be entertained for using it as a naval base in the later stages of a war when it may be urgently required.

(b) Apart from the air aspect, Malta must be prepared against the following forms of attack:–

Invasion,
Bombardment,
Submarine attack in the vicinity,
Mining operations.

(c) To defeat these various forms of enemy action it is considered the following minimum forces should be based on Malta:–
1 Destroyer Flotilla,
Submarines (see paragraph 5(b)(iii).),
24 Motor Torpedo Boats.
 (i) The destroyer flotilla is required to:–
  Act as a counter attacking force against enemy light forces operating in the vicinity of the island against our minesweepers, etc.
  Act as a local escort for shipping to and from Tunisia, etc.
  A/S operations.
  Attack of enemy convoys on passage to and from Tripoli.
 (ii) Submarines are required for local defence against landing operations and enemy heavy forces operating in the vicinity of the island. Submarines for this purpose can be required from those mentioned in paragraph 5 (b) (iii) above. There will always be a proportion resting in harbour and these must be relied on to carry out any defensive patrols required.
 (iii) Motor Torpedo Boats are required for night operations against enemy surface forces in the vicinity of the island; by day, they will be required for anti submarine work in conjunction with aircraft and destroyers.

5. TOTAL FORCES required for local defence in the Central and Eastern Mediterranean on the outbreak of war over and above those of the main fleet are as follows:–
12 Destroyers (9 Malta, 1 Haifa, 2 Alexandria)
1 Escort Vessel (Suez)
5 Patrol Vessels (1 Suez, 2 Port Said, 2 for escort duties between Alexandria and Port Said)
24 Motor Torpedo Boats.

Section III – MINING.

It is considered that the Italians will endeavour to cut the Mediterranean into two portions by mining in the area between Cape Bon and Marsala; in any case, they will notify it as a dangerous area in an attempt to force all shipping through the Straits of Messina.

On the principle that if an area is to be mined we should mine it first, it is proposed that the French should be asked to declare this area as dangerous except for a passage close in to Cape Bon and that they should be responsible for keeping this passage clear and for laying as many minefields as possible in the remainder of the area.

2. Offensive Mining. It is considered that an offensive mining policy can be best developed by carrying out attrition mining from the outbreak

of hostilities, in the vicinity of Brindisi, Otranto, Taranto, the east coast of Sicily, and the Libyan ports as far east as Benghazi. Malta is well situated to act as a base for these operations.

3. To enable this policy to be carried out, a flotilla of four minelaying submarines and a suitable mining depot from which to operate will be required. It is proposed that the mining depot should be situated at Malta and should have a capacity of five complete outfits for each minelaying submarine, i.e. 1250 Mark XVI mines. A protected mining dock will be necessary to enable the submarines to embark these mines without danger from air raids.

4. <u>Defensive mining</u>. In the Eastern Mediterranean it is proposed to declare the whole area from Alexandria to Port Said a dangerous area within the 200 fathom line to the extent that it is dangerous for vessels to fish or to anchor within this area. To implement this policy it is proposed to lay small deep minefields off Alexandria and Port Said, and in the vicinity of landfalls between these two ports. This should deter submarines from operating in this area. A defensive minefield will also be required off Marsa Matruh as a protection against raiders.

5. For these purposes, a fast surface minelayer with a mine depot at Haifa will be required at the commencement of hostilities. This vessel will also be available for laying an offensive minefield off Tobruk if this is considered necessary.

6. The capacity of the mining depot at Haifa should be capable of supplying the surface minelayer from an early period in the war. As the war develops it may become possible to convert a destroyer flotilla to minelaying duties and intensify the mining policy. It is proposed that Haifa should be used as the base from which the minelaying destroyers should operate. It is therefore of importance that the mining depot at Haifa should be capable of rapid expansion as the mining effort develops.

7. Haifa is therefore required to be developed as a minelaying depot on a skeleton basis in peace time, and to be rapidly expanded on the outbreak of hostilities into a base for <u>defensive</u> mining in the Eastern Mediterranean in the early days of the war, gradually evolving into a base for extensive <u>offensive</u> mining as destroyers become available as minelayers.

* * *

296. *Pound to Admiralty*

[ADM 116/3900] Office of Commander-in-Chief, Mediterranean
Station
At Malta, 14 November 1938

[Carbon]

MOST SECRET.

Med.01061/0708/9.

The Strategical aspect of the Situation in the Mediterranean on 1st October 1938

Be pleased to inform Their Lordships that, as a result of an exhaustive analysis of the position in the Mediterranean on 1st October 1938 (vide Med.01060/0708/8 of 14th November 1938) when it was necessary to take into consideration the requirements of a war against both Germany and Italy, I have come to the conclusion that a considerable revision of our strategical policy in the Mediterranean has become essential. The following remarks and proposals are therefore forwarded in case they may be of some assistance to Their Lordships if this matter is to be considered.

2.– From a study of Part II of Med.01060/0708/8 of 14th November 1938 in which the adequacy of the forces at my disposal are analysed (a summarised extract is attached)[1] it will be seen that even were it possible to provide the aircraft, light surface forces and submarines which are considered necessary, we shall, at any rate in the opening stages of a war, be operating in the face of much superior forces of this nature.

3.– It appears probable therefore that so long as the present situation persists and we are forced to operate from Alexandria it will not only be difficult to cut enemy communications between:–
  (a) Italy and Black Sea
  (b) Italy and Libya
but we shall also be subjected to a steady attrition of our forces in our efforts to cut these communications and carry out other operations.

Any such attrition will be difficult to make good owing to our lack of repair facilities in the Eastern Mediterranean. It is of great importance therefore that this unsatisfactory situation should not be allowed to persist longer than is absolutely necessary.

4.– I am unaware of the grand strategy of the Allies (France and Great Britain) but it is considered that at the earliest possible stage it should include an advance into Libya
  (a) from the East by British, Egyptian and possibly Dominion Forces.

---
[1] Not reproduced. See preceding document.

(b) from the West by French forces assisted by British and Dominion Forces.

5.– The advantages of such an offensive would be:–

(a) Offensive action by the Allies from the outset seems essential and an early success in the Mediterranean area would have far reaching effects in dissuading minor powers from throwing in their lot with the enemy. This would particularly apply to Roumania, Hungary, Bulgaria, Turkey and Greece.

(b) With Libya in our hands:–

(1) The sea route from Egypt to Malta would be much safer as regards air attack.

(2) Large air and land forces would be released.

(3) The air forces so released might be operated from Tunisia to obtain air superiority in the Central Mediterranean and thus make Malta sufficiently safe for use as a Fleet base from which to interrupt trade between Italian ports and between Italy and the Black Sea.

(4) The reinforcement of Italian East Africa by air would no longer be possible and if some of the troops released from Libya were used it should soon fall into our hands as it would be cut off from all supplies and reinforcements. The menace to our communications in the Red Sea would then be removed.

(5) The problem of the supply of Malta which requires on an average 5 ships per month would be greatly simplified.

6.– It is considered that effective use of the forces in the Mediterranean can only be made if some general plan is formulated on these lines and in which all three services can co-operate.

7.– With limited Naval Forces and material resources it is not possible either to attack or defend every vulnerable point, and we cannot afford to waste any of our strength in operations which are not vital to the general plan. This is particularly important in the case of both submarine and mining operations. Attrition of the enemy forces as an object is by itself unsatisfactory, and the policy to be followed must be laid down with some object in view.

8.– Apart from the recommendations included in my letter Med.01060/0708/8 of 14th November 1938 I am therefore of the opinion that the formulation of such a plan is the most vital and urgent need on the Mediterranean Station.

[Admiralty Minute on Docket]

Matters in connection with these papers will be discussed with the Commander-in-Chief, Mediterranean.[1]

[1] See Doc. No. 307.

2. No further action is required at present.
V.H. Danckwerts.
D. of P.
24. 11. 38.

297. *Backhouse to Pound*

[ADM 205/3] 8 December 1938

[Carbon]

I enclose a copy of a letter I have received from Cork, for what it is worth. I thought twice about sending it to you but decided that you would sooner have it than not, although there may be nothing in it. People do gossip frightfully, and women are apt to hear scraps of conversation about subjects which they do not understand and to draw all sorts of conclusions.

We know, of course, that the A/A defences are most inadequate, but these will certainly be rectified. If it would do any good by putting more heart into people I would have yet another drive about them. I have already said a great deal on the subject. How I should like to be a dictator for about a month!

[*Attachment*]

[Copy] Admiralty House, Portsmouth
7 December 1938

I am sorry to worry you with the following but I think it should be mentioned in case there is any foundation at all to the story.

On Sunday last I was out at Uppark and there was told of a woman (whom I knew) who had just recently come back from Malta and told them, J. & M.,[1] that all the young officers she had met (she was out there to see a Lieutenant) seemed to have their tails down, were saying they had no chance against the Italians, were outnumbered and if they got through one battle couldn't fight a second, etc.

[1] 'Jimmy' and Margaret Meade-Fetherstonhaugh. Uppark House, West Sussex, was the residence of the Meade-Fetherstonhaugh family. Adm the Hon. Sir Herbert Meade-Fetherstonhaugh (1875–1964). Known in the Service as 'Jimmy'. 3rd son of 4th Earl of Clanwilliam. Assumed additional name of Fetherstonhaugh by Royal Licence, 1931. Commanded destroyer *Goshawk* and a destroyer division at Heligoland, 1914; commanded destroyer *Meteor* in Dogger Bank action, 1915; commanded cruiser *Royalist* at Jutland, 1916; Naval Asst to 2nd Sea Lord, 1918–19; commanded battle cruiser *Renown*, 1921–23; commanded RN College, Dartmouth, 1923–25; RA (D), Med Fleet, 1926–28; commanded Royal Yacht, *Victoria and Albert*, 1931–34; retired, 1936; Cdre of Convoys, 1939–40; Lieut-Col, Home Guard, 1940–44; served as Lieut, RNVR, Small Vessels Pool, 1944–45. Sergeant-at-Arms, House of Lords, 1939–46.

The most serious statement was that an 'Admiral at Malta' had stated that they had not got the ammunition to fight with, and would have no chance, etc. There was a good deal more about being bombed out of existence, etc., no A.A. guns and so on.

I merely repeat what I was told and have no belief in it, and cannot credit it.

But if there is any truth at all it seems serious that anyone can express these ideas, and if any officers do repeat such views it must get to the lower deck. D.P. [Pound] would be the last to hear of it. I am perfectly sure neither Layton, Tovey[1] or Leatham[2] believe anything of the sort, and wouldn't utter it if they did.

I send this to you in case you think it worth taking up. Please do not bother to answer. I told my informants I should pass this on.

298. *Backhouse to Cork and Orrery*

[ADM 205/3] 12 December 1938

[Carbon]

Thanks for your letter about Malta. I sent some extracts from it to D.P. I heard much the same thing from another source, which is rather disturbing. I think it is thoroughly bad that the Fleet should always leave Malta when there is a scare, as if they were running away from danger, although actually there may be very good reasons for moving it.

I have been pressing hard for the last two months to get Malta adequately defended, and intend to keep up the pressure until I get it done. I hope this will effect some improvement.

I am sorry to say that there are other bases in much the same state, such as Aden, and Gibraltar is not properly defended either.

* * *

---

[1]RA [later AoF Sir] John Cronyn Tovey (1885–1971). Commanded destroyers *Jackal*, 1915–16, *Onslow* (including Jutland), 1916–17, *Ursa*, 1917–18, *Wolfhound*, 1918–19, *Sea Wolf*, 1922–24; Capt (D), 2nd DF, 1924, 8th DF, 1924–26, 6th DF, 1926–27; Naval Asst to 2nd Sea Lord, 1930–32; commanded battleship *Rodney*, 1932–34; Cdre, RN Barracks, Chatham, 1935–37; RA (D), Med Fleet, 1938–40; VA 2nd-in-command, Med Fleet, Jul–Oct 1940; VAC 1st BS, Home Fleet, Oct–Dec 1940; C-in-C Home Fleet, 1940–43; C-in-C Nore, 1943–46.

[2]RA [later Adm Sir] Ralph Leatham (1886–1954). Commanded cruisers *Yarmouth*, 1928, *Durban*, 1928–30; commanded battleships *Ramilles*, 1934–35, *Valiant*, 1935–36; COS and Maintenance Capt, Portsmouth, 1937–38; RA, 1st BS, 1938–39; C-in-C East Indies, 1939–41; Flag Officer-in-charge, Malta, 1942; Dep Gov, Malta, 1943; C-in-C Plymouth, 1943–45; Gov and C-in-C Bermuda, 1946–49.

299. Pound to Backhouse

[ADM 205/3] 14 December 1938

SECRET.
[Telegram] Addressed Admiralty, from C.-in-C. Mediterranean.
IMPORTANT. personal.
97. For First Sea Lord and C.-in-C.
Immediately on receipt of your letter I made exhaustive enquiries and have no hesitation in saying that the reported state of mind of the young officer is entirely without foundation. I hope it may be possible to take immediate steps to prevent the persons in question spreading such a damnable false impression. Am writing suggesting further action.
Admiralty please pass to C.-in-C. Portsmouth.
2053/14.

300. Halifax[1] to Stanhope.[2]

[ADM 116/4084] Foreign Office, S.W.1.
13 December 1938

I expect you will have read the telegram No. 395 to Burgos on the subject of attacks on the High Seas on the ships *Bramhill* and *Laverock* by aircraft presumably under the control of General Franco.[3] These attacks

---

[1] Viscount Halifax [Edward Frederick Lindley Wood] (1881–1959). Succeeded father as 3rd Viscount, 1934. Created Earl, 1944. MP (U) Ripon div, West Riding, Yorkshire, 1910–25; Pres, Board of Education, 1922–24, 1932–35; Min of Agriculture, 1924–25; Viceroy of India, 1926–31; Sec of State for War, 1935; Lord Privy Seal, 1935–37; Leader of the House of Lords, 1935–38; Lord Pres of the Council, 1937–38; Sec of State for Foreign Affairs, 1938–40; British amb in Washington, 1941–46.

[2] Earl Stanhope [James Richard Stanhope] (1880–1967). Succeeded father as 7th Earl, 1905. Civil Lord of the Admy, 1924–29; Parliamentary and Financial Sec to the Admy, Sept–Nov 1931; Under Sec of State for War, 1931–34; Parliamentary Under Sec of State for Foreign Affairs, 1934–36; First Commissioner of Works, 1936–37; Pres, Board of Education, 1937–38; 1st Lord of the Admy, Oct 1938–Sept 1939; Lord Pres of the Council, 1939–40; Leader of the House of Lords, 1938–40.

[3] The Master of the SS *Bramhill* reported that on 7 November when 80 miles off Valencia on a Nyon course he was compelled by a Nationalist float plane, which dropped a bomb close to him and opened machine gun fire, to alter his course to Pollenza. Two hours later the aircraft signalled him to continue his voyage. On the same day the Master of the SS *Laverock* reported that when 56 miles off Cape Tortosa and 39½ miles from Cape Dragonera his ship was twice circled by a three-engine monoplane and shortly afterwards another aircraft approached flying very low and opened fire with a machine gun mounted in the nose, the bullets hitting the water short and very close to the ship's side. During the evening the ship was circled by aircraft five times and examined by searchlight although no signals were made at any time.

take us back to the days preceding the Nyon Conference and you will have noticed that the attack on the *Bramhill* was made while she was 80 miles off Valencia on the Nyon course. You will appreciate the intense indignation which would be aroused in the House of Commons and in the public press were these incidents to receive publicity, and it is still not impossible that they will do so. At present our only reply to any criticism would be that we had taken the matter up 'strongly' with the Burgos Authorities. In the House and in the country 'protests' to General Franco are at a discount. I do not wish, any more than you do, to become involved in a war with him, but attacks on British ships on the High Seas are quite intolerable and I should like you to consider urgently the best method of coping in future with such incidents. If I am correctly informed the Nyon patrols are not at their full strength and I wonder whether as a first measure it might not be advisable to increase the number of His Majesty's Ships on these patrols? Could we not at the same time inform the Burgos Authorities that we have taken this action in reply to these quite unwarrantable interferences with our merchantmen? In addition, would it be possible to restore the air squadrons which were sent to the Mediterranean last year to take part in the Nyon patrols? If you have any other suggestions I should be only too happy to consider them, with you, since I do not feel at all happy in framing replies in the House of Commons, or in our own House, to these incidents, merely on the familiar lines of strong representations or verbal protests.

301. *Backhouse to Stanhope*

[ADM 116/4084] [13 December 1938]

I attach a note by D.C.N.S.[1] Of the three alternatives mentioned, I consider that (a) is much preferable to either (b) or (c) as it would be much less likely to lead to guns being fired.

It might lead, however, to trouble with the Italians as they use Palma, Majorca, as a naval base, and I believe that many of the aircraft which operate from Majorca are Italian. At one time they were nearly all Italian, and in fact the Italians were practically in control of the Island, but this condition has now changed a good deal.

As regards the Foreign Secretary's suggestion about a patrol, I do not think that this would have any effect unless in such strength as practically to guard the whole route, almost to establish a blockade. [*Added in holograph in margin*: 'In any case few of our present destroyers have any

---

[1] A.B. Cunningham.

effective A.A. armament.'] It would be most inconvenient if this were to be maintained for a length of time, and if it were withdrawn the attacks might start again. We might, therefore, be letting ourselves in for a continuous strain on our resources.

With regard to the aircraft, I see no purpose in having these on the route unless they are in a position to attack planes which attack our ships; mere patrol aircraft would be of no real help, also they would have to be in much larger numbers than had been contemplated previously. Another difficulty would be where they were to operate from, as the Spanish ports are a long way from Oran or Algiers, and at this time of the year the weather is frequently unfavourable for flying long distances over the sea.

I most strongly agree with D.C.N.S. that if we are to threaten reprisals we must be prepared to carry them out immediately if another incident occurs.

There is another line to take: we might establish a blockade, say, at Cadiz, and stop all Spanish Nationalists' ships entering, but I feel that Majorca, being the seat of the trouble, would be more likely to have an effect. If we were to undertake a blockade of the Balearic Islands, it would be necessary to inform the Italians previously that nothing at all would be allowed in, otherwise I feel sure that all necessary supplies would enter under the Italian Flag. Obviously we cannot stop Italian ships unless we are prepared to go a long way in this matter.

Yet another alternative is that we could put our ships into convoy, which would mean arranging convoy assembly ports, probably at Gibraltar, Marseilles, and possibly Malta and Algiers. As in the case of the patrol, however, once started this might have to be continued indefinitely, and would occupy a very large number of ships if convoys were frequent. As this would be merely defensive, it would necessarily have less effect than something positive, such as the blockade of Majorca.

I would suggest that, if it is decided to take stronger action, the matter should be further considered in conference between the Admiralty and the Foreign Office, so that we may be quite clear what it is likely to involve.

Perhaps, if Signor Mussolini were informed that we cannot stand these attacks on our ships by aircraft from Majorca, he might be induced to tell Franco that they must be stopped.

[*Attachment*]
  Protection of British Merchant Ships in Spanish Waters against Air Attack.
The forms of protection fall under the following broad headings:–
  (i) Direct protection.
  (ii) Threat of reprisals.

Direct protection, in order to be efficient, must be provided both at the Spanish ports concerned and on the high seas.

It has been accepted in the past and presumably still holds good that British ships enter Spanish ports at their own risk, consequently it will be necessary to establish safety areas off these ports in which British ships will be protected by one of H.M. Ships.

The ports in question are few and ships stationed at say, Barcelona, Valencia and Alicante would meet this requirement.

The protection of British ships on the high seas is more difficult and can only be carried out effectively by a large number of ships. The re-establishment of the Nyon routes and patrols on these routes appears to be the only satisfactory method of achieving this. At least one flotilla of destroyers would be required on patrol at a time.

This is rather a waste of effort as even with a large number of ships it cannot be ensured that one is present when an aircraft attack is carried out. Even less effective would be the re-establishment of the Nyon Air Patrol. The flying boats, even if present, which is highly improbable, could do little to hinder an attack on a British vessel. Their role was to look out for submarines.

Threat of reprisals. This must consist of informing the Nationalist authorities that unless attacks on British ships ceased reprisals would be carried out.

Reprisals could take the form of:–

(a) Blockade of Majorca, particularly the interception of fuel supplies.

(b) Capture of Spanish Nationalist warships and/or merchant vessels.

(c) Bombardment of Spanish Nationalist territory.

(a) would be comparatively simple and would in a short time bring the air operations from Majorca to a standstill.

(b) would also be quite simple and if sufficient force was used in the case of war vessels would I'm sure be achieved bloodlessly.

(c) is not recommended.

All the above might be interpreted by the Germans and Italians as intervention.

I feel myself that the mere threat would be enough but I would strongly urge that no threat be made unless with full intention of carrying it out in case of necessity.

A.B.C. [Cunningham]
13th December, 1938.

302. Stanhope to Halifax

[ADM 116/4084] 16 December 1938

[Carbon]

I have been thinking over your letter of the 13th December about the attacks by Franco's aircraft on British ships on the high seas.

The First Sea Lord whom I have consulted does not think that your suggestion to bring up the Nyon patrols to full strength would have any practical effect. Aircraft with their greatly superior speed can always keep away from warships and attack merchant ships when the warships are not actually present. The Nyon routes cover many hundreds of miles and the First Sea Lord does not think that the patrols would have real effect unless they were in such strength as practically to guard the whole route, in fact, almost to establish a blockade. This would clearly be a very big naval commitment.

As regards aircraft, these were used on the Nyon routes to look out for submarines. They could do little to hinder an attack on a British ship, and would have to be in much larger numbers than has been contemplated previously. Another difficulty would be to find an operational base for them, as the Spanish ports are a long way from Oran or Algiers and at this time of the year the weather is frequently unfavourable for flying long distances over the sea.

If it is felt necessary, for political reasons, to take action the least objectionable way in my view is that we should warn Franco that if any British ship is again interfered with on the high seas we shall seize one of his merchant ships and hold her until such time as we have received adequate guarantees as to the future and have been paid such compensation as is due for the interference caused to the British ship.

It is probable, I think, that such a threat would suffice, but we must be prepared to carry out our threat if it does not. You will know better than I whether Germany or Italy might consider such action on our part as direct intervention. I imagine that it would be difficult for either of them to do so although Italy might again feel free openly to send assistance to Franco.

The more likely event, however, is that Franco, having few merchant ships of his own, might intensify his action against British shipping [*added in holograph*: 'of which there are always a number in Franco ports'] and so cause a big disturbance in trade on which the Board of Trade would have something to say. Should Franco do so the Navy would of course have to take up the challenge and be faced with the big commitment of having to provide protection for all British shipping off the coasts of Nationalist Spain. If the Dictator Powers want to get us involved with Franco they might perhaps persuade him to exercise his nuisance value to the full in this way.

After reading the telegrams again I am not myself satisfied that the attacks on the *Laverock* and the *Bramhill* were deliberate attacks contrary to the principles of the London Naval Treaty. Although admittedly they may have been, I think it is more probable that they were attempts to send these ships in for examination, probably arising out of a mistake in identification by the aircraft. As you know we have always maintained that it is permissible for an aircraft to use a moderate degree of force (such as firing a machine gun or dropping a bomb nearby) in order to divert a merchant ship when belligerent rights are being exercised. Admittedly these rights do not exist in this case and therefore ought not to be put into operation, but that in fact is what Franco as regards air action is doing.

I dislike as much as you do seeing British ships on the high seas being interfered with and our counter-action being limited to futile protests. The difficulty, as I see it is that any adequate counter on our part may be taken by other Powers as direct interference in the Spanish war and the end of so-called non-intervention.

[*Added in holograph*;] This of course was written before our talk this morning.

303. Backhouse to Pound[1]

[ADM 205/3] 28 December 1938

[Carbon]

Most Secret.

I sent you a telegram shortly before Christmas about the Spring programme. You will have realised that the political outlook is very uncertain and that there is a feeling that it may become more so and to the extent that we may need to be more prepared for emergency. In these circumstances, I did not think it would be wise to announce a Home Fleet cruise programme right on to March in case we decide to bring the fleet or part of it home sooner.

There is nothing definite to go on but the times are not normal.

I should hope that after the P.M.'s visit to Rome we may be a little clearer about the outlook, but this does not follow and there may be other indications in January of what 1939 has in store. I will, of course, keep you as well informed as I am able. Nor do I want to give you an alarmist

---
[1] Copy to C-in-C Home Fleet.

view of the situation, only that we must not take chances. For this reason we are arranging that *Hood* shall not be allowed to go to extended notice after her return and we are stopping *London*'s reconstruction and some other work of a like nature. It is inconvenient but we can't afford to lay up any more important ships now.

I hope *Nelson* will be able to go to Malta as planned and if she does Charles Forbes will be able to tell you more as I shall see him again before he sails.

We intend, also, to try and speed up our arrangements for manning the Reserve Fleet as the present scheme is too inelastic and entails the use of Proclamation and of the word Mobilisation which everyone hates. If we can do this, we should be able to take more extensive precautionary measures.

There was a great rush of work before Christmas, principally C.I.D. and C.O.S., but I think this was due largely to all the Ministers wishing to go away for several weeks while Parliament is up.

We have completed our Estimates for 1939 but they have not yet gone to the Treasury or the Cabinet. I foresee a struggle but that is not unusual. Actually I think we have kept within reasonable limits, but the Chancellor is evidently uneasy about finance as such great demands are being made on him by the Air Force and for A.R.P. besides the lesser demands of the Navy and Army, although the latter is badly in need of more money after years of starvation.

Thank you for your letter about the state of feeling at Malta. I expected you would react strongly. I gather from C. and O. [Cork and Orrery] that the 'woman' in question has been rebuked, but I do not know who she is and I don't think he does.

304. *Muirhead-Gould to Pound*

[ADM 116/3892] H.M.S. *Devonshire*, at Sea
27 January 1939

CONFIDENTIAL

No.5184/191.
Report of Proceedings (For period 16th January to 27th January, 1939).

* * *

Sunday, 22nd January – Arrival at Marseilles and Departure for Caldetas.
17. At 0900 the pilot was taken onboard in L'Estaque Road and the ship berthed alongside Mole C (south) in National Basin, assisted by two small tugs, without difficulty.

18. Shortly after berthing I received a telegram from the Minister[1] asking me to return to Caldetas and stand by. I left Marseilles therefore at 1600 and proceeded to Caldetas at economical speed ordering *Greyhound* to remain there till my arrival.

Monday, 23rd January – Arrival at Caldetas.

19. During the Morning Watch great activity was apparent along the Beddalona–Tossa road and there seemed to be a certain amount of gunfire, probably A.A. One or two villages were observed to be on fire. I anchored off the Embassy at Caldetas at 0740. While approaching the anchorage, Barcelona had been twice heavily bombed from the air.

20. Lieutenant Commander Ouvry and the party I had left behind returned to the ship and the minister was good enough to say that they had been most useful in packing up the Embassy. During the day *Greyhound* was busy embarking the archives and baggage from the Embassy and working parties from *Devonshire* were employed fitting pinnaces for evacuating large numbers and in building pontoons on board. I also supplied two signalmen and one Telegraphist to the Embassy to assist in Communications. Commander D.K. Bain, H.M.S. *Greyhound* and I lunched with the Minister ashore and discussed the situation. It was agreed that all the Embassy female staff and such others as could be spared should proceed to Marseilles in H.M.S. *Greyhound* the following day. In addition to these, the Consul General was asked to send as many British subjects as were prepared to leave Barcelona. These were successfully embarked (10 Embassy staff and 10 refugees together with about one ton of archives and baggage) in *Greyhound* and sent to Marseilles on Tuesday 24th January.

21. Barcelona was bombed frequently during the day and raids were carried out on inland villages. During the night several bombs fell in the vicinity of Caldetas and the windows of the Swedish Legation were broken by a bomb which fell during dinner.

22. During my conversation with the Minister I urged him to leave Caldetas before it was too late and to erect his Embassy on board. He told me that his Consul General was obliged to be in Barcelona to look after British interests and he did not want to leave him yet.

Tuesday, 24th January – At Caldetas.

23. At 0800 *Glowworm* arrived from Palma having experienced very bad weather.

---

[1] Sir Ralph Claremont Skrine Stevenson (1895–1977). Chargé d'Affaires with local rank of Minister Plenipotentiary at Barcelona, Oct 1938–Feb 1939; Principal Private Sec to Sec of State for Foreign Affairs, 1939–41; Min at Montevideo, Uruguay, 1941–43; Amb to Yugoslavia, 1943–46; Amb to China, 1946–50; Amb to Egypt, 1950–55.

24. The junior staff of the Embassy having left in *Greyhound*, I supplied the Embassy with two Cypher Officers and a signal staff which was kept busy all day until dusk. The Cypher Officers then brought duplicate cypher books on board and were occupied during the greater part of the night in decyphering messages addressed to the Embassy.

25. Working parties from *Devonshire* were employed throughout the day in re-building the Arenys pier and in placing in position the pontoons constructed on board. This will be of great value as the original pier was extremely dangerous. The Embassy started to evacuate their stocks of provisions and another working party was engaged in loading and unloading lorries and transporting cases &c from the road to the pier.

26. The Minister went into Barcelona during the forenoon and when returning tried to avoid the coast road which was being bombed but at Grandollers was caught in an air raid which was so close to him that the telegraph wires fell on his car.

27. During the afternoon further heavy raids took place on Barcelona harbour and at about 1830 the British S.S. *Miocene* laden with petrol, was fired by incendiary bombs. The vessel burned for three days covering the port with a heavy pall of smoke.

28. During the day I endeavoured to re-organise the destroyer programme but this was made difficult by the delay in the arrivals of *Hostile* and *Hunter* due to bad weather. To meet requirements and possible contingencies I found it necessary to keep *Greyhound* and *Glowworm* another 24 hours by which time their reliefs would have arrived. I sailed *Glowworm* for Gandia in order to replenish the stocks of provisions there.

29. The American flagship *Omaha*, flag of Rear Admiral Lackey[1] and the French flagship *Suffren*, flag of Rear Admiral Kerdudo, arrived at 1830 and 1900 respectively. As I had hoisted all my boats I requested these Flag Officers that I might defer paying my respects until the next morning and this was arranged. I informed them that owing to air raids being made on the coast road near Caldetas I darkened ship and kept steam at short notice during the night.

30. During the night a message was received from the Minister that the seat of the Spanish Government had moved to the Province of Gerona and that Senor Negrín[2] had requested the Diplomatic Corps to leave the city. In spite of this he still gave assurances that the resistance would be

---

[1] RA Henry E. Lackey, USN (1876–1952). Cdr, Mine Force, Atlantic Fleet, 1920–21; commanded cruiser *Memphis*, 1924–27; Norfolk Naval Training Station, Virginia, 1927–30; battleship *California*, 1930–31 and Cruiser Division 4, 1933–35; Dir of Shore Establishments, 1935–37; commanded Squadron 40-T (neutrality patrol during Spanish Civil War), 1937–39; retired 1940.

[2] Dr Juan Negrín (1889–1956). Socialist Prime Minister of the Spanish Republic from October 1937 to the end of the Civil War.

intensified. It was the Minister's opinion however that the fall of Barcelona was imminent. At the same time the Foreign Office authorised the Minister to withdraw the Consul General from Barcelona if there were no longer any British subjects who wished to evacuate or if there were any danger of his being cut off.

Wednesday, 25th January – At Caldetas.

31. At 0725 the French destroyers *Le Fortune* and *Simoun* arrived and at 0800 the American destroyer *Badger* also arrived.

32. At 0800 I ordered *Hunter* to proceed with despatch to Caldetas as in view of the reports it seemed that communications by road with Barcelona would become impossible and I might need to send a ship to Barcelona to evacuate the Consul General and any refugees who had not been able to leave the city.

33. At 0830 bombers were reported overhead and a series of bomb attacks were carried out between St. Poldemar and Arenys. … Many of the bombs fell between the ship and the shore and the French destroyer *Le Fortune* who was between me and the shore was nearly hit. I cleared the Upper Deck except for A.A. guns' crews; but I am convinced that the bombs were not directed against ships and in subsequent raids I have generally sent everyone below. I was in some anxiety about a working party I had just sent to Arenys Pier in the pinnace but fortunately no-one was hit although bombs fell only 100 yards away.

34. Shortly after the raid the Minister asked me to come and see him urgently at Caldetas and informed me that he could not send the car to Arenys as the road had been cut by the bombs during the raid. It was therefore necessary for me to land on the beach in front of the Consulate through the surf. The Minister informed me that he had decided to evacuate with all the remainder of his staff except the Pro-Consul, Mr. Ambrose. At this time it was believed that the road to Arenys was completely blocked and a start was made to embark stores and baggage from the beach into the motor boat and pinnace. The French and American boats were engaged in a similar task and it was unfortunately noticeable that the American 'whaleboats' were much better suited for this rather hazardous operation than our whalers. Later in the forenoon news was received that the road was again open and I was very glad to be able to send the remainder of the stores and luggage to Arenys pier for embarkation. There the pontoons which had been built and placed by the ship proved invaluable for the work and were used also by the French and American boats.

35. These events precluded my calling on the American and French Admirals but they also were very busy and when I returned to the ship I found that they had asked me to consider all calls paid and returned. When

the U.S.S. *Omaha* sailed during the evening she made a very courteous message of regret that the Admiral had not been able to receive me. I called on Contre Amiral Kerdudo on Thursday 26th January and he intended to take tea with me when returning my call but unfortunately he was suddenly ordered to proceed to Port Vendres during the afternoon.

It may be mentioned here that the French Ambassador and a small staff left for Figueras by car about 1800 on Wednesday, 25th January.

36. During the afternoon of 25th January a large working party was employed in carrying innumerable cases of stores, baggage and personal belongings of the Embassy and staff, from the bluff above Arenys pier to the boats, a distance of at least four hundred yards, down some steps on a steep cliff.

At 1645 the Minister and most of his staff arrived onboard.

37. Although refugees are dealt with in an appendix which is not enclosed in the interim report, it can be said that refugees were embarked at intervals throughout 25th and 26th January whenever the Consul General at Barcelona was able to provide transport and get them through to Arenys.

38. A rather alarming air raid took place at 1830 when San Pol de Mar was hit several times and set on fire. Several bombs fell in the vicinity of Arenys and others in Caldetas, one of which partially destroyed the house of a British subject Mr. Witty and wounded him slightly. First aid was provided for him the following morning and he steadfastly refused to leave. Some civilians were killed in these raids.

39. Consul General Barcelona informed me that the A.A. defences of the city had ceased their activities during the afternoon and there is no doubt that they were already withdrawing northward. During the day several trains and hundreds of lorries streamed through Arenys moving to the north. There are also hundreds of unfortunate civilian refugees trudging along the road and many of these poor people have come from Tarragona.

40. Throughout the day I maintained a signal and wireless staff, two cypher officers and an officer and working party in the Consulate at Caldetas. All these came off with the Minister.

Thursday, 26th January – At Caldetas.

41. The Military Attaché landed on the beach at 0630 and was successful in evacuating some political refugees whose lives were in peril and whom the Minister was very anxious to take into safety. These included Judge Camin, Judge of the Supreme Court of Barcelona who in the course of his duties had succeeded in saving hundreds of Nationalist prisoners from the death sentence but who has unfortunately had to announce the death sentence in six cases and for which his life is known to be in forfeit if he falls into the hands of the Nationalists.

42. Also at 0630 I landed wireless ratings who dismantled the Embassy wireless equipment and got it (with the exception of the heaviest batteries) through the surf into the boats so that this rather valuable set has been saved.

43. During the forenoon the Secretary to the Minister and Mr. Cowan, Honorary Attaché and also a member of Sir Philip Chetwode's Commission,[1] landed on Caldetas beach in order to arrange the further evacuation of refugees.

Shortly afterwards the weather suddenly broke and a heavy swell and high sea made conditions on the beach extraordinarily difficult.

I regret that my whaler broached to and capsized in the surf and had to be left on the beach. She was hauled up clear of the surf and I hoped that H.M.S. *Hunter* would be able to salve her. *Hunter* has now reported that the boat is a complete wreck and has had to be abandoned.[2]

* * *

305. *Muirhead-Gould to Pound*

[ADM 116/3896] H.M.S. *Devonshire*, at Marseilles
12 February 1939

CONFIDENTIAL.
No.5787/191.
Report of Proceedings (For Period 3rd February to 10th February, 1939).

* * *

Friday 3rd February and Saturday 4th February. At Palma.
80. On receipt of Admiralty telegram timed 1330/4th February, the Consul arranged an interview with Admiral Moreno who had just returned to Palma and the Count San Luis, the Air Officer Commanding the Nationalist Air Force in Majorca.[3]

---

[1] Field Marshal Sir Philip Chetwode (1869–1950) was President of the International Commission for the Exchange of Prisoners which operated with limited success during the Spanish Civil War. Denis Cowan was his representative.

[2] At about 1600 Muirhead-Gould received word from the Consul General that Barcelona had fallen and Nationalist troops were entering the town without opposition. *Devonshire* sailed for Port Vendres and Marseilles at 1045 the following morning. HMS *Devonshire*, Report of Proceedings, 26 Jan–2 Feb 1939, ADM 116/3892.

[3] As Catalonia fell to the Nationalists the war seemed to be moving into its final stage. The Nationalists asked Hillgarth for British assistance in providing a ship to serve as a neutral meeting ground for negotiations to arrange the surrender of the island of Minorca, still in Republican hands. The Foreign Office approved, provided the Germans and Italians were not told and no foreign troops would be allowed on the island for a period of two years. Hugh Thomas, *The Spanish Civil War* (New York, 1963), p. 580 and n. 3.

Sunday 5th February – At Palma.

81. The interview referred to in paragraph 80 above took place at 0900 in the Naval Headquarters. Lieutenant Commander Alan Hillgarth had already explained the alternative procedure which had been suggested by His Majesty's Government but this did not find favour with either Moreno or San Luis. I accordingly informed them that His Majesty's Government were prepared to put one of H.M. Ships at their disposal to take San Luis to Minorca in order to negotiate a surrender and stipulated that the following points must be agreed to:–

(a) That I should inform Minorca W/T Station that I should arrive at Port Mahon at 0900 on Tuesday 7th February.

(b) That neither the Consul nor I should take any part in the negotiations.

(c) That no naval or air activity was to take place near Minorca after *Devonshire*'s departure until her return.

I decided to go in *Devonshire* instead of sending a destroyer for the following reasons:–

(a) The Commanding Officer H.M.S. *Hotspur* was unacquainted with the situation and did not know Hillgarth.

(b) Although in some ways it would have been convenient to have gone up harbour at Port Mahon in a destroyer, I foresaw the possibility of rioting in the town and considered that a destroyer in the inner harbour might be in a most dangerous position.

(c) Our destroyers are not dissimilar at long range from the Nationalist destroyers and I was anxious lest fire should be opened on the destroyer while approaching.

(d) It was suggested to me by the Consul that should the negotiations break down after a meeting had been held on board a destroyer in the inner harbour, we might be accused of taking a Nationalist officer up harbour for espionage purposes.

82. All our conditions were agreed to and in addition San Luis promised to give the Consul a signed guarantee as General Franco's representative, that no foreign troops would take part in the occupation of Minorca.

83. Moreno and San Luis informed me that in order to do everything possible to bring about a peaceful surrender, they were empowered to permit the escape of officers and others considered criminals, and asked me if in such cases I would give them passage to Marseilles. Admiral Moreno assured me that this escape would only be offered as a last resort if negotiations could not be completed by any other means. In these circumstances I felt that I could safely recommend that approval should be given if it became clear that negotiations would otherwise break down.

84. At the conclusion of the interview, on hearing that Admiral Moreno was leaving for Barcelona at 0930, I decided to deliver the protest ordered in Admiralty telegram timed 1148/3rd February. Admiral Moreno promised to order an inquiry.

85. Immediately after returning on board I commenced to call Minorca W/T Station on every probable wave including 600 metres (commercial) but was never able to communicate.

86. H.M.S. *Hostile* arrived from Gandia with one Italian prisoner of war who was disembarked in a boat from the Naval Base. Although Consul Valencia had been asked to collect 15 of the 40 prisoners remaining, he reported that he was unable to do so in time for *Hostile*. I have now asked him to concentrate the whole of the remaining prisoners in this exchange (about 40) at Gandia and to report when they are ready.

Monday, 6th February – At Palma and Departure for Port Mahon.

87. At 100 H.M. Ships *Hotspur* and *Havock* arrived to relieve *Hostile* and *Hunter* who sailed for Malta at 1100. At 1630 I sailed *Havock* for Gandia and the round trip, giving passage in her to Captain Williams, ex Master of the *Wintonia*.

88. I was rather disconcerted when Foreign Office refused permission for Consul Palma to visit Minorca as he was an acquaintance of the Governor of that Island who was believed to be General Brandaris, and also as he had already been to Minorca in H.M. Ships. In the event it was fortunate that he remained at Palma or I should have had no means of communicating from Port Mahon with the Nationalist authorities in Majorca.

89. At 2130 I sailed for Port Mahon with Count San Luis onboard as General Franco's representative.

Tuesday, 7th February – At Port Mahon.

90. As I had failed to communicate with Minorca W/T station, I had to consider the possibility that the Government batteries at Port Mahon might open fire as I approached though *Devonshire* did not resemble any Nationalist warship. I therefore streamed paravanes, prepared the ship for action and in order that as many people as possible should be down below I prepared to go to General Quarters. I approached the anchorage with my signal letters flying, large white ensigns at the fore and main topsail yards and displaying the International signal requesting a pilot. I was on the point of going to general quarters when the signal station answered my identity and signalled that a pilot was being sent.

At 0910 I anchored off the entrance to Port Mahon and waited for the pilot who came out shortly afterwards.

91. The pilot informed me that General Brandaris had left the island on Sunday 5th February for an unknown destination and that a Naval Captain J.L. Ubieta was new Governor of Minorca. He said that it would be perfectly in order for me to send an officer to enquire from the Governor when he would receive my call, as he was uncertain whether an officer would be sent out to meet me and volunteered to accompany my officer. He also told me that there was great anxiety in the town owing to the dropping of leaflets during the last two or three days and that he estimated at least 70 percent of the population were pro-Franco but that unfortunately the 30 percent who were pro-Government were Government officials, soldiers and commissars who had the arms and were in a position to drive the majority whichever way they wished.

I sent Lieutenant Commander J.J. Casement, R.N., and Instructor Lieutenant J.C. Gascoigne (who speaks Spanish) in the Jolly boat to call on the Governor and to ask when he would receive me. (I sent Instructor Lieutenant Gascoigne instead of the Interpreter as I thought it best to send a naval officer).

92. While these officers were ashore a Government naval officer arrived in a motor launch and invited me to call on the Governor between 1100 and 1130 G.M.T. which I promised to do. Shortly after he left my officers returned and reported that they had seen the Governor who would send a car for me at 1130 G.M.T. I proceeded at this time and was received by the Governor in due course. I explained to him exactly why I had come and asked whether he was prepared to come on board and see General Franco's representative. He refused point blank but after I had returned to the point several times asked who was the representative. I informed him that it was the Count San Luis whereupon he expressed more interest and said they had been shipmates together. He re-affirmed however that he could not possibly see San Luis without the permission of his Government which he promised to do within an hour. I expressed some doubt as to the possibility of this and informed him that his W/T station was not working which seemed to surprise him. He added that communication with Valencia and Alicante was quite good. Finally he agreed to come onboard at 1330 G.M.T. but would not promise to see San Luis. I then returned to the ship.

Captain Ubieta arrived onboard accompanied by an Engineer Officer San Martin. After some discussion in my cabin and a little stimulant I asked if Ubieta was prepared to see San Luis. He replied – 'NO'. He had failed to communicate with his Government and was not prepared to see anyone until he had received instructions. I pointed out that it seemed a pity not to take the opportunity for conversation with an old shipmate and urged that a meeting could do no possible harm and might do good. After some time I persuaded him to see San Luis in my presence but he reserved the right not to listen to any proposals. I informed San Luis who also begged me to remain present at the interview.

93. The interview was very interesting. There was Count San Luis, friend of the ex-King – friend of General Franco – a man who had twice been a member of the Senate in Madrid – once Lieutenant Governor of Seville – once Governor of Seville – and now occupying the position of Air Officer commanding in Majorca, wearing the uniform of a Lieutenant Commander in the Navy in which rank he retired many years ago. San Luis regarded anybody associated with the Spanish Government as a traitor and a 'Red'.

On the other side was Ubieta, who had either from choice or by accident found himself on the Spanish Government side at the beginning of the Civil War – who had led and directed the successful attack on *Baleares* and who wears the five pointed communist star as a reward – now a Captain in the Spanish Government Navy and Governor of Minorca, which post he had only held for two days. He considers anyone associated with Franco a rebel.

Unfortunately I do not speak Spanish but the substance of the interview was communicated to me afterwards by San Luis and Instructor Lieutenant Gascoigne, who, with the consent of both parties, was also present to interpret from time to time any remarks affecting my responsibilities.

Notes on the conversation which took place between the Spanish Nationalist San Luis and the Governor of Minorca, Captain Ubieta, form an enclosure to this report.[1]

94. After lasting one hour and forty minutes the meeting broke up, Ubieta saying that he would continue his efforts to communicate with his Government until 0700 G.M.T. the next day, at which time, failing instructions from them, he would himself decide one way or the other.

In these circumstances I decided I was justified in remaining at Port Mahon.

95. During the afternoon information was received from *Havock* that part of the Spanish Government were expected at Gandia during the night.

* * *

Wednesday, 8th February – At Port Mahon.

97. At about 1000 Ubieta came out to the ship accompanied by San Martin and a plain clothes political commissar. Much of the old ground was gone over again with greater and greater insistence by Ubieta on the guarantee of safety for such persons who felt their lives would be unsafe

---

[1] Not reproduced.

if they remained behind. At last San Luis informed me that he was unable to make further progress without promising that such persons would be given passage in H.M.S. *Devonshire*. He said that everything else had been accepted and that he was practically certain that if I gave an undertaking to give passage to these people to Marseilles, Ubieta would surrender Minorca.

I pointed out that I could not give a general promise but that I was prepared to take such persons as were vouched for by the Governor as being in genuine danger and provided always that they could show me some proof that they would be allowed to land in France. Ubieta replied that he could arrange this as there was a French Consul in Port Mahon. I again asked him how many people might be affected and was told thirty or forty, or perhaps a few more if they brought their families. I therefore promised that I would take such people provided they came off in their own boats and were vouched for by a pass or through their names being on the list which was to be provided and that they all exhibited a passport bearing the French Consul's visa.

98. During the meeting, which was a long one, two more Commissars had appeared with urgent messages for Ubieta which they delivered privately. On my promising to take these political refugees Ubieta informed San Luis that he and the Commissars were prepared to surrender the island but as their lives were now in danger they could not possibly go ashore again. I was asked whether I would agree to send a boat and a small armed party with an officer to bring off the wives and families of the ex-Governor and of the Commissars. I refused absolutely and again made it clear that I could only take people who came off in their own boats and who fulfilled the formalities I have already stated. San Martin, who showed more courage than the others, volunteered to go and fetch them himself.

San Luis took this opportunity to send letters into the town appointing as Governor a local Colonel (retired) Usaleti, in the name of General Franco, and directed him to keep order and prevent reprisals until the Nationalist troops arrived. At about this time at San Luis' request I informed Consul Palma that the Governor of Minorca was prepared to surrender and that he should ask the military command in Majorca to prepare to send the forces of occupation. I provided lunch for San Luis in my cabin and at their own request Ubieta and the two Commissars lunched alone in the Ward Room Guestroom. At about 1300 a second Commissar decided to go ashore and he assured San Luis that he was in a position to arrange for the quick surrender of the town and would take a further letter to Colonel Usaleti.

99. Shortly after 1330 my attention was called to Fort La Mola where a number of troops were observed to be watching intently some events

which were happening in the town out of sight of the ship. A few minutes later it was seen that three Nationalist fighters were 'roaring-up' the town of Mahon and villages in the vicinity. Their behaviour must have been intensely irritating to everyone in the town and villages, especially as they had been assured by San Luis that no flying would take place over Minorca whilst *Devonshire* was at anchor outside. Shortly afterwards it was observed that there were three bombers flying high over the island and one flying boat circling round at a lesser height.

Not until 1415 did the Government A.A. batteries pay any attention but unfortunately at last one or two rounds were fired from a battery beyond the town. The shells burst nowhere near the bombers or fighters but this was the signal for a series of unwarranted and brutal raids on Mahon and La Mola and in direct violation of the terms agreed upon in my interview with the Nationalist authorities on Sunday 5th February.

100. The first raid occurred at 1430 and was an attack on the batteries on La Mola point. The attack was carried out by three bombers who dropped three or four bombs each. Four fell on the fort (there appeared to be little damage caused, one fell on the cliff face, and several in the water between the cliff and the ship.

The upper deck was cleared and hands were kept below for some minutes.

101. Although I felt this raid must be a mistake and an isolated attack, I came to short notice for steam and prepared to weigh. Immediately the bombs dropped I telegraphed to *Hotspur* in plain language to protest strongly through the Consul at Palma against the bombing during the negotiations. San Luis associated himself with my protest.

102. Just at this time Consul Palma informed me that the military and naval commanders were preparing for the embarkation of the forces of occupation – that the garrison at Cuidadela had declared for Franco and that three battalions of infantry were being sent to support them. He also reported that aircraft would demonstrate over Minorca but would take no offensive action unless they met with resistance.

103. At about 1600 aircraft were again heard overhead and between 1600 and 1650 four or five deliberate raids were carried out, two of which were deliberate attacks on the town of Mahon, and Anti-Aircraft fire now became intense. A detailed account of these raids is contained in a separate report. ...

It is quite clear that the bombers were Italian.

104. After the first raid at 1600 I decided to proceed to sea for the safety of the Ship's Company and also to make a further protest through the Consul Palma. The raids were in such quick succession, however, and anti-aircraft fire was so intense that I could not get men on the forecastle

without grave risk. I therefore kept all hands below under cover until well about 1700 when A.A. fire ceased.

105. After weighing, I remained off the coast for some time and after dusk closed to within two miles of the harbour entrance where I remained stopped hoping that boats would come out. These raids had in the meantime caused San Luis to request me to communicate with Consul Palma asking him to telegraph General Franco offering his resignation as his task had been made impossible for him.

106. As I had hoped, a boat came off with the Commissar who reported that in spite of the bombing the local authorities ashore had maintained order and were still prepared to surrender on condition that there were no further attacks and that I would implement my promise to take off those people who wanted to leave and whose names were approved by the Governor. Having just previously received from Consul Palma confirmation that any action which had been taken by aircraft was directly contrary to orders and assurance that it would not be repeated, I gave the Commissar this information and reaffirmed my promise regarding the evacuation of approved persons. At 1925 I again anchored off the harbour entrance, darkening ship and burning only candle anchor lights.

107. The Commissar then went ashore but returned again at about 2100 with Colonel Usaleti, the Acting Governor. They jointly made a formal protest to San Luis and myself regarding the bombing but my assurance that it had not been ordered and was not a deliberate violation of the agreement seemed to satisfy them. Usaleti said he thought he would be able to maintain order during the night but would be much happier if H.M.S. *Devonshire* remained at anchor off the harbour. In fact he stated that the departure of the ship before the refugees had been taken on board would probably be disastrous. I therefore agreed to remain until all the persons mentioned in the Governor's list had been embarked.

108. Just before these gentlemen left the ship, there was a further outbreak of anti-aircraft fire, two shells burst on the water close to the ship and a loud explosion occurred on the point of La Mola. No aircraft were seen from the ship and it is doubtful whether the explosion had been a bomb or an exploding shell. (Reports obtained from refugees later suggest that three aircraft flew round the coast and dropped an occasional bomb. It was further stated by refugees that one aircraft dropped some more leaflets in the villages.) In spite of their apprehension of further raids Colonel Usaleti and the Commissar again went ashore promising that in any case he would prevent any further firing from the anti-aircraft batteries.

109. During the first watch there was intermittent machine and gunfire from various parts of the island.

Thursday 9th February – Departure for Marseilles.

110. My hopes for an organised evacuation of political refugees were dashed when about midnight a motor boat came alongside with several people who had passes from Usaleti but who were not on the Governor's list. These were followed during the middle watch by two large tugs, several more motor, sailing and rowing boats which clustered round the gangway. The persons whose names appeared in the Governor's list were inextricably mixed with all sorts of other people including some who ought never to have been given passage passes by Usaleti and many others who had no real reason whatever for leaving the island except for their present fear. Both San Luis and Ubieta did their best to persuade these people to remain in their boats and return to the shore but it was impossible to reason with them. I consulted both San Luis and Ubieta and we agreed that there was nothing else to be done but to take everybody on board.

The refugees were then searched for arms at the top of the gangway and three sacks full of revolvers were collected. They were assembled on the quarter deck until provision could be made for them under shelter forward. The stream continued unabated until after 0400.

111. At about this time San Luis told me in confidence that he felt unable to trust any further the promises made from Palma that there would be no further bombing. He was apprehensive that another raid would take place at daylight and strongly urged me to leave. I decided not to do so until all the persons on the list had been embarked but fortunately I was told soon afterwards that the list was cleared. I explained this to Ubieta and asked him formally if he were satisfied that I had carried out my promises. He agreed that he was perfectly satisfied and asgreed that we should leave at once. This was more easily said than done as there were still several boats round the gangway and when an effort was made to hoist the lower half of the ladder, these unfortunate people in the boats started to jump into the water to cling to the ship's side and were only rescued with difficulty. By this time I was actually under way but I obviously could not use my screws.

112. At 0500 there were no more people alongside in boats and there were no boats within a mile of the ship although two or three more could be seen coming slowly out of the harbour. I took advantage of this opportunity to proceed to sea at high speed.

113. I ascertained later that the artillery and machine gun fire, rumours of further air raids and the misbehaviour of armed bands of Nationalists who had started reprisals in town, had caused an absolute panic among those who were known not to be Nationalists.

Hundreds of these poor people had fled to the quays to embark in any boats they could find.

114. At about 0330 Consul Palma had informed me
  (i) that he had the most definite guarantee that Mahon would not be further bombed but if the authorities wished to surrender they must order opposition to stop.
  (ii) that fighting had broken out between the garrison at Cuidadela and Republican troops and that Cuidadela had asked for help.
  (iii) that troops from Majorca would land at Cuidadela early that morning (Thursday 9th February) and that a division of Navarrese troops from the mainland would arrive later in the day.
  (iv) that without the advice of San Luis the Nationalists could not send an expedition to Port Mahon and requested that San Luis might return to his own people as soon as possible.
  I therefore ordered *Hotspur* to raise steam and to rendezvous with me north of Majorca at 0900. At this time I transferred San Luis to *Hotspur* and he reached Palma at 1230.

115. I was now in a quandary as to where these 450 refugees could be landed. I telegraphed to Admiralty (my 0420/9th February) to ask whether arrangements could be made for France to receive them. My first intention had been to proceed to Port Vendres, the nearest French port, but after reconsideration I decided that it would be extremely difficult to disembark so many in ship's boats. I therefore decided to proceed to Marseilles. In order not to embarrass the British Consulate and French authorities by arriving late in the evening I reduced speed to arrive at 0830 Friday 10th February.

Friday 10th February – At Marseilles.

116. I arrived at Marseilles at 0830 and received every possible help from the Consulate general staff and French authorities. I was fortunate enough to persuade the civil and military authorities to extend preferential treatment to Captain Ubieta, his wife and their father and also Lieutenant Borras (A.D.C.) And his wife. They were allowed to leave in the care of the Spanish Consul.
  All the refugees had left the ship by 1430.

117. It was some consolation after these distressing events to receive assurances from San Luis and Captain Ubieta separately that they were extremely grateful for all that had been done for them by officers and ship's Company. Captain Ubieta handed me on leaving, the attached expression[1] of his satisfaction at the way he and his people had been treated.

* * *

---

[1] Not reproduced.

306. *Muirhead-Gould to Pound*

[ADM 116/3896] H.M.S. *Devonshire*, at Sea
20 February 1939

CONFIDENTIAL.
No. 5806/191.
Report of Proceedings (For Period 10th February to 19th February, 1939).

* * *

Friday 10th February – At Marseilles.
118. There were a large number of Press reporters and photographers on the quay who were extremely busy trying to get information from my berthing and brow parties and even calling up to the refugees on the upper deck. As I had already heard extraordinary rumours about what had happened at Minorca, I thought it best to prevent any further distorted reports being circulated and promised the press agents to give them a short authoritative account later on if they would cease from interrogating the Ship's Company. I accordingly prepared some notes from which I later addressed them but regret that I should have allowed myself to be questioned by them with the result that the extremely embarrassing personal interview was published in the press and broadcast on the radio. I need hardly say that nothing was further from my thoughts than to obtain such undesirable publicity at such a moment.

119. As a matter of interest, the men among the refugees who were of military age were first placed in a train but later in the evening they were transferred to a French ship which was proceeding to Port Vendres. From reports received I gather that they were not being very well looked after.

120. As soon as the refugees were clear the opportunity was taken to send the whole of the bedding and blankets etc., which had now been used by two lots of refugees, to be cleaned and disinfected.[1] This was accomplished by Sunday morning, 12th February. The ship was thoroughly cleaned and was in a normal condition by Sunday, when I held divisions and Divine Service in accordance with normal routine.

121. During the stay of the ship at Marseilles, leave was given to 659 men and although the ship, at Mole 'G', was five miles from the centre of the town, there were only three leave breaking offences and six men who returned on board drunk.

---

[1] In the period 26–28 January *Devonshire* had carried refugees from Caldetas to Marseilles. They included the Masters and crews of British ships sunk in Barcelona harbour.

THE APPROACH OF WAR, 1938–1939   487

* * *

131. On Monday 13th February, Consul Palma received a letter from Admiral Moreno expressing his personal regret at the bombing of Mahon on the afternoon of 8th Feburary for which he blames the pilots whom he states have been duly punished. He informed the Consul confidentially, with a request that this should not be quoted, that the real fault was with the Italian Air Force Brigadier. I understand that Moreno has in consequence of this incident asked officially for the withdrawal of all Italian airmen from Majorca and that General Kindelan[1] has agreed to this withdrawal gradually. General Kindelan is head of all the Air Forces on the Nationalist side.

I should add that I was very hospitably entertained to luncheon ashore [at Palma] by the Count San Luis on Friday 17th February – Admiral Moreu, commanding the cruisers, General Canovas, the Military Governor of Majorca, and many other senior officers were present with their wives and expressed themselves very grateful for His Majesty's Government part in bringing about the peaceful surrender of Minorca.

The Consul at Palma informed me also that a local newspaper had published a report that H.M.S. *Devonshire* had taken from Minorca 500 Red criminals. On Lieutenant-Commander Hillgarth protesting to Admiral Moreno a written apology was at once received from the paper and the account was corrected. It is interesting to note however, that no other references to *Devonshire*'s visit to Port Mahon have been allowed to appear in the local papers.

* * *

136. As a result of the bombing at Caldetas and Port Mahon, I have taken the following precautions for the safety of the ship and the personnel:–
(a) Action Stations have been revised so that only H.A. Control personnel and Captains and Phone Numbers of 4-inch guns are in exposed positions.
(b) Protective mattresses have been fitted round the bridges.
(c) Torpedo war-heads have been stowed below in the seamen's bathroom flat.
(d) Spare 4-inch H.E. shell have been cleared from the upper deck magazines and stowed below.

[1]General Alfredo Kindelán y Duany (1879–1962). Noted balloonist and pioneer of Spanish military aviation; a staunch monarchist, choose voluntary exile rather than serve republic, 1931–34; one of the leading conspirators in military revolt, 1936; head of Nationalist air force, 1936–39; Capt Gen of Balearics, 1939–41; Capt Gen of Catalonia, 1941–42; head Escuela Superior del Ejército, 1943–45; persistent critic of Franco identified with Monarchist causes, exiled to the Canary Islands, 1946; created marqués, 1961.

(e) Arrangements have been made and have been tried out for weighing the anchor from between decks to avoid exposing personnel.

* * *

[Admiralty Minutes]
Referred for early remarks. The papers touching the Minorcan question are attached.

It will be seen that we have received the thanks of the Duke of Alba,[1] and on the other hand something of a reprimand from the Spanish Embassy. It is, however, generally accepted that the Minorcan incident was very well managed by the Navy.

By way of being a sort of 'devil's advocate' M. Branch can only point to one slight blot. It was a principal intention of H.M. Govt. to avoid in any way being involved as intermediary in the negotiations at Minorca and it was made clear to S.N.O. that, while placing himself at the Consul's disposition, he was to do nothing more. In fact the Commanding Officer went a little beyond his brief. It was originally intended by the Board that a destroyer should go to Minorca with San Luis on board, but the Captain of the *Devonshire* decided for what seemed to be very good reasons to go himself. But he made certain stipulations to Admiral Moreno and San Luis which might perhaps have better been left to the Consul to make, i.e. he stipulated that neither the Consul nor himself should take any part in the negotiations and he also stipulated that no naval or air activity was to take place near Minorca after *Devonshire*'s departure and until her return. It is unfortunate that Hillgarth did not accompany San Luis, and the Foreign Office's insistence that he should not go probably made it inevitable that the Captain of the *Devonshire* should take a rather more active part than he otherwise would have done; though in this connection it must be remarked that he appears to have taken it upon himself to make his stipulations before he knew that Hillgarth would not accompany him.

The interview which the Captain of the *Devonshire* gave to the Press at Marseilles, and its result, has been dealt with. In the event no harm was done and he makes a graceful apology for any harm he may have caused.

His report of the proceedings is really most satisfactory.

J.H. James., for Head of M.
March, 1939.

* * *

[1] Jacobo Stuart Fitzjames y Falcó, 17th Duke of Alba (1878–1953). Representative of Franco in London and, after the latter was finally recognised, Spanish amb, 1939–45.

I do not altogether agree with M. Branch minute.

2. The stipulations which the Commanding Officer, *Devonshire*, made to Admiral Moreno were, in my opinion, very properly made by the Commanding Officer himself as Senior Naval Officer, Eastern Area.

3. M. Branch do not seem to be quite au fait with the position at Palma. There is never any question of the Senior Naval Officer acting through the consul; on proper occasions the Senior Naval Officer deals directly with the Spanish authorities.

4. I consider the Commanding Officer, *Devonshire*, though he may have slightly exceeded the letter of his instructions, showed enterprise and initiative and brought his mission to a very successful conclusion.

A.B.C. [Cunningham]
6th March, 1939.

I agree with D.C.N.S., para.4., and consider that taking a broad view, the C.O. of *Devonshire* carried out this 'operation' very successfully and that his actions were those best suited to the somewhat peculiar conditions with which he was confronted.

Please draft a letter to the C.-in-C. Med., copy to the R.A.C. 1st C.S., to the effect that Their Lordships have read this report with interest and that they consider that the C.O. H.M.S. *Devonshire* handled the situation with initiative and good judgement.

They have no doubt, and this is also the view of the Foreign Office, that it was mainly due to his actions that there were few casualties and little material damage in Minorca when the Island surrendered to General Franco's representative.

Their Lordships are pleased, therefore, to commend Captain Muirhead-Gould for the part he played, and request you to inform him accordingly.

R.B. [Backhouse]
9th March, 1939.

* * *

307. *Admiralty to Pound*

[ADM 116/3900] 13 February 1939

[Carbon]

MOST SECRET.

M.01155/39.

I am to refer to your submission Med.01061/0708/9 of the 14th November, 1938, concerning the strategic aspect of the situation in the

Mediterranean, in which you recommended an initial attack by British, French and Egyptian forces on Libya, in order to eliminate Italy from the war as soon as possible.

2. This question was discussed at the meeting with you held on 29th November, when it was shown that the limited land and air forces in Egypt were likely to be inadequate for such an operation, and I am to state, for purposes of record, that it was generally agreed that the most our forces could do would be to hold up an Italian attack on Egypt. It was explained also that at present the French plan as regards operations from Tunis is not known, although it was considered that they too might have to stand on the defensive, at any rate at first. It is possible that, as a result of staff conversations with the French, a different policy may be evolved, in which case a combined plan could be developed, but we are not yet in a position to proceed thus far. You will be informed of any change in the situation regarding this.

3. Their Lordships consider, however, that the Italians would have considerable anxiety as regards the maintenance of supplies and reinforcements between Italy and Libya even in face of our small force of submarines operating from Malta. Once the route by these ships is known, opportunities may occur also of attacking them in other ways.

4. With reference to the summary of recommendations made in Part III of your submission No. Med.01061/0708/8 of the 14th November, 1938, I am to forward for information the attached statement containing Their Lordships' observations in regard to Inter-Allied Staff Discussions and Red Sea Convoys.

* * *

[Attachment] <u>Inter-Allied Discussions</u>.

1. A summary of discussions with the French, which have taken place up to date, was handed to the Commander-in-Chief. Any further developments in staff conversations will be communicated. It was agreed to ask the French to provide Liaison Officers for Malta and Gibraltar.

<u>Red Sea Convoys</u>.

2. Steps are already being taken to improve the overland route via Iraq so that it can be used by lightly equipped troops, their stores and equipment being transported by sea. It is hoped that this will be possible by April, 1939. The use of the route, however, depends partly upon the attitude of the Arab countries and tribes, as well as the completion of the necessary shore arrangements. The Indian reinforcements may, therefore, still have to proceed by the Red Sea route, and in this event a division of 'Tribals' or destroyers must be detached to the Red Sea Force.

The greatest risk to Red Sea convoys and shipping at the commencement of war is considered to be by aircraft attack from Italian territory, the degree of which it is difficult to estimate. In any case, at first, it will necessitate convoys being escorted by ships possessing good anti-aircraft armament. In addition there will be some risk of submarine attack until the Italian submarines in this area have been dealt with.

308. *Pound to Backhouse*

[ADM 1/9930]   Commander-in-Chief, Mediterranean Station
21 February 1939

MOST SECRET.

I have been considering some of the outstanding points of joint interest to the French and ourselves in the Mediterranean in a war against Germany and Italy.

The extended basis of the staff conversations with the French, announced in your telegram 1044 of the 10th February, will no doubt present opportunities to raise some of the points with them.

2. The first is that of the line of demarcation between the Eastern and Western basins, which will be under British and French control respectively.

It has generally been assumed that this would be from CAPE BON to MARSALA (Sicily), and across the MESSINA STRAIT. But if, as foreshadowed in the enclosure to your letter, the war opens with an Italian attack on the French, we in the Eastern basin shall then find ourselves responsible for two-thirds of the coast of TUNISIA.

Whichever way the war opens TRIPOLI will be the African port which most threatens the security of TUNISIA, and it would seem to be to the French interest, as well as within the capabilities of their aircraft, to have TRIPOLI as an objective for attack. Offensive action by the French against TRIPOLI will also enable me to give more attention to BENGHAZI and ports further East which more directly threaten EGYPT.

It seems, therefore, that the dividing line should be to the East of TRIPOLI, and I suggest the meridian of LICATA – 13 DEGREES 57 MINUTES East – from Sicily to the African coast.

In the MESSINA STRAIT a definite line is also required, and I suggest one across the North Entrance from CAPE PELORO (Sicily) to PEZZO POINT (Italy) which will put both MESSINA and REGGIO in the British area of control.

3. The French policy with regard to the narrow waters between CAPE BON and MARSALA (Sicily) is naturally of great interest to us. Strong

patrols, both reconnaissance and offensive, would be most valuable, and possibly they may consider mining the shallower waters on the SICILY side and so diverting all through-shipping into the deep waters near CAPE BON where it could more easily be controlled.

4. In this connection too it would be interesting to learn the French reaction to a joint policy of declaring on the outbreak of war that the MESSINA STRAIT was a dangerous area closed to merchant shipping and that thereafter all shipping using the Strait would be liable to attack without warning. This would divert all neutral shipping to the channel off CAPE BON, and simplify the task of our submarine patrols off MESSINA.

5. The synchronising of French attacks on West Coast objectives with ours on the South East coast of Italy is another matter of importance; probably all that could be done at this stage would be to obtain their agreement in principle. The same would apply to the possible timing of their attacks on TRIPOLI with ours on BENGHAZI or other Libyan ports.

6. Finally, in considering the efficiency of our Gibraltar Straits patrol there is clearly a need for a daylight air patrol well to the Eastward, say between ORAN or ARZEU and CAPE PALOS (Spain). Had we any aircraft available for this patrol it would only be necessary to ask for the use of a base in ALGERIA, but under present conditions the French might possibly be persuaded to provide such a patrol, or perhaps have already contemplated something of the kind themselves.

It is presumed that the French would have no objection to a notice to mariners at the outbreak of war advising all shipping bound through the Gibraltar Straits to call at Gibraltar for a brief examination. Such a step would help to overcome the difficulties caused by the possibility of shipping making use of neutral territorial waters in the Straits.

309. *Captain A.R. Hammick[1] to Rear Admiral J.D. Cunningham*

[ADM 116/3896] H.M.S. *Sussex*, at San Raphael
22 March 1939

No. 564/191.

I have the honour to report the following proceedings of His Majesty's Ship under my command.

2. In accordance with the Commander-in-Chief's signal timed 0051/13, *Sussex* slipped and proceeded to sea from Gibraltar at 0900 on Monday, 13th March.

---

[1] Capt [later RA] Alexander Robert Hammick (1887–1969). Commanded cruiser *Sussex*, 1938–40; retired, 1941; Chf Staff Officer to Flag Officer-in-charge, Greenock, 1940–42.

THE APPROACH OF WAR, 1938-1939   493

3. At 0600 the following morning I met *Impulsive* and *Intrepid* in position 39° 05' N, 0° 05' E, and after transferring mail to *Impulsive* and receiving Eastern Area Temporary Memoranda, etc. took over patrol VG.

During the forenoon I sighted and passed the time of day with the Spanish Nationalist Armed Merchantman *Mar Negro*; this vessel was subsequently sighted at regular intervals and appeared to be on a patrol approximately similar to our own.

At evening quarters paravanes were recovered and streamed for exercise by the boys and Ordinary Seamen.

About 1900, the British ships *Stanhope* and *Stancor* were observed leaving Valencia harbour, the latter with a number of political refugees aboard, according to a report from the Consul, Valencia. Both these vessels were kept in sight by *Sussex* until 0100, when the *Stancor* steamed off to the southward. About 0500, when off Gandia, a light was observed in the distance aproaching the merchant ship, and on investigation this proved to be the Nationalist destroyer *Melilla*. However, on seeing the *Sussex* the destroyer disappeared to the southward, to return some time later with the *Mar Negro*. In the meantime the *Stanhope* had taken the opportunity to enter Gandia harbour. The *Melilla* and *Mar Negro* then continued to patrol together.

4. On Wednesday, 15th March, nothing of interest occurred. An air raid was heard to be in progress on Valencia about 1515, and was maintained for some little time. Seaboats were exercised at evening quarters.

5. At 1000 on Thursday, 16th March, off Gandia, received fresh vegetables and eggs from *Icarus*.

At 1030, Consul Valencia reported that he observed a British ship of the Stan line, apparently still outside the three mile limit, being approached by the *Canarias*. At this time paravanes were being recovered and streamed for exercise by the Boys and Ordinary seamen, and the ship was at the southward end of the patrol line VG. Course was at once set for Valencia and speed increased to eighteen knots.

At 1145 sighted *Stangate* and *Mar Negro* in company, off Valencia, about five miles from the coast. On my approach, however, the latter made off to the southward, and the *Stangate* continued on her course of 035°. No reply was given to my signal in international code asking where she was bound, and my semaphore message to the same effect was answered in a manner that was quite unintelligible. A moment afterwards a cap was seen being waved out of a for'd scuttle, followed by 'S.O.S.' flashed by a torch. I thereupon ordered the ship to stop engines, but it was sometime before the vessel was persuaded to obey, the final signal being made in morse code on the syren. (This delay was subsequently explained by the Captain of the Prize crew which was aboard, to be due to the fact that neither he nor any of his crew understood English, and could not in consequence understand the code books. The English master was not on the bridge and knew no Spanish in any case.)

While paravanes were being recovered, *Sussex* steamed round the *Stangate*, but no sooner had engines been stopped that *Stangate* was seen to get under way again. Again the order to stop was sounded on the syren and this time was obeyed immediately.

At 1330 the cutter was sent across with Commander Gilmour and Paymaster Sub-Lieutenant Rigge (as interpreter), and it was soon learned that the ship was in the hands of a prize crew from the *Mar Negro*, having been boarded inside the three mile limit. This crew consisted of an officer of Sub-Lieutenant's equivalent rank, one Petty Officer and about 14 seamen; all were armed, some with service type rifles but the majority carried automatic rifles of German manufacture.

As a result of detailed enquiries it was confirmed that the ship had indeed been stopped inside the three mile limit, and the English Master cheerfully admitted that he had been entering Valencia harbour when stopped, being at the time about two miles distant. He had no wireless transmitter but had a receiving set, and stated that he had no idea that a blockade was in force![1] His written statement is attached with a certificate on the reverse side, signed by the English Non-Intervention officer, to the effect that the ship was carrying neither cargo nor passengers when arrested. A report by the master of the previous movements of the ship is also attached.

6. The Captain of the Prize crew could not explain why he was steering 035° when his destination was Palma, nor why he was still flying the Red ensign and the Non-Intervention flag; but it is evident that he had hoped to evade the *Sussex*. No reasonable explanation could be given moreover why the *Mar Negro* had sheered off at my approach, but again it is apparent that she had hoped I should not prove too inquisitive and would allow the *Stangate* to continue on her course, in the normal manner.

7. On receiving information that the Master had signed a statement to the effect that he had been seized inside Spanish Territorial Waters, I decided that the ship was a lawful prize. I therefore decided to take no further action and informed the Captain of the prize crew that he should not be flying an ensign. The latter, on hearing this, did not manage to conceal his obvious delight and there followed much halloing between

---

[1] The British and French governments had recognised Franco on 28 February. On 6 March the Nationalists announced a blockade of the Spanish Mediterranean coast from Sagunto to Avra, that is most of the coastal territory still in the hands of the Republic. Territorial waters were closed to all navigation without authorisation.

him and his 'Comandante' in the *Mar Negro*, who by this time had returned and was lying close off his port quarter.

8. The cutter returned to the ship about 1515, and the *Stangate* proceeded on her journey to Palma, now steering the correct course.

Several signals were exchanged between the *Mar Negro* and *Sussex*, in Spanish, the final being 'Always gentleman, I thanks you' in reply to mine 'A quien madruga, Dios le ayuda'.

9. At 1815 I sighted a vessel about three miles outside Gandia. She turned out to be the *Stanhope* which had been reported to be due to leave Gandia at 1800 with 60 refugees on board. *Mar Negro* was near me at the time, and we both closed the merchant vessel. *Mar Negro* then ordered her to stop. I therefore instructed *Stanhope* to follow me, she being at this time about a mile outside territorial waters. *Mar Negro*, however, complained that the ship had been inside the three mile limit when ordered to stop, but I denied her claim and continued to escort *Stanhope* to the South East at her maximum speed of eight and a half knots. *Mar Negro* remained in company. I ascertained that the Master of the *Stanhope* had refused to take any passengers and had a cargo of oranges for Liverpool and London and was calling at Gibraltar en route. After some communication with *Mar Negro* I made 'Good night and Good Luck' to which she replied the same. I then considered the incident closed. At about 2030 *Mar Negro* attempted to pass from starboard to port between myself and *Stanhope*, having previously carried out a similar manoeuvre in the opposite direction, successfully. This time he was less successful and was rammed amidships by the *Stanhope*. He at once protested 'energetically' against the conduct of the merchant ship and said that he was ascertaining his damages; these were later reported to be 'of some importance'.[1] *Stanhope* suffered slight damage to the upper deck, most of the force of the impact having been taken by the bower anchor! My offer of assistance to the *Mar Negro* was refused with many thanks, and again a strong protest against the conduct of the *Stanhope*. Shortly afterwards the *Mar Negro* switched off Navigation lights and disappeared to the Northward, her arrival at Palma being reported the following day.

10. On Friday, 17th, during the forenoon, exercised General Quarters, with breakdowns.

11. At 1000 on Saturday, 18th, had a rendezvous with *Intrepid* thirteen miles off Gandia to transfer 9 boxes of aircraft parts for *Devonshire* and to collect six bags of mail for *Penelope*.

[1] When Hammick observed the *Mar Negro* anchored off Gandia on 30 March, he reported the damages 'consisted of a few dents and scratches on her port quarter'. H.M.S. *Sussex*, Report of Proceedings, 4 April 1939, ADM 116/3896.

During the afternoon *Intrepid* reported from Gandia that S.S. *Barrington Combe* was due to leave Valencia at 1830 with sixty refugees; arrived off Valencia about 1800 and found *Mar Cantabrico* (who apparently relieved *Mar Negro*) and *Melilla*, also closing the port. At 1830 *Sussex* was about a mile inshore of the Nationalist warships, steaming to the Southward, and a vessel was at this time observed leaving the port. A course parallel to that of the Spanish warships was maintained and speed adjusted so as to remain as long as possible directly between the *Mar Cantabrico* and the *Barrington Combe*. During this time the following signal was made to the *Mar Cantabrico*, with a 10″ signalling projector: 'Do you anticipate strong wind from the North tomorrow as we are expecting our mails?' Some time later the following reply was made: 'I am very sorry but I have not received the weather forecast for tomorrow.'

In the meantime the merchant ship had gained the safety of the open sea, so I made 'Good night' to the Nationalist warship and continued southwards on patrol VG.

12. On Sunday, 19th, an incident of minor interest occurred when the *Stanbrook* reported herself to be molested by a Nationalist destroyer, giving her position as four miles South East of Benidorme Island. I altered course and increased speed to investigate, but two hours later when asked whether she still required assistance the merchant ship replied asking for advice as to whether she should enter Alicante, she had been ordered by the destroyer not to do this, and the latter was still in company with her. I informed her that if she entered territorial waters the ship would be arrested, in which case I was not authorised to assist her. She thanked me for the advice upon which she evidently intended to act, and I resumed VG patrol.

13. The weather deteriorated considerably on Sunday and at 1945 *Penelope* reported that she had lost a man overboard and would be late relieving me; the man was never recovered, and *Penelope* relieved me at 2130. Mail and official correspondence, etc. were transferred in a cordite case secured to a grass hawser, this having originally been tailed on to a life buoy which was veered astern to *Penelope*. I was impressed by the manner in which *Penelope* was handled during this difficult manoeuvre.

Left *Penelope* at 2300 and proceeded for San Raphael.

* * *

310. Rear Admiral Tovey to Pound

[ADM 116/3896]　　　　　　　　　　　　　　H.M.S. *Galatea*, at Palma
　　　　　　　　　　　　　　　　　　　　　　　　　3 April 1939

SECRET

No. 288/26/2/6.

I have the honour to forward the following report of H.M.S. *Galatea* wearing my flag.

2. H.M.S. *Galatea* sailed from Gibraltar at 1400 on Friday, 17th March, and arrived at Palma at 0800 on Sunday, 19th March, 1939.

* * *

3. During the forenoon I took over the duties of Senior Naval Officer, Eastern Area from Captain G.C. Muirhead-Gould, D.S.C. and later met the Consul onboard *Devonshire*.

4. Mr. Waldron, the Interpreter, was transferred from *Devonshire* to *Galatea*. *Devonshire* sailed at 1730 for Monte Carlo.

* * *

6. At 1400 [20th March] I received information from the Consul, Valencia, informing me of an imminent attack on Cartagena by 25,000 Communist troops and requesting the visit of one of H.M. Ships.

As Cartagena is in the patrol area administered by the Rear Admiral-in-Charge, Gibraltar, I referred this news to him in my cypher message 1430/19th March, repeated to you, and informed the Consul Valencia that a decision would be communicated later.

Information was received on 20th March that H.M.S. *Wishart* had been instructed to call at Cartagena on 21st March, and her time of arrival was passed to the Consul, Valencia.

* * *

11. On Wednesday, 22nd March, with the permission of the authorities, my Squadron Gunnery Officer and the Consul visited the *Stangate*, and arrangements were made for the supply of more coal and of the few requirements of the Master and crew. The guard was carrying out his duties with tact and courtesy, and the Master and crew were entirely satisfied with their treatment.

12. At about 1900 on 22nd March, as reported in my 2009/22, *Mohawk*, on passage from Golfe Juan to patrol V.G., rescued the crew of a Spanish Government[1] aircraft from Majorca which had crashed in position 39° 38′

---

[1] With the recognition of Franco, units formerly designated as 'Nationalist' or 'Insurgent' were frequently referred to as 'Government', a term formerly reserved solely for Republican units.

N 00° 55' E. The crew of five Germans had been in the water for four hours. *Mohawk* sank the aircraft and brought the crew to Palma arriving at 2345 and sailing again for patrol.

13. At 2220 on Wednesday 22nd March a signal was received from *Penelope* reporting that the Spanish Government armed merchant cruiser *Mar Cantabrico* claimed to have captured S.S. *Stancor*. Although the Captain of the *Penelope* suggested that an element of doubt existed as to his estimate of the position of *Stancor* when he saw *Mar Cantabrico* signal to her to stop, it appeared that there was no doubt that the *Mar Cantabrico* had not established effective control of the *Stancor* whilst she was still in Territorial Waters and that the case was therefore covered by I.S.C.W. Section A.3 para. 18. I instructed *Penelope* to act accordingly.

14. In accordance with your 2355/22 I proceeded to sea in *Galatea* at 0600 on Thursday, 23rd March, to meet *Penelope*, who was escorting S.S. *Stancor* towards Palma, with *Mar Cantabrico* in company. These three ships were sighted at 0710 and the course of the *Stancor* was adjusted to keep well clear of Territorial Waters. Contrary to my expectations the weather and sea were suitable for boatwork. This was extremely fortunate, for shortly afterwards the weather so deteriorated that no further opportunity would have occurred for several days. At 0810 S.S. *Stancor* was instructed to stop engines and the Captain and Navigating Officer of *Penelope* came over to *Galatea*.

15. From the evidence of these two officers I came to the following conclusions:–

(a) That when *Mar Cantabrico* was first seen by *Penelope* to order *Stancor* to stop engines, the latter was definitely outside Territorial Waters.

(b) That *Mar Cantabrico* based her claim on having signalled 'stop' to *Stancor* by directional light some 20 minutes earlier, at which time *Stancor* was well within the three mile limit. I do not question this statement and though *Stancor* denies having seen this signal, his evidence is not considered to be entirely reliable. *Mar Cantabrico* however, took no further steps to enforce compliance with his order until *Stancor* was unquestionably outside Territorial Waters and later stated that he refrained from doing so owing to the presence of *Penelope*. This may well have been true, as *Penelope* passed between *Mar Cantabrico* and *Stancor* shortly after the alleged time of the first signal.

16. The evidence, in my opinion, established clearly that the *Mar Cantabrico*, very possibly owing to a commendable desire to avoid an 'incident' with a British warship, failed to take the necessary steps to gain effective control over *Stancor* and therefore, whatever the time of the first

signal the *Stancor* was entitled to protection under I.S.C.W. Section A.3 paragraph 20(d).

17. *Mar Cantabrico* had also complained that *Stancor* had made a deliberate attempt to ram her, but from the evidence of the Captain of the *Penelope* it is clear that what actually happened was that *Mar Cantabrico* then well outside territorial waters, altered course across *Stancor*'s bows in an endeavour to force her off her course and I am satisfied that *Stancor* was not to blame. It will be remembered that *Mar Negro* tried the same manoeuvre with *Stanhope* on 16th March, and brought about a collision (*Sussex*'s 2226/16).

18. At 1005 *Penelope* was instructed to shape course with *Stancor* towards Marseilles, well clear of territorial waters and in accordance with your 2355/22 I reported the results of my investigation to the Admiralty. *Galatea* and *Mar Cantabrico* remained in company with *Stancor*. At 1832, having received Admiralty's 1327/23, I instructed *Stancor* to proceed on her voyage, informing *Mar Cantabrico* that I had been ordered to do so. *Penelope* was detached to Malta. *Mar Cantabrico* parted company shortly afterwards, having made a signal 'I very much regret your resolution. I am notifying my Admiral. Goodbye.' The conduct of *Mar Cantabrico* was correct and most courteous throughout and her standard of signalling in English was high.

19. *Galatea* remained in company with *Stancor* until 2359 and then returned to Palma, arriving at 0700 on Friday, 24th March.

* * *

21. H.M. Consul had informed me that Admiral Moreno was very disturbed over the *Stancor* incident.

At my interview [25th March] I was at pains to explain to him that it was on account of the misunderstanding that had arisen that you had instructed me to go to sea in order personally to investigate the case, and that throughout my investigation I adopted an entirely impartial attitude. I explained to him at length all the circumstances of the case and he agreed that effective control had not been established within territorial waters. I further explained to him that when the *Stancor* was still within territorial waters there was no reason why the *Mar Cantabrico* should not have fired a blank round, followed if necessary by a shotted round across the bows of the *Stancor*, which would have entirely altered the case. Admiral Moreno demurred at the idea of firing a gun at all, but admitted that there was no reason why such action should not have been taken. There is no doubt that the attitude both of Admiral Moreno and of the Captain of the *Mar Cantabrico* was out of deference to H.M. Ships. Whilst endeavouring to avoid undue criticism of the Captain of the *Mar Cantabrico* I demonstrated to Admiral Moreno how easy it would have been for the *Mar Cantabrico* to have obtained control of the *Stancor* inside territorial waters if he had taken up a more suitable initial position. At the end of the interview Admiral Moreno was quite satisfied with the decision arrived at and was most friendly. The incident has in no way affected the pleasant relations existing between the two Services.

* * *

25. On Sunday, 26th March, the Consul at Valencia informed me by signal that 200 Italian prisoners-of-war were being concentrated at Valencia for exchange. The Republican authorities were not willing to release them unless they could be carried to a non-Spanish port and the Consul therefore asked whether they could be taken to Marseilles. Forty of these prisoners were to be exchanged against British prisoners in Nationalist hands, and the Consul requested that these should be given preference. I had been awaiting an opportunity of sending to this Consul his mails and the stores which *Penelope* had brought from Malta, so I decided at the same time to discuss the arrangements for the prisoners with him.

26. *Galatea* sailed from Porassa Bay at 1730 anchoring off Gandia at 0630 on Monday, 27th and landing the stores and mails.

27. The Consul, Mr. A. Gooden, came off to breakfast and told me the situation ashore. The S.S. *Stanland* which, with the *Atlantic Guide*, had been granted immunity by the Spanish Government to enter Valencia with a cargo of foodstuffs, was now in that harbour and I learnt from the Consul that she was due to sail about 1500. As the immunity did not cover departure, I deemed it advisable for *Mohawk*, now on patrol V.G., to be off Valencia when the *Stanland* came out. I therefore ordered *Mohawk* to close and collect the outgoing mails and she returned to her patrol during the forenoon.

28. *Stanland* sailed earlier than was expected, passing the three mile limit at 1426 without being molested, but 20 minutes later *Mohawk* reported that *Mar Negro* and a Spanish destroyer were closing her at high speed. *Mohawk* also closed and was joined by *Nubian*, who had arrived from St. Maxime to relieve *Mohawk*. The Spanish warships remained in company with *Stanland* for an hour before returning to their patrol off Valencia.

29. While *Stanland* was sailing, *Atlantic Guide* proceeded to enter harbour. *Mohawk* reported that a seaplane was attempting to stop *Atlantic Guide*, but the shore batteries opened fire and drove the seaplane off. It is thought that the action of the seaplane in endeavouring to stop the *Atlantic Guide* was a case of mistaken identity. After their failure with the

Stancor it is believed that the Spanish Naval Authorities decided to employ a seaplane to force the *Stanland* to acknowledge control inside territorial waters. Unfortunately for them the seaplane arrived half an hour too late and was probably misled into thinking the *Atlantic Guide* was the *Stanland*.

30. The Master of this ship had, during the forenoon, asked me for an assurance that the Republican Authorities at Valencia would prevent refugees from forcing their way onboard his ship. The Consul informed me that the harbour authorities had received orders to prevent any refugees from embarking, even if the ships were willing to take them. This surprising information was confirmed by Mr. Apfel who, as the representative authority at Gandia, had received a copy of the order. He stated that the order, which was signed by Colonel Casado,[1] directed that no Spanish subjects were to be allowed under any circumstances to embark in British ships leaving the country, owing to the instructions issued by the British Board of Trade. Harbour authorities were held responsible that the ban was enforced. Mr. Apfel agreed to provide me with a copy of the order, but the events of the next few days unfortunately made this impossible.

31. *Mohawk* left for Marseilles at 1700 after having closed *Galatea* and the Captain came onboard to make a verbal report of these incidents.

32. *Galatea* sailed at 1730 and arrived at Palma at 0700 on Tuesday, 28th. At 1730, the decyphering of a long and rather corrupt message from the Consul Valencia (2350/27) was completed. From this and subsequent signals and from the wireless news reports it was evident that it might shortly become necessary to embark not only the 200 Italian prisoners but also a large number of political refugees. The situation of the ships available was as follows:–

Galatea. At Palma. Complete with refugee stores. 30% short of fuel.

Mohawk. Arrived Marseilles 1630, 28th March, with mails and passengers, having been at sea for 7 days.

Nubian. Due to be relieved on patrol on the 29th, embark 50 Italian prisoners and proceed to Marseilles.

Sussex. Proceeding with moderate despatch to relieve *Nubian* on patrol at 1030, 29th March. No refugee stores.

Maine. At Golfe Juan. Accommodation for 300 refugees.

[1]Colonel Segismundo Casado (1893–1968). Officer commanding Republican Army of the Centre and leader of a group of officers who staged a coup in Madrid on 4 March against the Negrín government because they feared the latter had added too many Communists and this would prevent negotiation of a satisfactory peace with Franco. The officers established a 'Junta of National Defence'. The coup led to fighting in Madrid in what amounted to a civil war within a civil war in which the forces supporting the coup prevailed, but subsequent negotiations with Franco failed and the final collapse of Republican forces soon followed.

33. During the night I heard from the Consul Valencia that the Republican Authorities no longer objected to the Italian prisoners being taken to Palma and that the whole party, numbering 167, would be ready to embark in the morning. I therefore directed *Nubian* to embark them all for Palma. At 0748, Wednesday 29th *Nubian* reported that he could not get in touch with the Consul and that the situation at Gandia was chaotic. On receipt of the signal I ordered *Sussex* to proceed with despatch. Further signals from *Nubian* indicated that the Consul had arrived onboard. There appeared to be many armed refugees present, and at 1030 I ordered the embarkation to be restricted to Italian prisoners. Meanwhile a number of Italian bombers had been observed to leave Palma and I considered that they might be going to bomb Gandia. On receipt, therefore, of *Nubian*'s report (1035) that the situation was deteriorating and that it was doubtful whether the Italians could be embarked, I instructed her to confine her action to saving British lives.

34. I had asked the Consul, Palma, to come onboard as soon as possible in order that I might learn from him the views and intentions of the Nationalist Government authorities. It took longer than I expected for him to complete the necessary enquiries but, when he arrived, he informed me of the intense resentment that the evacuation of Republican leaders in H.M. Ships would cause. I reported this to you in my 1100. At 1143 *Galatea* weighed and proceeded at 25 knots to Gandia.

35. My last signal to *Nubian* had crossed another from her reporting an improvement in the situation at Gandia. The main body of armed refugees, who had arrived at Gandia in the hope of embarkation in British ships, had moved on Southwards to Alicante, under the false impression purposely given to them by Mr. Apfel, that there were some ships there waiting to evacuate them. As soon as the main body had passed, the embarkation of the Italians began and *Nubian* reported that it would probably be completed by 1430.

36. The Consul, Valencia, was now onboard *Sussex* and at 1222 *Sussex* requested approval to embark Colonel Casado and his staff. I approved their embarkation, provided that they clearly understood that their final disposal would depend on the decision of H.M. Government. At the same time I requested instructions as to their disposal.

37. *Galatea* arrived Gandia at 1650. After consultation with the Consul and with the Captains of *Sussex* and *Nubian*, I decided to transfer the 167 Italians from *Nubian* to *Sussex*, the latter sailing for Palma on completion of the transfer, and to sail *Nubian* for Marseilles with various persons whose passages had previously been arranged. *Galatea* would remain at Gandia to embark the political refugees. *Sussex* sailed at 1920 and *Nubian* at 2220, the latter being instructed to fuel at Marseilles before returning.

38. Colonel Casado himself had refused to embark under the conditions I had imposed, but 15 of his entourage, together with the wife and four children of one of them, were in *Sussex* when I arrived. These were transferred to *Galatea* before *Sussex* sailed. Colonel Casado and the five remaining members of his staff came off at 2300, having reconsidered their decision on the recommendation of the International Committee for the Co-ordination of Assistance to Republican Spain. Three members of this committee, M. Forcinal,[1] Sir George Young[2] and Lord Faringdon[3] had been onboard earlier in the evening and had represented that they were in a position to assist in selecting the Republicans qualified for evacuation under I.S.C.W.101 paragraph 3.

39. The first intimation I received of this Committee was when the Consul General Marseilles asked for a passage to Valencia for eight of its members (your 2139/24). In this signal they were described as 'The International Delegation for Relief and Refugees in Central Spain'. And this was all that was known of them when they appeared onboard on the evening of 29th March. It later became apparent that they were members not of an impartial charitable organisation but of one with very definite Left Wing tendencies. Some of them were in the habit of giving the Communist salute to their friends, Lord Faringdon going so far as to stand up in the stern sheets of *Galatea*'s boat carrying him to *Sussex* and give this salute to the refugees in *Galatea*. The same gentleman admitted that the organisation was, in some aspects of its work, acting directly contrary to the wishes of H.M. Government.

* * *

Although M. Forcinal, a French Deputy was the actual head of the Committee, Sir George Young conducted its dealings with me. I informed the Captain of *Sussex* and the Principal Medical Officer of *Maine* that the recommendations of this committee should be accepted with caution.

---

[1] Albert Forcinal (1887–1976). Socialist deputy (Eure) in Chamber of Deputies, 1928–42 and National Assembly, 1946–55; refused to participate in vote of full power to Petain in 1940 and subsequently joined the Resistance; arrested by Gestapo and imprisoned in Buchenwald, 1943–45; Sec of State for Veterans and War Victims, Oct–Nov, 1947.
[2] Sir George Young (1872–1952). 4th Baronet. Supporter of Labour party and author of *The New Spain*, 1933.
[3] A. Gavin Henderson (1902–77). Succeeded grandfather as 2nd Baron Faringdon, 1934. After what could be described as a gilded youth, joined the Labour party and sat in the House of Lords as a Labour peer; member of parliamentary pacificist group supporting appeasement but also strong supporter of Spanish Republic and active in finding homes for Basque refugee children; as a pacifist during World War II served in fire brigade rather than armed forces; on Executive Committee of Fabian Society, 1942–66; Vice Chairman, Fabian Society, 1959–60; Chairman, 1960–61; Vice President, 1970–77; Councillor, London County Council, 1958–61; Alderman, 1961–65.

40. Late in the evening of Wednesday 29th, the Consul returned onboard and I discussed with him the instructions addressed to him by the Foreign Office in Admiralty's 1933. These directed that the qualification for evacuation in H.M. Ships 'in imminent danger of their lives' was to be interpreted in as wide and generous a manner as possible. These instructions were largely responsible for my subsequent suggestion to send *Sussex* to Alicante. The Consul spent the night onboard *Galatea*, going ashore early in the morning to arrange for the embarkation of the main body of refugees. Admiralty signal 1604/30 which stated, in effect, that the Foreign Office telegram quoted above was meant for the Consul personally and did not imply any modification of the policy of H.M. Government, was received after all the refugees had been embarked. It is suggested that, if the modification of policy given in the Foreign Office telegram was not intended, it was most unfortunate that it should have been sent out, especially as the signal was sent through Naval channels. It is observed that the telegram ended with the words 'Admiralty informed', which led me to assume that the Admiralty concurred. Procedure of this kind does nothing to assist Naval and Consular Officers in their efforts to co-operate in difficult circumstances. The reflections on Mr. Gooden contained in this message are, in my opinion, entirely unjustified. As stated in my 1030/31, I have found him energetic, level-headed and most satisfactory to work with. He is comparatively new to Valencia, but in spite of this he handled a delicate and constantly changing situation with great tact and ability.

41. The embarkation of refugees began at 0600 on Thursday, 30th, and continued until 0930, by which time a further 143 men, 19 women and 2 children had been embarked. Their control at the point of embarkation in Gandia harbour was mainly in the hands of Mr. Apfel, the British Manager of the port. He was, as usual, completely in command of the situation and his assistance was invaluable. The refugees were instructed to give up their firearms before entering the boats, a large mound of assorted weapons being collected on the jetty. More pistols were thrown in the sea as soon as their owners had made certain that they, and not only their leaders, were safely on the way off. The few remaining arms were extracted after the refugees had arrived onboard. The type of person embarked varied from apparently respectable officers and officials of moderate tendencies to the lowest criminal types, and included a number of prominent Anarchists and members of the S.I.M. (The Spanish Cheka). With the refugees came eight members of the International Committee.

The Spanish Government armed merchant cruiser *Mar Negro*, which had been in the offing since the morning, anchored at 0900 but did not send in an armed party until after the last boatload of refugees had arrived alongside *Galatea*.

42. It was my intention that *Sussex* should return to Gandia immediately after disembarking the Italians at Palma, to take over the refugees from *Galatea* whose accommodation was being strained to the utmost. But *Sussex* experienced considerable delay at Palma and was not able to arrive back at Gandia till 1630 on 30th March. On receipt, therefore, of your 2340/29, directing that refugees should be transferred to *Maine*, it was decided to keep them onboard *Galatea* until *Maine*'s arrival. So long as the weather remained settled, their accommodation on the upper deck of *Galatea* would be satisfactory.

43. To implement the broader policy conveyed in Foreign Office telegram 1933/29, I proposed to you that I should send *Sussex* to Alicante. Pending your approval, I instructed *Sussex* to proceed to Gandia at 25 knots and, on her arrival at 1630, 30th March, transferred to her some of the International Committee, including M. Forcinal, now in French destroyer *Lynx*, which had arrived in the morning and which left for Toulon at 1730. Before *Sussex* could sail for Alicante, I received Admiralty signal 1604/30 (see paragraph 40), and also learnt that the Consular representative at Alicante was safe in Valencia, and that there were only three British residents in Alicante which was stated to be in Government hands. One of these was too old to be moved and the other two had expressed definite intentions of remaining. I therefore delayed, and later with your approval cancelled, *Sussex*'s departure for Alicante.

44. *Mohawk* arrived at 0530, Friday 31st March, bringing 25 tons of Red Cross stores and two American members of the International Committee for the Assistance of Child Refugees in Spain. These were all landed at Gandia, after which *Mohawk* sailed for Palma.

45. *Mar Negro* was now anchored off Gandia and I sent an officer to enquire from her Captain whether British merchant ships which had previously traded with Republican ports would be allowed to enter ports now in Government hands. This information had been asked for by Admiralty signal 1915/30. At the same time I informed the Captain of the *Mar Negro* of the identity of *Nubian*'s passengers and of the destination of the stores now being landed and assured him that no more refugees other than British subjects would be embarked. He was most cordial and promised to make the necessary enquiries of his Government and to inform me by signal of the reply. This, when received, was passed to Admiralty in my 2019/31 and 2230/31. It was to the effect that British ships might enter Spanish ports, but that the Spanish Government would like to know the names of the ships concerned; and that it was absolutely forbidden for any person to leave Spain in such ships.

46. The Spanish Government troops occupied Gandia during the afternoon, arriving both from Valencia and from Alicante. The telephones

had been taken over by the Army and no private calls were allowed; the roads were blocked with troops; so that I experienced some difficulty in getting into touch with the Consul at El Perello. When he finally managed to reach Gandia, the new local authorities tried to stop him from coming onboard, as no one was allowed to embark. I got in touch with the Captain of the *Mar Negro*, who kindly took steps that no obstacles were put in the way of the Consul coming and going as he wished.

47. At 0900, Saturday 1st April, *Maine* arrived and anchored. In view of the large proportion of undesirable characters among the refugees I had arranged, with your approval, to embark in *Maine* a Royal Marine guard of one officer and twenty men from *Sussex*. To avoid offending against the Geneva Convention the Marine guard carried no ammunition.[1] I also sent a list of the refugees with their previous occupations, for forwarding by air mail to the Admiralty, to arrive before those refugees who were to be allowed to enter the United Kingdom. The transfer was carried out by the boats of *Galatea* and *Sussex* and was completed by 1100, the International Committee being sent to *Maine* at the same time. *Maine* sailed at 1600 with instructions to arrive at Marseilles at 0630 on Monday, 3rd April.

48. *Nubian* arrived from Marseilles at 1100, bringing mails for all ships and stores for the International Red Cross. Some difficulty was experienced in obtaining permission to land these stores, but this was finally done on the following morning. It is understood that the Spanish Government authorities are likely to commandeer these stores, possibly for issue through their own relief organisation, and Consul General, Marseilles has been informed to this effect.

49. On Saturday 1st April, I reviewed the future requirements in the Eastern Area in view of the end of the Spanish Civil War, and, in my 1206/1, suggested that *Galatea* and two destroyers would be adequate, providing one ship at Palma, one at Gandia until normal communications are re-established and one maintaining communication with Marseilles.

This signal crossed your 1340/1, which I did not receive until late the same night.

50. On receipt of your 1911/1 early on Sunday, 2nd April, directing that *Sussex*, *Maine* and *Nubian* should proceed to Malta as soon as I could dispense with them, I ordered *Sussex* to sail for Malta. *Sussex* sailed at 0700.

51. Later in the forenoon I detached *Nubian* also, and she sailed for Malta at 1030. I also instructed *Maine* to proceed to Malta after disembarking refugees at Marseilles.

---

[1] Presumably this was because *Maine* was a hospital ship.

52. The Consul having informed me that he was quite content to remain without wireless communication for two days until the *Zulu* arrived and that he had established satisfactory relations with the new authorities at Valencia, I sailed for Palma at 1800.

53. *Galatea* arrived at Palma at 0700 today, Monday 3rd April. After a discussion with the Consul, I made my 1203/3 suggesting that the force maintained in the Eastern Area could be reduced to two ships.

311. Pound to Admiralty

[ADM 116/3844]　　　　　　　　　　　　　　　　　　　2 April 1939

SECRET.

[Telegram] Addressed Admiralty.

IMPORTANT.

659. Personal for D.C.N.S. for First Sea Lord if not away.

Part I.

First Sea Lord's private letter of 28th March agrees with my proposals for movements of Fleet during forthcoming month and may be summarised as follows:–

(A) In view of the international situation the Fleet should be as far as possible immediately ready for any emergency and should not have to make a large number of movements at last minute when there might be considerable political objections to them.

(B) In order to implement (A) that [the Fleet] should not carry out normal Summer Cruise to Adriatic or Italian ports and in fact that no cruise programme should be issued.

The abandonment of Summer Cruise would necessitate giving up visits to Athens and Roumanian and Bulgarian ports during July about which I have already been in communication with respective Ministers. I feel however that it is most important at present time not to have any commitment and First Sea Lord is in agreement that it is undesirable that ships should visit Black Sea ports whilst situation remains so uncertain.

(C) That Fleet should proceed to Eastern Mediterranean using Alexandria, Port Said and Haifa and should [undertake?] short cruises from these places to Greek and Cypriot ports should international situation allow.

(D) Advantages of above arrangements are:–

(i) Removal of possibility of air attack on Fleet at Malta.

(ii) Fleet can be kept tuned up particularly in anti-aircraft and anti-submarine work.

(iii) Our arrangements in Eastern Mediterranean would be perfected before emergency occurred.

(iv) Presence of Fleet in Eastern Mediterranean should have heartening effect on Eastern Mediterranean countries.

2140/1.

670. My 2140. Part II.

(E) It remains to decide when move to Eastern Mediterranean should be made. Unless it can be stated with certainty that an emergency will not arise shortly I am most strongly of opinion that it should commence as soon as possible subject to the visits to Italian ports which have already been arranged not being interfered with. Visit of H.M. Ships *Bulldog*, *Glorious*, *Inglefield* and *Mohawk* and 6th Division to French ports which takes place after 12th April to be cancelled. This would necessitate abandonment of most of second series of H.M.S. *Centurion* firings but ship would work up at Alexandria and fire at H.S.B.P.T. This disadvantage is slight compared with getting Fleet to their War Stations. Only other possible disadvantage is a political one but H.M. Government having declared their policy it would seem the more we show our readiness for any eventuality the better.

(F) Propose following movements as soon as they can be arranged after 12th April:

(i) H.M.S. *Ramilles*, one division of destroyers and one submarine for destroyers' anti-submarine training at Gibraltar. This force to carry out any duty in connection with Spain.

(ii) H.M.S. *Resource* refit at Alexandria.

(iii) H.M.S. *Maidstone* to disembark submarines' spare gear and torpedoes. Embark 202 Squadron stores and sail for Alexandria.

(iv) 202 Squadron to proceed to Alexandria.

(v) Four submarines and motor torpedo boats remain at Malta.

(vi) Remainder of Fleet proceed to Alexandria.

(G) So long as situation permits ships should proceed to Malta as necessary for docking.

2217/1.

312. *Vice Admiral Layton to Pound*

[ADM 116/3844] 6 April 1939

SECRET.

[Telegram] Addressed C.-in-C. Mediterranean, repeated Admiralty.

Reception by Naples Authorities both service and civilian markedly friendly. Discussion of present situation, however, avoided except for

occasional reference by locals to difficult times. Vague rumours heard today that Italian troops landed or Italian flag hoisted in Albania and also that more troops were called up last night but no confirmation. Populace generally friendly but somewhat puzzled at visit of ships in present situation in view of tone of Italian press. Criticism of present regime is more vocal than was expected probably due to war weariness after Spanish war and apparent dislike of Germans. Doubtful, however, how much importance should be attached to these facts or in view of strength of Fascist Regime and popularity of many of its achievements. First Lord's speech and subsequent Commons discussion main news subjects Naples *Mattino* this morning, Thursday.[1]
2337/6.

313. *Admiralty to Pound*

[ADM 116/3844] 7 April 1939

SECRET.

[Telegram] Addressed C.-in-C. Mediterranean. From Admiralty.

IMMEDIATE.

In view of Albanian situation question of retention of H.M. Ships in Italian waters is now under consideration. Hope to get decision this afternoon.
1240/7.

314. *Admiralty to Pound*

[ADM 116/3844] 7 April 1939

SECRET.

[Telegram] Addressed C.-in-C. Mediterranean. From Admiralty.

IMMEDIATE.

My 1240/7, decision is that ships are to remain where they are at normal notice.

Question will be reviewed tomorrow Saturday in the light of further information.
1806/7.

---

[1] This is probably a reference to the inadvertent reference in a speech at a public ceremony by the First Lord on the night of 4 April that the anti-aircraft guns of the Fleet were manned following what turned out to be a false report that an attack on the Fleet was imminent. The speech was picked up in the press and caused a sensation. See Viscount Cunningham of Hyndhope, *A Sailor's Odyssey* (London, 1951), pp. 200–201.

315. *Pound to Vice Admiral Ford*[1]

[ADM 116/3844]　　　　　　　　　　　　　　　　　　7 April 1939

SECRET.

[Telegram] Addressed V.A. Malta, repeated Admiralty.

IMPORTANT.

Muzzle the press as far as possible regarding all movements of ships.
2246/7.

316. *Pound to Layton et al.*

[ADM 116/3844]　　　　　　　　　　　　　　　　　　7 April 1939

SECRET.

[Telegram] Addressed V.A.C. First Battle Squadron, Captain(D) 2, H.M.S. *Malaya*, H.M.S. *Mohawk*. Repeated Admiralty.

699. Keep steam at 4 hours notice. Act as far as possible as if nothing had happened subject to following.

(a) Quietly recall all officers and men from extended leave.

(b) Give leave tomorrow Saturday from 1400 to 1800 to officers, petty officers, also to selected leading rates and men of able seamen rating.

(c) If officers or men have been invited to any evening entertainments they should attend unless you receive instructions to contrary but they are not to land before 1800.

(d) You may expect further instructions tomorrow, Saturday.
2304/7.

317. *Pound to Vice Admiral Ford*

[ADM 116/3844]　　　　　　　　　　　　　　　　　　7 April 1939

SECRET.

[Telegram] Addressed V.A. Malta, repeated R.A.C. First Cruiser Squadron, Rear Admiral (D) Mediterranean, Captain (D) First 'Tribal' Destroyer, Admiralty.

---

[1] VA [later Adm Sir] Wilbraham Tennyson Randle Ford (1880–1964). Commanded destroyer depot ship *Diligence*, 1922–23; cruiser *Calliope*, 1924–25; Dir of Physical Training and Sports, Admy, 1926; CO HMS *Ganges* (Training establishment, Shotley), 1927–28; commanded battleship *Royal Oak*, 1929–30; CO, Navigation School, 1930–32; RAC Australian Sqdn, 1934–36; VA-in-Charge and Adm Superintendent Dockyard, Malta, 1937–41; C-in-C Rosyth, 1942–44.

IMPORTANT.
704. Following steps are to be taken as unostentatiously as possible.
(a) Complete all ships with Naval Victualling and Central Stores. First Cruiser Squadron to complete last.
(b) Complete Supply Ship *Reliant*.
(c) H.M.S. *Resource* to be brought forward.
(d) Nothing is to be done in H.M.S. *Maidstone* or H.M.S. *Woolwich* which would increase time taken to get them ready for service. Plans to be prepared to get them ready for service irrespective of what armament they would have but no action to be taken without further instructions.
(e) Cease repairing H.M. Ships *Penelope*, *Afridi*, *Sikh* and *Gurkha* as much as possible.
(f) As soon as possible get H.M. Ships *Liverpool* and *Gloucester* in a condition in which they can steam.
(g) *Philomel* to proceed to Alexandria.
2349/7.

318. *Admiralty to Pound*

[ADM 116/3844]　　　　　　　　　　　　　　　　　　　　8 April 1939

SECRET.
[Telegram] From Admiralty.

IMMEDIATE.
There are no positive indications that Italian movements in Albania are part of a larger plan or that there is any need for British forces in the Mediterranean to take any special precautions. It would however be likely to raise public comment here if British ships remained in Italian ports under present circumstances. Ships at present in Italian ports should therefore leave tonight and return to Malta, or visit other ports as you think fit, so as to adhere to programme dates. This curtailment of visits to Italian ports should be ascribed to the obscurity of the present situation and the desirability of avoiding embarrassing incidents to which the presence of British ships might give rise. The movement indicated in your 2233/7 if solely due to the present situation should not take place.
1420/8.

319. *Pound to V.A. Malta et al.*

[ADM 116/3844] 9 April 1939

SECRET.

[Telegram] Addressed V.A. Malta, V.A.C. 1st Battle Squadron, R.A.(D) Mediterranean, R.A.C. 3rd Cruiser Squadron, R.A.C. 1st Cruiser Squadron, R.A. Gibraltar, Admiralty.

IMPORTANT. PERSONAL.

721. Minister Athens has informed me that Greek Government has received information from sources which have proved reliable in the past that Italians intend to occupy Corfu either tomorrow Monday or shortly afterwards. I have so far received no instructions from Admiralty but it is obvious that the Fleet must be prepared to leave Malta at short notice. All available ships are to be ready for war and at 4 repetition 4 hours notice for steam. V.A. Malta pass to Captain S(1).

1438/9.

320. *Admiralty to Pound*

[ADM 116/3844] 9 April 1939

SECRET.

[Telegram] From Admiralty.

IMMEDIATE. Part One.

Italian Chargé d'Affaires assures Foreign Secretary that Italy does not intend to infringe integrity or independence of Greece. On the other hand reports from other sources indicate that Italy intends some sort of occupation Corfu between 10th and 12th April. Italian Chargé d'Affaires informed that we should take very grave view of any such act. Both Greek and Yugoslavian Governments taking precautionary measures. Cabinet considering whole situation tomorrow morning. Part Two follows.

1710/9.

Part Two.

It is considered inadvisable that Fleet should concentrate in Malta. Ships on passage and those mentioned in your 1343/9 should rendezvous approximately 50 miles south of Malta until the situation clears. Units requiring fuel should proceed singly to Malta and store if necessary. Ships now at Malta should not leave as it is not desired that any movements should be made which might be construed as preparations

for hostilities. Further instructions may be expected after Cabinet meeting tomorrow.
1715/9.

321. *Pound to Admiralty*

[ADM 116/3844] 9 April 1939

SECRET.

[Telegram]

IMPORTANT.
728. Personal for D.C.N.S. [Cunningham] Reference 1501 from Admiralty. If information contained in Waterlow's[1] telegram No. 115 is correct it appears possible that we may deliver an ultimatum or note with a time limit. At any time after its delivery the Fleet may be attacked. I trust therefore that I may receive instructions to take the Fleet to sea before ultimatum is despatched. Waterlow's 116 not yet received.
2058/9.

322. *Admiralty to Pound*

[ADM 116/3844] 9 April 1939

SECRET.

[Telegram]

Personal from D.C.N.S.
Your 2058/9, this point has already been considered and will be kept in mind. Telegram 116 from Waterlow was slight addition to 115 and of no importance.
2140/9.

323. *Pound to Admiralty*

[ADM 116/3844] 10 April 1939

SECRET.
[Telegram] Addressed Admiralty, repeated V.A. Malta, R.A.C., 1st B.S., R.A.C., 1st C.S.

736. If situation develops so rapidly that it is undesirable to pass capital ships through Cape Bon area it is my intention to divert H.M.S. *Malaya*

[1] Sir Sydney Philip Waterlow (1878–1944). British Min in Athens, 1933–39.

to Gibraltar and to keep H.M.S. *Ramilles* with Eastern Mediterranean Force. I also intend to take all 8-inch cruisers with Fleet unless ordered to send one to Marseilles.[1]

008/10.

324. *Pound to Admiralty*

[ADM 116/3844] 10 April 1939

SECRET.

[Telegram]

IMPORTANT.

737. H.M.S. *Malaya* with H.M. Ships *Mohawk* and *Maori* in company pass Cape Bon at 0330 tomorrow Tuesday. 4th Division is proceeding to Malta. H.M. Ships *Warspite*, *Barham* and one division of destroyers will be concentrated south of Malta P.M. today Monday. Remainder of Fleet except H.M.S. *Penelope* in dockyard hands. H.M. Ships *Cossack* and *Aberdeen* and including H.M. Ships *Gloucester* and *Liverpool* are at 2 hours notice at Malta. I should like to get H.M.S. *Cossack* away for Gibraltar and sail H.M.S. *Arethusa* and one division of destroyers for Alexandria at earliest possible moment.

1151/10.

325. *Admiralty to Pound*

[ADM 116/3844] 10 April 1939

SECRET.

[Telegram]

No decision has yet been taken. It is hoped to be able to inform you definitely tonight Monday 10th after further meeting has been arranged. Admiralty 1715/9 should not now be taken as cancelling normal movements from Malta which should take place as you think fit.

1400/10.

---

[1] The battleship *Malaya* was at Menton. War plans called for mobilised reservists to travel across France by train to Marseilles and conveyed from Marseilles to the Fleet in a cruiser.

326. *Admiralty to Commanders-in-Chief*

[ADM 116/3844] 10 April 1939

SECRET.

[Telegram]

IMPORTANT.

Following is brief review of general situation in the Mediterranean.

France. In case of Italian attack on Greece assurance given that she will come in with us if we go to the assistance of Greece.

Italy. Formal assurance from Mussolini that he has no designs whatsoever on Greece and that he is anxious to maintain Anglo-Italian agreement. He also affirmed that Italian troops will be withdrawn from Spain. Italian Foreign Minister states this will be after victory march. Ciano[1] also gave assurances as to independence and integrity of Albania and respect for present frontiers and status quo in the Mediterranean. He has emphasised his desire to avoid any trouble with Great Britain. Withdrawal of our ships from Italian ports has been explained to him.

Greece. Mussolini's assurances have been sent direct to Greece and he has denied all rumours of hostile intentions of any sort. H.M. Government are inclined to take assurances of Italian Government as genuine.

1903/10.

327. *Admiralty to Pound*

[ADM 116/3844] 10 April 1939

SECRET.

[Telegram]

IMMEDIATE.

My 1903. Fleet can now assemble at Malta and normal docking, refits and practice programmes may be proceeded with. Precautions should be taken to the extent of keeping a proportion of A/A guns manned.

---

[1] Galeazzo Ciano (1903–44). Joined Italian foreign service, 1925; married Mussolini's daughter Edda, 1930; chief of Mussolini's press office, 1933; Undersecretary for Press and Propaganda, 1934–35; Min, 1935; served as bomber pilot in Ethiopian War, 1935–36; Min of Foreign Affairs, 1936–40; 1941–43; Amb to the Holy See, 1943; voted for the motion that led to the removal of Mussolini in the Fascist Grand Council, July 1943; subsequently arrested in the Fascist Italian Social Republic of German-held northern Italy, tried as a traitor and shot, Jan 1944.

Should you still wish, the force indicated in your 2333/7 may be sent to Alexandria, but the movement should not be made as a matter of urgency and should be announced as a normal visit.
1911/10.

328. *Admiralty to Pound*

[ADM 116/3844] 11 April 1939

SECRET.

[Telegram]

IMPORTANT.
Personal from D.C.N.S.
Admiralty 1903/10, although the Italian assurances to Greece and ourselves have undoubtedly eased the situation and it is considered improbable that any attack will be made on the Fleet at Malta, there is no doubt that tensions will continue during the discussions of the Albanian incident and until the question is settled.

In these circumstances I suggest that, while pushing ahead with the necessary dockings, refits, etc. the Mediterranean Fleet should be kept as far as possible in a state of preparedness to move to the Eastern Mediterranean.

I will keep you informed of any change in the situation and try to give you early notice should such a move appear to be imminent.
1130/11.

329. *Pound to Admiralty*

[ADM 116/3844] 13 April 1939

SECRET.

[Telegram]

IMPORTANT.
767. Experience in September crisis when Fleet was at Alexandria showed that there was considerable anxiety on the part of officers and men whose families were at Malta.

2. Apart from existing emergency Their Lordships have approved that the Fleet shall gradually concentrate at Alexandria commencing end of this month. As the Fleet will thus be away from Malta until the Autumn at least it is very desirable that as many of the families of officers and men

should leave Malta in the near future and I consider that the time has now come when I should recommend this.

3. The reaction to such a recommendation will be almost certainly that a considerable number of both officers and men will say they wish to send their families home but are not able to do so for financial reasons.

4. The importance of this is so great not only from the point of view of the efficiency of the Fleet but also from the point of view of the defence of Malta that I request Their Lordship's approval to inform all officers and men that they are recommended to send their families home now and that they will be given assisted passages either by transport or by ordinary means. It is important that all officers and men concerned should be treated alike.

5. The number concerned would be approximately 2700 and far the best method would be to provide a special transport for the purpose.

1403/13.

330. *Admiralty to Pound.*

[ADM 116/3844] 13 April 1939

SECRET.

[Telegram]

With reference to Admiralty telegram 2035 of 31st March, and your 2226/2, it seems practicable to transfer the Portsmouth floating dock to Alexandria at fairly short notice. This dock drawing 60 feet could probably be accommodated off the small mole to the east of the entrance without dredging.

It will take '*Queen Elizabeth*' and '*Resolution*' class battleships for ordinary dockings if lightened and it will take them in a damaged state up to 35 feet draft although in that condition one of these ships could probably not be raised more than about 10 feet, but subsequent action could be taken to lighten her and to proceed with repairs. The dock will also take damaged cruisers, aircraft carriers and smaller vessels. This transfer would only be carried out on the understanding that it was a temporary measure pending the completion of a new graving dock at Alexandria. Your views are requested prior to consultation with the Egyptian Government.

1655/13.

331. *Cunningham to Pound*

[ADM 1/9930] 14 April 1939

[Carbon]

MOST SECRET.

I am replying to your most secret letter of the 21st February, to the First Sea Lord, and am sorry that the answer has been delayed so long, but Naval Staff conversations with the French only commenced about a fortnight ago.

2. The line of demarcation between French and British Commands, drawn from Port Empedocle in Sicily to Ras Elmsel, 50 miles to the eastward of Tripoli, was agreed upon.

It was also agreed that it was not to be a hard and fast line never to be crossed, but only a division for general operational purposes.

3. The French minelaying facilities are very poor and they can only undertake very minor offensive operations by submarines at first.

4. It was agreed that we should not declare a dangerous area in the Messina Strait. Any action of this kind, unless it be the declaration of the mined area, would be against one of the principal tenets of International Law, which we aim at preserving.

5. The French agree in principle to the synchronisation of attacks in adjacent areas where simultaneous action would be advantageous. They point out that their Commander-in-Chief afloat is not in the same position as yourself. The French Commander-in-Chief has no discretion for undertaking any form of major operation, orders for which he receives from the French Admiralty. In their opinion, co-ordination of operations would be best arranged through the Admiralties.

6. The French intend to establish an air patrol approximately on the Oran–Cape Palos line and to the eastward of Bizerta.

7. The French agree in principle that their merchant ships should be subject to our Contraband Control in the Straits of Gibraltar. Measures to this end will be discussed.

8. In spite of paragraph 5 above, permission is now being obtained from the Committee of Imperial Defence for British and French Commanders-in-Chief of foreign stations to meet and prepare plans.

9. Your Chief of Staff, when at the Admiralty, received a summary of all conversations with the French naval authorities held up to date and will be able to give you information on these and other points.

332. *Pound to Admiralty*

[ADM 116/3844] 14 April 1939

SECRET.

[Telegram] Addressed Admiralty, repeated R.A. 1st Battle Squadron, V.A. Malta, R.A. Gibraltar.

IMPORTANT.

773. In view of presence of German submarine at Cadiz (see Consul Palma's 1112/12) it appears desirable that H.M.S. *Ramilles* should reach Gibraltar before emergency occurs. Propose therefore to sail H.M.S. *Ramilles*, H.M.S. *Severn* and second sub division for Gibraltar on completion of docking of H.M.S. *Ramilles* on 19th April.

H.M.S. *Ramilles* requires about another 10 days for working up owing to bad weather at Malta lately. Should movement of German ships as given in D.N.I.'s 1809 12 not require presence of H.M.S. *Ramilles* at Gibraltar it might even be desirable she should proceed to United Kingdom at once and complete working up at Portland so as to make passage before emergency occurs and thus avoid taking escorting destroyers away from Gibraltar. If H.M.S. *Ramilles* proceeds to Gibraltar before an emergency occurs she will provide crew for steaming H.M.S. *Active* from Malta to Gibraltar.

1149/14.

333. *Pound to Admiralty*

[ADM 116/3844] 14 April 1939

SECRET.

[Telegram]

IMPORTANT.

777. Your 1303 13th April. Much appreciate that question of providing some Mediterranean personnel in advance is being examined.

It would be most inconvenient to provide more than one 8-inch cruiser for Marseilles.[1]

This cruiser can take 800 and I hope that the number provided in advance may be such that this 800 will complete the total war requirements of the Station.

---

[1] The Admiralty informed Pound in this message that should an emergency arise it would be necessary to send at least one 8-inch cruiser to Marseilles to transport personnel to the fleet.

As regards this total number approximately 1800 were sent out in September 1938 emergency. Importance of reducing requirements was stressed at the meeting at the Admiralty in November 1938. Number to be sent was 1800 in A.L. M.00697/39 but has now been increased by Admiralty message 2000 to 2600.

Now that some mine-sweepers are commissioned and H.M.S. *Resource* is on the Station approximate total required before the outbreak of war for Malta and Eastern Mediterranean appears to be 1400 to 1500.

This number could be further reduced if a favourable decision could be given to Mediterranean 01219/0728/10F of 12th December 1938 and my 0903/11th March.[1]

1414/14.

334. *Pound to Admiralty*

[ADM 116/3844] 14 April 1939

SECRET.
[Telegram]

IMPORTANT. PERSONAL.

778. Personal for D.C.N.S. At the present time state of readiness of shore defences at the various fleet bases at Malta and in the Eastern Mediterranean is as follows:–

(A) Malta. About half the anti-aircraft guns are immediately ready, remainder having been removed to facilitate training of the crews. His Excellency the Governor ordered the above state of readiness when I informed him that the Fleet had been ordered to keep part of their anti-aircraft armament manned.

(B) Alexandria. Coast defence and anti-aircraft guns at 6 hours' notice. I have not asked for greater degree of readiness.

(C) Haifa. Coast defence guns manned at my request by improvised crews. Proper crews are still in United Kingdom.

Present state of affairs is considered unsatisfactory for the following reasons.

Neither Army nor R.A.F. appear to be kept as fully informed of the degree of tension as I am and hence may not consider it necessary to conform to the degree of readiness in the Fleet. Hence in the case of sudden attack Fleet may be defending the base instead of the base the Fleet.

Question of blackout also arises. Under the present arrangement I might broadcast information I had received from Admiralty [and] consider that

[1] Not reproduced.

blackout of whole island necessary for the safety of the Fleet but there is no certainty that this would be done on the information passed by me or S.N.O. present. During the present state of tension it appears desirable that Admiralty, War Office and Air Ministry who must be in possession of the best information as to any likelihood of a sudden attack should decide on what measures are necessary and that similar instructions should be sent to all three Services.

At the present moment I am in doubt whether the situation is sufficiently tense to require a proportion of the anti-aircraft guns of the fleet to be continuously manned. If it is then shore anti-aircraft guns should also continue to be manned.

H.E. the Governor and A.O.C. Mediterranean have seen this telegram.
1548/14.

335. *Admiralty to Pound*

[ADM 116/3844] 14 April 1939

SECRET.

[Telegram]

IMPORTANT.

Your 1403/13. It is not considered that immediate steps need be taken to remove families from Malta, vide A.T. 1303/13. Admiralty agree with you in principle and matter is being pursued with other Departments on Governor's letter to Secretary of State of 22nd February 1939 and a further communication will be made to you. You should take any steps which you consider desirable to encourage families to leave voluntarily and stop others coming observing that Mediterranean Fleet will be absent from Malta most of the summer. Special arrangements for assisted passages cannot be granted at this stage.
1851/14.

336. *Admiralty to Pound*

[ADM 116/3844] 15 April 1939

SECRET.

[Telegram] From D.C.N.S.

PERSONAL. IMPORTANT.

Your 1548/14. I am taking this up as an urgent matter. No special significance is attached at present to the movement of German units to the Spanish coast.[1] An emergency arising at an early date appears less probable. I request therefore that you will use your discretion as regards manning A/A armament.

1136/15.

337. *Pound to Admiralty.*

[ADM 116/3844] 18 April 1939

SECRET.

[Telegram]

IMPORTANT.

822. Movement of German Naval Forces to Spanish coast combined with deferring of Franco's victory march and consequent withdrawal of foreign troops from Spain gives one impression at this end that situation may arise in which H.M. Government will be loath to permit any large movement of Mediterranean Fleet. Also it appears quite possible that Axis may stage another stunt concurrently with Hitler's speech if as seems probable he will refuse to give any guarantee under existing status quo.

It appears to me desirable therefore that remainder of Mediterranean fleet should arrive at Alexandria Thursday 27th April. Under existing conditions it does not appear necessary to endeavour to camouflage where Fleet is going to and I should prefer to take Fleet straight to Alexandria so that no ship will be liable to unheralded attack by air or submarine in unprotected anchorages. I have no doubt that Italians are fully aware that it is our intention to move the fleet to Eastern Mediterranean shortly.

1458/18.

338. *Pound to Admiralty*

[ADM 1/9930] 19 April 1939

MOST SECRET.

Med. 0384/0708/50

POLICY AND CONDUCT OF WAR IN THE MEDITERRANEAN

Be pleased to lay before Their Lordships the questions of policy and conduct of war in the Mediterranean contained in the enclosures to this letter.

---

[1]The German government gave official notification that a substantial number of warships, including the *Deutschland*, *Graf Spee* and *Admiral Scheer*, would leave Germany on 18 April for exercises in Spanish waters lasting approximately four weeks. Admiralty to Cs-in-C, 12 April 1939, ADM 116/3844.

2. The importance of the earliest possible offensive action in war necessitates thorough investigation in peace of all possible war operations. Recent investigations have shown the need for answers to the several questions raised in these enclosures.

3. It is requested that I may be informed of Their Lordships views on the proposals made and that I may be supplied with copies of any draft notifications to foreign powers, or notices to mariners, which have already been or may now be drawn up in readiness for issue on the outbreak of war.

[Enclosure I]

Contraband Control at Gibraltar.

Investigation of the most effective patrols by armed boarding vessels in the Gibraltar Straits shows that this is not so straightforward a matter as it first appears.

2. Should SPAIN be hostile the selection of a patrol line or lines is a matter of avoiding the fire of any batteries sited on the Spanish European shore in addition to those at CEUTA.

3. If SPAIN however is neutral the territorial waters of both SPAIN and the Spanish colony of CEUTA must presumably be respected.

Merchant vessels hugging the European shore cannot obtain the shelter of neutral waters when east of CARNERO POINT, and they can therefore be intercepted by patrols to the Southward of GIBRALTAR.

On the African side of the Strait however the situation is governed by the degree of respect to be paid to the territorial waters of SPANISH MOROCCO. If these are respected, shipping by hugging the shore between say, TANGIER and CAPE NEGRO (TEUTAN BAY) could pass unintercepted through the Strait.

4. The provisions of the Treaty with Morocco 1856 and the Treaty of Algeciras 1906, which may be relevant to this question, are unfortunately not available to me. A recent opinion however, furnished by the Attorney General at Gibraltar to the Rear Admiral Gibraltar, concludes that there is no known treaty which deprives the SULTAN OF MOROCCO of his sovereign supremacy over the territorial waters of MOROCCO.

5. The contraband control work at GIBRALTAR would be facilitated by the issue on the outbreak of war of a notification by H.M. Government to all other powers engaged in Mediterranean and through-Mediterranean shipping advising them that a voluntary call by all merchant vessels at GIBRALTAR, when on passage through the Straits, would save them delay and inconvenience.

6. It is requested that a ruling may be given on the question of operations in war in the territorial waters of SPANISH MOROCCO, and that copies of any draft notification (para. 5 above) to be issued on the

outbreak of war may be held in peace by the Commander-in-Chief, Mediterranean and the Rear Admiral, Gibraltar.

[Enclosure II]
SUBMARINE ATTACK ON ENEMY CONVOYS TO LIBYA.

The importance of submarine patrols in the task of continuous interruption of the enemy's sea communications to LIBYA is apparent, and in fact no other type of vessel can maintain this continuous threat in the Central Mediterranean until the use of MALTA as a main base is again a possibility.

2. The effectiveness of submarine operations in the Central Mediterranean, however, and the chances of survival of the submarines themselves in that area, will be almost negligible if the ordinary system of visit and search of merchant ships has to be employed. Nor, if the patrols are to be effective, must any doubt be left in the minds of the submarine commanders as to what action they may or may not take against merchant shipping.

3. If, as is anticipated, the Italians are conducting a land campaign in North Africa it seems probable that their supplies and reinforcements to Libya will be run in escorted convoys; if they are sent in unescorted merchant ships it is clear that our submarines can take no action more drastic than visit and search, except in the case of a vessel clearly carrying troops, aeroplanes etc., or a recognised enemy fleet auxiliary (Articles 45 to 47 of CB 3012).

4. On the other hand it seems practically certain that some if not all of the vessels in an escorted enemy convoy which is without reasonable doubt bound for a Libyan port will come under one of the various categories described in article 39 of CB 3012, and that all vessels in that convoy may therefore be sunk without warning.

5. It is proposed therefore to issue instruction to all submarine commanders that in a war with Italy, when a land campaign is being conducted by the enemy in North Africa all ships in a South-bound escorted enemy convoy may be attacked without warning when encountered South of a line drawn between CAPE BON–CAPE PASSERO (Sicily) – and ELAPHONISI POINT (CRETE).

The Northern limit described above is inserted in order to prevent any possible confusion between convoys bound for LIBYA and those bound for BLACK SEA ports, which might result in unjustifiable action being taken against the latter convoys.

[Enclosure III]
OBSERVANCE OF GREEK TERRITORIAL WATERS.

Effective action against enemy and neutral merchant shipping on the routes between the DARDANELLES and ITALY is largely dependent on the degree of rigidity with which our Naval forces must respect Greek territorial waters.

2. There are broadly speaking two main routes to be considered
   (a) South of CAPE MATAPAN.
   (b) Through the CORINTH CANAL.

South of Cape Matapan.

3. The deep indentations in the South coast of MOREA fortunately provide several points at which vessels must leave Greek territorial waters if they are to avoid making prolonged detours involving by night some navigational risk. These points are, moreover, in sparsely populated districts and occasional encroachments inside the three-mile limit would probably cause little or no controversy.

Through the CORINTH CANAL.

4. Prolonged operations against shipping on this route present a more difficult problem and any scheme, such as is now projected, for the widening and deepening of the Canal is contrary to British interests from the point of view of a war with ITALY.

5. Encroachments into Greek territorial waters in or near the GULF OF ATHENS would quickly cause comment, whilst on leaving the GULF OF PATRAS shipping has a wide choice of routes through the Ionian Islands.

Before ordering operations in this latter area it would be necessary to know the views of H.M. Government as to whether the waters between the outlying islands of ITHACA, CEPHALONIA and ZANTE on the West and the mainland on the East should or should not be observed as Greek territorial waters.

6. If the attitude of GREECE seems likely to remain one of strict neutrality throughout the war, with no consequent prospects of our obtaining the use of Greek harbours, etc., it is for consideration whether more would not be gained by a policy from the outset of treating Greek territorial waters somewhat lightly and even of sinking a suitable merchant ship, or by other means stopping traffic, in the Canal when the landslide at present reported is cleared.

7. This would of course be a matter for H.M. Government to decide, and an early definition of their attitude towards GREECE would be essential to the Commander-in-Chief, Mediterranean at the outbreak of war.

[Enclosure IV]

THROUGH SHIPPING AND DANGEROUS AREAS.

The actual existence of British mines and potential minelaying craft on the Mediterranean Station makes it desirable that no time should be lost on the outbreak of war in making the most of the threat which their presence can exert.

2. The notification of dangerous areas will presumably be made by the Admiralty on the lines described in Section XI, paragraph 32 of A.L.M. 00643/38 (now superseded).

Such notification will naturally need previous discussion with the French Naval staff, but it is considered that it should be held ready in draft form in peace for immediate issue on the outbreak of war.

3. In the Eastern Mediterranean I consider the close approaches to the following enemy ports should be announced forthwith as dangerous areas, in order to interfere as much as possible with enemy movements and to impose the maximum possible mine-sweeping effort on the enemy from the outset:–

SYRACUSE, AUGUSTA, MESSINA, TARANTO, OTRANTO, BRINDISI, TRIPOLI, BENGHAZI, DERNA, TOBRUK, LEROS.

4. With regard to through-Mediterranean merchant shipping it is recommended that both the STRAITS OF MESSINA and approaches, and the MALTA Channel should be notified also to all powers as dangerous areas, and that shipping should be advised to use the deep water channel from off CAPE BON, to keep well to the South of PANTELLARIA and to pass to the Southward of Malta.

5. After a suitable interval from the issue of this notification it is proposed to instruct submarine Commanders on patrol off mineable waters off the ports in paragraph 3 above, and in the Southern approaches to the MESSINA Strait, that all enemy vessels in those areas may be attacked without warning.

6. If this policy is approved it is requested that I may be supplied with copies of the draft notifications in due course, and that I may be informed what is the length of the period of warning considered necessary before action may be taken as proposed in paragraph 5 above.[1]

---

[1] By the time these subjects had been discussed in the Admiralty and a reply prepared, Pound had been recalled to London to become First Sea Lord and Cunningham assumed command of the Mediterranean Fleet. See Docs Nos 356 and 359.

339. *Admiralty to Pound*

[ADM 116/3844]　　　　　　　　　　　　　　　　　21 April 1939

SECRET.

[Telegram]

PERSONAL.

Conversations have been authorised between the undermentioned authorities to discuss operations in the Mediterranean area and the Middle East:

(a) Commander-in-Chief, Mediterranean, with Commander-in-Chief, French Mediterranean Fleet.

(b) General Officer Commanding-in-Chief, Gibraltar; Staff Officer, representing General Officer Commanding-in-Chief, Egypt; Staff Officer, representing General Officer Commanding-in-Chief, Middle East (designate); Staff Officer, representing Commander-in-Chief, Mediterranean Fleet; Staff Officer, representing Air Officer Commanding-in-Chief, Middle East, with Commander of the French Forces in North Africa.

In order that you may start consideration of your part in these conversations, the following is the general line of action in the Mediterranean area and the Middle East which we propose to put before the French, and which we have good reason to suppose will be accepted:

(a) The general object of operations in the Mediterranean area and the Middle East is in co-operation with the French to render the Italian position in Libya and eventually in Ethiopia untenable.

(b) The general method to be adopted by the Allied Forces in achieving this object should be as follows:

(i) By sea. Operations to be designed to cut the communications between Italy and Libya whilst at the same time retaining control of the Red Sea and the entrances to the Mediterranean. Opportunities for exerting general pressure on Italy should be sought.

(ii) By land. Main offensive by French against:

(a) Libya.

(b) Spanish Morocco if Spain is hostile.

Operations by British Forces is to aim at defending Egypt and containing as many Italian Forces as possible on the Egyptian Front.

(iii) By air. Operations in defence of Egypt and in co-operation with the Royal Navy, against Italian communications to eastern Libya.

(iv) Organisation of insurrection among tribes in Libya and Ethiopia.

The actual date of the conversations under 1(b) above will be settled when agreement has been reached on the general strategical basis for operations in that area at the next stage of the Anglo-French Staff Conversations in London, due to begin on 24th April.

Further instructions in amplification of paragraph 2 will be sent as soon as discussion at the Anglo-French Conversations has reached the appropriate stage.

1740/21.

340. *Admiralty to Pound*

[ADM 116/3844] 22 April 1939

SECRET.

[Telegram]

IMMEDIATE.

Your 1700/21 and 1930/21.[1] It seems probable that the situation may become clearer after Hitler's speech on 28th April. In the meantime it is desirable that we should act on assumption that tension will not become more serious. For this reason it seems to be desirable that programmes should provide for later assembly at Alexandria of ships visiting other eastern Mediterranean ports. These visits could be cancelled at short notice if necessary.

Request therefore you will put forward revised programme providing for assembly of Fleet, other than ships mentioned at (a) of your 1708/21,[2] at Alexandria on or about 15th May.

This programme should be notified to the Press, explanation being given that it is desired to concentrate Fleet at Alexandria for exercises.

Steps to notify Greek Government and Ambassador, Cairo, will be taken by Admiralty as soon as revised programme is reported.

1210/22.

---

[1]Proposed programme for the Fleet.
[2]The ships listed as arriving at Alexandria on 28 April: *Warspite, Glorious, Bulldog* and First 'Tribal' Destroyer Flotilla.

341. *Pound to Admiralty*

[ADM 116/3844] 22 April 1939

SECRET.

[Telegram]

IMPORTANT.
880. Personal for D.C.N.S.
Your 1740/21.

(a) I lunched with Admiral Abrial[1] at Antibes and had a long talk with him and was left with the definite impression that further discussion with him would be of little value as naval strategy in Mediterranean is decided by either Admiral Esteva or Ministry of Marine.

(b) I do not consider it possible now to arrange synchronisation of operations against Eastern and Western coasts of Italy or ports in Libya. Exchange of information at the time as to intended operations is however desirable.

(c) No collaboration between myself and French authorities regarding operations mentioned in paragraph two (b) (i) [cutting communications between Italy and Libya] appears necessary so long as the dividing line between French and British areas is meridian Licata and a line across Straits of Messina as suggested in my personal letter of 21st February.

(d) The close co-operation necessary in Gibraltar and Bizerta–Malta area should be produced by the liaison officers provided for. This might be further improved by personal contact now if R.A. Gibraltar and V.A. Malta could meet their respective French authorities.

(e) I am not aware of existing operations of French cruisers, submarines and aircraft transport now in Eastern Mediterranean. If these are anything more than for local defence of Syria close co-operation with me will be necessary.

(f) If a meeting is considered necessary between myself and Admiral Esteva, it is suggested that latter who is very fond of flying should either come to see me at Malta before I leave or come to Alexandria.

(g) I am very reluctant to postpone my departure to Alexandria as apart from being divorced from fleet at what may be a critical time there are

---

[1]Vice Amiral Jean Marie Charles Abrial (1879–1962). Commanded cruiser *Tourville*, 1927–29; Contre amiral commanding 1re Division Légère, 1932–34; Sous Chef, Etat-Major Général, 1934–36; C-in-C Escadre de la Méditerranée, 1936–38; Préfet Maritime, Toulon, 1939; C-in-C Forces Maritimes du Nord, 1939–40; commanded Dunkirk sector during evacuation and thenceforth bitter against British for their 'flight' and supporter of Admiral Darlan and Vichy Government; Governor General Algeria, 1940–41; Secretary of State, Marine in Vichy Government, 1942–43; after liberation sentenced to 10 years' forced labour, 1946; pardoned, 1950.

many things to see to both as regards defence of Alexandria in harbour and organisation of the base.

2318/22.

342. *Pound to Admiralty*

[ADM 116/3844] 25 April 1939

SECRET.
[Telegram] Addressed Admiralty, repeated V.A. Malta, A.S. Malta, R.A.C. 3rd Cruiser Squadron, N.I.C. Port Said.

905. After assembly of Fleet at Alexandria 15th May all ships will complete from supply ship *Reliant* to full stowage of:
(a) frozen meat
(b) miscellaneous frozen products
(c) butter.

Desirable that supply ship *Reliant* should then be completed to full stowage in order to maintain supplies available in Eastern Mediterranean at highest possible level.

I wish to avoid sending *Reliant* to Malta to complete.

Request arrangements be made to freight replenishments to Port Said or Alexandria to arrive shortly after 15th May. Undesirable that (a) should be met from reserves held at Port Said in Army cold store. Understand (b) and (c) can be met from dues shortly expected at Malta.

R.A.C. 3rd Cruiser Squadron to report estimated quantities required by *Reliant* to complete after the Fleet has been supplied.

Anticipated subsequent monthly requirements in Eastern Mediterranean: (a) 100 tons; (b) 40 tons; (c) 20 tons.

2115/25.

343. *Pound to Admiralty*

[ADM 116/3844] 26 April 1939

SECRET.
[Telegram] Addressed Admiralty, repeated V.A. Malta.

911. On leaving Malta today Wednesday a submarine exercise was carried out. At 0925 H.M.S. *Afridi* gained contact with a submarine in position 085 degrees DELIMARA 7 miles. H.M.S. *Salmon*, the only British submarine in the vicinity, was on the surface and joined in the hunt. Contact with the submarine held continuous until 1200 when H.M.S.

Afridi was ordered to rejoin me as I did not wish to give what was obviously a foreign submarine further practice in evasion. Submarine gave whistling effect on four occasions and could be heard making bursts of speed which were very noisy. By plot submarine turned three circles one at slow speed and two at moderate speed. During the hunt one S.U.E. and one nine ounce charge were fired. Neither was answered nor was anything sighted throughout.
1953/26.

344. *Admiralty to Pound*

[ADM 116/3900] 1 May 1939

SECRET.

[Telegram]

Your 0053/19.[x] In view of probability that Turkey would be an ally in war against Italy, Their Lordships do not repeat not consider it necessary to make any further arrangements for this particular plan.
1854/1.

[x] Admiralty notation: 'Possibility of a British destroyer entering Black Sea before outbreak of hostilities in order to operate against Italian trade.'

345. *Admiralty to Pound*

[ADM 116/3845] 4 May 1939

SECRET.

[Telegram]

Your 2318/22nd April and my 1345/24th April.

(b) Concur. French Naval Staff consider that co-ordination of operations in Eastern and Western basins if desired must be arranged through the naval Staffs in Paris and London.

(c) Dividing line is Port Empedocle to Ras Elmsel. Concur that collaboration is not repetition not necessary.

(d) Concur. R.A. Gibraltar and V.A. Malta are authorised to confer personally with the French S.N.O.'s at Oran and Bizerta respectively. It is under consideration in war to appoint a British Liaison Officer to the Staff of the S.N.O. Bizerta as well as Oran.

(e) The only French naval forces to remain in the Eastern basin in war will be two submarines at Beyrouth. You are authorised to confer with the

French Rear Admiral Commanding in the Levant at Beyrouth on the employment of these submarines.

(f) Meeting with Admiral Esteva is not repetition not considered necessary. It is probable that operations of French forces in the Mediterranean will be controlled by the C.N.S. in Paris rather than by Admiral Esteva.

1111/4.

346. *Pound to Admiralty*

[ADM 116/3900]
Office of Commander-in-Chief, Mediterranean
Station
At Alexandria, 10 May 1939

MOST SECRET.

Med.0447/031/8.

STRATEGY IN THE MEDITERRANEAN.

Be pleased to lay before Their Lordships the following remarks which are forwarded in continuation of Med.01196/031/8 of 23rd December, 1938 in which the requirements of Royal Air Force aircraft for cooperation with the Fleet in a war with Italy were set out.

In Med.01061/07089 of 14th November, 1938, the essential need for a combined strategical plan was stressed, and it was proposed therein that the plan should take the form of a combined attack by the Naval, Military and Air Forces on Libya as soon after the outbreak of war as available resources made this possible.

2. It appears, however, that at the commencement of hostilities the relative strengths of the British and Italian Military and Air forces in Egypt and Libya respectively, will force us on the defensive in Egypt, and that we shall have to remain so until Egypt has been heavily reinforced, a process which will obviously take some time.

3. In the meantime, the only British force in the Mediterranean which can take the offensive against Italy is the Fleet. But in the nature of things the offensive activity of the Fleet must of necessity be limited to the exertion of economic pressure by the interruption of Italian sea communications and attacks on his Naval forces found at sea, supplemented by bombardments of Libyan and perhaps Sicilian ports, and possibly by Fleet Air Arm raids on enemy naval forces in harbour.

4. The success which may be expected to attend such operations and the period which will elapse before their effect begins to be decisive must depend largely on enemy counter measures. His surface forces may perhaps to some extent be discounted, but his large number of submarines

cannot be disregarded, whilst his air forces are obviously a very serious obstacle to sustained naval operations in the Eastern and Central Mediterranean.

5. The effect of Air Power on Naval operations is largely an unknown factor, at any rate so far as attacks on ships at sea are concerned, but events in the Spanish Civil War and to a lesser extent in the Sino-Japanese War, have shown that ships in harbour are far from immune from bombing attacks.

6. The Italians are in a position to develop heavy bombing attacks on our Fleet when at Alexandria, from Libyan airbases and from the Dodecanese, which can be readily reinforced from their Metropolitan Air Force, but we cannot bring to bear similar attacks on their Fleet. The Italians can bomb Malta with great intensity and in the opinion of some they can, and almost certainly will, wreck the dockyard and make the harbour untenable; but we have nothing with which to retaliate. Hence we shall undoubtedly be subject to attrition to a much greater extent than the Italians.

7. I do not know what the French can and intend to do to neutralise the Italian Metropolitan Air Force, but from a study of the 'Handbook on French Air Forces, S.D.132', which discloses the considerable numerical inferiority of the French Air Force compared with the Italian, it appears that the former will be quite inadequate for the object in view, having regard to the French commitments on their northern frontier in an Axis war.

8. It is, therefore, in my opinion, imperative for the early success of our arms in the Mediterranean, that plans should be made with the French now for moving a force of British bombers and fighters to Tunisia and Algeria whence they can bomb vital points in Italy and Sicily, including the Italian Fleet and dockyards at Taranto and Augusta. Even if the aircraft themselves cannot be sent till war has broken out, the ground arrangements could be progressed and it is considered that a start should be made as soon as possible to collect the stores, reserve equipment and bombs.

9. In view of the information contained in Admiralty Telegram 1854 of 1st May, 1939,[1] it is also considered that arrangements should be made for operating British bombers from the Turkish mainland against the Dodecanese.

10. I am of the opinion that the knowledge of the intention to base British bombers and fighters within comfortable range of Italy itself would act as a strong deterrent to the Italians embarking on a policy likely to lead to war with the British Empire, and would have a correspondingly heartening effect on the smaller Mediterranean Powers.

---

[1] That Turkey would be an ally. See Doc. No. 344.

347. *Admiralty to Pound*

[ADM 1/9946]   12 May 1939

SECRET.

[Telegram]

In connection with discussions with French Naval Staff it is desired to obtain from them their intentions as regards naval operations in the Mediterranean in the event of war. In order that reciprocal action may be taken by us request that you will forward at the earliest opportunity an outline of your intended operations.

1809/12.

348. *Pound to Admiralty*

[ADM 1/9946]   15 May 1939

SECRET.

[Telegram]

88. Your 1809 12th May. (a) Objects and dispositions already communicated to Their Lordships in Mediterranean War Orders.

(b) Initial round-up of enemy shipping in Eastern Mediterranean will be carried out.

(c) It is hoped that air reconnaissance will locate patrol positions of enemy submarines and if that is so and they are within reasonable distance of Alexandria intensive anti-submarine operations by Flotillas safeguarded by coastal reconnaissance flights will be carried out during opening day of war to endeavour to destroy morale of enemy submarines.

(d) Subsequent operations will be dependent to a certain extent on result of (c) and experience of air menace.

(e) Offensive submarine patrol off Messina, Otranto, Taranto, Benghazi as far as numbers permit.

(f) Continuous operations as far as number of ships permit to interrupt enemy communications with Libya and Black Sea. If Turks are with us latter will not be required.

(g) Other operations which it is proposed to carry out as opportunity offers and subject to (d) are as follows:

    (i) Carrier attack on warships in Taranto and Augusta unless as I hope shore based aircraft will be available for this.

    (ii) Bombardment, air attack and mining of Benghazi, Derna and Tobruk.

(iii) Defence of shipping between Port Said, Palestine, Syria, Cyprus [and] Alexandria.
(iv) Cooperation with Army if enemy invaded Egypt or we advance into Libya.
1532/15.

349. *Pound to Admiralty*

[ADM 116/3845] 15 May 1939

SECRET.

[Telegram]

IMPORTANT.
89. Situation regarding manning anti-aircraft guns and searchlights defences at Alexandria I consider most unsatisfactory. All 24 searchlights and 10 of the 22 guns have already been turned over to Egyptian troops. Four more guns are supposed to be turned over shortly. When this is done it is understood Officer Commanding Anti-Aircraft Brigade and the whole of the headquarters staff including majority of maintenance troops and possibly Instructors for searchlights are shortly to be withdrawn to United Kingdom leaving only one battery of eight guns manned by British troops.

It is common knowledge that Egyptians have as yet no efficient organisation for maintenance of either guns or searchlights and not much aptitude for this work. Experience is already showing that rapid loss of efficiency of guns and failure of searchlights will be the inevitable result unless considerable number of British maintenance specialists is left behind to continue present instruction for some months.

Taking into consideration the large part of vital Naval strength which is based on Alexandria I consider this gambling with our sea supremacy is quite unjustifiable. It is in any case very difficult to defend Alexandria against air attacks but from experience which has been gained it is clear that Fleet will not have the best protection possible in the circumstances when in harbour unless both anti-aircraft searchlights and guns are manned by British personnel, and I urge most strongly that this should be done without delay. At the very least General Officer Commanding British Troops in Egypt should be authorized to retain here for the next six months at Alexandria specialists necessary to continue instruction of Egyptian anti-aircraft units.

Fleet based on Alexandria is most vital target and I further urge that immediate steps should be taken to rearm anti-aircraft defences here with more modern guns and searchlights.
1701/15.

350. *Pound to Admiralty*

[ADM 116/3845] 15 May 1939

SECRET.

[Telegram] Addressed Admiralty, repeated A.O.C. Middle East.

IMPORTANT.

90. Your 1310 May 6th … R.D.F. set, I should install to west of Alexandria and should give adequate warning of aircraft from Libya which approach overland or oversea within 40 miles of coast. H.Q. Middle East considers sea patrol still necessary to locate aircraft approaching along routes more than 40 miles to seaward but if lattice masts for present set could be provided to increase its range, or better still a mains set provided for setting up at Dekdaria, necessity for sea patrol would not exist. In view of difficulties anticipated in taking up fishing boats and in maintaining an efficient patrol when they are taken up, most strongly urge that question of erecting lattice masts as an immediate requirement for the defence of Egypt and Fleet and early provision of a mains set may be taken up with Air Ministry. Mains set at Dekdaria will cover approaches from Dodecanese as well as from Libya.

1754/15.

351. *Admiralty to Pound*

[ADM 116/3845] 20 May 1939

SECRET.

[Telegram]

IMPORTANT.

Your 1701/15 has been forwarded to the War Office who are in agreement with your views. Matter is receiving immediate attention and decision is being communicated to you as soon as possible.

1217/20.

352. *Admiralty to Pound*

[ADM 116/3845] 21 May 1939

SECRET.

[Telegram]

146. Your 1217/20. I am very grateful. Request that provision of Bofors guns for anti-aircraft defence of Alexandria may also be considered.
1139/21.

353. *Rear Admiral Gibraltar*[1] *to Admiralty*

[ADM 116/3845] 22 May 1939

SECRET.

[Telegram] Addressed Admiralty, repeated C.-in-C. Mediterranean.

638. At 0915 today Monday a commercial aircraft with German markings and distinguishing letters D-AVUL flew along the south and detached Mole at height of about 70 feet distant 100 yards.

Similar flights against which Governor has protested were made by this aircraft on the 7th May and 17th May.

Aircraft appears to be making trips between Seville and Tetuan passing Gibraltar daily.[2]
1130/22.

354. *Admiralty to C.'s-in-C. Home Fleet and Mediterranean*

[ADM 1/9875] 6 June 1939

SECRET.

[Telegram] Addressed C.-in-C. Home Fleet and C.-in-C. Mediterranean, repeated R.A.(L).

It is expected that the Portsmouth Floating Dock will sail for Alexandria on 24th June and tow will occupy about five weeks.

---

[1] RA [later VA Sir] Alfred Englefield Evans (1884–1944). Dep DNI, Admy, 1929–30; Capt of the Fleet, Home Fleet, 1930–33; Cdre South America and CO cruiser *Exeter*, 1933–35; RA 2nd-in-Command, 1st CS, 1935–36; RA-in-Charge and Adm Superintendent Dockyard, Gibraltar, 1937–39; retired, June 1939; Cdre RNR of Convoys, 1939–40; Head of Naval Technical Mission, Ottawa and member of Naval Supply Council, North America, 1940–44.

[2] The registration D-AVUL belonged to a tri-motor Junkers JU52/3m in service with the German airline Deutsche Lufthansa, 1935–39. The Ju-52 was the airline's most numerous type and was one of approximately ten transferred to the Spanish airline Iberia at some date in 1939. R.E.G. Davies, *Lufthansa: An Airline and Its Aircraft* (New York, 1991), p. 44.

In order to guard against attack in the event of any deterioration of the political situation Commander-in-Chief Mediterranean is requested to take what precautions he considers necessary as regards dispositions during the passage through the Mediterranean.
1754/6.

355. Cunningham to Admiralty

[ADM 1/9875] 12 June 1939

SECRET.

[Telegram] Addressed Admiralty, repeated V.A. Malta, R.A. Alexandria.

325. Your 1754 6th June not to V.A. Malta, it is intended to have one division of 'TRIBALS' at Malta from 6th July available to move to Gibraltar or to meet dock on passage should situation require provision of close escort. One battleship and one 8-inch cruiser will also be available at Malta to provide cover, and main fleet will move westward if situation becomes serious. It is not intended to take a large force into the Western Mediterranean but to leave distant cover in their own sphere to the French. It is requested therefore that they may be informed of these dispositions and asked to take such measures as may be necessary should the situation deteriorate.
1816/12.

356. Cunningham to Admiralty

[ADM 1/9930] 22 June 1939

SECRET.

[Telegram]

404. Completion of plan for offensive action on outbreak of war is pretty dependent on reply to question put in Mediterranean letter 0384/0708/50 of 19th April.[1] In particular Their Lordships' views are needed on the matter in enclosures 2 and 3.[2] Early reply is therefore requested on those two questions. With regard to enclosure 3 it is further requested that I may be informed to what extent recent guarantees have affected situation as concerns use of Greek territorial waters at outset of hostilities having in view possibility of using secluded Greek harbours for fuelling light craft from oilers previously placed there.
2301/22.

[1] Doc. No. 338.
[2] Submarine attacks on convoys to Libya and observance of Greek territorial waters.

357. Cunningham to Admiralty

[ADM 116/3877]　　　　　　　　　　　　　　　　　　30 June 1939

SECRET.

[Telegram]

465. Intend to carry out trade attack exercise in Eastern Basin 0900 9th August to 2300 11th August.

Exercise is confined to chart 2606 and in addition ships will not proceed west of 031° East as far south as 036° North or to the north of a line from position 036° North 031° East to south point of Kaso Island thence to Santorin thence 270°.

Propose to use following anchorages: Gandia Bay and Gulf of Mirabella in Crete; Famagusta, Haifa, El Arish, Alexandria, Gulf of Kanais 031° 10′ North 028° 001′ East and Mersa Matruh.

Care will be taken not to fly over foreign territory except Crete. Request permission be obtained to fly over Crete. Request also that French, Greek, Italian and Turkish Governments be notified of exercises.

Ambassador Cairo and High Commissioner Palestine already informed. Governor Cyprus will be informed.

1209/30.

358. Admiralty to Cunningham

[ADM 116/3900]　　　　　　　　　　　　　　　　　　6 July 1939

[Carbon]

MOST SECRET.

With reference to your predecessor's submission of the 10th May, Med.0447/031/8, forwarding proposals for countering Italian air attack in the Mediterranean,[1] I am to inform you that this question was under consideration when the Chiefs of Staffs' European Appreciation 1939 was being drawn up, and has again been discussed during the Anglo-French Staff Conversations. So long, however, as our fighter strength is below the 800 envisaged as necessary for the Air Defence of Great Britain, and our bomber strength is only one-half of Germany's, the Air Staff view, supported by the Chiefs of Staffs, is that aircraft cannot be found for operation from Tunisia.

2. For similar reasons the French are not prepared to reinforce the air forces in North Africa, which consist of 81 bombers, 20 fighters, 1 or 2

[1] Doc. No. 346.

escadrilles of fighters for local defence and 1 escadrille of torpedo bombers for naval operations. The French have stated, during the Staff Conversations, that the continual menace to France constituted by the German Air Force, would prevent her from participating with large forces in any action which may be undertaken against continental Italy or Libya, and that she must retain the greater part of her air strength in the North-East in order to meet German air attacks. To enable France to release French units for service in the South-East and in Tunisia, the French would require this country to contribute a greater number of air units, especially fighters, for operations in the North-East of France, but for the reasons given in paragraph 1 above this country cannot entertain the proposal.

359. *Admiralty to Cunningham*

[ADM 1/9930] 8 July 1939

SECRET.

[Telegram]

Your 2301/22. Questions arising in your letter Med. 0384/0708/50 dated 19th April 1939 have been fully considered within the Admiralty. The following decisions are for your guidance in planning, but should not be considered final. A final reply will be made after questions raised have been discussed with the Foreign Office and other Departments.[1]

2. Enclosure I. Legal arguments in para. 4 are fully appreciated and their acceptance will be urged. Pending further information however your plans must be based on the assumption that all the territorial waters of the Spanish zone of Morocco must be respected as if the coast were legally part of Spanish territory. Reference para. 5 it is contemplated to urge all neutral shipping to call voluntarily at British contraband control bases.

3. Enclosure II. Para. 5 is not concurred in. Reference para. 4, it is considered that the correct interpretation of para. 4 of Article 39 of C.B. 3012 is that troopships, auxiliary vessels actually belonging to the enemy fighting forces or merchant ships known to be acting as auxiliaries in direct attendance on the enemy fleet may be sunk without warning provided they can be positively identified. *The only penalty incurred by an ordinary merchant ship in enemy convoy is that it cannot complain of chance damage resulting from a genuine attack on the escort or on*

---

[1] Italicised text is as amended in Admiralty telegram 1600/26 of 26 July 1939, ADM 1/9930.

another ship in the convoy belonging to the categories which may be lawfully sunk without warning. Submarines must, therefore, have strict orders to act as indicated above. Your para. 3 is concurred in as amplified above.

4. Enclosure III. Your paragraphs 3 and 6. It is considered that the disadvantages of a simple policy of treating Greek territorial waters lightly from the outbreak of war might well outweigh the advantages. The underlying assumption of territorial waters being immune is, however, that the territorial sovereign will prevent his waters from being abused by other belligerents. Greece would certainly be unable to ensure full control of all her waters and it is therefore probable that we should be able to find legal justification for any action inside her waters which became necessary. Nevertheless such action would only become possible on the actual failure of Greece to exercise due supervision and it must in any case depend upon a decision by H.M. Government taken at the time. Unless and until therefore, Admiralty orders were issued to the contrary, British Naval Patrols should respect the territorial waters of Greece if she is neutral. As regards the Corinth Canal, it is considered that it would be most undesirable to attempt to block it whether by sinking a merchant ship or otherwise. In any case, reasons for such operations are presumably much less important now that Turkey is our ally. Your paragraph 5, in no case do H.M. Government recognise a territorial belt exceeding 3 miles in width (this has a bearing also on Enclosure I). Waters to Eastwards of Cephalonia and Zante which are outside a belt of this width are therefore high seas, to which a right of passage through surrounding territorial waters exists.

5. Enclosure IV. Supposition in your para. 2 is confirmed. Admiralty will also state areas are dangerous on account of mines since there is no other legal method of creating a dangerous area. In considering mining policy the provision of the VIIIth Hague Convention Article 2 should be respected. Concur in paras. 3 and 4 with exception of Tripoli which is in French zone, but Trieste, Fiume, Venice and Pola could be added to list. Your para. 5 is not concurred in and in this connection please refer to personal most secret letter D.C.N.S. to C.-in-C. dated 14th April 1939 in reply to personal letter most secret C.-in-C. to C.N.S. dated 21st February 1939.[1]

6. Your 2301/22, the guarantee to Greece would not affect the position of Greek territorial waters so long as Greece remains neutral.

7. The policy of the Admiralty is in general directed at conducting war in the opening stages in strict conformity with international law in order

---

[1] Docs Nos 331 and 308, respectively.

that we can have a clear case for reprisals when and if international law is violated by the enemy.

1249/8.

360. *Cunningham to Admiralty*

[ADM 1/9946] 14 July 1939

SECRET.

[Telegram]

A. Plans are ready for bombarding military objectives in Libyan ports in some cases in conjunction with bombing by R.A.F. and plans for similar attacks on Sicily are being prepared. I am sure that such operations apart from material results will have considerable moral effect on Italians but it is important that they should be carried out immediately war breaks out in order to seize the initiative.

B. An important point of principle which appears also to affect the R.A.F. requires decision before I can order operations to be carried out. The policy of H.M. Government being to conduct war in accordance with international law (vide Para. 7 of Admiralty Message 1249 8th July) it needs to be quite clear whether attacks on shore objectives which involve danger to civilian life can only be undertaken as retaliatory actions. But the disregard of security of civilian life shown by totalitarian powers in Spain gives a clear pointer as to how they will act in the next war and it is evident that sooner or later we shall be forced to some extent to disregard it too or else be placed at a grave disadvantage by being unable to bomb or bombard many important objectives. In the case of Italy there are extremely few places which can be bombarded without risk to civilian life.

C. Para 148 of C.O.S. European appreciation 1939/40 would appear to permit attacks on defended ports without previous warning on outbreak of war and this is naturally what I would work on in order to retain initiative but if immediate offensives against enemy ports may not be undertaken it is suggested as a basis for action in this matter that if Malta is bombed and civilian casualties result that this should automatically untie my hands and leave me free without further instructions to bombard military objectives in defended ports notwithstanding risk to civilian life. Much as I dislike thus surrendering initiative to enemy this proposal appears to provide the only reasonable alternative. Part 2 follows.

1903/14.

Following is Part 2 of my 1903/14.

D. I feel strongly that this vital question requires decision before war breaks out and I urge that it is not left until the stress of moment which will inevitably result in valuable time being lost.

E. It would greatly assist conjoint planning with French and with R.A.F. if a ruling on this question could be given before meeting proposed in my 1542/6.

F. I wish to emphasize that if bombardments at earliest possible moment are ruled out, the initial operations that Fleet can carry out will be very tame affairs limited as they must be to anti-submarine operations and sweeps into Aegean and Central Mediterranean aiming at cutting off enemy surface forces and Libyan supply ships neither of which are likely to be at sea in early stages until Italians have seen the effect of their air and submarine attacks. Such operations will appear hesitant and lacking in vigour and decisiveness whereas striking of series of blows contemplated may well make enemy lose heart and quickly realize his vulnerability.

Moreover succession of attacks contemplated on enemy coast must tend to force his fleet to sea and thus give us an early chance of destroying it.

2038/14.

361. *Admiralty to Cunningham*

[ADM 1/9946] 29 July 1939

SECRET.

[Telegram]

Your 1903/14.

A. Most recent statement of policy of H.M.G. as regards bombardment of shore objectives is contained in Admiralty telegram 1430 of the 17th January 1936.[1]

B. Telegram referred to authorises commencement of offensive operations against military objectives immediately upon the outbreak of war.

C. Under Hague Convention No. IX warships may bombard defended towns (see Article 1, second paragraph, in this connection) and they may also bombard military objectives, as defined in Article 2, in undefended towns, subject only to observance of every precaution to avoid civilian casualties and damage. Although this Convention may be regarded as

[1] Doc. No. 91.

representing International Law on the subject, it is very desirable that naval bombardments should not be carried out with a greater degree of disregard for civilian life than air bombardment, and both in defended and undefended towns therefore attack should be limited to objectives regarded as legitimate air targets, and carried out subject to conditions explained in paragraph G below.

D. So far as concerns Air Bombardment, interpretation of Hague Draft Rules referred to in Admiralty telegram under reference has never been drawn up.

E. Pending Their Lordships' consultation with other interested departments following may be considered provisionally to be legitimate targets for planning purposes:

1. Military, Naval and Air Forces.
2. Coast Defence Works.
3. Naval, Military and Air Establishments.
4. Railways used for Military communications but not stations or trains unless known to be troop trains.
5. Wharves, etc., in commercial harbours actually being used for Military purposes may be attacked subject to every care being taken to avoid damage to merchant ships in harbour.

N.B. Factories not forming part of Military, Naval or Air Establishments should not repeat not be attacked without express Admiralty authority.

Question whether oil and petrol storage can be considered military objectives and so included in legitimate targets is under consideration and will be discussed with the French, but for the present they should be excluded.

F. If you consider additions to the list in paragraph E necessary for effective prosecution of initial operations which you have in mind you should acquaint Their Lordships of specific instances.

G. Possibly, however, the difficulty which you envisaged is rather the prohibition of indiscriminate bombing contained in Hague draft rules. Admiralty view is that this rules out attacks unless (a) it is possible to distinguish and identify the object in question, and (b) there is a reasonable expectation that damage will be restricted to legitimate objectives.

H. Above is tentative only, and matter is being taken up with other departments and French Government with a view to issue of comprehensive instructions. You will of course understand that if, as is probable, enemy commence indiscriminate attacks from the air, intention is to issue further instructions releasing you from legal restrictions.

1055/29.

362. *Cunningham to Admiralty*

[ADM 1/9905] Office of Commander-in-Chief, Mediterranean Station
2 August 1939

MOST SECRET
Med.0721/0700/11.
ANGLO-FRENCH CONVERSATIONS
Be pleased to inform Their Lordships that conversations were conducted during the recent visit of Vice Admiral Ollive,[1] Commander-in-Chief of the French Mediterranean Fleet to Malta. Three conferences were held on 27th, 28th and 29th July, at which the following were present:–
FRENCH
Vice Admiral Ollive,
Commandant-en-Chef, Flotte de Mediterranée.
Capitaine de Vaisseau Blehaut,[2]
Chief of Staff.
Capitaine de Corvette Laurin,
Ministry of Marine.
Capitaine de Corvette Reboul,
Staff Officer (Operations)
* Lieutenant de Vaisseau Boulanger.
BRITISH
Admiral Sir A.B. Cunningham, K.C.B., D.S.O.
Commander-in-Chief, Mediterranean Fleet.

---

[1] Vice Amiral d'escadre [later Amiral] Emmanuel Lucien Henri Ollive (1882–1950). Commanded battery of naval gunners on western front, 1914–17 commanded 2nd group of river gunboats, western front, 1917–18; commanded destroyers of 1re escadre, Toulon, 1935–36; commanded 3e escadre légère, 1936–37; 1st sous-chef, naval general staff, 1937–38; C-in-C Escadre de la Méditerranée, 1938; Préfet Maritime, Toulon, 1938–39; C-in-C Flotte de Méditerranée, May–Nov 1939; C-in-C Forces Maritimes de l'Atlantique Sud et d'Afrique, Nov 1939–July 1940; Préfet Maritime, Algiers, July 1940–Oct 1942; retired Jan 1943.
[2] Capitaine de Vaisseau [later Contre Amiral] Henri Paul Blehaut (1889–1962). Commanded submarines *Cigogne*, 1918, *Clorinde*, 1920–22, *Joëssel*, 1922–24, *Requin*, 1924–27; commanded 11th Torpedo Boat Division (Mediterranean), 1930–32; Chief of Staff in Far East Naval Forces, 1932–34; Chief of Staff, 3e division légère (cruisers), 1934–35; commanded 9e division légère (destroyers), 1936–38; Chief of Staff, Maritime Prefecture, Toulon, 1938–39; Chief of Staff to Admiral Ollive in the latter's functions as C-in-C Mediterranean Fleet and subsequently C-in-C South Atlantic and Africa, 1939–40; commanded 3rd Cruiser Division, 1941–42; 1st Cruiser Division, Sept–Nov 1942; Secretary of State for Marine and Colonies in Vichy Government, 1943–44; imprisoned in Germany, Aug 1944; arrested on return to France, tried in absentia by High Court and sentenced to 10 years, 1948; acquitted, 1955.

Rear Admiral H.R. Moore,[1] C.B., C.V.O., D.S.O.
(Representing the Vice Admiral, Malta).
Commodore A.U. Willis,[2] D.S.O.
Chief of Staff.
Air Commodore A.O. Leckie,[3] D.S.O., D.S.C., A.F.C.
Air Officer Commanding, Mediterranean.
Captain J.G.L. Dundas.[4]
(Plans Division, Admiralty).
Captain E.R. Gibson
Staff Officer (Operations).
Commander S.N. Blackburn.
Commander on Staff, H.M.S. *St. Angelo.*
Commander R.M. Dick, D.S.C.
Staff Officer (Plans).
* Lieutenant Commander J. Liddell.
Fleet W/T Officer.
Paymaster Lieutenant Commander H.S.P. Watch.
Secretary to Chief of Staff.

[1] RA [later Adm Sir] Henry Ruthven Moore (1886–1978). Naval Asst Sec to Cttee of Imperial Defence, 1921–24; commanded cruisers *Caradoc*, 1928–30, and *Dauntless*, 1930; Dep DP, Admy, 1930–32; DP, 1932–33; commanded cruiser *Neptune*, 1933–35; Cdre 1st Class and COS, Home Fleet, 1936–38; COS to C-in-C Portsmouth, 1938–39; RAC 3rd CS, 1939–40; Asst CNS (Trade), Admy, 1940–41; Vice CNS, 1941–43; VAC 2nd BS and 2nd-in-Command, Home Fleet, 1943–44; C-in-C Home Fleet, 1944–45; Head of British Admy Delegation, Washington, 1945–48; C-in-C The Nore, 1948–50; retired, 1951; Dep Lieut, Kent, 1957; High Sheriff of Kent, 1959–60.
[2] Cdre [later Adm of the Fleet Sir] Algernon Usborne Willis (1889–1976). Commanded destroyer *Warwick*, 1927–29; commanded cruiser *Kent* and Flag Capt, China Fleet, 1933–34; commanded battleship *Nelson* and Flag Capt, Home Fleet, 1934–35; commanded HMS *Vernon* (Torpedo School, Portsmouth), 1935–38; commanded battleship *Barham* and Flag Capt, 1st BS, Med Fleet, 1938–39; COS, Med Fleet, 1939–41; C-in-C Africa Station, 1941; C-in-C South Atlantic, 1941–42; commanded 3rd BS and 2nd-in-Command, Eastern Fleet, 1942–43; commanded Force H, Med, 1943; C-in-C Levant, Oct–Dec 1943; 2nd Sea Lord and Chief of Naval Personnel, 1944–46; C-in-C Med Fleet, 1946–48; C-in-C Portsmouth, 1948–50; retired, 1950.
[3] Air Cdre [later Air Marshal] Robert Leckie (1890–1975). Joined RNAS, 1915; Dir of Flying Operations, Canadian Air Board, 1919–22; officer commanding flying in carriers *Hermes* and *Courageous*, 1925–29; commanded RAF Bircham Newton, 1929–31; commanded No. 210 Flying Boat Squadron, 1931; commanded RAF Pembroke Dock, Southampton, 1931–33; commanded RAF Hendon, 1933–35; Dir of Training, Air Ministry, London, 1935–38; commanded RAF Med, 1938–40; Director of Training, Royal Canadian Air Force, 1940–44; Chief of Air Staff, Royal Canadian Air Force, 1944–47.
[4] Capt [later VA] John George Lawrence Dundas (1893–1952). Commanded sloop *Folkestone*, 1936–37; Asst DP, Admy, 1938–40; commanded cruiser *Nigeria*, 1940–42; COS to C-in-C Med, Feb–June 1943; COS to C-in-C Levant, June–Nov 1943; COS to C-in-C Med, 1943–44; Asst CNS, 1944–45.

THE APPROACH OF WAR, 1938–1939   547

Note 1. With reference to Admiralty Telegram 1157 of 22nd July, Group Captain Slessor,[1] after the British meeting preliminary to the conference stated that having communicated the information he brought with him to the Air Officer Commanding Mediterranean, he did not consider his presence necessary, and proceeded to Egypt to see the Air Officer Commanding-in-Chief, Middle East.

Note 2. * These officers formed a sub-committee on communication matters and did not attend the main discussions.

2. The discussions were conducted on cordial and frank terms and I greatly appreciate the readiness shown by Vice Admiral Ollive to collaborate to the utmost in any manner possible. The conversations were thus of considerable value in advancing the arrangements for collaboration of the naval and air forces working in the Mediterranean.

* * *

4. It is desired to draw particular attention to the following points:–

(a) The line of demarcation between French and British spheres has been re-arranged so as to allow more sea room for the French in their operations against Libyan traffic making for Tripoli from Italy. The line comes into force on 10th August.

(b) Since the conference concluded, the Admiralty message has been received on the subject of bombardment policy, which may lead to modification of paragraph 7 of the minutes, but as this matter is being discussed between the Admiralties, presumably both commands will receive the same instructions.

(c) It is observed that Admiralty preliminary discussions with the French as regards transport of personnel apparently only envisage movement of the reinforcements for Malta. It is requested that I may have further information on this subject particularly as to what is the intention as regards the remaining 1500 men for the Eastern Mediterranean.

(d) With reference to paragraph 14 of the minutes, arrangements are being made for Rear Admiral H.R. Moore, C.B., C.V.O., D.S.O., the Rear Admiral Commanding, Third Cruiser Squadron, to visit Bizerta during the week commencing 31st July.

---

[1]Group Capt [later Marshal of the RAF Sir] John Cotesworth Slessor (1897–1979). Dep Dir of Plans, Air Ministry, 1937–38; Dir of Plans, Air Ministry, 1938–41; AOC No. 5 (Bomber) Group, 1941–42; Asst CAS (Policy), 1942–43; AOC Coastal Command, Feb 1943–Jan 1944; C-in-C RAF Med and Middle East, 1944–45; Air Member for Personnel, 1945–47; Commandant of the Imperial Defence College, 1948–50; CAS, 1950–52; retired 1953.

(e) With reference to paragraph 11(b) of the minutes, it is understood that such a code is in fact in production and it is felt that this should be in force in the two navies at the earliest possible moment.

(f) Item 11 of the agenda was not proceeded with as the French Commander-in-Chief was fully aware of the arrangements.

6. I am sending copies of this letter & enclosures to Vice Admiral Ollive, the Air Officer Commanding-in-Chief, Middle East and the Air Officer Commanding, Mediterranean.[1]

* * *

[Enclosure No. 2]   28th July 1939.

REPORT OF CONFERENCE BETWEEN COMMANDER-IN-CHIEF, MEDITERRANEAN AND THE VICE ADMIRAL COMMANDANT, FRENCH MEDITERRANEAN FLEET.

The Commander-in-Chief opened the meeting by welcoming the French Admiral and his Staff and said that he much appreciated Vice Admiral Ollive's presence in view of his very recent assumption of command. The French Commander-in-Chief replied that he was only too glad of the opportunity of making this contact.

2. Mutual explanation of Chain of Command.

The system of command was mutually made clear and certain details as to limitation of station areas explained.

3. Allied Strategical Policy in the Mediterranean.

The Commander-in-Chief stated that the policy was 'to exert in co-operation with the French such pressure as to make the Italian position in Libya and eventually Ethiopia untenable'. The French Admiral agreed that this policy was in accordance with his instructions and that it was in accordance with that agreed upon in London.

4. Allied Naval Object.

The Commander-in-Chief stated the Allied Naval Object as 'to cut all Italian sea communications, at the same time retaining control of the entrance to the Mediterranean and exerting such other pressure on Italy as may be possible by naval means'. The French Admiral agreed that this object was in accordance with his instructions and added that his mission also included the protection of the North African lines of communication.

5. The Commander-in-Chief, Mediterranean, referred to his tasks as given in the Naval War Memorandum, namely:–

---

[1]Confirmation of arrangements made at meetings with Rear Admiral Commanding Division du Levant.

(1) To bring enemy naval forces to action wherever found.

(2) To ensure safe passage of reinforcements to the Eastern Mediterranean.

(3) To obtain command of the Eastern Basin.

The French Admiral noted this, saying that each ally had much the same tasks whilst between them lay the forces to be destroyed.

6. The best method of Applying Combined Pressure on the Enemy.

The Commander-in-Chief described generally the operations which he had in mind on the outbreak of war (these are given in Appendix I).

It was agreed that bombardment of enemy bases would have great moral effect and might well achieve our ultimate aim of destroying the enemy's surface forces by forcing them to sea.

It was agreed that while directly synchronised action was not feasible, collaboration could take place; for instance, operations might be designed some time ahead to occur on the same day, or even to follow each other at a short interval, so that the second operation reaped the benefit of the dispersion of force caused by the first. Such concerted action is also calculated to have the effect of drawing off forces from one zone, thus facilitating action by the other ally in his zone.

7. Bombarding Enemy Ports.

It appeared that the instructions likely to be in force regarding bombarding action on the outbreak of war were substantially the same for each country in that neither ally could initiate operations likely to endanger civil life until so authorised by their governments as a retaliatory measure. The French Admiral observed that his only direct instructions on this matter at present, were to adhere to the international law as regards attack on enemy communications.

8. French Plan.

The French Admiral then briefly described the French Naval Plans (see Appendix II). He further stated his view that enemy shipping would not pass to the westward of Sicily on its way to Libya even if loaded in North Italy, he said that as from some as yet unconfirmed intelligence, he understood that the Italians were constructing a number of fast ships for passage to Libya. Thus, French action against enemy shipping was likely to take place chiefly well to the eastward of Cape Bon. The French Admiral then emphasized the vital importance of reconnaissance reports of enemy forces and shipping emerging from the Straits of Messina or ports in Southern Italy or Sicily, since such knowledge was essential to the successful operation of the light forces at Bizerte.

9. The Line of Demarcation.

The operations of French and British light forces raised the question of the line of demarcation between French and British spheres of responsibility. It was considered the present line was unsatisfactory and it was agreed to make the following recommendations:–

(a) A line drawn from Cape Empedocle to Ras Elmsel down to the latitude of 35° 30′ N, then in a direction 110° to the meridian 15° 34′ E then down that meridian to the North African coast. This line, it was agreed, should come into force on 10th August.

(b) This line was not rigid for surface forces who could cross it during operations against the enemy, it being understood that information of such action should be passed at once to the Command concerned.

(c) The line was to be regarded as rigid for submarines but in view of possible errors in position keeping in aircraft, French submarines should be further instructed not to come within a range of 50 miles of Malta except in emergency and after giving previous warning.

(d) With regard to aircraft, these may cross the line when engaged in reconnaissance operations prior information having been given, it being understood that they will in no case attack submarines when in the zone of the other ally unless enemy identity has been established beyond possible doubt.

(e) It was also agreed that aircraft working from Bizerta and Malta may mutually cross the line and land on each other's aerodromes, and recognition arrangements and lines of approach will be interchanged for that purpose.

10. Expected Disposition of Allied Forces.

The Commander-in-Chief showed the British dispositions and explained them (Appendix III). The French Admiral expressed concern at the shortage of fast and powerful aircraft at Malta, particularly when it was explained that none would be available until next December. He said that he sincerely hoped that this date might be advanced and that he would represent to his authorities the necessity of more fast and powerful aircraft in Tunisia if such an increase is possible, having in view the important reconnaissance duties needed in Southern Italy and Sicily the need of which has already been stressed.

11. Interchange of Plans.

(a) Plan 1.

After describing the Plan, the Commander-in-Chief stated that attack on Augusta by Fleet Air Arm was dependent on sufficient good ship targets being present, and pointed out that the shallow waters was an obstacle to the use of aircraft torpedoes.

With regard to Force 'W', he pointed out that he did not expect that Malta could be used except by night in the early stages of the war, and then probably only for refuelling. And it must therefore be expected that periods might occur when Force 'W' might have had to return to

Alexandria. The intention was, as far as possible, however, always to retain a force in the Central Mediterranean.

(b) <u>Reconnaissance</u>.

The system of Flying Boat reconnaissance between Malta and Cephalonia was explained, showing a continuous patrol daily by four Flying Boats. It was explained that this patrol could be maintained initially only for fourteen days.

The methods of passing information obtained by this patrol to the French were then discussed, and it was explained that communication arrangements for this purpose were already being worked out between the Malta and Bizerta commands. The importance of a rapid operational code for the interchange of enemy reports and similar messages was again emphasized.

(c) <u>Submarine Dispositions</u>.

The Commander-in-Chief confirmed that the submarine patrols off Italian ports were offensive as well as for reconnaissance. He pointed out that these might not always be off South Italian ports, e.g. a patrol might be required off Benghazi.

(d) <u>Plan 2</u>.

Plan 2 was explained (see Appendix I). The French had no observations on this plan. It was emphasized that the arrangements were such that it could be changed over to Plan 1 at any time should circumstances allow.

12. <u>Measures for Passage of British Re-inforcements</u>.

The Commander-in-Chief explained that it was the intention to embark re-inforcements at Marseilles prior to the outbreak of war, and that he was concerned over the question of their safe passage through the central Mediterranean, should war break out during the operation. The French Admiral observed that a scheme was already agreed to in principle between the two Admiralties for transferring a force of 1100 men either from Casablanca or Algiers and sending them by rail to Tunis or Sousse to embark in British light craft for onward passage. As full details of this had not yet been received, further information will be sought from the Admiralty.

13. <u>Exchange of Liaison Officers</u>.

It was decided after discussion that this proposal was not at the moment practicable, and that it should therefore be dropped.

14. <u>Further Conferences with Subordinate Commands</u>.

The Commander-in-Chief stated that it was hoped to arrange a further conference shortly between the Admiral Commanding, Force 'W', the Admiral Commanding Force Leger d'Attaque, and also between the Vice Admiral, Malta and Prefet Maritime, Bizerta. The French Admiral agreed that this was desirable. These authorities would then be able to exchange ideas in connection with the general way in which they intended to employ the forces under their command.

15. <u>Exchange of Current Disposition of the Fleet</u>.

The desirability of such an exchange was fully agreed upon, and it was considered it should be effected by communication between the Commanders-in-Chief, details being settled by their staffs.

These communications of movements will take place only in periods of tension on the initiative of either Commander-in-Chief. Proposed movements will be signalled if possible before they take place. Information will be passed by methods already decided upon.

[Enclosures]

APPENDIX I[1]
GENERAL DESCRIPTION OF PLANS ON THE OUTBREAK OF WAR

1. It is hoped that the Fleet may have been able to reach the Central Mediterranean by the time war is declared in which case the plans described below can be put into operation more quickly.

2. Two typical plans are given.

PLAN I.   Is the one which will be carried out if permission has been given to bombard and bomb defended harbours. (Note:– This permission is only expected after the enemy have shown disregard of civilian life).

PLAN II.   Is the alternative plan and is much less effective than Plan I but is the maximum permissible until we have a free hand to attack enemy ports.

PLAN I.   Note:– Z represents day of leaving harbour.

The fleet leaves ALEXANDRIA so as to be well to the west of CRETE at dark on Z+1; it then divides into three forces.

FORCE 'A'.   1 Battleship, 2 8″ Cruisers.
                2 6″ Cruisers, Destroyer screen.
     'B'   Aircraft Carrier and screen.
     'C'   2 Battleships, 1 8″ Cruiser.
             1 6″ Cruiser, Destroyer screen.

FORCE 'A'.   Arrives off DERNA and TOBRUK at dawn on Z+2 day and bombards. This attack being supported by bombing by Royal Air Force from Egypt.

FORCE 'B'.   (Aircraft Carrier) keeps well to the north and provides reconnaissance for Forces 'A' and 'C' and has a striking force ready for enemy surface forces.

[1] Appendix I is stamped 'enclosure 3' in original.

FORCE 'C'.   Arrives off BENGHAZI at dawn on Z+2 day and bombards. It may be possible for the Royal Air Force from Egypt to co-operate by bombing at the same time.

PHASE II.   Forces 'A', 'B' and 'C' retire to northwest. Force 'C' steers for MALTA but after dark turns north and attacks CATANIA and AUGUSTA at dawn on Z+3 day.

Meanwhile Force 'B' (Aircraft Carrier) protected by Force 'A' moves far enough northwest to launch a dawn bombing attack on ships in AUGUSTA.

PHASE III.   All forces retire towards CRETE (unless Phase II has brought the enemy heavy forces to sea). On arrival off CRETE, 2 6" Cruisers and 4 'Tribal' Class destroyers are detached and sweep back to MALTA from which base they will then continue to operate as Force 'W'. The rest of the fleet returns to ALEXANDRIA to replenish, possibly carrying out an anti-submarine sweep between LIBYA and CRETE.

Reconnaissance.   Throughout operations a continuous Flying Boat Patrol is kept by day between MALTA and CEPHALONIA to report enemy surface forces and shipping.

Submarines.   Will be placed off TARANTO, STRAITS OF MESSINA and AUGUSTA.

PLAN II.

Fleet sails at dawn on Z day and sweeps as wide an area as possible between NILE DELTA and EASTERN CRETE.

Fleet concentrates just before dark. After dark 2 6" Cruisers and 4 Destroyers are detached. These turn back and pass round eastern end of CRETE and sweep out the area to the north of CRETE. This force passes through the KITHERA Channels eventually joining fleet on Z+2 day for Phase III.

PHASE II.

Fleet steers west during the night and at dawn on Z+2 day is close to the coast of CRETE (MESSARA BAY). During the night anti-submarine striking forces have been detached and these take up positions 30 miles apart. *Glorious* flies off air patrols and a concentrated submarine hunt is carried out all Z+1 day between west CRETE and LIBYA.

PHASE III.

At dark on Z+1 day fleet concentrates and moves to a line southeast from NAVARIN on which it again spreads and sweeps all Z+2 day in a northwesterly direction up the coast of GREECE.

In the evening of Z+2 day the fleet concentrates and just before dark turns to steer for MALTA. After dark a force of Cruisers and Destroyers is detached which sweeps up through the STRAITS OF OTRANTO. This force, on returning from its sweep will go to MALTA from which it will continue to operate as Force 'W'.

PHASE IV.

The fleet steers southeast during the night, remaining in general support in the hope that the OTRANTO raid may bring surface forces to sea in pursuit.

If this is ineffective, the fleet will turn and sweep eastward across LIBYAN trade routes eventually returning to ALEXANDRIA to replenish.

Reconnaissances and Submarines.   As for Plan I.

Note:– Plan II can easily be converted to Plan I should circumstances arise to allow of this.

APPENDIX II.

Translation of Note 65F of the French General Staff.

OPERATIONS OF THE FRENCH NAVAL FORCES IN THE MEDITERRANEAN ON THE OUTBREAK OF HOSTILITIES

On the outbreak of war the French Naval Forces in the Mediterranean would be employed as follows:–

1. The protection of troop transports (to Corsica and between France and North Africa).

2. Offensive action by submarines:–

(a) In the Ligurian Sea (the forces available in the northeastern area operating in the zone GENOA – SPEZIA and between Corsica and France.

(b) In the lower Tyrrenian Sea and between Sicily and Libya (traffic with Tripolitania (the forces available in the southeast operating in particular Trapani, Marsala and south of Sardinia, off Cagliari and San Antiocco.

3. Attack on enemy communications between Sicily and Libya (traffic with Tripoli). Attacks by light surface forces and air forces based at Bizerta, operating from an anchorage situated south of Sfax, and by submarines.

It would be of advantage to combine these attacks with the operations of British naval forces in their zone of operations.

4. Attack on enemy coast by surface vessels in the Ligurian Sea. Objectives proposed in this region are:–

(a) Region VADO–SAVONA
   Petrol tanks

Industrial works
Port of SAVONA and steel works
(b) Region of GENOA
VOLTRI and PRA
SESTRI PONENTE
POLCEVERA
Port of GENOA

Attacks on these objectives would be carried out according to circumstances:–

By strong forces comprising cruisers and contre torpilleurs supported by battleships (in the course of such an operation all the objectives could be attacked).

Or by light forces in small numbers (contre torpilleurs) on a small number of objectives.

The above objectives could be combined with air bombardments.

5. Besides the operations specified above, we will make provision for air reconnaissance and attack on submarines off the coasts of Provence, Corsica and Tunisia especially in the Sicily channel and Sardinia channel.

6. Other operations which we have considered carrying out eventually according to circumstances are as follows:–

Bombardment of PALERMO by surface vessels and aircraft.
Bombardment of TRIPOLI by aircraft.
Bombardment of PANTELLARIA by aircraft.
Co-operation with land forces engaged on the Tunisian frontier.

  by [order of] the Admiral of the Fleet
   Chief of the General Staff[1]
    Signed: Bourragué[2]

Authenticated translation.
 (Sd). S.N. Blackburn
  Commander. 28th July 1939.

---

[1] Amiral de la Flotte François Xavier Darlan (1881–1942). Chief of Naval General Staff, 1937–40; C-in-C French Maritime Forces, 1939–40; Minister of Marine, 1940–41; Vice President of the Council, Minister of Foreign Affairs and of the Interior, 1941–42; C-in-C French Armed Forces, 1942; present in North Africa at moment of Allied landings, he concluded an armistice ending fighting between Vichy forces and Allies and became High Commissioner in French Africa, Nov 1942; assassinated 24 Dec 1942.

[2] Contre Amiral [later Vice Amiral] Célestin Jean Léon Bourragué (1886–1955). Sous Chef, Naval General Staff, 1937–39; commanded 4th CS in Atlantic Squadron, 1939–40; commanded Force Y (reinforcements for Dakar), 1940; President of Cttee for Coordination of Imperial Telecommunications, 1940–41, 1943–45; Chief of Gen Staff for National Defence, 1941–42; Dir of Armistice Services, 1942–43.

APPENDIX III.
DISPOSITION OF BRITISH FORCES ON OUTBREAK OF WAR.
Based on expected situation – end of August.

1. ALEXANDRIA.
Eastern Mediterranean Fleet.
3 Battleships – *Warspite*, *Barham*, *Malaya*.
1 Aircraft Carrier *Glorious* & attendant destroyer *Bulldog*.
　3 Reconnaissance
　Squadrons and　　} 48 [aircraft].
　1 Fighter Squadron
3 8″ Cruisers – *Devonshire*, *Shropshire*, *Sussex* – First Cruiser Squadron.
2 6″ Cruisers – *Arethusa*, *Penelope* – Third Cruiser Squadron
1 6″ Cruiser – *Galatea* (Rear Admiral Destroyers).
2½ Flotillas of Destroyers – 'H', 'I' Class.
　　　　　　　　　　　4 'Tribal' Class.
　　　　　　　　　　　4 'G' Class.
　　　　　　　(1st, 2nd, 3rd and 4th Flotillas).
1 Submarine
Half a squadron R.A.F. Flying Boats (Sunderlands).
4 Minesweepers – 'Sutton' Class.
Notes:– (a) one or more of the 8″ Cruisers may be bringing reinforcements from Marseilles.
　　　(b) 2 Destroyers will be stationed in the Suez Canal Zone.
　　　(c) 'I' Class Destroyers are being replaced by "K" Class.

2. RED SEA.
1 Flotilla of Destroyers – 4 of 'G' Class, 4 'Tribal' Class.
1 Anti-aircraft cruiser – *Curlew*.

3. HAIFA.
1 Escort Vessel.

4. MALTA.
6 Submarines – 2 Minelayers and 4 'S' Class.
12. Motor Torpedo Boats
3 Minesweepers – 'Sutton' Class.
1½ R.A.F. Flying Boat Squadrons (Londons and Sunderlands).
Note:– It is the intention, very early, to detach two 6″ Cruisers and four 'Tribal' Class to operate in Central Mediterranean, based on Malta.

5. GIBRALTAR.
1 Flotilla of Destroyers – 13th Flotilla of 'V' and 'W' Class, chiefly from reserve.
2 Minesweepers – will come from England.

General Note.
These dispositions show the forces available on the outbreak of war. An increase will take place when reserves are mobilised, particularly in Anti-submarine and Minesweeping craft.[1]

* * *

363. *Admiralty to C.'s-in-C.*

[ADM 1/10188] 30 August 1939

SECRET.

[Telegram] Addressed C.-in-C. America and West Indies, North Atlantic, Mediterranean, China, East Indies, South Atlantic.

IMMEDIATE.

If it becomes necessary to despatch the Warning Telegram to you it seems likely that Italy as well as Germany may be specified as a potential enemy. In view of the alliance between Germany and Italy the possibility of Italian hostility cannot be excluded though it is expected that Italy will do her utmost to avoid becoming involved.

2. H.M.G. regard Italian neutrality if it can by any means be assured, as decidedly preferable to her active hostility. The French Government agree with this policy. Consequently your action should be governed by the general principles that follow.

3. If you receive a Warning Telegram specifying both Germany and Italy as potential enemies, your defensive precautions against attack by Italy should as far as possible be non-provocative. If subsequently you receive the War Telegram in which only Germany is specified as the enemy, you should avoid initiating any action against Italy which is likely to bring her in against us. This of course does not relieve you of your responsibility for taking necessary precautions against the possibility of

---

[1] The French provided the British with a similar but less specific list of forces available and their dispositions. This was Appendix IV in the original. It may be summarised as follows: Forces effected to the C-in-C High Sea Forces: older battleships *Lorraine* and *Bretagne*; 6 10,000-ton cruisers; 12 destroyers and 9 1,500-ton torpedo boats. The Maritime Prefect of the 3rd Region (Toulon): 3 destroyers; 3 610-ton torpedo boats; 3 gunboats (*avisos*); approximately 26 large and small submarines; 57 naval aircraft (3–4 reconnaissance); and 32 air force planes. The Maritime Prefect of the 4th Region (Bizerta): cruiser *Emile Bertin* and 3 7,700-ton cruisers; 6 destroyers; 3 610-ton torpedo boats; 2 gunboats; minelayer *Castor*; 17 large and small submarines (3 detached to the Levant and the anticipation the latter would be reinforced); 12 naval aircraft; and 24 air force planes (12 seaplanes). At Oran: 3 1,500-ton torpedo boats; 14–16 small submarines; seaplane carrier *Commandant Teste* and 30–40 naval aircraft. There was also a small number of torpedo boats, submarines and naval aircraft based in Morocco.

sudden attack by Italy nor of making such defensive dispositions as you think advisable. Nor does it debar you from immediate local retaliation if attacked by Italian forces.

4. Should the development of the situation demand any modification of these instructions, you will be notified immediately.

1115/30.

SOURCES AND DOCUMENTS

Sources

The National Archives, Kew
Admiralty Papers   ADM
Cabinet Papers   CAB
Foreign Office Papers FO

The National Maritime Museum, Greenwich
Chatfield Papers   CHT

Numerical List of Documents

Part I: The Mediterranean Fleet from 1930 to the Ethiopian Crisis

1	Summary of Mediterranean Fleet Exercises (1)	Nov 1930	ADM 186/147
2	Summary of Mediterranean Fleet Exercises (2)	Nov 1930	ADM 186/148
3	Chatfield to Admiralty	24 April 1931	ADM 1/8752/198
4	Hallifax to Tyrrell	8 May 1931	ADM 1/8752/198
5	Admiralty to Chatfield	23 Nov 1931	ADM 116/2860
6	Chatfield to Admiralty	21 Dec 1931	ADM 116/2860
7	Admiralty to Chatfield	1 Mar 1932	ADM 116/2860
8	Chatfield to Admiralty	25 Aug 1932	ADM 1/8761/245
9	Fisher to Admiralty	10 April 1933	ADM 116/3473
10	Fisher to Chatfield	29 May 1933	CHT/4/5
11	Admiralty to Fisher	2 June 1933	ADM 116/3473
12	Fisher to Admiralty	20 June 1933	ADM 116/3473
13	Fisher to Chatfield	21 July 1933	CHT/4/5
14	Fisher to Admiralty	29 Aug 1933	ADM 116/3473
15	Admiralty to Fisher	3 Oct 1933	ADM 116/3473
16	Fisher to Admiralty	6 Oct 1933	ADM 116/3473
17	Fisher to Chatfield	17 Nov 1933	CHT/4/5
18	Fisher to Admiralty	23 Jan 1934	ADM 116/3473
19	Chatfield to Fisher	11 May 1934	CHT/4/5
20	Chatfield to Fisher	14 June 1934	CHT/4/5
21	Chatfield to Fisher	2 Aug 1934	CHT/4/5
22	Board of Admiralty Minutes	4 Oct 1934	ADM 167/90
23	Admiralty to Fisher	5 Jan 1935	ADM 116/3473

560   THE MEDITERRANEAN FLEET, 1930–1939

24	Chatfield to Fisher	24 Jan 1935	ADM 116/3489
25	Admiralty to Fisher	15 Mar 1935	ADM 116/3536
26	Fisher to Chatfield	30 Mar 1935	ADM 116/3489

Part II: The Abyssinian Crisis, 1935–1936

27	Admiralty to Fisher	7 Aug 1935	ADM 116/3536
28	Little to Fisher	10 Aug 1935	ADM 116/3038
29	Fisher to Admiralty	20 Aug 1935	ADM 116/3038
30	Fisher to Admiralty	21 Aug 1935	ADM 116/3038
31	Fisher to Admiralty	22 Aug 1935	ADM 116/3038
32	Admiralty to Fisher	22 Aug 1935	ADM 116/3536
33	Admiralty to Fisher	22 Aug 1935	ADM 116/3038
34	Rear Admiral Gibraltar to Fisher	23 Aug 1935	ADM 116/3038
35	Fisher to Admiralty	23 Aug 1935	ADM 116/3040
36	Admiralty to Fisher	24 Aug 1935	ADM 116/3040
37	Fisher to Admiralty	25 Aug 1935	ADM 116/3040
38	Vice Admiral Malta to Fisher	25 Aug 1935	ADM 116/3038
39	Admiralty to Vice Admiral Malta	26 Aug 1935	ADM 116/3038
40	Admiralty to Fisher	26 Aug 1935	ADM 116/3038
41	Rear Admiral Gibraltar to Fisher	27 Aug 1935	ADM 116/3038
42	Admiralty to Fisher	28 Aug 1935	ADM 116/3038
43	Backhouse to MacKenzie	28 Aug 1935	ADM 116/3039
44	Fisher to Chatfield	29 Aug 1935	CHT/4/5
45	Admiralty to Fisher	29 Aug 1935	ADM 116/3038
46	Fisher to Admiralty	31 Aug 1935	ADM 116/3038
47	Fisher to Admiralty	1 Sept 1935	ADM 116/3040
48	Fisher to Admiralty	1 Sept 1935	ADM 116/3038
49	Admiralty to Fisher	2 Sept 1935	ADM 116/3038
50	Memorandum by Chatfield	3 Sept 1935	CAB 16/138
51	Admiralty to Fisher	3 Sept 1935	ADM 116/3040
52	Admiralty to Fisher	4 Sept 1935	ADM 116/3038
53	Fisher to Admiralty	4 Sept 1935	ADM 116/3038
54	Fisher to Admiralty	5 Sept 1935	ADM 116/3038
55	Admiralty to C-in-C East Indies	11 Sept 1935	ADM 116/3038
56	Admiralty to Fisher	11 Sept 1935	ADM 116/3038
57	Fisher to Admiralty	15 Sept 1935	ADM 116/3038
58	Chatfield to Dreyer	16 Sept 1935	CHT/4/4
59	Admiralty to Fisher	18 Sept 1935	ADM 116/3536
60	Admiralty to Fisher and Backhouse	18 Sept 1935	ADM 116/3038
61	Admiralty to Fisher	21 Sept 1935	ADM 116/3038
62	Admiralty to Fisher	3 Oct 1935	ADM 116/3038
63	Admiralty to Fisher, Backhouse and Rose	4 Oct 1935	ADM 116/3487
64	Fisher to Admiralty	8 Oct 1935	ADM 116/3038
65	Backhouse to Admiralty	9 Oct 1935	ADM 116/3038
66	Fisher to Admiralty	19 Oct 1935	ADM 116/3038
67	Backhouse to Admiralty	25 Oct 1935	ADM 116/3038
68	Fisher to Chatfield	25 Oct 1935	CHT/4/5
69	Pound to Chatfield	26 Oct 1935	CHT/4/10
70	Admiralty to Fisher and French	26 Oct 1935	ADM 1/8804

71	Fisher to Admiralty	29 Oct 1935	ADM 116/3050
72	Admiralty to Fisher	1 Nov 1935	ADM 116/3487
73	Admiralty to Fisher	2 Nov 1935	ADM 116/3038
74	Admiralty to Fisher	2 Nov 1935	ADM 116/3050
75	Admiralty to Fisher	2 Nov 1935	ADM 116/3038
76	Admiralty to Fisher and Backhouse	5 Nov 1935	ADM 116/3050
77	Fisher to Chatfield	8 Nov 1935	CHT/4/5
78	Admiralty to Fisher	18 Nov 1935	ADM 116/3046
79	Fisher to Admiralty	3 Dec 1935	ADM 116/3046
80	Fisher to Admiralty	6 Dec 1935	ADM 116/3398
81	Admiralty to Fisher	7 Dec 1935	ADM 116/3046
82	Admiralty to Fisher, Rose and Bailey	18 Dec 1935	ADM 116/3046
83	Chatfield to Fisher	20 Dec 1935	CAB 53/5
84	Admiralty to Fisher	21 Dec 1935	ADM 116/3398
85	Fisher to Chatfield	23 Dec 1935	CAB 53/5
86	Mountbatten to Vice Admiral Malta	24 Dec 1935	ADM 116/3398
87	Revised Mediterranean War Plan (Summary)	Jan 1936	ADM 116/3476
88	Fisher to Chatfield	11 Jan 1936	CHT/4/5
89	Admiralty to Fisher	13 Jan 1936	CAB 53/5
90	Admiralty to Chatfield and Backhouse	15 Jan 1936	ADM 116/3038
91	Admiralty to Fisher, Backhouse and Rose	17 Jan 1936	ADM 1/9946
92	Backhouse to Fisher	5 Feb 1936	CHT/4/5
93	Fisher to Admiralty	7 Feb 1936	ADM 116/3038
94	Admiralty to Fisher	19 Feb 1936	ADM 116/3058
95	Fisher to Chatfield	25 Feb 1936	CHT/4/5
96	Fisher to Chatfield	6 Mar 1936	CHT/4/5
97	Fisher to Admiralty	19 Mar 1936	ADM 116/3468
98	Pound to Chatfield	31 Mar 1936	CHT/4/10
99	Minutes by James and Chatfield	22–27 April 1936	ADM 116/3042
100	Pound to Chatfield	30 April 1936	CHT/4/10
101	Admiralty to Backhouse, Pound et al.	1 May 1936	ADM 116/3042
102	Admiralty to Backhouse, Pound and Rose	1 May 1936	ADM 116/3042
103	Admiralty to Backhouse and Pound	14 May 1936	ADM 116/3042
104	Admiralty to C-in-Cs Mediterranean Home Fleet and East Indies	21 May 1936	ADM 116/3042
105	Pound to Chatfield	30 May 1936	CHT/4/10
106	Admiralty to Pound et al.	19 June 1936	ADM 116/3042
107	Pound to Admiralty	22 June 1936	ADM 116/3042
108	Memorandum by Fisher	29 June 1936	CAB 16/147
109	Hoare to Eden	3 July 1936	FO 371/20382
110	Chatfield to Vansittart	6 July 1936	ADM 116/3042
111	Admiralty to Pound	6 July 1936	ADM 116/3043
112	Admiralty to Pound et al.	8 July 1936	ADM 116/3042

113	Admiralty to Pound et al.	8 July 1936	ADM 116/3042
114	Admiralty Message Home and Abroad	11 July 1936	ADM 116/3043
115	Admiralty Message Home and Abroad	14 July 1936	ADM 116/3043
116	Pound to Chatfield	23 July 1936	CHT/4/10

Part III: The Spanish Civil War and the Nyon Agreements, 1936–1937

117	Pound to Forces Western Mediterranean	23 July 1936	ADM 116/3534
118	Godfrey to Rear Admiral 1st Cruiser Squadron	31 July 1936	ADM 116/3051
119	Pound to Chatfield	8 Aug 1936	CHT/4/10
120	Godfrey to Rear Admiral 1st Battle Squadron	29 Aug 1936	ADM 116/3051
121	Pound to Somerville	2 Sept 1936	FO 371/20537
122	Somerville to Pound	3 Sept 1936	FO 371/20537
123	Pound to Chatfield	18 Sept 1936	CHT/4/10
124	Pound to Mediterranean Fleet	23 Sept 1936	ADM 116/3051
125	Goolden to Blake	10 Oct 1936	ADM 116/3051
126	Admiralty Memorandum	22 Dec 1936	ADM 116/3917
127	Courage to Pound	15 Feb 1937	ADM 116/3534
128	Admiralty to Cs-in-C Home Fleet, Mediterranean and Vice Admiral Gibraltar	20 Feb 1937	ADM 116/9170
129	Pound to Admiralty	21 Feb 1937	ADM 116/9170
130	Beckett to Vice Admiral Gibraltar	10 April 1937	ADM 116/3534
131	Pound to Blake	6 April 1937	ADM 116/3534
132	Pound to Blake	6 April 1937	ADM 116/3534
133	Tait to Blake	8 April 1937	ADM 116/3534
134	Tait to Blake	8 April 1937	ADM 116/3534
135	Admiralty to Blake	10 April 1937	ADM 116/3534
136	Minute by Vice Admiral Gibraltar	11 April 1937	ADM 116/3917
137	Miles to Somerville	28 April 1937	ADM 116/3681
138	Deverell to Vice Admiral Gibraltar	30 April 1937	ADM 116/3518
139	Vice Admiral Gibraltar to Admiralty	13 May 1937	ADM 116/3521
140	Vice Admiral Gibraltar to Pound	13 May 1937	ADM 116/3521
141	HMS *Hunter* to Admiralty	13 May 1937	ADM 116/3521
142	Admiralty to Captain (D), 2nd Destroyer Flotilla	14 May 1937	ADM 116/3521
143	Wells to Admiralty	14 May 1937	ADM 116/3521
144	Admiralty to Pound	14 May 1937	ADM 116/3521
145	Wells to Vice Admiral Gibraltar	14 May 1937	ADM 116/3521
146	Admiralty to Pound	14 May 1937	ADM 116/3521
147	Wells to Admiralty	14 May 1937	ADM 116/3521
148	Pound to Ships in Western Mediterranean	15 May 1937	ADM 116/3521
149	Vice Admiral Gibraltar to Pound	15 May 1937	ADM 116/3521
150	Pound to Admiralty	16 May 1937	ADM 116/3521
151	Vice Admiral Gibraltar to Pound	17 May 1937	ADM 116/3521

SOURCES AND DOCUMENTS 563

152	Vice Admiral Gibraltar to Admiralty	17 May 1937	ADM 116/3521
153	Vice Admiral Gibraltar to Admiralty	17 May 1937	ADM 116/3521
154	Pound to Vice Admiral Gibraltar	18 May 1937	ADM 116/3521
155	Pound to Admiralty	22 May 1937	ADM 116/3521
156	Admiralty to Pound	25 May 1937	ADM 116/3521
157	Pound to Admiralty	27 May 1937	ADM 116/3521
158	Gordon to Rear Admiral Gibraltar	7 June 1937	ADM 116/3519
159	Arliss to Rear Admiral Gibraltar	12 June 1937	ADM 116/3519
160	Pound to Admiralty	13 June 1937	ADM 116/3521
161	Admiralty to Pound	14 June 1937	ADM 116/3521
162	Pound to Admiralty	24 June 1937	ADM 116/3522
163	Pound to Vice Admiral Gibraltar	25 June 1937	ADM 116/3521
164	Tait to Pound	26 July 1937	ADM 116/3520
165	Pound to Mediterranean Station	18 Aug 1937	ADM 116/3522
166	Somerville to Pound	24 Aug 1937	ADM 116/3522
167	Pound to Somerville	24 Aug 1937	ADM 116/3522
168	Admiralty to Pound	26 Aug 1937	ADM 116/3522
169	Pound to Somerville	26 Aug 1937	ADM 116/3522
170	Richmond to Somerville	27 Aug 1937	ADM 116/3519
171	Pound to Admiralty	28 Aug 1937	ADM 116/3522
172	Admiralty to Pound	28 Aug 1937	ADM 116/3522
173	Pound to Somerville	29 Aug 1937	ADM 116/3522
174	Somerville to Admiralty	4 Sept 1937	ADM 116/3534
175	Pound to Somerville	1 Sept 1937	ADM 116/3534
176	Admiralty to Pound	1 Sept 1937	ADM 116/3917
177	Somerville to Pound	1 Sept 1937	ADM 116/3917
178	Somerville to 2nd DF	3 Sept 1937	ADM 116/3534
179	Somerville to Pound	3 Sept 1937	ADM 116/3522
180	Pound to Somerville	3 Sept 1937	ADM 116/3522
181	Pound to Admiralty	5 Sept 1937	ADM 116/3522
182	Somerville to Pound	5 Sept 1937	ADM 116/3522
183	Admiralty to Pound	6 Sept 1937	ADM 116/3522
184	Admiralty to Pound	7 Sept 1937	ADM 116/3522
185	Somerville to Pound	8 Sept 1937	ADM 116/3522
186	Pound to Admiralty	8 Sept 1937	ADM 116/3522
187	Pound to Admiralty	8 Sept 1937	ADM 116/3522
188	Admiralty to Pound	9 Sept 1937	ADM 116/3522
189	Pound to Backhouse	10 Sept 1937	ADM 116/3522
190	Pound to Chatfield	10 Sept 1937	ADM 116/3522
191	Chatfield to Pound	11 Sept 1937	ADM 116/3522
192	Chatfield to Pound	11 Sept 1937	ADM 116/3522
193	Chatfield to Pound	12 Sept 1937	ADM 116/3522
194	Pound to Chatfield	12 Sept 1937	ADM 116/3522
195	Chatfield to Pound	12 Sept 1937	ADM 116/3522
196	Chatfield to Pound	13 Sept 1937	ADM 116/3522
197	Admiralty to Pound	13 Sept 1937	ADM 116/3522
198	Chatfield to Pound	13 Sept 1937	ADM 116/3522
199	Chatfield to Pound	13 Sept 1937	ADM 116/3522
200	Pound to Chatfield	14 Sept 1937	ADM 116/3522
201	Pound to Admiralty	14 Sept 1937	ADM 116/3522

202	Director of Naval Intelligence to Chatfield	14 Sept 1937	ADM 116/3525
203	Pound to Chatfield	14 Sept 1937	ADM 116/3522
204	Chatfield to Pound	15 Sept 1937	ADM 116/3522
205	Chatfield to Pound	15 Sept 1937	ADM 116/3522
206	Pound to Admiralty	17 Sept 1937	ADM 116/3525
207	Pound to Admiralty	17 Sept 1937	ADM 116/3525
208	Pound to Admiralty	17 Sept 1937	ADM 116/3525
209	Pound to Admiralty	17 Sept 1937	ADM 116/3525
210	Admiralty to Pound	18 Sept 1937	ADM 116/3525
211	Pound to Admiralty	19 Sept 1937	ADM 116/3525
212	Pound to Admiralty	20 Sept 1937	ADM 116/3525
213	Admiralty to Pound	20 Sept 1937	ADM 116/3525
214	Pound to Chatfield	20 Sept 1937	CHT/4/10
215	Pound to Mediterranean Station	21 Sept 1937	ADM 116/3524
216	Muirhead-Gould to Pound	22 Sept 1937	ADM 116/3681
217	Admiralty to Pound	1 Oct 1937	ADM 116/3524
218	Admiralty to Pound	1 Oct 1937	ADM 116/3524
219	Chatfield to Pound	5 Oct 1937	ADM 116/3529
220	Pound to Chatfield	6 Oct 1937	ADM 116/3529
221	A.B. Cunningham to Pound	7 Oct 1937	ADM 116/3529
222	Pound to Chatfield	8 Oct 1937	ADM 116/3529
223	Admiralty to Pound	9 Oct 1937	ADM 116/3529
224	Pound to A.B. Cunningham	11 Oct 1937	ADM 116/3524
225	Pound to Admiralty	14 Oct 1937	ADM 116/3530
226	Godfrey to A.B. Cunningham	14 Oct 1937	ADM 116/3681
227	Pound to Admiralty	16 Oct 1937	FO 371/21169
228	Admiralty to Pound	18 Oct 1937	FO 371/21169
229	Admiralty to Pound	20 Oct 1937	ADM 116/3530
230	Pound to A.B. Cunningham	21 Oct 1937	ADM 116/3525
231	Admiralty to Pound	22 Oct 1937	ADM 116/3533
232	Admiralty to Pound	23 Oct 1937	ADM 116/3834
233	Binney to Pound	23 Oct 1937	ADM 116/3525
234	Pound to Binney	25 Oct 1937	ADM 116/3525
235	Binney to Pound	25 Oct 1937	ADM 116/3525
236	Admiralty to Pound	25 Oct 1937	ADM 116/3525
237	Pound to Admiralty	30 Oct 1937	ADM 116/3525
238	Admiralty to Pound	3 Nov 1937	ADM 116/3525
239	A.B. Cunningham to Pound	4 Nov 1937	ADM 116/9948
240	Admiralty to Pound	10 Nov 1937	ADM 116/3533
241	Pound to Somerville	11 Nov 1937	ADM 116/3533
242	Somerville to Pound	11 Nov 1937	ADM 116/3533
243	Pound to HMS *Repulse*	12 Nov 1937	ADM 116/3530
244	Godfrey to A.B. Cunningham	29 Nov 1937	ADM 116/3681
245	Pound to Chatfield	17 Nov 1937	CHT/4/10
246	Chatfield to Pound	23 Nov 1937	CHT/4/10
247	Somerville to Pound	24 Nov 1937	ADM 116/3892
248	Somerville to Admiralty	29 Nov 1937	ADM 116/3533
249	Admiralty to Pound	3 Dec 1937	ADM 116/3525
250	Pound to Admiralty	4 Dec 1937	ADM 116/3525

SOURCES AND DOCUMENTS 565

251	Admiralty to Pound	6 Dec 1937	ADM 116/3530
252	Admiralty to Pound	9 Dec 1937	ADM 116/3529
253	Pound to Binney	14 Dec 1937	ADM 116/3533
254	Pound to Chatfield	22 Dec 1937	CHT/4/10
255	Chatfield to Pound	30 Dec 1937	CHT/4/10

Part IV: The Approach of War, 1938–1939

256	Admiralty to Pound	1 Jan 1938	ADM 116/3530
257	Pound to British Ambassador Rome	2 Jan 1938	ADM 116/3525
258	Bernotti to Pound	5 Jan 1938	ADM 116/3530
259	Admiralty to Pound	25 Jan 1938	ADM 1/9948
260	A.B. Cunningham to Pound	4 Feb 1938	ADM 116/3679
261	Admiralty to Pound, Backhouse and Calvert	5 Feb 1938	ADM 116/3530
262	Calvert to Admiralty	6 Feb 1938	ADM 116/3532
263	Pound to Chatfield	7 Feb 1938	CHT/4/10
264	Calvert to Admiralty	12 Feb 1938	ADM 116/4084
265	Hutton to Pound	10 Mar 1938	ADM 116/3530
266	Admiralty to Pound	11 Mar 1938	ADM 116/3530
267	Pound to Wells	12 Mar 1938	ADM 116/3530
268	Admiralty to Pound	19 Mar 1938	ADM 116/3530
269	Pound to Rear Admiral Gibraltar	20 Mar 1938	ADM 116/3530
270	Somerville to Pound	27 Mar 1938	ADM 1/9949
271	Pound to Chatfield	4 April 1938	ADM 1/9534
272	Pound to A.B. Cunningham	14 April 1938	ADM 116/3530
273	Pound to Admiralty	17 June 1938	ADM 116/3530
274	Admiralty to Pound	21 June 1938	ADM 116/3530
275	Admiralty to Pound	24 June 1938	ADM 116/3530
276	Pound to Admiralty	25 June 1938	ADM 116/3530
277	Rear Admiral 3rd Cruiser Squadron to Pound	25 June 1938	ADM 116/3530
278	Pound to Admiralty	26 June 1938	ADM 116/3530
279	Admiralty to Pound	27 June 1938	ADM 116/3530
280	Pound to Admiralty	23 Aug 1938	ADM 1/9543
281	Pound to Chatfield	24 Aug 1938	CHT/4/10
282	Admiralty to Pound	27 Aug 1938	ADM 1/9543
283	Admiralty to Pound	31 Aug 1938	ADM 1/9542
284	Pound to Admiralty	31 Aug 1938	ADM 1/9542
285	Pound to Admiralty	31 Aug 1938	ADM 116/4068
286	Pound to Admiralty	1 Sept 1938	ADM 1/9542
287	Admiralty to Forbes and Pound	1 Sept 1938	ADM 1/9543
288	Admiralty to Pound	6 Sept 1938	ADM 1/9542
289	Admiralty to Cs-in-C	6 Sept 1938	ADM 116/4068
290	Pound to Admiralty	21 Sept 1938	ADM 116/4161
291	Admiralty to Pound and Forbes	21 Sept 1938	ADM 116/4161
292	Muirhead-Gould to Pound	23 Sept 1938	ADM 116/3903
293	Layton to Pound	3 Oct 1938	ADM 116/3903
294	Backhouse to Pound	11 Oct 1938	ADM 205/3
295	Pound to Admiralty (Excerpts)	14 Nov 1938	ADM 116/3900
296	Pound to Admiralty	14 Nov 1938	ADM 116/3900

566   THE MEDITERRANEAN FLEET, 1930–1939

297	Backhouse to Pound	8 Dec 1938	ADM 205/3
298	Backhouse to Cork and Orrery	12 Dec 1938	ADM 205/3
299	Pound to Backhouse	14 Dec 1938	ADM 205/3
300	Halifax to Stanhope	13 Dec 1938	ADM 116/4084
301	Backhouse to Stanhope	13 Dec 1938	ADM 116/4084
302	Stanhope to Halifax	16 Dec 1938	ADM 116/4084
303	Backhouse to Pound	28 Dec 1938	ADM 205/3
304	Muirhead-Gould to Pound	27 Jan 1939	ADM 116/3892
305	Muirhead-Gould to Pound	12 Feb 1939	ADM 116/3896
306	Muirhead-Gould to Pound	20 Feb 1939	ADM 116/3896
307	Admiralty to Pound	13 Feb 1939	ADM 116/3900
308	Pound to Backhouse	21 Feb 1939	ADM 1/9930
309	Hammick to J.D. Cunningham	22 Mar 1939	ADM 116/3896
310	Tovey to Pound	3 April 1939	ADM 116/3896
311	Pound to Admiralty	2 April 1939	ADM 116/3844
312	Layton to Pound	6 April 1939	ADM 116/3844
313	Admiralty to Pound	7 April 1939	ADM 116/3844
314	Admiralty to Pound	7 April 1939	ADM 116/3844
315	Pound to Ford	7 April 1939	ADM 116/3844
316	Pound to Layton et al.	7 April 1939	ADM 116/3844
317	Pound to Ford	7 April 1939	ADM 116/3844
318	Admiralty to Pound	8 April 1939	ADM 116/3844
319	Pound to Vice Admiral Malta et al.	9 April 1939	ADM 116/3844
320	Admiralty to Pound	9 April 1939	ADM 116/3844
321	Admiralty to Pound	9 April 1939	ADM 116/3844
322	Admiralty to Pound	9 April 1939	ADM 116/3844
323	Pound to Admiralty	10 April 1939	ADM 116/3844
324	Pound to Admiralty	10 April 1939	ADM 116/3844
325	Admiralty to Pound	10 April 1939	ADM 116/3844
326	Admiralty to Cs-in-C	10 April 1939	ADM 116/3844
327	Admiralty to Pound	10 April 1939	ADM 116/3844
328	Admiralty to Pound	11 April 1939	ADM 116/3844
329	Pound to Admiralty	13 April 1939	ADM 116/3844
330	Admiralty to Pound	13 April 1939	ADM 116/3844
331	A.B. Cunningham to Pound	14 April 1939	ADM 1/9930
332	Pound to Admiralty	14 April 1939	ADM 116/3844
333	Pound to Admiralty	14 April 1939	ADM 116/3844
334	Pound to Admiralty	14 April 1939	ADM 116/3844
335	Admiralty to Pound	14 April 1939	ADM 116/3844
336	Admiralty to Pound	15 April 1939	ADM 116/3844
337	Pound to Admiralty	18 April 1939	ADM 116/3844
338	Pound to Admiralty	19 April 1939	ADM 1/9930
339	Admiralty to Pound	21 April 1939	ADM 116/3844
340	Admiralty to Pound	22 April 1939	ADM 116/3844
341	Pound to Admiralty	22 April 1939	ADM 116/3844
342	Pound to Admiralty	25 April 1939	ADM 116/3844
343	Pound to Admiralty	26 April 1939	ADM 116/3844
344	Admiralty to Pound	1 May 1939	ADM 116/3900
345	Admiralty to Pound	4 May 1939	ADM 116/3845
346	Pound to Admiralty	10 May 1939	ADM 116/3900

347	Admiralty to Pound	12 May 1939	ADM 1/9946
348	Pound to Admiralty	15 May 1939	ADM 1/9946
349	Pound to Admiralty	15 May 1939	ADM 116/3845
350	Pound to Admiralty	15 May 1939	ADM 116/3845
351	Admiralty to Pound	20 May 1939	ADM 116/3845
352	Admiralty to Pound	21 May 1939	ADM 116/3845
353	Rear Admiral Gibraltar to Admiralty	22 May 1939	ADM 116/3845
354	Admiralty to Cs-in-C Home Fleet and Mediterranean	6 June 1939	ADM 1/9875
355	A.B. Cunningham to Admiralty	12 June 1939	ADM 1/9875
356	A.B. Cunningham to Admiralty	22 June 1939	ADM 1/9930
357	A.B. Cunningham to Admiralty	30 June 1939	ADM 116/3877
358	Admiralty to A.B. Cunningham	6 July 1939	ADM 116/3900
359	Admiralty to A.B. Cunningham	8 July 1939	ADM 1/9930
360	A.B. Cunningham to Admiralty	14 July 1939	ADM 1/9946
361	Admiralty to A.B. Cunningham	29 July 1939	ADM 1/9946
362	A.B. Cunningham to Admiralty	2 Aug 1939	ADM 1/9905
363	Admiralty to Cs-in-C	30 Aug 1939	ADM 1/10188

INDEX

Officers are usually listed under the ranks they held when first mentioned.

Abrail, Vice Amiral Jean Marie Charles, 529, 529 n. 1
Abyssinia *see* Ethopia
Aden, 106, 161, 172–3, 177, 186, 361, 363
Admiralty: and Malta, 7–8, 34–5, 51–2, 82; strength of Mediterranean Fleet, 9, 16–17, 20–22, 50–51, 71; and Gibraltar, 28, 153–4, 245–6; and Turkey, 55–6, 531; potential conflict with Italy (1935), 64–5, 68–70, 74, 78 n. 3, 91, 93, 112–15, 126, 148–50; reversion to normal conditions (1936), 74, 75, 170–73, 176, 183–7; and Spanish Civil War, 194, 196–7, 197, 199–201, 204–5, 224–31; and non-intervention committee, 233–5, 244–5, 261–2, 266; and incidents with British warships, 255–6, 259 n. 1, 288, 290; attacks on merchant shipping, 278, 283, 336 n. 1, 351–2, 366–71; and "pirate" submarine attacks, 296–8, 302–4, 335; and Nyon agreements, 313–15, 317, 320–21, 350–51; and German warships in Spain, 322, 410, 521–2; and relaxation of Nyon patrols, 353, 369, 383, 396, 420; and Czechoslovak crisis, 384–5, 427–9, 432, 434–5, 437–8; and renewal of attacks on shipping, 399, 403–4, 420–21, 424; and Italian occupation of Albania, 386, 389, 509, 511–14; and Spanish Morocco, 393, 540; and Greek territorial waters, 393, 541; and Mediterranean strategy, 489–91; 515–16, 539–40; staff talks with French, 527–8, 531, 534; and 1939 cruise programme, 528; and Hague Treaty rules for conduct of war, 540–43; and Italy, 394–5, 557–8.

See also Chatfield, Fisher, Pound, Backhouse, Cunningham, A.B.
Aegean Sea, 201–2, 205, 374
Air Council, 7, 30
Alba, Duke of, 488, 488 n. 1
Albania, 389, 509–11
Alexandria: and potential conflict with Italy, 65, 72, 97, 113, 145; and concentration of fleet, 67, 70, 74, 385, 389, 433, 444–8; British forces at, 118, 122, 158; base defences, 125, 135, 146, 158–9, 164, 520; and end of Ethiopian crisis, 182–3, 185; and floating dock, 363, 372, 390, 517, 537–8; and AA defences, 390, 446, 535–7; mentioned, 10 n. 2
Algeria, 392, 492, 533
Almeria, 198, 254, 257–8, 263
Amsden, Captain William F. (USN), 210
Anderson, Captain Walter Stratton (USN), 167
Anglo-French Agreement (1937), 311–13, 318, 320–21
Anglo-French staff talks, 384, 391, 490. *See also* Pound, Cunningham, A.B., France, Navy
Anglo-German Naval Agreement (1935), 64
Apfel, Mr. Edwin, 501–2, 504
Arliss, Commander Stephen H.T., 264
Arzeu, 202, 204, 319, 323, 359
Atlantic Fleet, 3 n. 2, 5, 9, 12, 17, 18 n. 1, 21. *See also* Home Fleet
Aubert, Capitaine de frégate Yves, 272–3
Augusta: in British war plans, 392, 526, 533–4, 550, 552; mentioned, 66, 68, 78, 104, 140, 150–51

Backhouse, Vice Admiral Roger: career, 46 n. 2; C-in-C Home Fleet, 69–70, 87–8, 116–17, 119–20, 150–53, 155; designated 1SL, 376, 406–7;

569

Backhouse, Vice Admiral Roger: (*cont'd*) and Czechoslovak crisis, 385, 441–2; and alleged pessimism in Mediterranean Fleet, 387, 463, 471; and attacks on British shipping, 466–7; programme for 1939, 470–71; mentioned, 3, 46, 49, 89, 324, 358, 393, 489. *See also* Admiralty, Home Fleet
Badoglio, Marshal Pietro, 175
Bailey, Rear Admiral Sidney Robert, 134
Bain, Commander D.K., 472
Balbo, Maresciallo dell'Aria Italo, 33
Baldwin, Stanley, 163 n. 1
Balearic Islands: 192. *See also* Palma, Majorca, Minorca
Barcelona: situation, 219–24, 270; evacuation, 472–6
Bastarreche, Rear Admiral Francisco: port admiral at Palma, 199, 274, 277, 278; and attacks on British merchant ships, 351; at Cadiz, 364–5; mentioned, 283–4
Beatty, Admiral of the Fleet 1st Earl, 3
Beckett, Lieut. Commander Roger Caton, 239–41
Bernotti, Ammiraglio di Squadra Romeo, 349–50, 358, 396, 398
Bevan, Captain Richard Hugh Lorraine, 14
Binney, Rear Admiral Thomas Hugh: career, 89 n. 2; in Aegean, 338, 343, 347–8, 355, 357; mentioned, 211, 218, 407
Bizerta, 133, 137, 139
Black, Vice Consul R.A., 364
Blake, Rear Admiral Geoffrey, 47, 360, 374
Blehaut, Capitaine de vaisseau Henri Paul, 545
Bonham-Carter, Lieut. General Charles, 188, 360
Borghese, Lieutenant Junio, 200, 204 n. 1
Bourragué, Contre Amiral Célestin Jean Léon, 555
Boyd, Captain Denis William, 169
Boyle, Admiral Sir William Henry Dudley, 39, 46, 156, 161, 463–4
Brandaris, General J., 478–9
Brownrigg, Vice Admiral Henry John, 89, 146

Cables, undersea, 70, 129–30
Cadiz, 363–5
Cagliari, 69, 114, 116–18, 120
Calvert, Rear Admiral Thomas Frederick Parker, 403–5

Camin, Judge, 475
Campbell, Major General David Gtaham, 25 n. 1
Casado, Colonel Segismundo, 501–3
Casement, Lieut. Colonel J.J., 479
Cazenave, Enseigne de vaisseau H.L.R., 138–40
Chamberlin, Neville, 48, 205
Chatfield, Admiral Alfred Ernle Montacute: career, 14 n. 1; commands Mediterranean Fleet, 3–9, 14–16; and defence of Suez Canal, 10, 44–5; and strength of Mediterranean Fleet, 17–20, 42–4, 383, 418; and visit of Prince of Wales, 22–5; and Defence Requirements Committee, 38 n. 2 and n3, 48; and Silver Jubilee Review, 45–6; and Samos incident, 47–8; programme, 48–9; and Keyes, 49–50; and Ethiopian crisis, 67–8, 73, 74, 93–102, 162–6, 182; critical of British foreign policy, 109, 362–3; and Spanish Civil War, 194, 196, 246, 330–31, 347–8; and Noyon agreements, 201, 203, 306–7, 421 n. 1; and German threat, 206, 360–61; and Japan, 206, 375–6; and potential American cooperation, 375–6; and defence of Egypt, 361; and MTBs, 362. *See also* Admiralty, Fisher, Pound
Chilton, Sir Henry, 367 n. 1
Chetwode, Vice Admiral George, 33
Chetwode, Field Marshal Philip, 476
Chiefs of Staff Committee: and potential war with Italy, 69, 72, 94, 96, 111, 135–6, 147; and Libya, 205, 344–6; mentioned, 206, 360
Churchill, Winston S., 359
Ciano, Galeazzo, 515
Committee of Imperial Defence: and Gibraltar, 6, 28; and defence requirements, 38 n. 2; and Mediterranean area, 52–3, 361, priorities, 206, 372
Companys, Don Luis, 222
Cooper, Alfred Duff, 343, 421 n. 1
Corfu, 21–3, 389, 512. *See also* Greece
Corinth Canal, 391, 451–2, 525, 541. *See also* Greece
Cork and Orrey, Earl *see* Boyle, Admiral Sir William
Courage, Lieut. Commander R.E., 231, 285

INDEX 571

Cox, Reverend Hubert, 275
Cunningham, Rear Admiral Andrew Browne: career, 122 n. 4; and Fisher, 66, 122, 161; and Spanish Civil War, 331–4, 381–2, 387, 400–403, 467–8, 489; as Mediterranean C-in-C, 393–4, 426, 526 n. 1, 538–9; as DCNS, 516; meeting with French, 394, 518, 545–52; plans for war, 538, 552–7; and International Law, 542–4; mentioned, 168, 407. *See also* Admiralty, Mediterranean Fleet
Cunningham, Captain John Henry Dacres, 121
Cyprus, 10 n. 2, 175, 362–3
Czechoslovak Crisis (1938), 384–5, 388, 427, 429–30, 436–42,

Daily Mail, 281
Danckwerts, Captain Victor Hilary, 169
Dangerfield, Commander, 331
Darlan, Amiral de la Flotte François Xavier, 555
Defence Requirements Committee, 48
Deverell, Commander G.R., 249–53
Dreyer, Admiral Frederic Charles, 109 n. 1, 156
Dundas, Captain John George Lawrence, 546

Eden, Robert Anthony, 78–9, 181, 343
Edward, Prince of Wales: visits fleet, 22–5; as King Edward VIII, 187
Egypt: defence of, 70, 72, 205, 343–5, 359–61, 490; and Czechoslovak crisis, 433; mentioned, 65, 177, 184, 188, 324. *See also* Alexandria, Suez Canal, Sollum, Mersa Matruh
Esteva, Vice Amiral Jean Pierre: career, 311 n. 1; meeting with Pound, 202–3, 311, 315, 318, 320, 349–50, 358; mentioned, 396, 397, 401, 529, 532
Ethiopia, 11, 63–4. *See also* Italy, League of Nations, Haile Selassie
Evans, Rear Admiral Alfred Englefield, 178, 185, 537
Eyres-Monsell, Bolton Meredith, 165 n. 1

"Fairey Queen" (drone), 26–7
Faringdon, 2nd Baron, 503
Field, Admiral Sir Frederick L., 3
Firth, Commander Charles Leslie, 187
Fischel, Konteradmiral Hermann von, 266–7

Fisher, Neville, 124
Fisher, Admiral William Wordsworth: career, 26 n. 1; as C-in-C Mediterranean Fleet, 3, 6, 10, 26–8, 33, 39–42, 46, 131; and Malta defences, 7–8, 29–31, 37–9, 80–81, 89–90, 108; and Turkey, 47 n. 2, 55 n. 3, 56 n. 1; and passage of fleet to the Far East, 56–9; and potential conflict with Italy, 64–8, 70, 72–3, 78–9, 91; 392 n. 2; and strength of Mediterranean Fleet, 71, 132–3; and French Navy, 72; on vulnerability of capital ships to air, 73, 179–80; and Home Fleet, 75–80; and Port "X", 91–2, 135 n. 1; planned operations, 103–4, 117, 155; on situation at Alexandria, 121, 130–31; and Backhouse, 152, 155; to be C-in-C Portsmouth, 156; provides summary of Ethiopian crisis, 157–9; mentioned, 123–4, 159–61. *See also* Alexandria, Mediterranean Fleet, Admiralty, Chatfield
Flandin, Pierre Etienne, 162
Fleet Air Arm: and potential conflict with Italy, 66, 84, 392; 802 Squadron, 83–4; 812 Squadron, 83–4; 820 Squadron, 178; 823 Squadron, 83–4; 825 Squadron, 83–4; mentioned, 49, 83, 91, 131, 447, 454
Forbes, Vice Admiral Charles Morton, 122, 131, 188, 376, 471
Forcinal, Albert, 503, 505
Ford, Vice Admiral Wilbraham Tennyson Randle, 510
Fouad I, King of Egypt, 168–9
France: as potential enemy, 7; and Ethiopian crisis, 64–5, 71–2, 76, 79 n. 2, 98, 101, 105, 162–3; discussions with British, 110–11, 127–8, 136–7; and Spanish Civil War, 192, 201, 306–7; and Tunisia, 393; numerical inferiority of air force, 533, 539–40. *See also* France, Navy, Nyon Agreements
France, Navy: large destroyers, 272; officers in Spain, 273; and Nyon agreement, 298, 311–14, 318, 323, 396; visit to Malta, 358; and potential war with Germany, 384, 431; role in war with Italy, 391, 394, 414; arrangements for liaison officers, 427, 527–8, 531–2,

France, Navy: (cont'd)
539–40, 545–52; staff talks with British, 491–2, 518, 529; planned operations, 554–5; forces available, 557 n. 1. See also Laborde, Michelier, Estiva, Ollive

Franco y Bahamonde, Generalissimo Francisco: career, 272 n. 1; and attacks on British ships, 465 n. 1, 466, 469–70; and San Luis, 480, 483; recognized by British and French, 494 n. 1, 497 n. 1; mentioned, 191, 206, 272, 387, 403, 412, 467, 477. See also Spain, Nationalist

Franco y Bahamonde, Colonel Ramon, 242–3

French, Vice Admiral Wilfred Frankland, 85, 89–90

Gambardella, Ammiraglio di squadra Fausto, 14–15
Gandia, 500, 502, 504–6
Gascoigne, Instructor Lieut. J.C., 479–80
Genoa, 69–70, 108, 114, 116–18, 120, 555
George V, King of England, 10, 25
George II, King of the Hellenes, 426
George, Prince, 4, 22–5
Germany: and Spanish Civil War, 191, 195, 264 n. 3; and Great Britain, 197–8, 202; aggressive policies, 383–4, strength of air power, 442. See also Germany, Navy
Germany, Navy: and Great Britain, 222; and Spanish Civil War, 322, 409, 519; and Czechoslovak crisis, 427–8, 441; on Spanish coast, 521–2
Gibraltar: poor air defences, 6, 26, 28; submarine defences, 82, 86; and British naval operations, 148, 152–4; and Spanish Civil War, 196, 245–6; and German threat, 384, 537; and Czechoslovak crisis, 429–30, 436–40, 450–51; and war plans, 523–4, 556. See also Mediterranean Fleet, Home Fleet
Glinsky, Captain, 281
Godfrey, Captain John Henry: career, 207 n. 1; and Spanish Civil War, 207–10, 212–14, 275 n. 2; proceedings in Aegean, 337–43, 355–7; and Palestine, 426
Gooden, Mr. A., 500, 503
Goolden, Captain Francis Hugh, 219–24

Gordon, Lieut. Commander R.C., 262–4
Great Britain: and Spanish Civil War, 193–6, 198–9. See also Great Britain, Board of Trade, Great Britain, Foreign Office
Great Britain, Army: Gordon Highlanders, 188, 207–8, 212; Worcestershire Regiment, 27; 11th Hussars, 146
Great Britain, Board of Trade, 421 n. 1, 501
Great Britain, Cabinet, 69, 73–4, 81, 163, 167, 434
Great Britain, Foreign Office: and Ethiopian crisis, 74, 126, 129, 163 n. 1, 165, 182; and Turkey, 9, 48, 55; and Spanish Civil War, 303, 369, 382, 387, 407, 409, 421 n. 1; and Hillgarth, 303; and surrender of Minorca, 476 n. 3, 478; and Spanish refugees, 504; mentioned, 55, 79, 86, 107
Greece: and British use of ports, 91, 93–4, 98, 136, 145, 147, 201, 317, 320; protest over Skyros, 205, 399–40; use of territorial waters, 391, 393, 524–5, 538, 541; and Italian threat to Corfu, 389, 512. See also Aegean, Skyros, Corinth Canal, Corfu
Greece, Navy, 358, 389, 391, 426

Hague Convention (1907), 149–50, 541, 543–4
Haifa: and Ethiopian crisis, 67, 74, 130, 157; defences, 164, 183, 185, 520; use in war, 448, 458, 460; 556
Haile Selassie, 175 n. 3
Haining, Lieut. General Robert Hadden, 426
Halifax, Viscount (Edward Frederick Lindley Wood), 387, 465–6
Hallifax, Captain Guy W., 15
Hammick, Captain Alexander Robert, 492–6
Hemsted, Paymaster Captain James Rustat, 176
Henderson, Vice Admiral Reginald Guy Hannam, 363
Hillgarth, Lieut. Commander Alan Hugh: career, 209 n. 1; at Palma, 194, 208–9, 411; and attacks on British ships, 244, 382, 407–8, 499; praised, 273–5, 413; and Minorca, 387, 476–8, 482–4, 487–8; mentioned, 413
Hitler, Adolph, 7, 163, 198, 441–2, 528

INDEX   573

Hoare, Sir Samuel, 63–4, 72, 74, 181, 217–18
Home Fleet: combined exercises with Mediterranean Fleet, 39; situation, 43, 46, 50–51; passage to Far East, 53, 58; and potential conflict with Italy, 66–7, 69, 71, 116–17, 127, 152; reinforces Mediterranean, 77, 87–8, 93, 132, 165; question of leave, 134–5; and easing of tension, 176, 184; and Spanish Civil War, 193, 200, 204, 234; destroyers for Mediterranean, 300, 302–3; and Nyon patrols, 419–22. *See also* Atlantic Fleet, Backhouse
Horan, Captain Henry Edward, 374, 406
Horton, Rear Admiral Max, 155, 207 n. 2
Hutton, Captain Fitzroy Evelyn Patrick, 408 n. 4

Iachino, Contrammiraglio Angelo, 250
im Thurn, Vice Admiral John Knowles, 32, 39, 55 n. 3
Ingersoll, Captain Royal E. (USN), 375 n. 1
International Delegation for Relief and Refugees in Central Spain, 503–5
International Red Cross, 269, 505–6
Invergordon Mutiny, 5, 33 n. 2, 49, 193
Italy: threat of air force, 7, 29–30, 33, 67, 76, 443, 533; and Ethiopia, 11, 63, 64, 69, 110, 163–4, 166; and Yugoslavia, 35; and belligerent rights, 71, 115, 126; strategic disadvantages, 94–5, 97, 101; geographical advantages, 96–7, 100–101, 443; and Spanish Civil War, 188, 191–3, 195, 214–15, 411–12, 482, 487; and attacks on British ships, 244, 466–7; and Nyon agreements, 297, 323, 327–8, 349–50; in Chatfield's opinion, 362–3; and Great Britain, 372, 383; and Czechoslovak crisis, 433–4; and Red Sea, 441; and Albania, 509, 511–12, 515. *See also* Italy, Navy, League of Nations
Italy, Navy: improvement, 15; and Spanish Civil War, 192, 195, 198, 201–2, 398, 407, 459; and Great Britain, 197, 222, covert submarine activity, 316, 381; and Nyon agreements, 202–3; visit to Malta, 425–6; strength, 95, 413–14, 452–3. *See also* Ricci, Bernotti

James, Mr. J.H., 488

James, Vice Admiral William Milbourne, 73, 162–6, 183, 194, 246, 416–18, 421 n. 1
Japan: potential threat to Suez Canal, 10, 53–4, 57–8; situation in Far East, 64, 206, 363, 375–6, 383–4
Jerram, Paymaster Commander Rowland, 183
Junod, Dr. Marcel, 269
Jutland, Battle of, 4, 248

Kennedy-Purvis, Vice Admiral Charles Edward, 400, 407
Kerdudo, Contre Amiral, 473, 475
Keyes, Admiral of the Fleet Roger John Brownlow, 49–50, 358
Kindelán y Duany, General Alfredo, 487
Kinloch, Lieut. David Charles, 150

Laborde, Vice Amiral Jean Joseph, Comte de, 72, 138–42
Lackey, Rear Admiral Henry E. (USN), 473
Lampson, Sir Miles Wedderburn, 115, 155, 169, 361, 363
Laval, Pierre, 63, 72
Layton, Rear Admiral Geoffrey, 385, 407, 436–40, 463, 508–9
League of Nations: and Italy, 8, 11, 63–5, 71, 74–7; votes sanctions, 69, 93–4, 101–2, 105, 110, 126, 162–3, 173–4, 182; and Spanish Civil War, 198; mentioned, 36, 165, 167, 170–72
Leatham, Rear Admiral Ralph, 464
Leckie, Air Commodore Robert, 546
Libya: and Italian forces, 205, 343, 345; and British war plans, 386, 391, 461–2, 490, 524, 533, 548. *See also* Tobruk
Little, Vice Admiral Charles James Colebrooke, 77–9, 109
Locarno, Treaty, 7
London Naval Treaty (1930), 48, 95, 308, 325, 351, 398–9, 403, 470
Lyon, Rear Admiral George Hamilton D'Oyly, 59, 122, 130, 160, 175
Lyster, Captain Arthur, 392

MacDonald, James Ramsey, 49
Mackenzie, Captain K.H.L., 87
Maclean, Air Vice Marshal Cuthbert T., 115
Maddalena, 69, 118, 120, 128
Majorca, 214–16, 467–8, 487. *See also* Palma, Balearic Islands

Malaga, 216–18, 249–53
Malta: air defences lamentable, 6–7, 29–32, 34–5, 37–8, 83; viability as base, 8, 97, 99, 108, 458–9; combined operations at, 27, 33–4; political crisis, 38–9; in war with Italy, 65–6, 76, 456, 533; defences evaluated, 82, 84–5, 145, 157, 387, 390, 520; naval families, 124–5, 390, 516–17, 521; return of fleet, 176, 187; visit of Turkish fleet, 10, 192; submarine shelters, 372; and Italian occupation of Albania, 389, 511–13, 515–16; visit of Italian warships, 425; and Czechoslovak crisis, 449–50; British forces at, 556. *See also* Mediterranean Fleet, Fisher, Pound
Marriott, Captain J.P.R., 122–3, 156, 174
Meade-Fetherstonhaugh, Admiral Sir Herbert, 463
Mediterranean Fleet: strength of, 3, 9, 16–17, 20–22, 42–5, 50–51, 71, 74, 132, 134, 181–3; and a war with Japan, 3, 52–9; annual routine, 3–4, 10, 15–16; visit of royal princes, 4, 22–5; exercises, 4–5, 12–13, 39, 160–61, 192; redistribution of forces, 5, 16–17; strategical investigations, 35–7; vulnerability to air attack, 6, 8, 96; 179–80; and "Fairey Queen" drones, 8; and silver anniversary review, 10; preparedness for war, 39–42, 102–3; and potential conflict with Italy, 64–9, 127, 144–6; easing of tensions over Ethiopia, 176–7, 182, 184–5; and Spanish Civil War, 192–4, 196, 218–19, 231–41, 381, 388; and coronation review, 193; role in war with Germany, 384; and Czechoslovak crisis, 384, 433; alleged pessimism in fleet, 386–7; and Italian occupation of Albania, 389, 510–13; on outbreak of war, 394–5; comparison with Italian fleet, 413–15; adequacy of fleet, 451–7; false rumors, 508 n. 1. *See also* Fisher, Pound, Cunningham, Admiralty, Alexandria, Malta
Mersa Matruh, 122, 135, 146, 345 n. 1
Metaxas, General Ioannis, 339
Meyrick, Rear Admiral Sidney Julius, 160
Miaja, General José, 270, 274

Miles, Captain Geoffrey John Audley, 247–9
Minorca, 387–8, 476–86, 488. *See also* Port Mahon, Muirhead-Gould
Mobil Naval Base, 65, 67–8, 74, 104, 135, 183, 186. *See also* Port "X"
Moore, Rear Admiral Henry Ruthven, 546–7
Moreno, Vice Admiral Francisco: and attacks on neutral ships, 351–2, 366–8, 370, 381–2, 399, 401–4, 407–8, 411–12, 499–500; and surrender of Minorca, 476–8, 487–9; mentioned, 409, 413. *See also* Spain, Navy (Nationalist)
Moriondo di Morenco, Contrammiraglio Alberto, 411
Mountbatten, Commander Louis, 72, 138–44, 151, 202
Mudros, 300, 310, 338, 355–7
Muirhead-Gould, Captain Gerard C.: career, 326 n. 1; and perceived submarine attack, 326–7; and Czechoslovak crisis, 385, 434–5; and surrender of Minorca, 387, 476–86; and press interview, 486, 488; and evacuation of embassy staff, 471–6; mentioned, 489, 497
Muselier, Contre Amiral Emile Henri Désiré, 142–3
Mussolini, Benito: career, 78 n. 1; and bellicose statements, 7, 63–4, 78, remains neutral, 394, and Czechoslovak crisis, 433, 441; and Greece, 515; mentioned, 4, 109, 146, 205, 395, 467

Negrin, Dr. Juan, 473, 501 n. 1
Neville, Lieut. Commander E., 138–40, 144
News Chronicle, 281
Non Intervention Committee, 192, 195, 233–5, 249, 251, 253, 265
Norman, Commander Horace Geoffrey, 406
Nyon Agreements: proposals, 306–9; Anglo-French agreement, 201–2, 311–13, 320–21; and Germany, 322; and Mediterranean Fleet, 325–6, 331, 351; joined by Italy, 327–30, 341–50, 381; reduction of patrols, 396–7; recurrence of attacks, 398, 404, 419–23, 465–6, 468, 470; mentioned, 338–9, 383

Okengit, Lieutenant Hilmi, 340

Ollive, Vice Amiral Emmanuel, 394, 545, 547–8

Packer, Captain Herbert Annesley, 339–40
Palestine, 188, 192, 344, 426. *See also* Haifa
Palma, 208–10, 466. *See also* Majorca
Palmer, Sir Herbert Richmond, 175
Phillips, Captain Thomas Spencer Vaughn, 245–6, 375 n. 1
Pipon, Rear Admiral Murray, 82, 197, 239, 245, 254, 258–9
Porcel, Dr. Jose, 411
Port Mahon, 477–85, 487. *See also* Minorca
Port Said, 10, 52, 54, 59, 107, 113, 118
Port "X": in British plans, 65, 67–72, 91–2, 98–9, 101, 104, 111; and Royal Marines, 125–6; temporary abandonment, 135, 144; mentioned, 89, 112–14, 123, 128, 131, 147. *See also* Mobil Naval Base
Pound, Vice Admiral Alfred Dudley Pickman Rogers: career, 46 n. 1; as Chief of Staff to Fisher, 70, 121, 123–4, 159–60; assumes command of Mediterranean Fleet, 73–4, 159–61, 165, 193; and end of Ethiopian crisis, 167–70, 174, 177, 187; aand Spanish Civil War, 194, 196–202, 204–5, 207, 275–6; meeting with Esteva, 202–3, 318, 320; and German threat, 206, 322, 384, and Japan, 206, 383–4, 405; visits to Spanish ports, 211, 214–18; and non intervention committee, 236–8, 260–62, 265, 292; and attacks on British ships, 241–2, 266–7, 276–9, 368–9, 382; and mining of *Hunter*, 257–60, 267; and question of belligerent rights, 282–4; and attacks by unknown submarines, 290, 293–6, 299–302, 305–6; and Nyon agreements, 310–11, 315–16, 319–21, 335; and Italians, 323, 382–3, 413–15; and situation in *Warspite*, 324, 406; and *Basilisk* affair, 331–2, 334–5, 359; and reinforcement of western Mediterranean, 336–7, 346; and defence of Egypt, 343–4, 372; and reduction of Nyon patrols, 345–6, 353–5, 371, 383, 396–7; and *Ilex* affair, 347–8, 359; meeting with French and Italian admirals, 349–50, 358; and MTBs, 372–3; and Czechoslovak crisis, 384–6, 426, 428–31; and situation in Mediterranean (Oct. 1938), 443–60; and alleged pessimism in Mediterranean Fleet, 387, 465; and Albanian crisis, 389, 510–14; and AA defences of Malta and Alexandria, 390, 535–7; and Nyon patrols, 405, 419–24; and Spanish Nationalist submarines, 409–10; staff talks with French, 391, 491–2, 529–30; plans for war, 392, 424–5, 519–26, 531, 534–5; and Italian visit to Malta, 425–6; fleet programme (1939), 507–8, 530; and naval families at Malta, 516–17; becomes First Sea Lord, 393; mentioned, 112, 463. *See also* Admiralty, Mediterranean Fleet
Pouritch, Bojidar, 181 n. 3
Pridham, Captain Arthur Francis, 426

"Queen Bee" drones, 73, 154, 156, 160, 179–80

Raikes, Rear Admiral Robert Henry Taunton, 122, 155
Ramsay, Vice Admiral Alexander Robert Maude, 131, 154, 187
Ransome, Lieut. Commander P.C., 232 n. 1
Red Sea: and undersea cables, 70, 74; and Italian Navy, 100, 105–6, 172; and British forces, 105–6, 556; British plans in event of war, 107, 361, 455–6, 490–91
Riccardi, Ammiraglio di Squadra Arturo, 425
Richmond, Commander Maxwell, 279–82
Robert, Vice Amiral Georges, 16 n. 1
Roosevelt, Franklin Delano, 375 n. 1
Rose, Vice Admiral Frank Forrester, 105, 155
Rossi, Conde (Aconovaldo Bonacorsi), 216
Royal Air Force: and Malta, 8, 84; at Arzeu, 202, 313, 315, 319, 323, 359; 80(F) Squadron, 447; 202 Squadron, 30, 300, 310, 425, 447, 508; 205 Squadron, 106; 210 Squadron, 319, 32
Royal Marines, 125, 506
Royal Navy: 1st Battle Squadron, 40, 121, 145, 171, 310; 3rd Battle Squadron, 18; Battle Cruiser Squadron, 36, 112; 1st Cruiser

Royal Navy: (cont'd)
  Squadron, 33, 57, 145, 161, 337, 415, 429–30; 2nd Cruiser Squadron, 36, 73, 78 n. 2, 108, 112, 114, 116, 118, 133, 148, 152–3, 160–61, 403; 3rd Cruiser Squadron, 10, 36, 56–8, 67, 73, 104, 145, 324, 382, 414–15; 3rd Light Cruiser Squadron, 148, 152–3, 156–7, 160–61, 178; 1st Destroyer Flotilla, 23, 59, 145, 171–2, 176, 182, 184, 237–8, 265, 300, 310–19, 436, 556; 2nd Destroyer Flotilla, 17, 93 n. 2, 154, 161, 172, 238, 265, 286, 289, 291–2, 300, 310, 323, 334, 419, 556; 3rd Destroyer Flotilla, 10, 25, 56, 59, 161, 197, 213, 247–9, 300, 310, 396, 428, 430, 436–7, 450, 556; 4th Destroyer Flotilla, 23, 36, 169–70, 176, 178, 184, 353, 556; 5th Destroyer Flotilla, 36, 87, 93 n. 2, 132, 145–6, 154, 353–4; 6th Destroyer Flotilla, 93 n. 2, 103–4, 108, 112, 114, 116, 133, 152, 154, 160, 168–9, 178, 292, 396; 13th Destroyer Flotilla, 556; 19th Destroyer Flotilla, 132, 145; 20th Destroyer Flotilla, 114, 116–17, 133, 170–71; 21st Destroyer Flotilla, 114, 116–18, 133, 169, 171–2; 1st "Tribal" Destroyer Flotilla, 528 n. 1; 2nd Destroyer Division, 356; 5th Destroyer Division, 338, 430; 6th Destroyer Division, 436, 439–40, 450, 508; 7th Destroyer Division, 354; 8th Destroyer Division, 321, 354; 11th Destroyer Division, 200, 292, 402; 12th Destroyer Division, 321; 15th Destroyer Division, 185; 1st Submarine Flotilla, 145; 2nd Submarine Flotilla, 36, 78 n. 2, 87; 2nd Anti Submarine Flotilla, 176, 178, 184; 1st Minesweeping Flotilla, 178, 197, 260, 265–6; 2nd Minesweeping Flotilla, 184–5; 5th Minesweeping Flotilla, 176, 178, 184; 1st MTB Flotilla, 310
Runciman, Sir Walter Viscount, 428

Sakallarion, Admiral, 340
Samos Affair, 47, 55 n. 2
San Luis, Lieut. Commander Count: and Minorca, 476–85; mentioned, 487–8

San Martin, Engineer Officer, 479–81
Scurfield, Lieut. Commander B.G., 256 n. 1, 258 n. 1
Secret Anchorage "M" (Kamaran Bay, Red Sea), 57
Secret Anchorage "T" (Addu Atoll), 57–8
Servaes, Captain Reginald Maxwell, 360
Ships (British): *Atlantic Guide*, 500–501; *African Mariner*, 356–7, 423; *Atlantic Guide*, 500–501; *Alcira*, 403; *Aquitania*, 385–6, 438–40, 444, 446, 450; *Atreus*, 185; *Barrington Combe*, 496; *Benarty*, 183, 185; *Bramhill*, 465–6, 470; *Burlington*, 288; *Carthage*, 125; *Cervantes*, 204, 336–7; *Chitral*, 125; *Ciscar*, 103; *City of Marseilles*, 280; *Clonlara*, 399, 401–2; *Cutty Sark*, 122; *Derbyshire*, 208; *Dominia*, 129; *Endymion*, 402, 408; *Euphorbia*, 356, 364–5; *Gibel Zerjon*, 212–14; *Hawkinge*, 103; *Heminge*, 103; *Jan Weems*, 204, 350–52; *Jenny*, 268; *Lady Dennison Pender*, 129; *Lake Geneva*, 401–2; *Levant*, 129; *Laverock*, 465, 470; *Marvia*, 352; *Miocene*, 473; *Montcalm*, 440; *Nahlin*, 187 n. 2; *Noemijulia*, 199, 268, 277, 279–81; *Nuddea*, 185; *Rajputana*, 125; *Retriever*, 129; *Rio Azul*, 185; *Somersetshire*, 337; *Spanker*, 103; *Stanbreak*, 496; *Stancor*, 493, 498–501; *Stangate*, 493–5, 497; *Stanhope*, 493, 495; *Stanland*, 500–501; *Thistlegarth*, 344; *Thorpness*, 401, 408; *Wintonia*, 478; *Woodford*, 200
Ships (Dutch): *Hannah*, 400, 403
Ships (French): *Imaréthie*, 271; *L'Obstine*, 210
Ships (German): *Schwabenland*, 439; *Sexta*, 251
Ships (Italian): *Ausonia*, 123
Ships (Spanish Republican): *Caleelcofauba*, 324; *Campeador*, 198
Sidi Barrani, 72, 147
Simon, John Alisebrook, 55
Singapore, 7, 38, 43, 376
Skyros, 205, 316, 338–40, 342, 355, 357
Slessor, Group Captain John Cotesworth, 547
Sollum, El, 70, 72, 115–16, 137, 146–7
Somerville, Rear Admiral James: career, 90 n. 1; and Spanish Civil War,

INDEX

199, 200, 215–16, 277–9, 292; and *Havock* incident, 284–9, 291; and Nyon patrols, 296, 298–9, 354; at Cadiz, 363–5; and Morocco, 366; proceedings, 410–13; mentioned, 67, 90, 168 n. 2
Soviet Union, 191, 297
Spain, 188, 191–4, 381, 387, 523–4. *See also* Spain (Nationalist), Spain (Republican)
Spain (Nationalist): potential enemy, 7; and attacks on shipping, 199, 231–3, 236–44, 277–8, 336–7, 350 n. 4, 352, 387, 465–6; and blockade, 204–5, 494 n. 1; and Royal Navy, 196–7, 249–51; use of submarines, 381, 398–9; naval activity, 271, 400, 419, 421, 436, 439–40, 493–6, 498, 500, 505
Spain (Republican), 196, 271, 440 n. 1. *See also* Barcelona
Spanish Morocco, 391, 393, 523, 540
Stanhope, James Richard, 7th Earl, 465–6, 469–70
Stevenson, Ralph Clement Skrine, 472–5
Suez Canal: defence of, 10, 44–5, 52–9, 384, 415–18, 457–8; and Ethiopian crisis, 64–5, 76–7; mentioned, 145, 185

Tait, Captain William Eric Campbell, 242–4, 268–75
Talbot, Captain Arthur George, 347, 359
Taranto: as potential target, 104, 392, 460, 526, 533–4; mentioned, 68, 78
Tobruk, 145, 157, 552
Tomkinson, Vice Admiral Wilfred, 49
Tovey, Rear Admiral John Cronyn, 388, 464, 497–507
Troup, Rear Admiral James Andrew Gardiner, 316 n. 1
Trucy, Capitaine de vaisseau P.E., 139, 142–3
Tunisia, 393, 449, 458, 462, 533, 539
Turkey: and Samos incident, 9, 46, 55 n. 2; naval visit to Malta, 10, 55–6, 192; and Ethiopian crisis, 136; and potential British use of harbours, 201, 205, 316–17, 340–42, 359; use of air bases, 533; remains neutral, 392; assumed to be ally, 531, 541. *See also* Samos Affair
Tyrrell, William George, 1st Baron, 15

Ubieta, Captain J.L., 479–81, 484–5
Usalete, Colonel, 481, 483–4

United States, 375–6

Valencia, 500–503
Vansittart, Sir Robert Gilbert, 79, 182

Warships (American):
Cruisers:
Omaha, 473, 475; *Quincy*, 210, 221
Destroyer:
Badger, 474
Gunboat:
Panay, 375 n. 1
Warships (British):
Battleships:
Barham, 50, 92, 102–3, 118, 161, 178, 187, 202, 262, 306, 310, 315, 324–5, 371, 373, 406, 556; *Malaya*, 43, 50, 201, 262, 310, 320, 323, 325, 337–8, 342–3, 355, 357, 371, 406, 429, 513, 556; *Nelson*, 16, 50, 116, 118, 152, 170–71, 376, 471; *Queen Elizabeth*, 9, 14–16, 22–4, 42, 44, 47 n. 1, 49–50, 65, 123, 171, 187, 372 n. 1, 377, 406; *Ramilles*, 10 n. 2, 16, 44, 50, 116, 171, 508, 514, 519; *Resolution*, 23, 39, 44, 50, 65, 121; *Revenge*, 23–4, 27–8, 36, 44, 50, 439; *Rodney*, 50, 108, 116, 118, 152, 170–71, 177; *Royal Oak*, 17, 23, 42, 50, 439; *Royal Sovereign*, 27–8, 50, 116, 171; *Valiant*, 42, 49–50, 161, 171, 178, 372 n. 1, 376; *Warspite*, 42, 44, 49–50, 262, 324, 372 n. 1, 375–6, 394, 406, 528 n. 1, 556
Battlecruisers:
Hood, 43, 50, 133, 150, 152, 169–71, 176, 178, 183, 262, 310, 375, 385, 400, 405, 426, 428–30, 436–40, 450, 471; *Renown*, 42, 49–50, 133, 155, 161, 166, 171, 372 n. 1, 376; *Repulse*, 43, 50, 171, 188 n. 1, 194, 201, 204–5, 207–10, 212–14, 220, 262, 275 n. 2, 300, 310, 316, 320, 323, 337–44, 355–7, 375, 428, 430
Aircraft Carriers:
Albatross, 427; *Courageous*, 36, 67, 87, 132–3, 154, 183, 187, 344; *Eagle*, 17, *Furious*, 69, 108, 114, 116–17, 133, 344; *Glorious*, 23–4, 40, 57, 65–6, 83–4, 89–90, 133, 160–61, 187, 205, 299, 310, 319, 323–4, 344, 346, 392, 414–15, 428, 430, 432, 447, 508, 528 n. 1, 553, 556; *Hermes*, 313

Warships (British): (*cont'd*)
Cruisers:
  Achilles, 153, 174, 176, 185; *Adventure*, 27, 69, 176–7, 185; *Ajax*, 70–71, 131, 153, 155, 160–61, 176, 178, 185; *Arethusa*, 160–61, 256, 382–3, 413 n. 1, 415–16, 418, 429, 444, 556; *Australia* (RAN), 95, 130, 169, 174, 176, 184; *Berwick*, 69, 176, 185; *Cairo*, 133, 171, 310; *Capetown*, 175; *Ceres*, 23; *Colombo*, 23, 105; *Coventry*, 23, 155, 160–61, 178, 184 187, 327, 361; *Curlew*, 23, 135, 160–61, 178, 184, 187, 361, 556; *Cumberland*, 177; *Delhi*, 237, 310, 416–17; *Despatch*, 338, 416–17; *Devonshire*, 47 n. 1, 130, 208, 220, 241, 326–7, 385, 387–8, 400, 428, 430, 434–5, 472–3, 476–8, 481–3, 486–9, 497, 556; *Diomede*, 171; *Durban*, 130; *Emerald*, 105, 416; *Enterprise*, 175 n. 3, 416; *Exeter*, 160, 174, 176, 178, 185, 188 n. 1; *Frobisher*, 55 n. 3; *Galatea*, 69, 122, 161, 187, 199, 211, 285–9, 310, 321, 363–4, 388, 410, 413–14, 497, 501–7, 556; *Gloucester*, 511; *Leander*, 153; *Liverpool*, 511; *London*, 23–4, 27, 36, 39, 47 n. 1, 55–6, 177, 211, 219–23, 241, 430–31, 437, 440, 450, 471; *Manchester*, 439, 444; *Norfolk*, 105; *Penelope*, 237, 274, 286, 288–9, 306, 310, 382, 413 n. 1, 415–16, 418, 444, 495–6, 498–500, 511, 556; *Shropshire*, 23, 25, 77, 88 n. 1, 211, 219, 241–2, 268–9, 271–5, 430–31, 444, 448–9, 556; *Southampton*, 403; *Sussex*, 23, 27–8, 69, 177, 388, 430, 492–5, 556; *Sydney* (RAN), 153, 155, 160–61, 169, 174, 176, 184
Destroyers:
  Achates, 247; *Acheron*, 247; *Active*, 286–9, 519; *Afridi*, 385, 390, 421–2, 428, 434, 511, 530–31; *Anthony*, 247; *Arrow*, 196, 249, 252, 411; *Basilisk*, 203–4, 330–35; *Boreas*, 203, 331, 333, 382, 411; *Bulldog*, 508, 528 n. 1, 556; *Campbell*, 171; *Codrington*, 23, 194, 212–13; *Cossack*, 385, 421–2, 428, 434; *Douglas*, 23, 161, 185, 310, 444; *Faulknor*, 70; *Firedrake*, 337 n. 1; *Gallant*, 196, 239–44; *Gipsy*, 195, 231–3, 250–51; *Glowworm*, 356, 371, 440 n. 1, 472–3; *Grafton*, 187, 252; *Grenade*, 356; *Greyhound*, 356, 371, 440 n. 1, 472, 473; *Griffin*, 371; *Gurkha*, 511; *Hardy*, 253–6, 258, 288; *Hasty*, 288–9; *Havock*, 195, 199–200, 202–3, 231–2, 284–90, 330–32, 334–5, 478, 480; *Hereward*, 198, 263–5, 286–8; *Hero*, 264; *Hostile*, 199, 279–82, 286, 289, 473, 478; *Hotspur*, 277, 286–8, 477–8, 482, 485; *Hunter*, 196–7, 253–60, 265, 267, 473–4, 476, 478; *Hyperion*, 253–4, 288; *Icarus*, 340, 355, 493; *Ilex*, 205, 339, 347, 355; *Imogen*, 338–9, 342, 347, 437; *Imperial*, 437; *Impulsive*, 493; *Inglefield*, 338, 343, 347, 430, 437, 508; *Intrepid*, 493, 495–6; *Isis*, 355; *Ivanhoe*, 437; *Keith*, 23; *Kempenfelt*, 87, 103, 354, 382, 411; *Mackay*, 87, 132, 171; *Mohawk*, 497–8, 500–501, 505, 508; *Nubian*, 500–502, 505–6; *Rowena*, 118, 160; *Shamrock*, 185; *Sikh*, 511; *Torrid*, 118, 160; *Thruster*, 118, 160; *Vanoc*, 238, 261, 321, 420–22, 438; *Wessex*, 171; *Westminster*, 171; *Windsor*, 171; *Wishart*, 72, 138, 141, 143, 457, 497; *Wolsey*, 161, 185, 212, 214, 444, 449; *Worcester*, 217; *Zulu*, 507
Submarines:
  Clyde, 449; *L.23*, 88; *Narwhal*, 178, 185; *Pandora*, 168; *Porpoise*, 69; *Proteus*, 168, 185; *Salmon*, 530; *Seahorse*, 184; *Sealion*, 436; *Severn*, 338, 343, 519; *Starfish*, 184; *Thames*, 436–8, 449–50
Sloops and Patrol Vessels:
  Aberdeen, 352, 394, 457; *Bideford*, 449; *Bryony*, 22–3; *Chrysanthemum*, 22–3; *Cornwallis* (RIN), 107; *Enchantress;* 343; *Kingfisher*, 261; *Mallard*, 261; *Ormonde*, 185; *Penzance*, 176, 185; *Puffin*, 261
Minelayers and Minesweepers:
  Elgin, 266; *Lydd*, 266; *Plover*, 450; *Saltash*, 266
Trawlers:
  Agate, 262; *Amethyst*, 262; *Dee*, 178, 184; *Garry*, 178, 184; *Hawthorn*, 262; *Holly*, 262; *Jasper*, 262; *Kate*

INDEX 579

Lewis, 178, 185; *Laurel*, 262;
Liffey, 178, 184; *Lilac*, 262;
Magnolia, 262; *Willow*, 262
Hospital Ship:
   Maine, 22–3, 219, 237, 254 n. 1, 269, 310, 501, 503, 505–6
Depot Ships and Other Auxiliaries:
   Bacchus, 146; *Brigand*, 446; *Centurion*, 67, 80, 89, 358–9, 405, 419, 508; *Cherryleaf*, 449; *Cyclops*, 22–3, 202, 237, 313, 315–16, 318–19; *Churraca*, 327, 344; *Guardian*, 67–8, 87–8, 104, 178; *Lucia*, 87–8; *Maidstone*, 444, 447, 454, 508, 511; *Perthshire*, 22–3; *Petrella*, 318–19, 449; *Philomel*, 511; *Plumleaf*, 318; *Protector*, 326–7, 344, 430, 445; *Reliant*, 511, 530; *Resource*, 22–3, 89, 91, 130, 194, 212, 237, 254, 324, 360, 416–17, 444–5, 508, 511, 520; *Roysterer*, 446; *St. Day*, 185; *St. Omer*, 256; *Sandhurst*, 22–3; *Viscol*, 292; *Woolwich*, 68, 89, 91, 104, 130, 194, 310, 318, 511
Warships (British Classes): "A" (destroyers), 42, 44; "Acasta" (destroyers), 247; "County" (cruisers), 194; "D" (cruisers), 73, 160, 382, 416; "Dido" (cruisers), 418; "E" (destroyers), 151; "G" (destroyers), 152, 154, 556; "H" (destroyers), 178, 183, 556; "I" (destroyers), 441, 556; "Leaf" (oilers), 449; "London" (cruisers), 155; "Porpoise" (submarines), 454; "Queen Elizabeth", 9, 21, 41–2, 51, 517; "Resolution" (battleships), 517; "River" (submarines), 453–4; "Royal Sovereign", 9, 41, 51, 117; "Tribal" (destroyers, 324, 383, 394, 414–17, 456–7
Warships (French):
Battleships:
   Bretagne, 557 n. 1; *Lorraine*, 557 n. 1; *Provence*, 15–16, 394
Cruisers:
   Algérie, 358; *Emile Bertin*, 557 n. 1; *Suffren*, 473
Destroyers and Torpedo Boats:
   Le Fortune, 474; *Lynx*, 505; *Simoun*, 474; *Verdun*, 272–3
Submarine:
   Diamante, 280–81
Seaplane Carrier:
   Commandant Teste, 221 n. 1, 557 n. 1

Minelayer:
   Castor, 557 n. 1
Warships (German):
Armoured Ships:
   Admiral Graf Spee, 522 n. 1; *Admiral Scheer*, 198, 263, 522 n. 1; *Deutschland*, 195, 197, 202, 263–4, 385, 438–9, 522 n. 1
Cruisers:
   Emden, 436; *Leipzig*, 195, 198, 266, 335
Torpedo Boats:
   Albatross, 263; *Leopard*, 263; *Luchs*, 263; *Seeadler*, 263
Submarines:
   U.26, 264; *U.27*, 436, 438–9; *U.30*, 436, 438–9; *U.35*, 410 n. 1
Warships (German Classes):
   "Deutschland", 375
Warships (Greek)
Cruiser:
   Averoff, 358 n. 3
Warships (Italian):
Cruisers:
   Bari, 105; *Emanuele Filiberto Duca d'Aosta*, 350 n. 2; *Quarto*, 299, 411; *Taranto*, 105; *Trento*, 36; *Trieste*, 14, 36
Destroyers and Torpedo Boats:
   Audace, 105; *Falco*, 250–51; *Impavido*, 105; *Maestrale*, 208; *Nicoloso da Recco*, 250; *Palestro*, 105; *Pantera*, 105; *Saetta*, 198; *Tigre*, 105
Submarines:
   Diaspro, 200; *Iride*, 200, 204 n. 1; *Jalea*, 200 n. 1; *Septembrinni*, 105; *Settimo*, 105; *Torricelli*, 198 n. 3
Sloops and Patrol Vessels:
   Axio, 105; *C. Berta*, 105; *Oetia*, 105; *P. Corsinni*, 105
Depot Ships and Other Auxiliaries:
   Ausonia, 105; *Cherso*, 105; *Città di Milano*, 130; *Città di Siracusa*, 130; *Glasone*, 130; *Lussin*, 105; *Volta*, 105
Warships (Portuguese):
Destroyer:
   Douro, 221
Warships (Spanish Nationalist):
Cruisers:
   Almirante Cervera, 439; *Baleares*, 192, 271, 382, 408, 410–11, 480; *Canarias*, 192, 271, 274, 382, 411, 413, 439, 493; *Navarra*, 439

Warships (Spanish Nationalist): (*cont'd*)
  Destroyer:
    Melilla, 493, 496
  Submarines:
    Gonzales Lopez, 204 n. 1; *Mola*, 407; *Sanjurjo*, 407–8
  Armed Merchantmen:
    Mar Cantabrico, 496, 498–9; *Mar Negro*, 493–6, 499–500, 504–6
Warships (Spanish Republican):
  Battleship:
    Jaime I, 208, 254, 271
  Cruisers:
    Libertad, 208; *Miguel de Cervantes*, 198 n. 3, 212–13
  Destroyers:
    Churruca, 200 n. 1; *Jose Luis Diez*, 439; *Lazaga*, 253; *Lepanto*, 251–2; *Sanchey Barcaiztequi*, 251–2
  Submarine:
    B.5, 213
Warships (Turkish):
  Cruiser:
    Hamidiye, 435
Washington Naval Treaty, 95
Waterlow, Sir Sydney Philip, 513
Wauchope, Major General Arthur, 175
Weir, Major General George A., 115, 301
Wells, Rear Admiral Gerald Alymer, 122–3, 423
Wells, Rear Admiral Lionel Victor, 255–7, 338, 409 n. 1
Westminster, 2nd Duke of, 122
Willis, Commodore Algernon Usborne, 546
Wodehouse, Captain Norman Atherton, 374, 406

Young, Sir George, 503

NAVY RECORDS SOCIETY – LIST OF VOLUMES
(as at 1 October 2015)

Members wishing to order any volumes should write to Robin Brodhurst, The Mill, Stanford Dingley, Reading, RG7 6LS or email him at robinbrodhurst@gmail.com.

Those volumes marked OP will be printed from scanned copies or original discs. For other titles the Society still retains some original copies.

1. *State Papers relating to the Defeat of the Spanish Armada, 1588.* Vol. I. Ed. Professor J.K. Laughton. (£15.00)
2. *State Papers relating to the Defeat of the Spanish Armada, 1588.* Vol. II. Ed. Professor J.K. Laughton. (£15.00)
1 + 2. Combined Volumes. (£20.00)
3. *Letters of Lord Hood, 1781–1783.* Ed. D. Hannay. **OP**
4. *Index to James's Naval History, 1886,* by C.G. Toogood. Ed. Hon. T.A. Brassey. **OP**
5. *Life of Captain Stephen Martin, 1666–1740.* Ed. Sir Clements R. Markham. **OP**
6. *Journal of Rear Admiral Bartholomew James, 1752–1828.* Ed. Professor J.K. Laughton and Cdr J.Y.F. Sullivan. **OP**
7. *Hollond's Discourses of the Navy, 1638 and 1659 and Slyngesbie's Discourse on the Navy, 1660.* Ed. J.R. Tanner. **OP**
8. *Naval Accounts and Inventories of the Reign of Henry VII, 1485–1488 and 1495–1497.* Ed. M. Oppenheim. **OP**
9. *The Journal of Sir George Rooke, 1700–1702.* Ed. O. Browning. **OP**
10. *Letters and Papers relating to the War with France, 1512–1513.* Ed. A. Spont. **OP**
11. *Papers relating to the Navy during The Spanish War, 1585–1587.* Ed. J.S. Corbett. (£15.00)
12. *Letters and Papers of Admiral of the Fleet Sir Thomas Byam Martin, 1733–1854,* Vol. II (see Vol. 24). Ed. Admiral Sir Richard Vesey Hamilton. **OP**
13. *Letters and Papers relating to the First Dutch War, 1652–1654,* Vol. I. Ed. S.R. Gardiner. **OP**

14. *Dispatches and Letters relating to the Blockade of Brest, 1803–1805*, Vol. I. Ed. J. Leyland. **OP**
15. *History of the Russian Fleet during the reign of Peter The Great, by a Contemporary Englishman, 1724.* Ed. Vice-Admiral Sir Cyprian A.G. Bridge. **OP**
16. *Logs of the Great Sea Fights, 1794–1805*, Vol. I. Ed. Rear Admiral Sir T. Sturges Jackson. **OP**
17. *Letters and Papers relating to the First Dutch War, 1652–1654*, Vol. II. Ed. S.R. Gardiner. **OP**
18. *Logs of the Great Sea Fights, 1794–1805*, Vol. II. Ed. Rear Admiral Sir T. Sturges Jackson. (£25.00)
19. *Letters and Papers of Admiral of the Fleet Sir Thomas Byam Martin, 1773–1854*, Vol. III (see Vol. 24). Ed. Admiral Sir R. Vesey Hamilton. **OP**
20. *The Naval Miscellany*, Vol. I. Ed. Professor J.K. Laughton. (£15.00)
21. *Dispatches and Letters relating to the Blockade of Brest, 1803–1805.* Vol. II. Ed. J. Leyland. **OP**
22. *The Naval Tracts of Sir William Monson*, Vol. I. Ed. M. Oppenheim. **OP**
23. *The Naval Tracts of Sir William Monson*, Vol. II. Ed. M. Oppenheim. **OP**
24. *Letters and Papers of Admiral of the Fleet Sir Thomas Byam Martin, 1773–1854*, Vol. I. Ed. Admiral Sir R. Vesey Hamilton. **OP**
25. *Nelson and the Neapolitan Jacobins.* Ed. H.G. Gutteridge. (£25.00)
26. *A Descriptive Catalogue of the Naval Mss. in the Pepysian Library*, Vol. I. Ed. J.R. Tanner. **OP**
27. *A Descriptive Catalogue of the Naval Mss. in the Pepysian Library*, Vol. II. Ed. J.R. Tanner. **OP**
28. *The Correspondence of Admiral John Markham, 1801–1807.* Ed. Sir Clements R. Markham. **OP**
29. *Fighting Instructions, 1530–1816.* Ed. J.S. Corbett. **OP**
30. *Letters and Papers relating to the First Dutch War, 1652–1654*, Vol. III. Ed. S.R. Gardiner and C.T. Atkinson. **OP**
31. *The Recollections of James Anthony Gardner, 1775–1814.* Ed. Admiral Sir R. Vesey Hamilton and Professor J.K. Laughton. (£15.00)
32. *Letters and Papers of Charles, Lord Barham, 1758–1813*, Vol. I. Ed. Professor Sir J.K. Laughton. (£15.00)
33. *Naval Songs and Ballads.* Ed. Professor C.H. Firth. **OP**

34. *Views of the Battles of the Third Dutch War.* Ed. J.S. Corbett. **OP**
35. *Signals and Instructions, 1776–1794.* Ed. J.S. Corbett. **OP**
36. *A Descriptive Catalogue of the Naval Mss. in the Pepysian Library,* Vol. III. Ed. J.R. Tanner. **OP**
37. *Letters and Papers relating to the First Dutch War, 1652–1654,* Vol. IV. Ed. C.T. Atkinson. **OP**
38. *Letters and Papers of Charles, Lord Barham, 1758–1813,* Vol. II. Ed. Professor Sir J.K. Laughton. (£15.00)
39. *Letters and Papers of Charles, Lord Barham, 1758–1813,* Vol. III. Ed. Professor Sir J.K. Laughton. (£15.00)
40. *The Naval Miscellany,* Vol. II. Ed. Professor Sir J.K. Laughton. (£15.00).
41. *Letters and Papers relating to the First Dutch War, 1652–1654.* Vol. V. Ed. C.T. Atkinson. (£15.00)
42. *Papers relating to the Loss of Minorca, 1756.* Ed. Capt H.W. Richmond. **OP**
43. *The Naval Tracts of Sir William Monson,* Vol. III. Ed. M. Oppenheim. (£15.00)
44. *The Old Scots Navy, 1689–1710.* Ed. J. Grant. **OP**
45. *The Naval Tracts of Sir William Monson,* Vol. IV. Ed. M. Oppenheim. (£15.00)
46. *Private Papers of George, 2nd Earl Spencer, 1794–1801,* Vol. I. Ed. J.S. Corbett. **OP**
47. *The Naval Tracts of Sir William Monson,* Vol. V. Ed. M. Oppenheim. (£15.00)
48. *Private Papers of George, 2nd Earl Spencer, 1794–1801,* Vol. II. Ed. J.S. Corbett. **OP**
49. *Documents relating to the Law and Custom of the Sea,* Vol. I, 1205–1648. Ed. R.G. Marsden. **OP**
50. *Documents relating to the Law and Custom of the Sea,* Vol. II, 1649–1767. Ed. R.G. Marsden. (£15.00)
51. *The Autobiography of Phineas Pett.* Ed. W.G. Perrin. **OP**
52. *The Life of Admiral Sir John Leake,* Vol. I. Ed. G.A.R. Callender. (£15.00)
53. *The Life of Admiral Sir John Leake,* Vol. II. Ed. G.A.R. Callender. (£15.00)
54. *The Life and Works of Sir Henry Mainwaring,* Vol. I. Ed. G.E. Manwaring. (£15.00)
55. *The Letters of Lord St. Vincent, 1801–1804,* Vol. I. Ed. D. Bonner-Smith. **OP**
56. *The Life and Works of Sir Henry Mainwaring,* Vol. II. Ed. G.E. Manwaring and W.G. Perrin. **OP**

57. *A Descriptive Catalogue of the Naval Mss. in the Pepysian Library*, Vol. IV. Ed. J.R. Tanner. **OP**
58. *Private Papers of George, 2nd Earl Spencer, 1794–1801*, Vol. III. Ed. Rear Admiral H.W. Richmond. **OP**
59. *Private Papers of George, 2nd Earl Spencer, 1794–1801*, Vol. IV. Ed. Rear Admiral H.W. Richmond. **OP**
60. *Samuel Pepys's Naval Minutes*. Ed. Dr J.R. Tanner. (£15.00)
61. *The Letters of Earl St. Vincent, 1801–1804*, Vol. II. Ed. D. Bonner-Smith. **OP**
62. *The Letters and Papers of Admiral Viscount Keith*, Vol. I. Ed. W.G. Perrin. **OP**
63. *The Naval Miscellany*, Vol. III. Ed. W.G. Perrin. **OP**
64. *The Journal of the 1st Earl of Sandwich, 1659–1665*. Ed. R.C. Anderson. **OP**
65. *Boteler's Dialogues*. Ed. W.G. Perrin. (£15.00)
66. *Letters and Papers relating to the First Dutch War, 1652–1654*, Vol. VI (& index). Ed. C.T. Atkinson. (£15.00)
67. *The Byng Papers*, Vol. I. Ed. W.C.B. Tunstall. **OP**
68. *The Byng Papers*, Vol. II. Ed. W.C.B. Tunstall. (£15.00)
69. *The Private Papers of John, Earl of Sandwich, 1771–1782*. Vol. I, *1770–1778*. Ed. G.R. Barnes and J.H. Owen. **OP**
    Corrigenda to Papers relating to the First Dutch War, 1652–1654. Ed. Capt A. Dewar. **(free)**
70. *The Byng Papers*, Vol. III. Ed. W.C.B. Tunstall. (£15.00)
71. *The Private Papers of John, Earl of Sandwich, 1771–1782*, Vol. II, *1778–1780*. Ed. G.R. Barnes and J.H. Owen. **OP**
72. *Piracy in the Levant, 1827–1828*. Ed. Lt Cdr C.G. Pitcairn Jones R.N. **OP**
73. *The Tangier Papers of Samuel Pepys*. Ed. E. Chappell. (£15.00)
74. *The Tomlinson Papers*. Ed. J.G. Bullocke. (£15.00)
75. *The Private Papers of John, Earl of Sandwich, 1771–1782*, Vol. III, *1780–1782*. Ed. G.R.T. Barnes and Cdr J.H. Owen. **OP**
76. *The Letters of Robert Blake*. Ed. Rev. J.R. Powell. **OP**
77. *Letters and Papers of Admiral the Hon. Samuel Barrington*, Vol. I. Ed. D. Bonner-Smith. (£15.00)
78. *The Private Papers of John, Earl of Sandwich*, Vol. IV. Ed. G.R.T. Barnes and Cdr J.H. Owen. **OP**
79. *The Journals of Sir Thomas Allin, 1660–1678*, Vol. I. Ed. R.C. Anderson. (£15.00)
80. *The Journals of Sir Thomas Allin, 1660–1678*, Vol. II. Ed. R.C. Anderson (£15.00)

81. *Letters and Papers of Admiral the Hon. Samuel Barrington*, Vol. II. Ed. D. Bonner-Smith. **OP**
82. *Captain Boteler's Recollections, 1808–1830*. Ed. D. Bonner-Smith. **OP**
83. *The Russian War, 1854: Baltic and Black Sea*. Ed. D. Bonner-Smith and Capt A.C. Dewar R.N. **OP**
84. *The Russian War, 1855: Baltic*. Ed. D. Bonner-Smith. **OP**
85. *The Russian War, 1855: Black Sea*. Ed. Capt. A.C. Dewar. **OP**
86. *Journals and Narratives of the Third Dutch War*. Ed. R.C. Anderson. **OP**
87. *The Naval Brigades of the Indian Mutiny, 1857–1858*. Ed. Cdr W.B. Rowbotham. **OP**
88. *Patee Byng's Journal, 1718–1720*. Ed. J.L. Cranmer-Byng. **OP**
89. *The Sergison Papers, 1688–1702*. Ed. Cdr R.D. Merriman. (£15.00)
90. *The Keith Papers*, Vol. II. Ed. C. Lloyd. **OP**
91. *Five Naval Journals, 1789–1817*. Ed. Rear Admiral H.G. Thursfield. **OP**
92. *The Naval Miscellany*, Vol. IV. Ed. C. Lloyd. **OP**
93. *Sir William Dillon's Narrative of Professional Adventures, 1790–1839*, Vol. I, *1790–1802*. Ed. Professor M. Lewis. **OP**
94. *The Walker Expedition to Quebec, 1711*. Ed. Professor G.S. Graham. **OP**
95. *The Second China War, 1856–1860*. Ed. D. Bonner-Smith and E.W.R. Lumby. **OP**
96. *The Keith Papers*, Vol. III. Ed. C.C. Lloyd. (£15.00)
97. *Sir William Dillon's Narrative of Professional Adventures, 1790–1839*, Vol. II, *1802–1839*. Ed. Professor M. Lewis. **OP**
98. *The Private Correspondence of Admiral Lord Collingwood*. Ed. Professor E. Hughes. **OP**
99. *The Vernon Papers, 1739–1745*. Ed. B.McL. Ranft. **OP**
100. *Nelson's Letters to his Wife and Other Documents, 1785–1831*. Ed. G.P.B. Naish. (£25.00)
101. *A Memoir of James Trevenen, 1760–1790*. Ed. Professor C.C. Lloyd. **OP**
102. *The Papers of Admiral Sir John Fisher*, Vol. I. Ed. Lt Cdr P.K. Kemp R.N. **OP**
103. *Queen Anne's Navy*. Ed. Cdr R.D. Merriman R.I.N. **OP**
104. *The Navy and South America, 1807–1823*. Ed. Professor G.S. Graham and Professor R.A. Humphreys. (£15.00)
105. *Documents relating to the Civil War*. Ed. Rev. J.R. Powell and E.K. Timings. **OP**

106. *The Papers of Admiral Sir John Fisher*, Vol. II. Ed. Lt. Cdr P.K. Kemp R.N. **OP**
107. *The Health of Seamen*. Ed. Professor C.C. Lloyd. **OP**
108. *The Jellicoe Papers*, Vol. I, *1893–1916*. Ed. A Temple Patterson. (£15.00)
109. *Documents relating to Anson's Voyage Round the World, 1740–1744*. Ed. Dr G. Williams. **OP**
110. *The Saumarez Papers: The Baltic 1808–1812*. Ed. A.N. Ryan. **OP**
111. *The Jellicoe Papers*, Vol. II, *1916–1935*. Ed. A. Temple Patterson. (£15.00)
112. *The Rupert and Monk Letterbook, 1666*. Ed. Rev. J.R. Powell and E.K. Timings. **OP**
113. *Documents relating to the Royal Naval Air Service*, Vol. I, *1908–1918*. Ed. Capt S.W. Roskill. (£15.00)
114. *The Siege and Capture of Havana, 1762*. Ed. Professor D. Syrett. **OP**
115. *Policy and Operations in the Mediterranean, 1912–1914*. Ed. E.W.R. Lumby. **OP**
116. *The Jacobean Commissions of Enquiry, 1608 and 1618*. Ed. A.P. McGowan. (£15.00)
117. *The Keyes Papers*, Vol. I, *1914–1918*. Ed. Professor P.G. Halpern. (£15.00)
118. *The Royal Navy and North America: The Warren Papers, 1736–1752*. Ed. Dr J. Gwyn. **OP**
119. *The Manning of the Royal Navy: Selected Public Pamphlets, 1693–1873*. Ed. Professor J.S. Bromley. (£15.00)
120. *Naval Administration, 1715–1750*. Ed. Professor D.A. Baugh. (£15.00)
121. *The Keyes Papers*, Vol. II, *1919–1938*. Ed. Professor P.G. Halpern. (£15.00)
122. *The Keyes Papers*, Vol. III, *1939–1945*. Ed. Professor P.G. Halpern. (£15.00)
123. *The Navy of the Lancastrian Kings: Accounts and Inventories of William Soper, Keeper of the King's Ships, 1422–1427*. Ed. Dr S. Rose. (£15.00)
124. *The Pollen Papers: The Privately Circulated Printed Works of Arthur Hungerford Pollen, 1901–1916*. Ed. Dr J.T. Sumida. (£15.00)
125. *The Naval Miscellany*, Vol. V. Ed. N.A.M. Rodger. (£15.00)
126. *The Royal Navy in the Mediterranean, 1915–1918*. Ed. Professor P.G. Halpern. (£15.00)

127. *The Expedition of Sir John Norris and Sir Francis Drake to Spain and Portugal, 1589.* Ed. Professor R.B. Wernhan. (£15.00)
128. *The Beatty Papers*, Vol. I. *1902–1918*. Ed. Professor B.McL. Ranft. (£15.00)
129. *The Hawke Papers, A Selection: 1743–1771.* Ed. Dr R.F. Mackay. (£15.00)
130. *Anglo-American Naval Relations, 1917–1919.* Ed. M. Simpson. (£15.00)
131. *British Naval Documents 1204–1960.* Ed. Professor J.B. Hattendorf, Dr R.J.B. Knight, A.W.H. Pearsall, Dr N.A.M. Rodger and Professor G. Till, (£25.00)
132. *The Beatty Papers*, Vol. II, *1916–1927*. Ed. Professor B.McL. Ranft. (£25.00)
133. *Samuel Pepys and the Second Dutch War.* Ed. R. Latham. (£25.00)
134. *The Somerville Papers.* Ed. M. Simpson with assistance from J. Somerville. (£25.00)
135. *The Royal Navy in the River Plate, 1806–1807.* Ed. J.D. Grainger. (£25.00)
136. *The Collective Naval Defence of the Empire, 1900–1940.* Ed. Professor N. Tracy. (£25.00)
137. *The Defeat of the Enemy Attack on Shipping, 1939–1945.* Ed. Dr E.J. Grove. (£25.00)
138. *Shipboard Life and Organisation, 1731–1815.* Ed. B. Lavery. (£25.00)
139. *The Battle of the Atlantic and Signals Intelligence: U-boat Situations and Trends, 1941–1945.* Ed. Professor D. Syrett. (£25.00)
140. *The Cunningham Papers*, Vol. I: *The Mediterranean Fleet, 1939–1942*. Ed. M. Simpson. (£25.00)
141. *The Channel Fleet and the Blockade of Brest, 1793–1801.* Ed. Dr R. Morriss. (£25.00)
142. *The Submarine Service, 1900–1918.* Ed. N.A. Lambert. (£25.00)
143. *Letters and Papers of Professor Sir John Knox Laughton, 1830–1915.* Ed. Professor A.D. Lambert. (£25.00)
144. *The Battle of the Atlantic and Signals Intelligence: U-boat Tracking Papers, 1941–1947.* Ed. Professor D. Syrett. (£25.00)
145. *The Maritime Blockade of Germany in the Great War: The Northern Patrol, 1914–1918.* Ed. J.D. Grainger.
146. *The Naval Miscellany*, Vol. VI. Ed. Dr M. Duffy. (£25.00)
147. *The Milne Papers. Papers of Admiral of the Fleet Sir Alexander Milne 1806–1896*, Vol. I, *1820–1859*. Ed. Professor J. Beeler. (£25.00)

148. *The Rodney Papers*, Vol. I, *1742–1763*. Ed. Professor D. Syrett. (£25.00)
149. *Sea Power and the Control of Trade: Belligerent Rights from the Russian War to the Beira Patrol, 1854–1970*. Ed. N. Tracy. (£25.00)
150. *The Cunningham Papers*, Vol. II: *The Triumph of Allied Sea Power, 1942–1946*. Ed. M. Simpson. (£25.00)
151. *The Rodney Papers*, Vol. II, *1763–1780*. Ed. Professor D. Syrett. (£25.00)
152. *Naval Intelligence From Berlin: The Reports of the British Naval Attachés in Berlin, 1906–1914,* Ed. Dr M.S. Seligmann. (£25.00)
153. *The Naval Miscellany*, Vol. VII. Ed. Dr S. Rose. (£25.00)
154. *Chatham Dockyard, 1815–1865. The Industrial Transformation.* Ed. P. MacDougal. (£25.00)
155. *Naval Courts Martial, 1793–1815*. Ed. Dr J. Byrn. (£25.00)
156. *Anglo-American Naval Relations, 1919–1939*. Ed. M. Simpson. (£30.00)
157. *The Navy of Edward VI and Mary*. Ed. Professor D.M Loades and Dr C.S. Knighton. (£30.00)
158. *The Royal Navy and the Mediterranean, 1919–1929*. Ed. Professor P. Halpern. (£30.00)
159. *The Fleet Air Arm in the Second World War*, Vol. I, *1939–1941*. Ed. Dr B. Jones. (£30.00)
160. *Elizabethan Naval Administration.* Ed. Professor D.M. Loades and Dr C.S. Knighton. (£40.00)
161. *The Naval Route to the Abyss: The Anglo-German Naval Race, 1895–1914*. Ed. Dr M.S. Seligmann, Dr F. Nägler and Professor M. Epkenhans. (£40.00)
162. *The Milne Papers. Papers of Admiral of the Fleet Sir Alexander Milne 1806–1896*, Vol. II, *1860–1862*. Ed. Professor J. Beeler. (£40.00)

OCCASIONAL PUBLICATIONS.

O.P. 1 *The Commissioned Sea Officers of the Royal Navy, 1660–1815.* Ed. Professor D. Syrett and Professor R.L. DiNardo. (£25.00)

O.P. 2 *The Anthony Roll of Henry VIII's Navy.* Ed. Dr C.S. Knighton and Professor D.M. Loades. (£25.00)

Robin Brodhurst
The Mill,
Stanford Dingley,
Reading, RG7 6LS.

robinbrodhurst@gmail.com